MATHEMATICAL ANALYSIS

SECOND EDITION

MATHEMATICAL ANALYSIS

SECOND EDITION

TOM M. APOSTOL
California Institute of Technology

ADDISON-WESLEY PUBLISHING COMPANY

Reading, Massachusetts

Menlo Park, California · London · Amsterdam · Don Mills, Ontario · Sydney

This book is in the
ADDISON-WESLEY SERIES IN MATHEMATICS

Consulting Editor: Lynn Loomis

ISBN 0-201-00288-4
    23 24 25 26 27 28 MA 9594

To my parents

PREFACE

A glance at the table of contents will reveal that this textbook treats topics in analysis at the "Advanced Calculus" level. The aim has been to provide a development of the subject which is honest, rigorous, up to date, and, at the same time, not too pedantic. The book provides a transition from elementary calculus to advanced courses in real and complex function theory, and it introduces the reader to some of the abstract thinking that pervades modern analysis.

The second edition differs from the first in many respects. Point set topology is developed in the setting of general metric spaces as well as in Euclidean n-space, and two new chapters have been added on Lebesgue integration. The material on line integrals, vector analysis, and surface integrals has been deleted. The order of some chapters has been rearranged, many sections have been completely rewritten, and several new exercises have been added.

The development of Lebesgue integration follows the Riesz-Nagy approach which focuses directly on functions and their integrals and does not depend on measure theory. The treatment here is simplified, spread out, and somewhat rearranged for presentation at the undergraduate level.

The first edition has been used in mathematics courses at a variety of levels, from first-year undergraduate to first-year graduate, both as a text and as supplementary reference. The second edition preserves this flexibility. For example, Chapters 1 through 5, 12, and 13 provide a course in differential calculus of functions of one or more variables. Chapters 6 through 11, 14, and 15 provide a course in integration theory. Many other combinations are possible; individual instructors can choose topics to suit their needs by consulting the diagram on the next page, which displays the logical interdependence of the chapters.

I would like to express my gratitude to the many people who have taken the trouble to write me about the first edition. Their comments and suggestions influenced the preparation of the second edition. Special thanks are due Dr. Charalambos Aliprantis who carefully read the entire manuscript and made numerous helpful suggestions. He also provided some of the new exercises. Finally, I would like to acknowledge my debt to the undergraduate students of Caltech whose enthusiasm for mathematics provided the original incentive for this work.

Pasadena   T.M.A.
September 1973

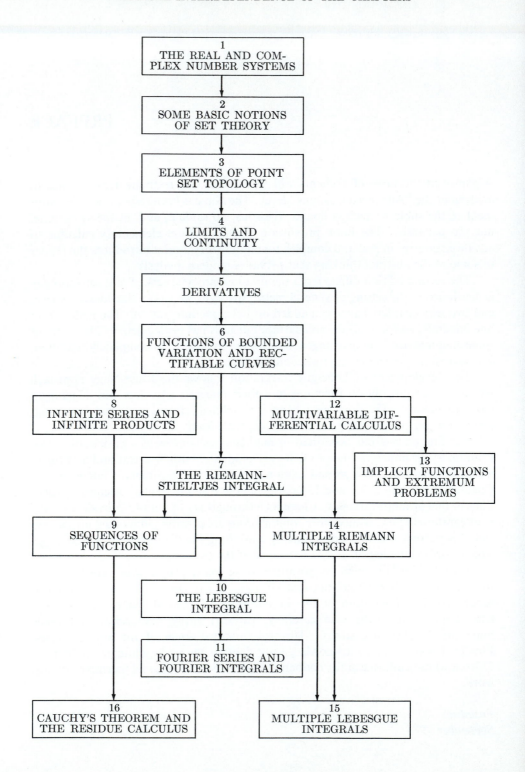

CONTENTS

Chapter 7   The Riemann–Stieltjes Integral

Chapter 8   Infinite Series and Infinite Products

CHAPTER 1

THE REAL AND
COMPLEX NUMBER SYSTEMS

1.1 INTRODUCTION

Mathematical analysis studies concepts related in some way to real numbers, so we begin our study of analysis with a discussion of the real-number system.

Several methods are used to introduce real numbers. One method starts with the positive integers 1, 2, 3, ... as undefined concepts and uses them to build a larger system, the positive *rational numbers* (quotients of positive integers), their negatives, and zero. The rational numbers, in turn, are then used to construct the *irrational numbers*, real numbers like $\sqrt{2}$ and π which are not rational. The rational and irrational numbers together constitute the real-number system.

Although these matters are an important part of the foundations of mathematics, they will not be described in detail here. As a matter of fact, in most phases of analysis it is only the *properties* of real numbers that concern us, rather than the methods used to construct them. Therefore, we shall take the real numbers themselves as undefined objects satisfying certain axioms from which further properties will be derived. Since the reader is probably familiar with most of the properties of real numbers discussed in the next few pages, the presentation will be rather brief. Its purpose is to review the important features and persuade the reader that, if it were necessary to do so, all the properties could be traced back to the axioms. More detailed treatments can be found in the references at the end of this chapter.

For convenience we use some elementary set notation and terminology. Let S denote a set (a collection of objects). The notation $x \in S$ means that the object x is in the set S, and we write $x \notin S$ to indicate that x is not in S.

A set S is said to be a *subset* of T, and we write $S \subseteq T$, if every object in S is also in T. A set is called *nonempty* if it contains at least one object.

We assume there exists a nonempty set $\mathbf{R}$ of objects, called real numbers, which satisfy the ten axioms listed below. The axioms fall in a natural way into three groups which we refer to as the *field axioms*, the *order axioms*, and the *completeness axiom* (also called the *least-upper-bound axiom* or the *axiom of continuity*).

1.2 THE FIELD AXIOMS

Along with the set $\mathbf{R}$ of real numbers we assume the existence of two operations, called *addition* and *multiplication*, such that for every pair of real numbers x and y

the *sum* $x + y$ and the *product* xy are real numbers uniquely determined by x and y satisfying the following axioms. (In the axioms that appear below, x, y, z represent arbitrary real numbers unless something is said to the contrary.)

Axiom 1. $x + y = y + x$, $xy = yx$ (*commutative laws*).

Axiom 2. $x + (y + z) = (x + y) + z$, $x(yz) = (xy)z$ (*associative laws*).

Axiom 3. $x(y + z) = xy + xz$ (*distributive law*).

Axiom 4. *Given any two real numbers x and y, there exists a real number z such that $x + z = y$. This z is denoted by $y - x$; the number $x - x$ is denoted by 0. (It can be proved that 0 is independent of x.) We write $-x$ for $0 - x$ and call $-x$ the negative of x.*

Axiom 5. *There exists at least one real number $x \neq 0$. If x and y are two real numbers with $x \neq 0$, then there exists a real number z such that $xz = y$. This z is denoted by y/x; the number x/x is denoted by 1 and can be shown to be independent of x. We write x^{-1} for $1/x$ if $x \neq 0$ and call x^{-1} the reciprocal of x.*

From these axioms all the usual laws of arithmetic can be derived; for example, $-(-x) = x$, $(x^{-1})^{-1} = x$, $-(x - y) = y - x$, $x - y = x + (-y)$, etc. (For a more detailed explanation, see Reference 1.1.)

1.3 THE ORDER AXIOMS

We also assume the existence of a relation $<$ which establishes an ordering among the real numbers and which satisfies the following axioms:

Axiom 6. *Exactly one of the relations $x = y$, $x < y$, $x > y$ holds.*

NOTE. $x > y$ means the same as $y < x$.

Axiom 7. *If $x < y$, then for every z we have $x + z < y + z$.*

Axiom 8. *If $x > 0$ and $y > 0$, then $xy > 0$.*

Axiom 9. *If $x > y$ and $y > z$, then $x > z$.*

NOTE. A real number x is called *positive* if $x > 0$, and *negative* if $x < 0$. We denote by $\mathbf{R}^+$ the set of all positive real numbers, and by $\mathbf{R}^-$ the set of all negative real numbers.

From these axioms we can derive the usual rules for operating with inequalities. For example, if we have $x < y$, then $xz < yz$ if z is positive, whereas $xz > yz$ if z is negative. Also, if $x > y$ and $z > w$ where both y and w are positive, then $xz > yw$. (For a complete discussion of these rules see Reference 1.1.)

NOTE. The symbolism $x \leq y$ is used as an abbreviation for the statement:

$$\text{``}x < y \qquad \text{or} \qquad x = y.\text{''}$$

Thus we have $2 \leq 3$ since $2 < 3$; and $2 \leq 2$ since $2 = 2$. The symbol $\geq$ is similarly used. A real number x is called *nonnegative* if $x \geq 0$. A pair of simultaneous inequalities such as $x < y$, $y < z$ is usually written more briefly as $x < y < z$.

The following theorem, which is a simple consequence of the foregoing axioms, is often used in proofs in analysis.

Theorem 1.1. *Given real numbers a and b such that*

$$a \leq b + \varepsilon \qquad \text{for every } \varepsilon > 0. \tag{1}$$

Then $a \leq b$.

Proof. If $b < a$, then inequality (1) is violated for $\varepsilon = (a - b)/2$ because

$$b + \varepsilon = b + \frac{a - b}{2} = \frac{a + b}{2} < \frac{a + a}{2} = a.$$

Therefore, by Axiom 6 we must have $a \leq b$.

Axiom 10, the completeness axiom, will be described in Section 1.11.

1.4 GEOMETRIC REPRESENTATION OF REAL NUMBERS

The real numbers are often represented geometrically as points on a line (called the *real line* or the *real axis*). A point is selected to represent 0 and another to represent 1, as shown in Fig. 1.1. This choice determines the scale. Under an appropriate set of axioms for Euclidean geometry, each point on the real line corresponds to one and only one real number and, conversely, each real number is represented by one and only one point on the line. It is customary to refer to the *point x* rather than the point representing the real number x.

Figure 1.1

The order relation has a simple geometric interpretation. If $x < y$, the point x lies to the left of the point y, as shown in Fig. 1.1. Positive numbers lie to the right of 0, and negative numbers to the left of 0. If $a < b$, a point x satisfies the inequalities $a < x < b$ if and only if x is *between* a and b.

1.5 INTERVALS

The set of all points between a and b is called an *interval*. Sometimes it is important to distinguish between intervals which include their endpoints and intervals which do not.

NOTATION. The notation $\{x : x \text{ satisfies } P\}$ will be used to designate the set of all real numbers x which satisfy property P.

Definition 1.2. *Assume a < b. The open interval (a, b) is defined to be the set*

$$(a, b) = \{x : a < x < b\}.$$

The closed interval $[a, b]$ *is the set* $\{x : a \leq x \leq b\}$. *The half-open intervals* $(a, b]$ *and* $[a, b)$ *are similarly defined, using the inequalities* $a < x \leq b$ *and* $a \leq x < b$, *respectively. Infinite intervals are defined as follows:*

$$(a, +\infty) = \{x : x > a\}, \qquad [a, +\infty) = \{x : x \geq a\},$$

$$(-\infty, a) = \{x : x < a\}, \qquad (-\infty, a] = \{x : x \leq a\}.$$

The real line **R** is sometimes referred to as the open interval $(-\infty, +\infty)$. A single point is considered as a "degenerate" closed interval.

NOTE. The symbols $+\infty$ and $-\infty$ are used here purely for convenience in notation and are not to be considered as being real numbers. Later we shall extend the real-number system to include these two symbols, but until this is done, the reader should understand that all real numbers are "finite."

1.6 INTEGERS

This section describes the *integers*, a special subset of **R**. Before we define the integers it is convenient to introduce first the notion of an *inductive set*.

Definition 1.3. *A set of real numbers is called an inductive set if it has the following two properties:*

a) *The number 1 is in the set.*

b) *For every x in the set, the number x + 1 is also in the set.*

For example, **R** is an inductive set. So is the set $\mathbf{R}^+$. Now we shall define the positive integers to be those real numbers which belong to every inductive set.

Definition 1.4. *A real number is called a positive integer if it belongs to every inductive set. The set of positive integers is denoted by* $\mathbf{Z}^+$.

The set $\mathbf{Z}^+$ is itself an inductive set. It contains the number 1, the number $1 + 1$ (denoted by 2), the number $2 + 1$ (denoted by 3), and so on. Since $\mathbf{Z}^+$ is a subset of every inductive set, we refer to $\mathbf{Z}^+$ as the *smallest* inductive set. This property of $\mathbf{Z}^+$ is sometimes called the *principle of induction*. We assume the reader is familiar with proofs by induction which are based on this principle. (See Reference 1.1.) Examples of such proofs are given in the next section.

The negatives of the positive integers are called the *negative integers*. The positive integers, together with the negative integers and 0 (zero), form a set **Z** which we call simply the *set of integers*.

1.7 THE UNIQUE FACTORIZATION THEOREM FOR INTEGERS

If n and d are integers and if $n = cd$ for some integer c, we say d is a *divisor* of n, or n is a *multiple* of d, and we write $d|n$ (read: d divides n). An integer n is called a *prime* if $n > 1$ and if the only positive divisors of n are 1 and n. If $n > 1$ and n is not prime, then n is called *composite*. The integer 1 is neither prime nor composite.

This section derives some elementary results on factorization of integers, culminating in the *unique factorization theorem*, also called *the fundamental theorem of arithmetic*.

The fundamental theorem states that (1) every integer $n > 1$ can be represented as a product of prime factors, and (2) this factorization can be done in only one way, apart from the order of the factors. It is easy to prove part (1).

Theorem 1.5. *Every integer $n > 1$ is either a prime or a product of primes.*

Proof. We use induction on n. The theorem holds trivially for $n = 2$. Assume it is true for every integer k with $1 < k < n$. If n is not prime it has a positive divisor d with $1 < d < n$. Hence $n = cd$, where $1 < c < n$. Since both c and d are $<n$, each is a prime or a product of primes; hence n is a product of primes.

Before proving part (2), uniqueness of the factorization, we introduce some further concepts.

If $d|a$ and $d|b$ we say d is a *common divisor* of a and b. The next theorem shows that every pair of integers a and b has a common divisor which is a linear combination of a and b.

Theorem 1.6. *Every pair of integers a and b has a common divisor d of the form*

$$d = ax + by$$

where x and y are integers. Moreover, every common divisor of a and b divides this d.

Proof. First assume that $a \geq 0$, $b \geq 0$ and use induction on $n = a + b$. If $n = 0$ then $a = b = 0$, and we can take $d = 0$ with $x = y = 0$. Assume, then, that the theorem has been proved for $0, 1, 2, \ldots, n - 1$. By symmetry, we can assume $a \geq b$. If $b = 0$ take $d = a$, $x = 1$, $y = 0$. If $b \geq 1$ we can apply the induction hypothesis to $a - b$ and b, since their sum is $a = n - b \leq n - 1$. Hence there is a common divisor d of $a - b$ and b of the form $d = (a - b)x + by$. This d also divides $(a - b) + b = a$, so d is a common divisor of a and b and we have $d = ax + (y - x)b$, a linear combination of a and b. To complete the proof we need to show that every common divisor divides d. Since a common divisor divides a and b, it also divides the linear combination $ax + (y - x)b = d$. This completes the proof if $a \geq 0$ and $b \geq 0$. If one or both of a and b is negative, apply the result just proved to $|a|$ and $|b|$.

NOTE. If d is a common divisor of a and b of the form $d = ax + by$, then $-d$ is also a divisor of the same form, $-d = a(-x) + b(-y)$. Of these two common divisors, the nonnegative one is called the *greatest common divisor* of a and b, and is denoted by $\gcd(a, b)$ or, simply by (a, b). If $(a, b) = 1$ then a and b are said to be *relatively prime*.

Theorem 1.7 (Euclid's Lemma). *If $a|bc$ and $(a, b) = 1$, then $a|c$.*

Proof. Since $(a, b) = 1$ we can write $1 = ax + by$. Therefore $c = acx + bcy$. But $a|acx$ and $a|bcy$, so $a|c$.

Theorem 1.8. *If a prime p divides ab, then $p|a$ or $p|b$. More generally, if a prime p divides a product $a_1 \cdots a_k$, then p divides at least one of the factors.*

Proof. Assume $p|ab$ and that p does not divide a. If we prove that $(p, a) = 1$, then Euclid's Lemma implies $p|b$. Let $d = (p, a)$. Then $d|p$ so $d = 1$ or $d = p$. We cannot have $d = p$ because $d|a$ but p does not divide a. Hence $d = 1$. To prove the more general statement we use induction on k, the number of factors. Details are left to the reader.

Theorem 1.9 (Unique factorization theorem). *Every integer $n > 1$ can be represented as a product of prime factors in only one way, apart from the order of the factors.*

Proof. We use induction on n. The theorem is true for $n = 2$. Assume, then, that it is true for all integers greater than 1 and less than n. If n is prime there is nothing more to prove. Therefore assume that n is composite and that n has two factorizations into prime factors, say

$$n = p_1 p_2 \cdots p_s = q_1 q_2 \cdots q_t. \qquad (2)$$

We wish to show that $s = t$ and that each p equals some q. Since p_1 divides the product $q_1 q_2 \cdots q_t$, it divides at least one factor. Relabel the q's if necessary so that $p_1|q_1$. Then $p_1 = q_1$ since both p_1 and q_1 are primes. In (2) we cancel p_1 on both sides to obtain

$$\frac{n}{p_1} = p_2 \cdots p_s = q_2 \cdots q_t.$$

Since n is composite, $1 < n/p_1 < n$; so by the induction hypothesis the two factorizations of n/p_1 are identical, apart from the order of the factors. Therefore the same is true in (2) and the proof is complete.

1.8 RATIONAL NUMBERS

Quotients of integers a/b (where $b \neq 0$) are called *rational numbers*. For example, $1/2$, $-7/5$, and 6 are rational numbers. The set of rational numbers, which we denote by $\mathbf{Q}$, contains $\mathbf{Z}$ as a subset. The reader should note that all the field axioms and the order axioms are satisfied by $\mathbf{Q}$.

We assume that the reader is familiar with certain elementary properties of rational numbers. For example, if a and b are rational, their average $(a + b)/2$ is also rational and lies between a and b. Therefore between any two rational numbers there are infinitely many rational numbers, which implies that if we are given a certain rational number we cannot speak of the "next largest" rational number.

1.9 IRRATIONAL NUMBERS

Real numbers that are not rational are called *irrational*. For example, the numbers $\sqrt{2}$, e, π and e^π are irrational.

Ordinarily it is not too easy to prove that some particular number is irrational. There is no simple proof, for example, of the irrationality of e^π. However, the irrationality of certain numbers such as $\sqrt{2}$ and $\sqrt{3}$ is not too difficult to establish and, in fact, we easily prove the following:

Theorem 1.10. *If n is a positive integer which is not a perfect square, then $\sqrt{n}$ is irrational.*

Proof. Suppose first that n contains no square factor > 1. We assume that $\sqrt{n}$ is rational and obtain a contradiction. Let $\sqrt{n} = a/b$, where a and b are integers having no factor in common. Then $nb^2 = a^2$ and, since the left side of this equation is a multiple of n, so too is a^2. However, if a^2 is a multiple of n, a itself must be a multiple of n, since n has no square factors > 1. (This is easily seen by examining the factorization of a into its prime factors.) This means that $a = cn$, where c is some integer. Then the equation $nb^2 = a^2$ becomes $nb^2 = c^2n^2$, or $b^2 = nc^2$. The same argument shows that b must also be a multiple of n. Thus a and b are both multiples of n, which contradicts the fact that they have no factor in common. This completes the proof if n has no square factor > 1.

If n has a square factor, we can write $n = m^2k$, where $k > 1$ and k has no square factor > 1. Then $\sqrt{n} = m\sqrt{k}$; and if $\sqrt{n}$ were rational, the number $\sqrt{k}$ would also be rational, contradicting that which was just proved.

A different type of argument is needed to prove that the number e is irrational. (We assume familiarity with the exponential e^x from elementary calculus and its representation as an infinite series.)

Theorem 1.11. *If $e^x = 1 + x + x^2/2! + x^3/3! + \cdots + x^n/n! + \cdots$, then the number e is irrational.*

Proof. We shall prove that e^{-1} is irrational. The series for e^{-1} is an alternating series with terms which decrease steadily in absolute value. In such an alternating series the error made by stopping at the nth term has the algebraic sign of the first neglected term and is less in absolute value than the first neglected term. Hence, if $s_n = \sum_{k=0}^{n} (-1)^k/k!$, we have the inequality

$$0 < e^{-1} - s_{2k-1} < \frac{1}{(2k)!},$$

from which we obtain

$$0 < (2k - 1)! \, (e^{-1} - s_{2k-1}) < \frac{1}{2k} \leq \frac{1}{2}, \tag{3}$$

for any integer $k \geq 1$. Now $(2k - 1)! \, s_{2k-1}$ is always an integer. If e^{-1} were rational, then we could choose k so large that $(2k - 1)! \, e^{-1}$ would also be an integer. Because of (3) the difference of these two integers would be a number between 0 and $\frac{1}{2}$, which is impossible. Thus e^{-1} cannot be rational, and hence e cannot be rational.

NOTE. For a proof that π is irrational, see Exercise 7.33.

The ancient Greeks were aware of the existence of irrational numbers as early as 500 B.C. However, a satisfactory theory of such numbers was not developed until late in the nineteenth century, at which time three different theories were introduced by Cantor, Dedekind, and Weierstrass. For an account of the theories of Dedekind and Cantor and their equivalence, see Reference 1.6.

1.10 UPPER BOUNDS, MAXIMUM ELEMENT, LEAST UPPER BOUND (SUPREMUM)

Irrational numbers arise in algebra when we try to solve certain quadratic equations. For example, it is desirable to have a real number x such that $x^2 = 2$. From the nine axioms listed above we cannot prove that such an x exists in **R** because these nine axioms are also satisfied by **Q** and we have shown that there is no rational number whose square is 2. The completeness axiom allows us to introduce irrational numbers in the real-number system, and it gives the real-number system a property of continuity that is fundamental to many theorems in analysis.

Before we describe the completeness axiom, it is convenient to introduce additional terminology and notation.

Definition 1.12. *Let S be a set of real numbers. If there is a real number b such that $x \leq b$ for every x in S, then b is called an upper bound for S and we say that S is bounded above by b.*

We say *an* upper bound because every number greater than b will also be an upper bound. If an upper bound b is also a member of S, then b is called the *largest member* or the *maximum element* of S. There can be at most one such b. If it exists, we write

$$b = \max S.$$

A set with no upper bound is said to be *unbounded above*.

Definitions of the terms *lower bound, bounded below, smallest member* (or *minimum element*) can be similarly formulated. If S has a minimum element we denote it by min S.

Examples

1. The set $\mathbf{R}^+ = (0, +\infty)$ is unbounded above. It has no upper bounds and no maximum element. It is bounded below by 0 but has no minimum element.

2. The closed interval $S = [0, 1]$ is bounded above by 1 and is bounded below by 0. In fact, max $S = 1$ and min $S = 0$.

3. The half-open interval $S = [0, 1)$ is bounded above by 1 but it has no maximum element. Its minimum element is 0.

For sets like the one in Example 3, which are bounded above but have no maximum element, there is a concept which takes the place of the maximum element. It is called the *least upper bound* or *supremum* of the set and is defined as follows:

Definition 1.13. *Let S be a set of real numbers bounded above. A real number b is called a least upper bound for S if it has the following two properties:*

a) *b is an upper bound for S.*

b) *No number less than b is an upper bound for S.*

Examples. If $S = [0, 1]$ the maximum element 1 is also a least upper bound for S. If $S = [0, 1)$ the number 1 is a least upper bound for S, even though S has no maximum element.

It is an easy exercise to prove that a set cannot have two different least upper bounds. Therefore, if there is a least upper bound for S, there is *only* one and we can speak of *the* least upper bound.

It is common practice to refer to the least upper bound of a set by the more concise term *supremum*, abbreviated *sup*. We shall adopt this convention and write

$$b = \sup S$$

to indicate that b is the supremum of S. If S has a maximum element, then $\max S = \sup S$.

The *greatest lower bound*, or *infimum* of S, denoted by inf S, is defined in an analogous fashion.

1.11 THE COMPLETENESS AXIOM

Our final axiom for the real number system involves the notion of supremum.

Axiom 10. *Every nonempty set S of real numbers which is bounded above has a supremum; that is, there is a real number b such that $b = \sup S$.*

As a consequence of this axiom it follows that every nonempty set of real numbers which is bounded below has an infimum.

1.12 SOME PROPERTIES OF THE SUPREMUM

This section discusses some fundamental properties of the supremum that will be useful in this text. There is a corresponding set of properties of the infimum that the reader should formulate for himself.

The first property shows that a set with a supremum contains numbers arbitrarily close to its supremum.

Theorem 1.14 (Approximation property). *Let S be a nonempty set of real numbers with a supremum, say $b = \sup S$. Then for every $a < b$ there is some x in S such that*

$$a < x \leq b.$$

Proof. First of all, $x \le b$ for all x in S. If we had $x \le a$ for every x in S, then a would be an upper bound for S smaller than the least upper bound. Therefore $x > a$ for at least one x in S.

Theorem 1.15 (Additive property). *Given nonempty subsets A and B of* **R**, *let C denote the set*

$$C = \{x + y : x \in A, \quad y \in B\}.$$

If each of A and B has a supremum, then C has a supremum and

$$\sup C = \sup A + \sup B.$$

Proof. Let $a = \sup A$, $b = \sup B$. If $z \in C$ then $z = x + y$, where $x \in A$, $y \in B$, so $z = x + y \le a + b$. Hence $a + b$ is an upper bound for C, so C has a supremum, say $c = \sup C$, and $c \le a + b$. We show next that $a + b \le c$. Choose any $\varepsilon > 0$. By Theorem 1.14 there is an x in A and a y in B such that

$$a - \varepsilon < x \quad \text{and} \quad b - \varepsilon < y.$$

Adding these inequalities we find

$$a + b - 2\varepsilon < x + y \le c.$$

Thus, $a + b < c + 2\varepsilon$ for every $\varepsilon > 0$ so, by Theorem 1.1, $a + b \le c$.

The proof of the next theorem is left as an exercise for the reader.

Theorem 1.16 (Comparison property). *Given nonempty subsets S and T of* **R** *such that $s \le t$ for every s in S and t in T. If T has a supremum then S has a supremum and*

$$\sup S \le \sup T.$$

1.13 PROPERTIES OF THE INTEGERS DEDUCED FROM THE COMPLETENESS AXIOM

Theorem 1.17. *The set* $\mathbf{Z}^+$ *of positive integers* $1, 2, 3, \ldots$ *is unbounded above.*

Proof. If $\mathbf{Z}^+$ were bounded above then $\mathbf{Z}^+$ would have a supremum, say $a = \sup \mathbf{Z}^+$. By Theorem 1.14 we would have $a - 1 < n$ for some n in $\mathbf{Z}^+$. Then $n + 1 > a$ for this n. Since $n + 1 \in \mathbf{Z}^+$ this contradicts the fact that $a = \sup \mathbf{Z}^+$.

Theorem 1.18. *For every real x there is a positive integer n such that $n > x$.*

Proof. If this were not true, some x would be an upper bound for $\mathbf{Z}^+$, contradicting Theorem 1.17.

1.14 THE ARCHIMEDEAN PROPERTY OF THE REAL NUMBER SYSTEM

The next theorem describes the Archimedean property of the real number system. Geometrically, it tells us that any line segment, no matter how long, can be covered by a finite number of line segments of a given positive length, no matter how small.

Theorem 1.19. *If $x > 0$ and if y is an arbitrary real number, there is a positive integer n such that $nx > y$.*

Proof. Apply Theorem 1.18 with x replaced by y/x.

1.15 RATIONAL NUMBERS WITH FINITE DECIMAL REPRESENTATION

A real number of the form

$$r = a_0 + \frac{a_1}{10} + \frac{a_2}{10^2} + \cdots + \frac{a_n}{10^n},$$

where a_0 is a nonnegative integer and $a_1, \ldots, a_n$ are integers satisfying $0 \le a_i \le 9$, is usually written more briefly as follows:

$$r = a_0 . a_1 a_2 \cdots a_n.$$

This is said to be a *finite decimal representation* of r. For example,

$$\frac{1}{2} = \frac{5}{10} = 0.5, \qquad \frac{1}{50} = \frac{2}{10^2} = 0.02, \qquad \frac{29}{4} = 7 + \frac{2}{10} + \frac{5}{10^2} = 7.25.$$

Real numbers like these are necessarily rational and, in fact, they all have the form $r = a/10^n$, where a is an integer. However, not all rational numbers can be expressed with finite decimal representations. For example, if $\frac{1}{3}$ could be so expressed, then we would have $\frac{1}{3} = a/10^n$ or $3a = 10^n$ for some integer a. But this is impossible since 3 does not divide any power of 10.

1.16 FINITE DECIMAL APPROXIMATIONS TO REAL NUMBERS

This section uses the completeness axiom to show that real numbers can be approximated to any desired degree of accuracy by rational numbers with finite decimal representations.

Theorem 1.20. *Assume $x \ge 0$. Then for every integer $n \ge 1$ there is a finite decimal $r_n = a_0 . a_1 a_2 \cdots a_n$ such that*

$$r_n \le x < r_n + \frac{1}{10^n}.$$

Proof. Let S be the set of all nonnegative integers $\le x$. Then S is nonempty, since $0 \in S$, and S is bounded above by x. Therefore S has a supremum, say $a_0 = \sup S$. It is easily verified that $a_0 \in S$, so a_0 is a nonnegative integer. We call a_0 the *greatest integer* in x, and we write $a_0 = [x]$. Clearly, we have

$$a_0 \le x < a_0 + 1.$$

Now let $a_1 = [10x - 10a_0]$, the greatest integer in $10x - 10a_0$. Since $0 \leq 10x - 10a_0 = 10(x - a_0) < 10$, we have $0 \leq a_1 \leq 9$ and

$$a_1 \leq 10x - 10a_0 < a_1 + 1.$$

In other words, a_1 is the largest integer satisfying the inequalities

$$a_0 + \frac{a_1}{10} \leq x < a_0 + \frac{a_1 + 1}{10} .$$

More generally, having chosen $a_1, \ldots, a_{n-1}$ with $0 \leq a_i \leq 9$, let a_n be the largest integer satisfying the inequalities

$$a_0 + \frac{a_1}{10} + \cdots + \frac{a_n}{10^n} \leq x < a_0 + \frac{a_1}{10} + \cdots + \frac{a_n + 1}{10^n}. \tag{4}$$

Then $0 \leq a_n \leq 9$ and we have

$$r_n \leq x < r_n + \frac{1}{10^n} ,$$

where $r_n = a_0 . a_1 a_2 \cdots a_n$. This completes the proof. It is easy to verify that x is actually the supremum of the set of rational numbers $r_1, r_2, \ldots$.

1.17 INFINITE DECIMAL REPRESENTATIONS OF REAL NUMBERS

The integers $a_0, a_1, a_2, \ldots$ obtained in the proof of Theorem 1.20 can be used to define an infinite decimal representation of x. We write

$$x = a_0 . a_1 a_2 \cdots$$

to mean that a_n is the largest integer satisfying (4). For example, if $x = \frac{1}{8}$ we find $a_0 = 0$, $a_1 = 1$, $a_2 = 2$, $a_3 = 5$, and $a_n = 0$ for all $n \geq 4$. Therefore we can write

$$\tfrac{1}{8} = 0.125000 \cdots$$

If we interchange the inequality signs $\leq$ and $<$ in (4), we obtain a slightly different definition of decimal expansions. The finite decimals r_n satisfy $r_n < x \leq r_n + 10^{-n}$ although the digits $a_0, a_1, a_2, \ldots$ need not be the same as those in (4). For example, if we apply this second definition to $x = \frac{1}{8}$ we find the infinite decimal representation

$$\tfrac{1}{8} = 0.124999 \cdots$$

The fact that a real number might have two different decimal representations is merely a reflection of the fact that two different sets of real numbers can have the same supremum.

1.18 ABSOLUTE VALUES AND THE TRIANGLE INEQUALITY

Calculations with inequalities arise quite frequently in analysis. They are of particular importance in dealing with the notion of absolute value. If x is any real number, the absolute value of x, denoted by $|x|$, is defined as follows:

$$|x| = \begin{cases} x, & \text{if } x \geq 0, \\ -x, & \text{if } x \leq 0. \end{cases}$$

A fundamental inequality concerning absolute values is given in the following:

Theorem 1.21. *If $a \geq 0$, then we have the inequality $|x| \leq a$ if, and only if, $-a \leq x \leq a$.*

Proof. From the definition of $|x|$, we have the inequality $-|x| \leq x \leq |x|$, since $x = |x|$ or $x = -|x|$. If we assume that $|x| \leq a$, then we can write $-a \leq -|x| \leq x \leq |x| \leq a$ and thus half of the theorem is proved. Conversely, let us assume $-a \leq x \leq a$. Then if $x \geq 0$, we have $|x| = x \leq a$, whereas if $x < 0$, we have $|x| = -x \leq a$. In either case we have $|x| \leq a$ and the theorem is proved.

We can use this theorem to prove the *triangle inequality*.

Theorem 1.22. *For arbitrary real x and y we have*

$$|x + y| \leq |x| + |y| \qquad \text{(the triangle inequality)}.$$

Proof. We have $-|x| \leq x \leq |x|$ and $-|y| \leq y \leq |y|$. Addition gives us $-(|x| + |y|) \leq x + y \leq |x| + |y|$, and from Theorem 1.21 we conclude that $|x + y| \leq |x| + |y|$. This proves the theorem.

The triangle inequality is often used in other forms. For example, if we take $x = a - c$ and $y = c - b$ in Theorem 1.22 we find

$$|a - b| \leq |a - c| + |c - b|.$$

Also, from Theorem 1.22 we have $|x| \geq |x + y| - |y|$. Taking $x = a + b$, $y = -b$, we obtain

$$|a + b| \geq |a| - |b|.$$

Interchanging a and b we also find $|a + b| \geq |b| - |a| = -(|a| - |b|)$, and hence

$$|a + b| \geq ||a| - |b||.$$

By induction we can also prove the generalizations

$$|x_1 + x_2 + \cdots + x_n| \leq |x_1| + |x_2| + \cdots + |x_n|$$

and

$$|x_1 + x_2 + \cdots + x_n| \geq |x_1| - |x_2| - \cdots - |x_n|.$$

1.19 THE CAUCHY–SCHWARZ INEQUALITY

We shall now derive another inequality which is often used in analysis.

Theorem 1.23 (Cauchy–Schwarz inequality). *If* $a_1, \ldots, a_n$ *and* $b_1, \ldots, b_n$ *are arbitrary real numbers, we have*

$$\left(\sum_{k=1}^{n} a_k b_k \right)^2 \le \left(\sum_{k=1}^{n} a_k^2 \right) \left(\sum_{k=1}^{n} b_k^2 \right).$$

Moreover, if some $a_i \ne 0$ *equality holds if and only if there is a real* x *such that* $a_k x + b_k = 0$ *for each* $k = 1, 2, \ldots, n$.

Proof. A sum of squares can never be negative. Hence we have

$$\sum_{k=1}^{n} (a_k x + b_k)^2 \ge 0$$

for every real x, with equality if and only if each term is zero. This inequality can be written in the form

$$Ax^2 + 2Bx + C \ge 0,$$

where

$$A = \sum_{k=1}^{n} a_k^2, \qquad B = \sum_{k=1}^{n} a_k b_k, \qquad C = \sum_{k=1}^{n} b_k^2.$$

If $A > 0$, put $x = -B/A$ to obtain $B^2 - AC \le 0$, which is the desired inequality. If $A = 0$, the proof is trivial.

NOTE. In vector notation the Cauchy–Schwarz inequality takes the form

$$(\mathbf{a} \cdot \mathbf{b})^2 \le \|\mathbf{a}\|^2 \|\mathbf{b}\|^2,$$

where $\mathbf{a} = (a_1, \ldots, a_n)$, $\mathbf{b} = (b_1, \ldots, b_n)$ are two n-dimensional vectors,

$$\mathbf{a} \cdot \mathbf{b} = \sum_{k=1}^{n} a_k b_k,$$

is their dot product, and $\|\mathbf{a}\| = (\mathbf{a} \cdot \mathbf{a})^{1/2}$ is the length of $\mathbf{a}$.

1.20 PLUS AND MINUS INFINITY AND THE EXTENDED REAL NUMBER SYSTEM R*

Next we extend the real number system by adjoining two "ideal points" denoted by the symbols $+\infty$ and $-\infty$ ("plus infinity" and "minus infinity").

Definition 1.24. *By the extended real number system* $\mathbf{R}^*$ *we shall mean the set of real numbers* $\mathbf{R}$ *together with two symbols* $+\infty$ *and* $-\infty$ *which satisfy the following properties:*

a) *If* $x \in \mathbf{R}$, *then we have*

$$x + (+\infty) = +\infty, \qquad\qquad x + (-\infty) = -\infty,$$
$$x - (+\infty) = -\infty, \qquad\qquad x - (-\infty) = +\infty,$$
$$x/(+\infty) = x/(-\infty) = 0.$$

b) *If $x > 0$, then we have*

$$x(+\infty) = +\infty, \qquad\qquad x(-\infty) = -\infty.$$

c) *If $x < 0$, then we have*

$$x(+\infty) = -\infty, \qquad\qquad x(-\infty) = +\infty.$$

d) $(+\infty) + (+\infty) = (+\infty)(+\infty) = (-\infty)(-\infty) = +\infty,$
$(-\infty) + (-\infty) = (+\infty)(-\infty) = -\infty.$

e) *If $x \in \mathbf{R}$, then we have $-\infty < x < +\infty$.*

NOTATION. We denote $\mathbf{R}$ by $(-\infty, +\infty)$ and $\mathbf{R}^*$ by $[-\infty, +\infty]$. The points in $\mathbf{R}$ are called "finite" to distinguish them from the "infinite" points $+\infty$ and $-\infty$.

The principal reason for introducing the symbols $+\infty$ and $-\infty$ is one of convenience. For example, if we define $+\infty$ to be the sup of a set of real numbers which is not bounded above, then every nonempty subset of $\mathbf{R}$ has a supremum in $\mathbf{R}^*$. The sup is finite if the set is bounded above and infinite if it is not bounded above. Similarly, we define $-\infty$ to be the inf of any set of real numbers which is not bounded below. Then every nonempty subset of $\mathbf{R}$ has an inf in $\mathbf{R}^*$.

For some of the later work concerned with limits, it is also convenient to introduce the following terminology.

Definition 1.25. *Every open interval $(a, +\infty)$ is called a neighborhood of $+\infty$ or a ball with center $+\infty$. Every open interval $(-\infty, a)$ is called a neighborhood of $-\infty$ or a ball with center $-\infty$.*

1.21 COMPLEX NUMBERS

It follows from the axioms governing the relation $<$ that the square of a real number is never negative. Thus, for example, the elementary quadratic equation $x^2 = -1$ has no solution among the real numbers. New types of numbers, called *complex numbers*, have been introduced to provide solutions to such equations. It turns out that the introduction of complex numbers provides, at the same time, solutions to general algebraic equations of the form

$$a_0 + a_1 x + \cdots + a_n x^n = 0,$$

where the coefficients $a_0, a_1, \ldots, a_n$ are arbitrary real numbers. (This fact is known as the *Fundamental Theorem of Algebra*.)

We shall now define complex numbers and discuss them in further detail.

Definition 1.26. *By a complex number we shall mean an ordered pair of real numbers which we denote by (x_1, x_2). The first member, x_1, is called the real part of the complex number; the second member, x_2, is called the imaginary part. Two complex numbers $x = (x_1, x_2)$ and $y = (y_1, y_2)$ are called equal, and we write $x = y$, if,* *and only if,* $x_1 = y_1$ *and* $x_2 = y_2$. *We define the sum* $x + y$ *and the product* xy *by the equations*

$$x + y = (x_1 + y_1, x_2 + y_2), \qquad xy = (x_1 y_1 - x_2 y_2, x_1 y_2 + x_2 y_1).$$

NOTE. The set of all complex numbers will be denoted by **C**.

Theorem 1.27. *The operations of addition and multiplication just defined satisfy the commutative, associative, and distributive laws.*

Proof. We prove only the distributive law; proofs of the others are simpler. If $x = (x_1, x_2)$, $y = (y_1, y_2)$, and $z = (z_1, z_2)$, then we have

$$\begin{aligned} x(y + z) &= (x_1, x_2)(y_1 + z_1, y_2 + z_2) \\ &= (x_1 y_1 + x_1 z_1 - x_2 y_2 - x_2 z_2, x_1 y_2 + x_1 z_2 + x_2 y_1 + x_2 z_1) \\ &= (x_1 y_1 - x_2 y_2, x_1 y_2 + x_2 y_1) + (x_1 z_1 - x_2 z_2, x_1 z_2 + x_2 z_1) \\ &= xy + xz. \end{aligned}$$

Theorem 1.28.

$$(x_1, x_2) + (0, 0) = (x_1, x_2), \qquad (x_1, x_2)(0, 0) = (0, 0),$$

$$(x_1, x_2)(1, 0) = (x_1, x_2), \qquad (x_1, x_2) + (-x_1, -x_2) = (0, 0).$$

Proof. The proofs here are immediate from the definition, as are the proofs of Theorems 1.29, 1.30, 1.32, and 1.33.

Theorem 1.29. *Given two complex numbers* $x = (x_1, x_2)$ *and* $y = (y_1, y_2)$, *there exists a complex number* z *such that* $x + z = y$. *In fact,* $z = (y_1 - x_1, y_2 - x_2)$. *This* z *is denoted by* $y - x$. *The complex number* $(-x_1, -x_2)$ *is denoted by* $-x$.

Theorem 1.30. *For any two complex numbers* x *and* y, *we have*

$$(-x)y = x(-y) = -(xy) = (-1, 0)(xy).$$

Definition 1.31. *If* $x = (x_1, x_2) \neq (0, 0)$ *and* y *are complex numbers, we define* $x^{-1} = [x_1/(x_1^2 + x_2^2), -x_2/(x_1^2 + x_2^2)]$, *and* $y/x = yx^{-1}$.

Theorem 1.32. *If* x *and* y *are complex numbers with* $x \neq (0, 0)$, *there exists a complex number* z *such that* $xz = y$, *namely,* $z = yx^{-1}$.

Of special interest are operations with complex numbers whose imaginary part is 0.

Theorem 1.33. $\qquad (x_1, 0) + (y_1, 0) = (x_1 + y_1, 0),$

$$(x_1, 0)(y_1, 0) = (x_1 y_1, 0),$$

$$(x_1, 0)/(y_1, 0) = (x_1/y_1, 0), \qquad if \ y_1 \neq 0.$$

NOTE. It is evident from Theorem 1.33 that we can perform arithmetic operations on complex numbers with zero imaginary part by performing the usual real-number operations on the real parts alone. Hence the complex numbers of the form $(x, 0)$ have the same arithmetic properties as the real numbers. For this reason it is

Fig. 1.3   **Geometric Representation**   **17**

convenient to think of the real number system as being a special case of the complex number system, and we agree to identify the complex number $(x, 0)$ and the real number x. Therefore, we write $x = (x, 0)$. In particular, $0 = (0, 0)$ and $1 = (1, 0)$.

1.22 GEOMETRIC REPRESENTATION OF COMPLEX NUMBERS

Just as real numbers are represented geometrically by points on a line, so complex numbers are represented by points in a plane. The complex number $x = (x_1, x_2)$ can be thought of as the "point" with coordinates (x_1, x_2). When this is done, the definition of addition amounts to addition by the *parallelogram law*. (See Fig. 1.2.)

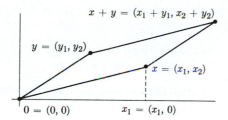

$$x + y = (x_1 + y_1, x_2 + y_2)$$

$$y = (y_1, y_2)$$

$$x = (x_1, x_2)$$

$$0 = (0, 0) \qquad x_1 = (x_1, 0)$$

Figure 1.2

The idea of expressing complex numbers geometrically as points on a plane was formulated by Gauss in his dissertation in 1799 and, independently, by Argand in 1806. Gauss later coined the somewhat unfortunate phrase "complex number." Other geometric interpretations of complex numbers are possible. Instead of using points on a plane, we can use points on other surfaces. Riemann found the sphere particularly convenient for this purpose. Points of the sphere are projected from the North Pole onto the tangent plane at the South Pole and thus there corresponds to each point of the plane a definite point of the sphere. With the exception of the North Pole itself, each point of the sphere corresponds to exactly one point of the plane. This correspondence is called a *stereographic projection*. (See Fig. 1.3.)

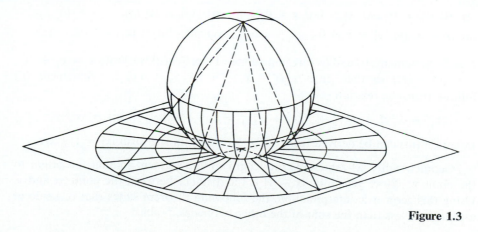

Figure 1.3

1.23 THE IMAGINARY UNIT

It is often convenient to think of the complex number (x_1, x_2) as a two-dimensional vector with components x_1 and x_2. Adding two complex numbers by means of Definition 1.26 is then the same as adding two vectors component by component. The complex number $1 = (1, 0)$ plays the same role as a unit vector in the horizontal direction. The analog of a unit vector in the vertical direction will now be introduced.

Definition 1.34. *The complex number* $(0, 1)$ *is denoted by* i *and is called the imaginary unit.*

Theorem 1.35. *Every complex number* $x = (x_1, x_2)$ *can be represented in the form* $x = x_1 + ix_2$.

Proof. $x_1 = (x_1, 0)$, $ix_2 = (0, 1)(x_2, 0) = (0, x_2)$,
$\qquad x_1 + ix_2 = (x_1, 0) + (0, x_2) = (x_1, x_2)$.

The next theorem tells us that the complex number i provides us with a solution to the equation $x^2 = -1$.

Theorem 1.36. $i^2 = -1$.

Proof. $i^2 = (0, 1)(0, 1) = (-1, 0) = -1$.

1.24 ABSOLUTE VALUE OF A COMPLEX NUMBER

We now extend the concept of absolute value to the complex number system.

Definition 1.37. *If* $x = (x_1, x_2)$, *we define the modulus, or absolute value, of* x *to be the nonnegative real number* $|x|$ *given by*

$$|x| = \sqrt{x_1^2 + x_2^2} .$$

Theorem 1.38.

i) $|(0, 0)| = 0$, *and* $|x| > 0$ *if* $x \neq 0$.   ii) $|xy| = |x| \, |y|$.

iii) $|x/y| = |x|/|y|$, *if* $y \neq 0$.   iv) $|(x_1, 0)| = |x_1|$.

Proof. Statements (i) and (iv) are immediate. To prove (ii), we write $x = x_1 + ix_2$, $y = y_1 + iy_2$, so that $xy = x_1 y_1 - x_2 y_2 + i(x_1 y_2 + x_2 y_1)$. Statement (ii) follows from the relation

$$|xy|^2 = x_1^2 y_1^2 + x_2^2 y_2^2 + x_1^2 y_2^2 + x_2^2 y_1^2 = (x_1^2 + x_2^2)(y_1^2 + y_2^2) = |x|^2 |y|^2.$$

Equation (iii) can be derived from (ii) by writing it in the form $|x| = |y| \, |x/y|$.

Geometrically, $|x|$ represents the length of the segment joining the origin to the point x. More generally, $|x - y|$ is the distance between the points x and y. Using this geometric interpretation, the following theorem states that one side of a triangle is less than the sum of the other two sides.

Theorem 1.39. *If x and y are complex numbers, then we have*

$$|x + y| \leq |x| + |y| \qquad (triangle\ inequality).$$

The proof is left as an exercise for the reader.

1.25 IMPOSSIBILITY OF ORDERING THE COMPLEX NUMBERS

As yet we have not defined a relation of the form $x < y$ if x and y are arbitrary complex numbers, for the reason that it is impossible to give a definition of $<$ for complex numbers which will have all the properties in Axioms 6 through 8. To illustrate, suppose we were able to define an order relation $<$ satisfying Axioms 6, 7, and 8. Then, since $i \neq 0$, we must have either $i > 0$ or $i < 0$, by Axiom 6. Let us assume $i > 0$. Then taking, $x = y = i$ in Axiom 8, we get $i^2 > 0$, or $-1 > 0$. Adding 1 to both sides (Axiom 7), we get $0 > 1$. On the other hand, applying Axiom 8 to $-1 > 0$ we find $1 > 0$. Thus we have both $0 > 1$ and $1 > 0$, which, by Axiom 6, is impossible. Hence the assumption $i > 0$ leads us to a contradiction. [Why was the inequality $-1 > 0$ not already a contradiction?] A similar argument shows that we cannot have $i < 0$. Hence the complex numbers cannot be ordered in such a way that Axioms 6, 7, and 8 will be satisfied.

1.26 COMPLEX EXPONENTIALS

The exponential e^x (x real) was mentioned earlier. We now wish to define e^z when z is a complex number in such a way that the principal properties of the real exponential function will be preserved. The main properties of e^x for x real are the law of exponents, $e^{x_1}e^{x_2} = e^{x_1+x_2}$, and the equation $e^0 = 1$. We shall give a definition of e^z for complex z which preserves these properties and reduces to the ordinary exponential when z is real.

If we write $z = x + iy$ (x, y real), then for the law of exponents to hold we want $e^{x+iy} = e^x e^{iy}$. It remains, therefore, to define what we shall mean by e^{iy}.

Definition 1.40. *If $z = x + iy$, we define $e^z = e^{x+iy}$ to be the complex number*
$$e^z = e^x (\cos y + i \sin y).$$

This definition* agrees with the real exponential function when z is real (that is, $y = 0$). We prove next that the law of exponents still holds.

---

* Several arguments can be given to motivate the equation $e^{iy} = \cos y + i \sin y$. For example, let us write $e^{iy} = f(y) + ig(y)$ and try to determine the real-valued functions f and g so that the usual rules of operating with real exponentials will also apply to complex exponentials. Formal differentiation yields $e^{iy} = g'(y) - if'(y)$, if we assume that $(e^{iy})' = ie^{iy}$. Comparing the two expressions for e^{iy}, we see that f and g must satisfy the equations $f(y) = g'(y), f'(y) = -g(y)$. Elimination of g yields $f(y) = -f''(y)$. Since we want $e^0 = 1$, we must have $f(0) = 1$ and $f'(0) = 0$. It follows that $f(y) = \cos y$ and $g(y) = -f'(y) = \sin y$. Of course, this argument *proves* nothing, but it strongly suggests that the definition $e^{iy} = \cos y + i \sin y$ is reasonable.

Theorem 1.41. *If* $z_1 = x_1 + iy_1$ *and* $z_2 = x_2 + iy_2$ *are two complex numbers, then we have*

$$e^{z_1}e^{z_2} = e^{z_1+z_2}.$$

Proof.

$$e^{z_1} = e^{x_1}(\cos y_1 + i \sin y_1), \qquad e^{z_2} = e^{x_2}(\cos y_2 + i \sin y_2),$$

$$e^{z_1}e^{z_2} = e^{x_1}e^{x_2}[\cos y_1 \cos y_2 - \sin y_1 \sin y_2$$

$$+ i(\cos y_1 \sin y_2 + \sin y_1 \cos y_2)].$$

Now $e^{x_1}e^{x_2} = e^{x_1+x_2}$, since x_1 and x_2 are both real. Also,

$$\cos y_1 \cos y_2 - \sin y_1 \sin y_2 = \cos (y_1 + y_2)$$

and

$$\cos y_1 \sin y_2 + \sin y_1 \cos y_2 = \sin (y_1 + y_2),$$

and hence

$$e^{z_1}e^{z_2} = e^{x_1+x_2}[\cos (y_1 + y_2) + i \sin (y_1 + y_2)] = e^{z_1+z_2}.$$

1.27 FURTHER PROPERTIES OF COMPLEX EXPONENTIALS

In the following theorems, z, z_1, z_2 denote complex numbers.

Theorem 1.42. e^z *is never zero.*

Proof. $e^z e^{-z} = e^0 = 1$. Hence e^z cannot be zero.

Theorem 1.43. *If x is real, then* $|e^{ix}| = 1$.

Proof. $|e^{ix}|^2 = \cos^2 x + \sin^2 x = 1$, and $|e^{ix}| > 0$.

Theorem 1.44. $e^z = 1$ *if, and only if, z is an integral multiple of* $2\pi i$.

Proof. If $z = 2\pi i n$, where n is an integer, then

$$e^z = \cos (2\pi n) + i \sin (2\pi n) = 1.$$

Conversely, suppose that $e^z = 1$. This means that $e^x \cos y = 1$ and $e^x \sin y = 0$. Since $e^x \neq 0$, we must have $\sin y = 0$, $y = k\pi$, where k is an integer. But $\cos (k\pi) = (-1)^k$. Hence $e^x = (-1)^k$, since $e^x \cos (k\pi) = 1$. Since $e^x > 0$, k must be even. Therefore $e^x = 1$ and hence $x = 0$. This proves the theorem.

Theorem 1.45. $e^{z_1} = e^{z_2}$ *if, and only if, $z_1 - z_2 = 2\pi i n$ (where n is an integer).*

Proof. $e^{z_1} = e^{z_2}$ if, and only if, $e^{z_1-z_2} = 1$.

1.28 THE ARGUMENT OF A COMPLEX NUMBER

If the point $z = (x, y) = x + iy$ is represented by polar coordinates r and θ, we can write $x = r \cos \theta$ and $y = r \sin \theta$, so that $z = r \cos \theta + ir \sin \theta = re^{i\theta}$.

The two numbers r and θ uniquely determine z. Conversely, the positive number r is uniquely determined by z; in fact, $r = |z|$. However, z determines the angle θ only up to multiples of 2π. There are infinitely many values of θ which satisfy the equations $x = |z| \cos \theta$, $y = |z| \sin \theta$ but, of course, any two of them differ by some multiple of 2π. Each such θ is called an *argument* of z but *one* of these values is singled out and is called the *principal argument* of z.

Definition 1.46. *Let* $z = x + iy$ *be a nonzero complex number. The unique real number* θ *which satisfies the conditions*

$$x = |z| \cos \theta, \qquad y = |z| \sin \theta, \qquad -\pi < \theta \le +\pi$$

is called the principal argument of z, *denoted by* $\theta = \arg(z)$.

The above discussion immediately yields the following theorem:

Theorem 1.47. *Every complex number* $z \ne 0$ *can be represented in the form* $z = re^{i\theta}$, *where* $r = |z|$ *and* $\theta = \arg(z) + 2\pi n$, *n being any integer.*

NOTE. This method of representing complex numbers is particularly useful in connection with multiplication and division, since we have

$$(r_1 e^{i\theta_1})(r_2 e^{i\theta_2}) = r_1 r_2 e^{i(\theta_1 + \theta_2)} \qquad \text{and} \qquad \frac{r_1 e^{i\theta_1}}{r_2 e^{i\theta_2}} = \frac{r_1}{r_2} e^{i(\theta_1 - \theta_2)}.$$

Theorem 1.48. *If* $z_1 z_2 \ne 0$, *then* $\arg(z_1 z_2) = \arg(z_1) + \arg(z_2) + 2\pi n(z_1, z_2)$, *where*

$$n(z_1, z_2) = \begin{cases} 0, & \text{if } -\pi < \arg(z_1) + \arg(z_2) \le +\pi, \\ +1, & \text{if } -2\pi < \arg(z_1) + \arg(z_2) \le -\pi, \\ -1, & \text{if } \pi < \arg(z_1) + \arg(z_2) \le 2\pi. \end{cases}$$

Proof. Write $z_1 = |z_1| e^{i\theta_1}$, $z_2 = |z_2| e^{i\theta_2}$, where $\theta_1 = \arg(z_1)$ and $\theta_2 = \arg(z_2)$. Then $z_1 z_2 = |z_1 z_2| e^{i(\theta_1 + \theta_2)}$. Since $-\pi < \theta_1 \le +\pi$ and $-\pi < \theta_2 \le +\pi$, we have $-2\pi < \theta_1 + \theta_2 \le 2\pi$. Hence there is an integer n such that $-\pi < \theta_1 + \theta_2 + 2n\pi \le \pi$. This n is the same as the integer $n(z_1, z_2)$ given in the theorem, and for this n we have $\arg(z_1 z_2) = \theta_1 + \theta_2 + 2\pi n$. This proves the theorem.

1.29 INTEGRAL POWERS AND ROOTS OF COMPLEX NUMBERS

Definition 1.49. *Given a complex number* z *and an integer* n, *we define the* nth *power of* z *as follows:*

$$z^0 = 1, \qquad z^{n+1} = z^n z, \qquad \text{if } n \ge 0,$$

$$z^{-n} = (z^{-1})^n, \qquad \text{if } z \ne 0 \text{ and } n > 0.$$

Theorem 1.50, which states that the usual laws of exponents hold, can be proved by mathematical induction. The proof is left as an exercise.

Theorem 1.50. *Given two integers m and n, we have, for $z \neq 0$,*

$$z^n z^m = z^{n+m} \quad and \quad (z_1 z_2)^n = z_1^n z_2^n.$$

Theorem 1.51. *If $z \neq 0$, and if n is a positive integer, then there are exactly n distinct complex numbers $z_0, z_1, \ldots, z_{n-1}$ (called the nth roots of z), such that*

$$z_k^n = z, \quad for\ each\ k = 0, 1, 2, \ldots, n - 1.$$

Furthermore, these roots are given by the formulas

$$z_k = Re^{i\phi_k}, \quad where \quad R = |z|^{1/n},$$

and

$$\phi_k = \frac{\arg(z)}{n} + \frac{2\pi k}{n} \quad (k = 0, 1, 2, \ldots, n - 1).$$

NOTE. The n nth roots of z are equally spaced on the circle of radius $R = |z|^{1/n}$, center at the origin.

Proof. The n complex numbers $Re^{i\phi_k}$, $0 \leq k \leq n - 1$, are distinct and each is an nth root of z, since

$$(Re^{i\phi_k})^n = R^n e^{in\phi_k} = |z| e^{i[\arg(z) + 2\pi k]} = z.$$

We must now show that there are no other nth roots of z. Suppose $w = Ae^{i\alpha}$ is a complex number such that $w^n = z$. Then $|w|^n = |z|$, and hence $A^n = |z|$, $A = |z|^{1/n}$. Therefore, $w^n = z$ can be written $e^{in\alpha} = e^{i[\arg(z)]}$, which implies

$$n\alpha - \arg(z) = 2\pi k \quad for\ some\ integer\ k.$$

Hence $\alpha = [\arg(z) + 2\pi k]/n$. But when k runs through all integral values, w takes only the distinct values $z_0, \ldots, z_{n-1}$. (See Fig. 1.4.)

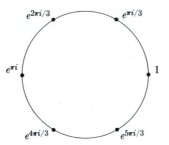

Figure 1.4

1.30 COMPLEX LOGARITHMS

By Theorem 1.42, e^z is never zero. It is natural to ask if there are other values that e^z cannot assume. The next theorem shows that zero is the only exceptional value.

Theorem 1.52. *If z is a complex number $\neq 0$, then there exist complex numbers w such that $e^w = z$. One such w is the complex number*

$$\log |z| + i \arg (z),$$

and any other such w must have the form

$$\log |z| + i \arg (z) + 2n\pi i,$$

where n is an integer.

Proof. Since $e^{\log |z| + i \arg (z)} = e^{\log |z|}e^{i \arg (z)} = |z|e^{i \arg (z)} = z$, we see that $w = \log |z| + i \arg (z)$ is a solution of the equation $e^w = z$. But if w_1 is any other solution, then $e^w = e^{w_1}$ and hence $w - w_1 = 2n\pi i$.

Definition 1.53. *Let $z \neq 0$ be a given complex number. If w is a complex number such that $e^w = z$, then w is called a logarithm of z. The particular value of w given by*

$$w = \log |z| + i \arg (z)$$

is called the principal logarithm of z, and for this w we write

$$w = \text{Log } z.$$

Examples

1. Since $|i| = 1$ and $\arg (i) = \pi/2$, $\text{Log } (i) = i\pi/2$.
2. Since $|-i| = 1$ and $\arg (-i) = -\pi/2$, $\text{Log } (-i) = -i\pi/2$.
3. Since $|-1| = 1$ and $\arg (-1) = \pi$, $\text{Log } (-1) = \pi i$.
4. If $x > 0$, $\text{Log } (x) = \log x$, since $|x| = x$ and $\arg (x) = 0$.
5. Since $|1 + i| = \sqrt{2}$ and $\arg (1 + i) = \pi/4$, $\text{Log } (1 + i) = \log \sqrt{2} + i\pi/4$.

Theorem 1.54. *If $z_1 z_2 \neq 0$, then*

$$\text{Log } (z_1 z_2) = \text{Log } z_1 + \text{Log } z_2 + 2\pi i n(z_1, z_2),$$

where $n(z_1, z_2)$ is the integer defined in Theorem 1.48.

Proof.

$$\text{Log } (z_1 z_2) = \log |z_1 z_2| + i \arg (z_1 z_2)$$
$$= \log |z_1| + \log |z_2| + i [\arg (z_1) + \arg (z_2) + 2\pi n(z_1, z_2)].$$

1.31 COMPLEX POWERS

Using complex logarithms, we can now give a definition of complex powers of complex numbers.

Definition 1.55. *If $z \neq 0$ and if w is any complex number, we define*

$$z^w = e^{w \text{ Log } z}.$$

Examples

1. $i^i = e^{i \operatorname{Log} i} = e^{i(i\pi/2)} = e^{-\pi/2}$.

2. $(-1)^i = e^{i \operatorname{Log}(-1)} = e^{i(i\pi)} = e^{-\pi}$.

3. If n is an integer, then $z^{n+1} = e^{(n+1)\operatorname{Log} z} = e^{n \operatorname{Log} z} e^{\operatorname{Log} z} = z^n z$, so Definition 1.55 does not conflict with Definition 1.49.

The next two theorems give rules for calculating with complex powers:

Theorem 1.56. $z^{w_1} z^{w_2} = z^{w_1 + w_2}$ if $z \neq 0$.

Proof. $z^{w_1 + w_2} = e^{(w_1 + w_2) \operatorname{Log} z} = e^{w_1 \operatorname{Log} z} e^{w_2 \operatorname{Log} z} = z^{w_1} z^{w_2}$.

Theorem 1.57. If $z_1 z_2 \neq 0$, then

$$(z_1 z_2)^w = z_1^w z_2^w e^{2\pi i w \, n(z_1, z_2)},$$

where $n(z_1, z_2)$ is the integer defined in Theorem 1.48.

Proof. $(z_1 z_2)^w = e^{w \operatorname{Log}(z_1 z_2)} = e^{w [\operatorname{Log} z_1 + \operatorname{Log} z_2 + 2\pi i \, n(z_1, z_2)]}$.

1.32 COMPLEX SINES AND COSINES

Definition 1.58. *Given a complex number z, we define*

$$\cos z = \frac{e^{iz} + e^{-iz}}{2}, \qquad \sin z = \frac{e^{iz} - e^{-iz}}{2i}.$$

NOTE. When z is real, these equations agree with Definition 1.40.

Theorem 1.59. *If $z = x + iy$, then we have*

$$\cos z = \cos x \cosh y - i \sin x \sinh y,$$
$$\sin z = \sin x \cosh y + i \cos x \sinh y.$$

Proof.

$$\begin{aligned} 2 \cos z &= e^{iz} + e^{-iz} \\ &= e^{-y}(\cos x + i \sin x) + e^{y}(\cos x - i \sin x) \\ &= \cos x(e^y + e^{-y}) - i \sin x(e^y - e^{-y}) \\ &= 2 \cos x \cosh y - 2i \sin x \sinh y. \end{aligned}$$

The proof for $\sin z$ is similar.

Further properties of sines and cosines are given in the exercises.

1.33 INFINITY AND THE EXTENDED COMPLEX PLANE C*

Next we extend the complex number system by adjoining an ideal point denoted by the symbol ∞.

Definition 1.60. *By the extended complex number system $\mathbf{C}^*$ we shall mean the complex plane $\mathbf{C}$ along with a symbol ∞ which satisfies the following properties:*

a) *If $z \in \mathbf{C}$, then we have $z + \infty = z - \infty = \infty$, $z/\infty = 0$.*

b) *If $z \in \mathbf{C}$, but $z \neq 0$, then $z(\infty) = \infty$ and $z/0 = \infty$.*

c) $\infty + \infty = (\infty)(\infty) = \infty$.

Definition 1.61. *Every set in $\mathbf{C}$ of the form $\{z : |z| > r \geq 0\}$ is called a neighborhood of ∞, or a ball with center at ∞.*

The reader may wonder why two symbols, $+\infty$ and $-\infty$, are adjoined to $\mathbf{R}$ but only one symbol, ∞, is adjoined to $\mathbf{C}$. The answer lies in the fact that there is an ordering relation $<$ among the real numbers, but no such relation occurs among the complex numbers. In order that certain properties of real numbers involving the relation $<$ hold without exception, we need two symbols, $+\infty$ and $-\infty$, as defined above. We have already mentioned that in $\mathbf{R}^*$ every nonempty set has a sup, for example.

In $\mathbf{C}$ it turns out to be more convenient to have just one ideal point. By way of illustration, let us recall the stereographic projection which establishes a one-to-one correspondence between the points of the complex plane and those points on the surface of the sphere distinct from the North Pole. The apparent exception at the North Pole can be removed by regarding it as the geometric representative of the ideal point ∞. We then get a one-to-one correspondence between the extended complex plane $\mathbf{C}^*$ and the total surface of the sphere. It is geometrically evident that if the South Pole is placed on the origin of the complex plane, the exterior of a "large" circle in the plane will correspond, by stereographic projection, to a "small" spherical cap about the North Pole. This illustrates vividly why we have defined a neighborhood of ∞ by an inequality of the form $|z| > r$.

EXERCISES

Integers

1.1 Prove that there is no largest prime. (A proof was known to Euclid.)

1.2 If n is a positive integer, prove the algebraic identity

$$a^n - b^n = (a - b) \sum_{k=0}^{n-1} a^k b^{n-1-k}.$$

1.3 If $2^n - 1$ is prime, prove that n is prime. A prime of the form $2^p - 1$, where p is prime, is called a *Mersenne prime*.

1.4 If $2^n + 1$ is prime, prove that n is a power of 2. A prime of the form $2^{2^m} + 1$ is called a *Fermat prime*. *Hint.* Use Exercise 1.2.

1.5 The *Fibonacci numbers* 1, 1, 2, 3, 5, 8, 13, ... are defined by the recursion formula $x_{n+1} = x_n + x_{n-1}$, with $x_1 = x_2 = 1$. Prove that $(x_n, x_{n+1}) = 1$ and that $x_n = (a^n - b^n)/(a - b)$, where a and b are the roots of the quadratic equation $x^2 - x - 1 = 0$.

1.6 Prove that every nonempty set of positive integers contains a smallest member. This is called the *well-ordering principle*.

Rational and irrational numbers

1.7 Find the rational number whose decimal expansion is 0.3344444 ...

1.8 Prove that the decimal expansion of x will end in zeros (or in nines) if, and only if, x is a rational number whose denominator is of the form 2^n5^m, where m and n are non-negative integers.

1.9 Prove that $\sqrt{2} + \sqrt{3}$ is irrational.

1.10 If a, b, c, d are rational and if x is irrational, prove that $(ax + b)/(cx + d)$ is usually irrational. When do exceptions occur?

1.11 Given any real $x > 0$, prove that there is an irrational number between 0 and x.

1.12 If $a/b < c/d$ with $b > 0$, $d > 0$, prove that $(a + c)/(b + d)$ lies between a/b and c/d.

1.13 Let a and b be positive integers. Prove that $\sqrt{2}$ always lies between the two fractions a/b and $(a + 2b)/(a + b)$. Which fraction is closer to $\sqrt{2}$?

1.14 Prove that $\sqrt{n - 1} + \sqrt{n + 1}$ is irrational for every integer $n \geq 1$.

1.15 Given a real x and an integer $N > 1$, prove that there exist integers h and k with $0 < k \leq N$ such that $|kx - h| < 1/N$. *Hint.* Consider the $N + 1$ numbers $tx - [tx]$ for $t = 0, 1, 2, \ldots, N$ and show that some pair differs by at most $1/N$.

1.16 If x is irrational prove that there are infinitely many rational numbers h/k with $k > 0$ such that $|x - h/k| < 1/k^2$. *Hint.* Assume there are only a finite number $h_1/k_1, \ldots, h_r/k_r$ and obtain a contradiction by applying Exercise 1.15 with $N > 1/\delta$, where δ is the smallest of the numbers $|x - h_i/k_i|$.

1.17 Let x be a positive rational number of the form

$$x = \sum_{k=1}^{n} \frac{a_k}{k!},$$

where each a_k is a nonnegative integer with $a_k \leq k - 1$ for $k \geq 2$ and $a_n > 0$. Let $[x]$ denote the greatest integer in x. Prove that $a_1 = [x]$, that $a_k = [k! \, x] - k[(k - 1)! \, x]$ for $k = 2, \ldots, n$, and that n is the smallest integer such that $n! \, x$ is an integer. Conversely, show that every positive rational number x can be expressed in this form in one and only one way.

Upper bounds

1.18 Show that the sup and inf of a set are uniquely determined whenever they exist.

1.19 Find the sup and inf of each of the following sets of real numbers:

    a) All numbers of the form $2^{-p} + 3^{-q} + 5^{-r}$, where p, q, and r take on all positive integer values.

    b) $S = \{x : 3x^2 - 10x + 3 < 0\}$.

    c) $S = \{x : (x - a)(x - b)(x - c)(x - d) < 0\}$, where $a < b < c < d$.

1.20 Prove the comparison property for suprema (Theorem 1.16).

1.21 Let A and B be two sets of positive numbers bounded above, and let $a = \sup A$, $b = \sup B$. Let C be the set of all products of the form xy, where $x \in A$ and $y \in B$. Prove that $ab = \sup C$.

1.22 Given $x > 0$ and an integer $k \geq 2$. Let a_0 denote the largest integer $\leq x$ and, assuming that $a_0, a_1, \ldots, a_{n-1}$ have been defined, let a_n denote the largest integer such that

$$a_0 + \frac{a_1}{k} + \frac{a_2}{k^2} + \cdots + \frac{a_n}{k^n} \leq x.$$

a) Prove that $0 \leq a_i \leq k - 1$ for each $i = 1, 2, \ldots$

b) Let $r_n = a_0 + a_1 k^{-1} + a_2 k^{-2} + \cdots + a_n k^{-n}$ and show that x is the sup of the set of rational numbers $r_1, r_2, \ldots$

NOTE. When $k = 10$ the integers $a_0, a_1, a_2, \ldots$ are the digits in a decimal representation of x. For general k they provide a representation in the scale of k.

Inequalities

1.23 Prove Lagrange's identity for real numbers:

$$\left(\sum_{k=1}^{n} a_k b_k \right)^2 = \left(\sum_{k=1}^{n} a_k^2 \right) \left(\sum_{k=1}^{n} b_k^2 \right) - \sum_{1 \leq k < j \leq n} (a_k b_j - a_j b_k)^2.$$

Note that this identity implies the Cauchy–Schwarz inequality.

1.24 Prove that for arbitrary real a_k, b_k, c_k we have

$$\left(\sum_{k=1}^{n} a_k b_k c_k \right)^4 \leq \left(\sum_{k=1}^{n} a_k^4 \right) \left(\sum_{k=1}^{n} b_k^2 \right)^2 \left(\sum_{k=1}^{n} c_k^4 \right).$$

1.25 Prove Minkowski's inequality:

$$\left(\sum_{k=1}^{n} (a_k + b_k)^2 \right)^{1/2} \leq \left(\sum_{k=1}^{n} a_k^2 \right)^{1/2} + \left(\sum_{k=1}^{n} b_k^2 \right)^{1/2}.$$

This is the triangle inequality $\|\mathbf{a} + \mathbf{b}\| \leq \|\mathbf{a}\| + \|\mathbf{b}\|$ for n-dimensional vectors, where $\mathbf{a} = (a_1, \ldots, a_n)$, $\mathbf{b} = (b_1, \ldots, b_n)$ and

$$\|\mathbf{a}\| = \left(\sum_{k=1}^{n} a_k^2 \right)^{1/2}.$$

1.26 If $a_1 \geq a_2 \geq \cdots \geq a_n$ and $b_1 \geq b_2 \geq \cdots \geq b_n$, prove that

$$\left(\sum_{k=1}^{n} a_k \right) \left(\sum_{k=1}^{n} b_k \right) \leq n \sum_{k=1}^{n} a_k b_k.$$

Hint. $\sum_{1 \leq j \leq k \leq n} (a_k - a_j)(b_k - b_j) \geq 0.$

Complex numbers

1.27 Express the following complex numbers in the form $a + bi$.

a) $(1 + i)^3$,

b) $(2 + 3i)/(3 - 4i)$,

c) $i^5 + i^{16}$,

d) $\frac{1}{2}(1 + i)(1 + i^{-8})$.

1.28 In each case, determine all real x and y which satisfy the given relation.

a) $x + iy = |x - iy|$,

b) $x + iy = (x - iy)^2$,

c) $\sum_{k=0}^{100} i^k = x + iy.$

1.29 If $z = x + iy$, x and y real, the complex conjugate of z is the complex number $\bar{z} = x - iy$. Prove that:

a) $\overline{z_1 + z_2} = \bar{z}_1 + \bar{z}_2$,   b) $\overline{z_1 z_2} = \bar{z}_1 \bar{z}_2$,   c) $z\bar{z} = |z|^2$,

d) $z + \bar{z} =$ twice the real part of z,

e) $(z - \bar{z})/i =$ twice the imaginary part of z.

1.30 Describe geometrically the set of complex numbers z which satisfies each of the following conditions:

a) $|z| = 1$,   b) $|z| < 1$,   c) $|z| \leq 1$,

d) $z + \bar{z} = 1$,   e) $z - \bar{z} = i$,   f) $\bar{z} + z = |z|^2$.

1.31 Given three complex numbers z_1, z_2, z_3 such that $|z_1| = |z_2| = |z_3| = 1$ and $z_1 + z_2 + z_3 = 0$. Show that these numbers are vertices of an equilateral triangle inscribed in the unit circle with center at the origin.

1.32 If a and b are complex numbers, prove that:

a) $|a - b|^2 \leq (1 + |a|^2)(1 + |b|^2)$.

b) If $a \neq 0$, then $|a + b| = |a| + |b|$ if, and only if, b/a is real and nonnegative.

1.33 If a and b are complex numbers, prove that

$$|a - b| = |1 - \bar{a}b|$$

if, and only if, $|a| = 1$ or $|b| = 1$. For which a and b is the inequality $|a - b| < |1 - \bar{a}b|$ valid?

1.34 If a and c are real constants, b complex, show that the equation

$$az\bar{z} + b\bar{z} + \bar{b}z + c = 0 \qquad (a \neq 0, z = x + iy)$$

represents a circle in the xy-plane.

1.35 Recall the definition of the inverse tangent: given a real number t, $\tan^{-1}(t)$ is the unique real number θ which satisfies the two conditions

$$-\frac{\pi}{2} < \theta < +\frac{\pi}{2}, \qquad \tan\theta = t.$$

If $z = x + iy$, show that

a) $\arg(z) = \tan^{-1}\left(\dfrac{y}{x}\right)$,   if $x > 0$,

b) $\arg(z) = \tan^{-1}\left(\dfrac{y}{x}\right) + \pi$,   if $x < 0$, $y \geq 0$,

c) $\arg(z) = \tan^{-1}\left(\dfrac{y}{x}\right) - \pi$,   if $x < 0$, $y < 0$,

d) $\arg(z) = \dfrac{\pi}{2}$ if $x = 0$, $y > 0$; $\arg(z) = -\dfrac{\pi}{2}$ if $x = 0$, $y < 0$.

1.36 Define the following "pseudo-ordering" of the complex numbers: we say $z_1 < z_2$ if we have either

   i) $|z_1| < |z_2|$ or   ii) $|z_1| = |z_2|$ and arg $(z_1) <$ arg (z_2).

Which of Axioms 6, 7, 8, 9 are satisfied by this relation?

1.37 Which of Axioms 6, 7, 8, 9 are satisfied if the pseudo-ordering is defined as follows? We say $(x_1, y_1) < (x_2, y_2)$ if we have either

   i) $x_1 < x_2$ or   ii) $x_1 = x_2$ and $y_1 < y_2$.

1.38 State and prove a theorem analogous to Theorem 1.48, expressing arg (z_1/z_2) in terms of arg (z_1) and arg (z_2).

1.39 State and prove a theorem analogous to Theorem 1.54, expressing Log (z_1/z_2) in terms of Log (z_1) and Log (z_2).

1.40 Prove that the nth roots of 1 (also called the nth roots of unity) are given by α, $\alpha^2, \ldots, \alpha^n$, where $\alpha = e^{2\pi i/n}$, and show that the roots $\neq 1$ satisfy the equation

$$1 + x + x^2 + \cdots + x^{n-1} = 0.$$

1.41 a) Prove that $|z^i| < e^\pi$ for all complex $z \neq 0$.

   b) Prove that there is no constant $M > 0$ such that $|\cos z| < M$ for all complex z.

1.42 If $w = u + iv$ (u, v real), show that

$$z^w = e^{u \log |z| - v \arg (z)} e^{i[v \log |z| + u \arg (z)]}.$$

1.43 a) Prove that Log $(z^w) = w$ Log $z + 2\pi in$, where n is an integer.

   b) Prove that $(z^w)^\alpha = z^{w\alpha} e^{2\pi in\alpha}$, where n is an integer.

1.44   i) If θ and a are real numbers, $-\pi < \theta \leq +\pi$, prove that

$$(\cos \theta + i \sin \theta)^a = \cos (a\theta) + i \sin (a\theta).$$

   ii) Show that, in general, the restriction $-\pi < \theta \leq +\pi$ is necessary in (i) by taking $\theta = -\pi, a = \frac{1}{2}$.

   iii) If a is an integer, show that the formula in (i) holds without any restriction on θ. In this case it is known as DeMoivre's theorem.

1.45 Use DeMoivre's theorem (Exercise 1.44) to derive the trigonometric identities

$$\sin 3\theta = 3 \cos^2 \theta \sin \theta - \sin^3 \theta,$$
$$\cos 3\theta = \cos^3 \theta - 3 \cos \theta \sin^2 \theta,$$

valid for real θ. Are these valid when θ is complex?

1.46 Define $\tan z = (\sin z)/(\cos z)$ and show that for $z = x + iy$, we have

$$\tan z = \frac{\sin 2x + i \sinh 2y}{\cos 2x + \cosh 2y}.$$

1.47 Let w be a given complex number. If $w \neq \pm 1$, show that there exist two values of $z = x + iy$ satisfying the conditions $\cos z = w$ and $-\pi < x \leq +\pi$. Find these values when $w = i$ and when $w = 2$.

1.48 Prove Lagrange's identity for complex numbers:

$$\left| \sum_{k=1}^{n} a_k b_k \right|^2 = \sum_{k=1}^{n} |a_k|^2 \sum_{k=1}^{n} |b_k|^2 - \sum_{1 \le k < j \le n} |a_k \bar{b}_j - a_j \bar{b}_k|^2.$$

Use this to deduce a Cauchy–Schwarz inequality for complex numbers.

1.49 a) By equating imaginary parts in DeMoivre's formula prove that

$$\sin n\theta = \sin^n \theta \left\{ \binom{n}{1} \cot^{n-1} \theta - \binom{n}{3} \cot^{n-3} \theta + \binom{n}{5} \cot^{n-5} \theta - + \cdots \right\}.$$

b) If $0 < \theta < \pi/2$, prove that

$$\sin (2m + 1)\theta = \sin^{2m+1}\theta \, P_m(\cot^2 \theta)$$

where P_m is the polynomial of degree m given by

$$P_m(x) = \binom{2m+1}{1} x^m - \binom{2m+1}{3} x^{m-1} + \binom{2m+1}{5} x^{m-2} - + \cdots$$

Use this to show that P_m has zeros at the m distinct points $x_k = \cot^2 \{\pi k/(2m + 1)\}$ for $k = 1, 2, \ldots, m$.

c) Show that the sum of the zeros of P_m is given by

$$\sum_{k=1}^{m} \cot^2 \frac{\pi k}{2m + 1} = \frac{m(2m - 1)}{3},$$

and that the sum of their squares is given by

$$\sum_{k=1}^{m} \cot^4 \frac{\pi k}{2m + 1} = \frac{m(2m - 1)(4m^2 + 10m - 9)}{45}.$$

NOTE. These identities can be used to prove that $\sum_{n=1}^{\infty} n^{-2} = \pi^2/6$ and $\sum_{n=1}^{\infty} n^{-4} = \pi^4/90$. (See Exercises 8.46 and 8.47.)

1.50 Prove that $z^n - 1 = \prod_{k=1}^{n} (z - e^{2\pi i k/n})$ for all complex z. Use this to derive the formula

$$\prod_{k=1}^{n-1} \sin \frac{k\pi}{n} = \frac{n}{2^{n-1}} \qquad \text{for } n \ge 2.$$

SUGGESTED REFERENCES FOR FURTHER STUDY

1.1 Apostol, T. M., *Calculus*, Vol. 1, 2nd ed. Xerox, Waltham, 1967.

1.2 Birkhoff, G., and MacLane, S., *A Survey of Modern Algebra*, 3rd ed. Macmillan, New York, 1965.

1.3 Cohen, L., and Ehrlich, G., *The Structure of the Real-Number System*. Van Nostrand, Princeton, 1963.

1.4 Gleason, A., *Fundamentals of Abstract Analysis*. Addison–Wesley, Reading, 1966.

1.5 Hardy, G. H., *A Course of Pure Mathematics*, 10th ed. Cambridge University Press, 1952.

1.6 Hobson, E. W., *The Theory of Functions of a Real Variable and the Theory of Fourier's Series*, Vol. 1, 3rd ed. Cambridge University Press, 1927.

1.7 Landau, E., *Foundations of Analysis*, 2nd ed. Chelsea, New York, 1960.

1.8 Robinson, A., *Non-standard Analysis*. North–Holland, Amsterdam, 1966.

1.9 Thurston, H. A., *The Number System*. Blackie, London, 1956.

1.10 Wilder, R. L., *Introduction to the Foundations of Mathematics*, 2nd ed. Wiley, New York, 1965.

CHAPTER 2

SOME BASIC NOTIONS
OF SET THEORY

2.1 INTRODUCTION

In discussing any branch of mathematics it is helpful to use the notation and terminology of set theory. This subject, which was developed by Boole and Cantor in the latter part of the 19th century, has had a profound influence on the development of mathematics in the 20th century. It has unified many seemingly disconnected ideas and has helped reduce many mathematical concepts to their logical foundations in an elegant and systematic way.

We shall not attempt a systematic treatment of the theory of sets but shall confine ourselves to a discussion of some of the more basic concepts. The reader who wishes to explore the subject further can consult the references at the end of this chapter.

A collection of objects viewed as a single entity will be referred to as a *set*. The objects in the collection will be called *elements* or *members* of the set, and they will be said to *belong to* or to be *contained in* the set. The set, in turn, will be said to *contain* or to be *composed of* its elements. For the most part we shall be interested in sets of mathematical objects; that is, sets of numbers, points, functions, curves, etc. However, since much of the theory of sets does not depend on the nature of the individual objects in the collection, we gain a great economy of thought by discussing sets whose elements may be objects of any kind. It is because of this quality of generality that the theory of sets has had such a strong effect in furthering the development of mathematics.

2.2 NOTATIONS

Sets will usually be denoted by capital letters:

$$A, B, C, \ldots, X, Y, Z,$$

and *elements* by lower-case letters: $a, b, c, \ldots, x, y, z$. We write $x \in S$ to mean "x is an element of S," or "x belongs to S." If x does not belong to S, we write $x \notin S$. We sometimes designate sets by displaying the elements in braces; for example, the set of positive even integers less than 10 is denoted by $\{2, 4, 6, 8\}$. If S is the collection of all x which satisfy a property P, we indicate this briefly by writing $S = \{x : x \text{ satisfies } P\}$.

From a given set we can form new sets, called *subsets* of the given set. For example, the set consisting of all positive integers less than 10 which are divisible by 4, namely, $\{4, 8\}$, is a subset of the set of even integers less than 10. In general, we say that a set A is a subset of B, and we write $A \subseteq B$ whenever every element of A also belongs to B. The statement $A \subseteq B$ does not rule out the possibility that $B \subseteq A$. In fact, we have both $A \subseteq B$ and $B \subseteq A$ if, and only if, A and B have the same elements. In this case we shall call the sets A and B equal and we write $A = B$. If A and B are not equal, we write $A \neq B$. If $A \subseteq B$ but $A \neq B$, then we say that A is a *proper subset* of B.

It is convenient to consider the possibility of a set which contains no elements whatever; this set is called the *empty set* and we agree to call it a subset of every set. The reader may find it helpful to picture a set as a box containing certain objects, its elements. The empty set is then an empty box. We denote the empty set by the symbol $\emptyset$.

2.3 ORDERED PAIRS

Suppose we have a set consisting of two elements a and b; that is, the set $\{a, b\}$. By our definition of equality this set is the same as the set $\{b, a\}$, since no question of order is involved. However, it is also necessary to consider sets of two elements in which order *is* important. For example, in analytic geometry of the plane, the coordinates (x, y) of a point represent an *ordered pair* of numbers. The *point* (3, 4) is different from the point (4, 3), whereas the *set* $\{3, 4\}$ is the same as the set $\{4, 3\}$. When we wish to consider a set of two elements a and b as being *ordered*, we shall enclose the elements in parentheses: (a, b). Then a is called the first element and b the second. It is possible to give a purely set-theoretic definition of the concept of an ordered pair of objects (a, b). One such definition is the following:

Definition 2.1. $(a, b) = \{\{a\}, \{a, b\}\}$.

This definition states that (a, b) is a set containing two elements, $\{a\}$ and $\{a, b\}$. Using this definition, we can prove the following theorem:

Theorem 2.2. $(a, b) = (c, d)$ *if, and only if, $a = c$ and $b = d$.*

This theorem shows that Definition 2.1 is a "reasonable" definition of an ordered pair, in the sense that the object a has been distinguished from the object b. The proof of Theorem 2.2 will be an instructive exercise for the reader. (See Exercise 2.1.)

2.4 CARTESIAN PRODUCT OF TWO SETS

Definition 2.3. *Given two sets A and B, the set of all ordered pairs (a, b) such that $a \in A$ and $b \in B$ is called the cartesian product of A and B, and is denoted by $A \times B$.*

Example. If **R** denotes the set of all real numbers, then $\mathbf{R} \times \mathbf{R}$ is the set of all complex numbers.

2.5 RELATIONS AND FUNCTIONS

Let x and y denote real numbers, so that the ordered pair (x, y) can be thought of as representing the rectangular coordinates of a point in the xy-plane (or a complex number). We frequently encounter such expressions as

$$xy = 1, \qquad x^2 + y^2 = 1, \qquad x^2 + y^2 \le 1, \qquad x < y. \qquad \text{(a)}$$

Each of these expressions defines a certain set of ordered pairs (x, y) of real numbers, namely, the set of all pairs (x, y) for which the expression is satisfied. Such a set of ordered pairs is called a *plane relation*. The corresponding set of points plotted in the xy-plane is called the *graph* of the relation. The graphs of the relations described in (a) are shown in Fig. 2.1.

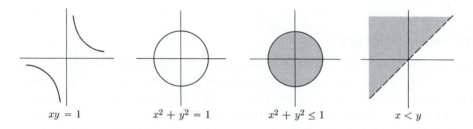

$$xy = 1 \qquad\qquad x^2 + y^2 = 1 \qquad\qquad x^2 + y^2 \le 1 \qquad\qquad x < y$$

Figure 2.1

The concept of relation can be formulated quite generally so that the objects x and y in the pairs (x, y) need not be numbers but may be objects of any kind.

Definition 2.4. *Any set of ordered pairs is called a relation.*

If S is a relation, the set of all elements x that occur as first members of pairs (x, y) in S is called the *domain* of S, denoted by $\mathcal{D}(S)$. The set of second members y is called the *range* of S, denoted by $\mathcal{R}(S)$.

The first example shown in Fig. 2.1 is a special kind of relation known as a *function*.

Definition 2.5. *A function F is a set of ordered pairs (x, y), no two of which have the same first member. That is, if $(x, y) \in F$ and $(x, z) \in F$, then $y = z$.*

The definition of function requires that for every x in the domain of F there is exactly one y such that $(x, y) \in F$. It is customary to call y the *value of F at x* and to write

$$y = F(x)$$

instead of $(x, y) \in F$ to indicate that the pair (x, y) is in the set F.

As an alternative to describing a function F by specifying the pairs it contains, it is usually preferable to describe the domain of F, and then, for each x in the domain, to describe how the function value $F(x)$ is obtained. In this connection, we have the following theorem whose proof is left as an exercise for the reader.

Theorem 2.6. *Two functions F and G are equal if and only if*

a) $\mathscr{D}(F) = \mathscr{D}(G)$ (*F and G have the same domain*), *and*

b) $F(x) = G(x)$ *for every x in $\mathscr{D}(F)$.*

2.6 FURTHER TERMINOLOGY CONCERNING FUNCTIONS

When the domain $\mathscr{D}(F)$ is a subset of $\mathbf{R}$, then F is called a *function of one real variable*. If $\mathscr{D}(F)$ is a subset of $\mathbf{C}$, the complex number system, then F is called a *function of a complex variable*.

If $\mathscr{D}(F)$ is a subset of a cartesian product $A \times B$, then F is called a *function of two variables*. In this case we denote the function values by $F(a, b)$ instead of $F((a, b))$. A function of two real variables is one whose domain is a subset of $\mathbf{R} \times \mathbf{R}$.

If S is a subset of $\mathscr{D}(F)$, we say that F is *defined on S*. In this case, the set of $F(x)$ such that $x \in S$ is called the *image of S under F* and is denoted by $F(S)$. If T is any set which contains $F(S)$, then F is also called a *mapping from S to T*. This is often denoted by writing

$$F : S \rightarrow T.$$

If $F(S) = T$, the mapping is said to be *onto T*. A mapping of S into itself is sometimes called a *transformation*.

Consider, for example, the function of a complex variable defined by the equation $F(z) = z^2$. This function maps every sector S of the form $0 \le \arg(z) \le \alpha \le \pi/2$ of the complex z-plane onto a sector $F(S)$ described by the inequalities $0 \le \arg[F(z)] \le 2\alpha$. (See Fig. 2.2.)

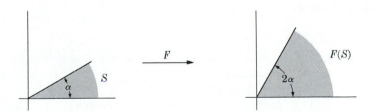

Figure 2.2

If two functions F and G satisfy the inclusion relation $G \subseteq F$, we say that G is a *restriction* of F or that F is an *extension* of G. In particular, if S is a subset of $\mathscr{D}(F)$ and if G is defined by the equation

$$G(x) = F(x) \qquad \text{for all } x \text{ in } S,$$

then we call G the restriction of F to S. The function G consists of those pairs $(x, F(x))$ such that $x \in S$. Its domain is S and its range is $F(S)$.

2.7 ONE-TO-ONE FUNCTIONS AND INVERSES

Definition 2.7. *Let F be a function defined on S. We say F is one-to-one on S if, and only if, for every x and y in S,*

$$F(x) = F(y) \quad implies \quad x = y.$$

This is the same as saying that a function which is one-to-one on S assigns distinct function values to distinct members of S. Such functions are also called *injective*. They are important because, as we shall presently see, they possess *inverses*. However, before stating the definition of the inverse of a function, it is convenient to introduce a more general notion, that of the *converse* of a relation.

Definition 2.8. *Given a relation S, the new relation $\check{S}$ defined by*

$$\check{S} = \{(a, b) : (b, a) \in S\}$$

is called the converse of S.

Thus an ordered pair (a, b) belongs to $\check{S}$ if, and only if, the pair (b, a), with elements interchanged, belongs to S. When S is a *plane relation*, this simply means that the graph of $\check{S}$ is the reflection of the graph of S with respect to the line $y = x$. In the relation defined by $x < y$, the converse relation is defined by $y < x$.

Definition 2.9. *Suppose that the relation F is a function. Consider the converse relation $\check{F}$, which may or may not be a function. If $\check{F}$ is also a function, then $\check{F}$ is called the inverse of F and is denoted by F^{-1}.*

Figure 2.3(a) illustrates an example of a function F for which $\check{F}$ is not a function. In Fig. 2.3(b) both F and its converse are functions.

The next theorem tells us that a function which is one-to-one on its domain always has an inverse.

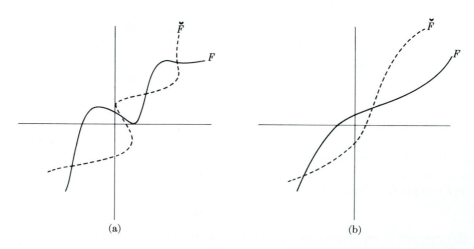

(a)   (b)

Figure 2.3

Theorem 2.10. *If the function F is one-to-one on its domain, then $\check{F}$ is also a function.*

Proof. To show that $\check{F}$ is a function, we must show that if $(x, y) \in \check{F}$ and $(x, z) \in \check{F}$, then $y = z$. But $(x, y) \in \check{F}$ means that $(y, x) \in F$; that is, $x = F(y)$. Similarly, $(x, z) \in \check{F}$ means that $x = F(z)$. Thus $F(y) = F(z)$ and, since we are assuming that F is one-to-one, this implies $y = z$. Hence, $\check{F}$ is a function.

NOTE. The same argument shows that if F is one-to-one on a subset S of $\mathscr{D}(F)$, then the restriction of F to S has an inverse.

2.8 COMPOSITE FUNCTIONS

Definition 2.11. *Given two functions F and G such that $\mathscr{R}(F) \subseteq \mathscr{D}(G)$, we can form a new function, the composite $G \circ F$ of G and F, defined as follows: for every x in the domain of F, $(G \circ F)(x) = G[F(x)]$.*

Since $\mathscr{R}(F) \subseteq \mathscr{D}(G)$, the element $F(x)$ is in the domain of G, and therefore it makes sense to consider $G[F(x)]$. In general, it is not true that $G \circ F = F \circ G$. In fact, $F \circ G$ may be meaningless unless the range of G is contained in the domain of F. However, the associative law,

$$H \circ (G \circ F) = (H \circ G) \circ F,$$

always holds whenever each side of the equation has a meaning. (Verification will be an interesting exercise for the reader. See Exercise 2.4.)

2.9 SEQUENCES

Among the important examples of functions are those defined on subsets of the integers.

Definition 2.12. *By a finite sequence of n terms we shall understand a function F whose domain is the set of numbers $\{1, 2, \ldots, n\}$.*

The range of F is the set $\{F(1), F(2), F(3), \ldots, F(n)\}$, customarily written $\{F_1, F_2, F_3, \ldots, F_n\}$. The elements of the range are called *terms* of the sequence and, of course, they may be arbitrary objects of any kind.

Definition 2.13. *By an infinite sequence we shall mean a function F whose domain is the set $\{1, 2, 3, \ldots\}$ of all positive integers. The range of F, that is, the set $\{F(1), F(2), F(3), \ldots\}$, is also written $\{F_1, F_2, F_3, \ldots\}$, and the function value F_n is called the nth term of the sequence.*

For brevity, we shall occasionally use the notation $\{F_n\}$ to denote the infinite sequence whose nth term is F_n.

Let $s = \{s_n\}$ be an infinite sequence, and let k be a function whose domain is the set of positive integers and whose range is a subset of the positive integers.

Assume that k is "order-preserving," that is, assume that

$$k(m) < k(n), \qquad \text{if } m < n.$$

Then the composite function $s \circ k$ is defined for all integers $n \geq 1$, and for every such n we have

$$(s \circ k)(n) = s_{k(n)}.$$

Such a composite function is said to be a *subsequence* of s. Again, for brevity, we often use the notation $\{s_{k(n)}\}$ or $\{s_{k_n}\}$ to denote the subsequence of $\{s_n\}$ whose nth term is $s_{k(n)}$.

Example. Let $s = \{1/n\}$ and let k be defined by $k(n) = 2^n$. Then $s \circ k = \{1/2^n\}$.

2.10 SIMILAR (EQUINUMEROUS) SETS

Definition 2.14. *Two sets A and B are called similar, or equinumerous, and we write $A \sim B$, if and only if there exists a one-to-one function F whose domain is the set A and whose range is the set B.*

We also say that F establishes a *one-to-one correspondence* between the sets A and B. Clearly, every set A is similar to itself (take F to be the "identity" function for which $F(x) = x$ for all x in A). Furthermore, if $A \sim B$ then $B \sim A$, because if F is a one-to-one function which makes A similar to B, then F^{-1} will make B similar to A. Also, if $A \sim B$ and if $B \sim C$, then $A \sim C$. (The proof is left to the reader.)

2.11 FINITE AND INFINITE SETS

A set S is called *finite* and is said to contain n elements if

$$S \sim \{1, 2, \ldots, n\}.$$

The integer n is called the *cardinal number* of S. It is an easy exercise to prove that if $\{1, 2, \ldots, n\} \sim \{1, 2, \ldots, m\}$ then $m = n$. Therefore, the cardinal number of a finite set is well defined. The empty set is also considered finite. Its cardinal number is defined to be 0.

Sets which are not finite are called *infinite sets*. The chief difference between the two is that an infinite set must be similar to some proper subset of itself, whereas a finite set cannot be similar to any proper subset of itself. (See Exercise 2.13.) For example, the set $\mathbf{Z}^+$ of all positive integers is similar to the proper subset $\{2, 4, 8, 16, \ldots\}$ consisting of powers of 2. The one-to-one function F which makes them similar is defined by $F(x) = 2^x$ for each x in $\mathbf{Z}^+$.

2.12 COUNTABLE AND UNCOUNTABLE SETS

A set S is said to be *countably infinite* if it is equinumerous with the set of all positive integers; that is, if

$$S \sim \{1, 2, 3, \dots\}.$$

In this case there is a function f which establishes a one-to-one correspondence between the positive integers and the elements of S; hence the set S can be displayed as follows:

$$S = \{f(1), f(2), f(3), \dots\}.$$

Often we use subscripts and denote $f(k)$ by a_k (or by a similar notation) and we write $S = \{a_1, a_2, a_3, \dots\}$. The important thing here is that the correspondence enables us to use the positive integers as "labels" for the elements of S. A countably infinite set is said to have cardinal number $\aleph_0$ (read: *aleph nought*).

Definition 2.15. *A set S is called countable if it is either finite or countably infinite. A set which is not countable is called uncountable.*

The words *denumerable* and *nondenumerable* are sometimes used in place of *countable* and *uncountable*.

Theorem 2.16. *Every subset of a countable set is countable.*

Proof. Let S be the given countable set and assume $A \subseteq S$. If A is finite, there is nothing to prove, so we can assume that A is infinite (which means S is also infinite). Let $s = \{s_n\}$ be an infinite sequence of distinct terms such that

$$S = \{s_1, s_2, \dots\}.$$

Define a function on the positive integers as follows:

Let $k(1)$ be the smallest positive integer m such that $s_m \in A$. Assuming that $k(1), k(2), \dots, k(n-1)$ have been defined, let $k(n)$ be the smallest positive integer $m > k(n-1)$ such that $s_m \in A$. Then k is order-preserving: $m > n$ implies $k(m) > k(n)$. Form the composite function $s \circ k$. The domain of $s \circ k$ is the set of positive integers and the range of $s \circ k$ is A. Furthermore, $s \circ k$ is one-to-one, since

$$s[k(n)] = s[k(m)],$$

implies

$$s_{k(n)} = s_{k(m)},$$

which implies $k(n) = k(m)$, and this implies $n = m$. This proves the theorem.

2.13 UNCOUNTABILITY OF THE REAL NUMBER SYSTEM

The next theorem shows that there are infinite sets which are not countable.

Theorem 2.17. *The set of all real numbers is uncountable.*

Proof. It suffices to show that the set of x satisfying $0 < x < 1$ is uncountable. If the real numbers in this interval were countable, there would be a sequence $s = \{s_n\}$ whose terms would constitute the whole interval. We shall show that this is impossible by constructing, in the interval, a real number which is not a term of this sequence. Write each s_n as an infinite decimal:

$$s_n = 0.u_{n,1}u_{n,2}u_{n,3}\cdots,$$

where each $u_{n,i}$ is $0, 1, \ldots,$ or 9. Consider the real number y which has the decimal expansion

$$y = 0.v_1v_2v_3\cdots,$$

where

$$v_n = \begin{cases} 1, & \text{if } u_{n,n} \neq 1, \\ 2, & \text{if } u_{n,n} = 1. \end{cases}$$

Then no term of the sequence $\{s_n\}$ can be equal to y, since y differs from s_1 in the first decimal place, differs from s_2 in the second decimal place, $\ldots$, from s_n in the nth decimal place. (A situation like $s_n = 0.1999\ldots$ and $y = 0.2000\ldots$ cannot occur here because of the way the v_n are chosen.) Since $0 < y < 1$, the theorem is proved.

Theorem 2.18. *Let $\mathbf{Z}^+$ denote the set of all positive integers. Then the cartesian product $\mathbf{Z}^+ \times \mathbf{Z}^+$ is countable.*

Proof. Define a function f on $\mathbf{Z}^+ \times \mathbf{Z}^+$ as follows:

$$f(m, n) = 2^m 3^n, \qquad \text{if } (m, n) \in \mathbf{Z}^+ \times \mathbf{Z}^+.$$

Then f is one-to-one on $\mathbf{Z}^+ \times \mathbf{Z}^+$ and the range of f is a subset of $\mathbf{Z}^+$.

2.14 SET ALGEBRA

Given two sets A_1 and A_2, we define a new set, called the *union* of A_1 and A_2, denoted by $A_1 \cup A_2$, as follows:

Definition 2.19. *The union $A_1 \cup A_2$ is the set of those elements which belong either to A_1 or to A_2 or to both.*

This is the same as saying that $A_1 \cup A_2$ consists of those elements which belong to at least one of the sets A_1, A_2. Since there is no question of order involved in this definition, the union $A_1 \cup A_2$ is the same as $A_2 \cup A_1$; that is, set addition is commutative. The definition is also phrased in such a way that set addition is associative:

$$A_1 \cup (A_2 \cup A_3) = (A_1 \cup A_2) \cup A_3.$$

The definition of union can be extended to any finite or infinite collection of sets:

Definition 2.20. *If F is an arbitrary collection of sets, then the union of all the sets in F is defined to be the set of those elements which belong to at least one of the sets in F, and is denoted by*

$$\bigcup_{A \in F} A.$$

If F is a finite collection of sets, $F = \{A_1, \ldots, A_n\}$, we write

$$\bigcup_{A \in F} A = \bigcup_{k=1}^{n} A_k = A_1 \cup A_2 \cup \cdots \cup A_n.$$

If F is a countable collection, $F = \{A_1, A_2, \ldots\}$, we write

$$\bigcup_{A \in F} A = \bigcup_{k=1}^{\infty} A_k = A_1 \cup A_2 \cup \cdots$$

Definition 2.21. *If F is an arbitrary collection of sets, the intersection of all sets in F is defined to be the set of those elements which belong to every one of the sets in F, and is denoted by*

$$\bigcap_{A \in F} A.$$

The intersection of two sets A_1 and A_2 is denoted by $A_1 \cap A_2$ and consists of those elements common to both sets. If A_1 and A_2 have no elements in common, then $A_1 \cap A_2$ is the empty set and A_1 and A_2 are said to be *disjoint*. If F is a finite collection (as above), we write

$$\bigcap_{A \in F} A = \bigcap_{k=1}^{n} A_k = A_1 \cap A_2 \cap \cdots \cap A_n,$$

and if F is a countable collection, we write

$$\bigcap_{A \in F} A = \bigcap_{k=1}^{\infty} A_k = A_1 \cap A_2 \cap \cdots$$

If the sets in the collection have no elements in common, their intersection is the empty set. Our definitions of union and intersection apply, of course, even when F is not countable. Because of the way we have defined unions and intersections, the commutative and associative laws are automatically satisfied.

Definition 2.22. *The complement of A relative to B, denoted by B − A, is defined to be the set*

$$B - A = \{x : x \in B, \text{ but } x \notin A\}.$$

Note that $B - (B - A) = A$ whenever $A \subseteq B$. Also note that $B - A = B$ if $B \cap A$ is empty.

The notions of union, intersection, and complement are illustrated in Fig. 2.4.

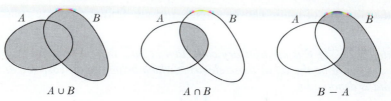

$A \cup B$ $A \cap B$ $B - A$

Figure 2.4

Theorem 2.23. *Let F be a collection of sets. Then for any set B, we have*

$$B - \bigcup_{A \in F} A = \bigcap_{A \in F} (B - A),$$

and

$$B - \bigcap_{A \in F} A = \bigcup_{A \in F} (B - A).$$

Proof. Let $S = \bigcup_{A \in F} A$, $T = \bigcap_{A \in F} (B - A)$. If $x \in B - S$, then $x \in B$, but $x \notin S$. Hence, it is not true that x belongs to at least one A in F; therefore x belongs to no A in F. Hence, for every A in F, $x \in B - A$. But this implies $x \in T$, so that $B - S \subseteq T$. Reversing the steps, we obtain $T \subseteq B - S$, and this proves that $B - S = T$. To prove the second statement, use a similar argument.

2.15 COUNTABLE COLLECTIONS OF COUNTABLE SETS

Definition 2.24. *If F is a collection of sets such that every two distinct sets in F are disjoint, then F is said to be a collection of disjoint sets.*

Theorem 2.25. *If F is a countable collection of disjoint sets, say $F = \{A_1, A_2, \ldots\}$, such that each set A_n is countable, then the union $\bigcup_{k=1}^{\infty} A_k$ is also countable.*

Proof. Let $A_n = \{a_{1,n}, a_{2,n}, a_{3,n} \ldots\}$, $n = 1, 2, \ldots$, and let $S = \bigcup_{k=1}^{\infty} A_k$. Then every element x of S is in at least one of the sets in F and hence $x = a_{m,n}$ for some pair of integers (m, n). The pair (m, n) is uniquely determined by x, since F is a collection of disjoint sets. Hence the function f defined by $f(x) = (m, n)$ if $x = a_{m,n}$, $x \in S$, has domain S. The range $f(S)$ is a subset of $\mathbf{Z}^+ \times \mathbf{Z}^+$ (where $\mathbf{Z}^+$ is the set of positive integers) and hence is countable. But f is one-to-one and therefore $S \sim f(S)$, which means that S is also countable.

Theorem 2.26. *If $F = \{A_1, A_2, \ldots\}$ is a countable collection of sets, let $G = \{B_1, B_2, \ldots\}$, where $B_1 = A_1$ and, for $n > 1$,*

$$B_n = A_n - \bigcup_{k=1}^{n-1} A_k.$$

Then G is a collection of disjoint sets, and we have

$$\bigcup_{k=1}^{\infty} A_k = \bigcup_{k=1}^{\infty} B_k.$$

Proof. Each set B_n is constructed so that it has no elements in common with the earlier sets $B_1, B_2, \ldots, B_{n-1}$. Hence G is a collection of disjoint sets. Let $A = \bigcup_{k=1}^{\infty} A_k$ and $B = \bigcup_{k=1}^{\infty} B_k$. We shall show that $A = B$. First of all, if $x \in A$, then $x \in A_k$ for some k. If n is the smallest such k, then $x \in A_n$ but $x \notin \bigcup_{k=1}^{n-1} A_k$, which means that $x \in B_n$, and therefore $x \in B$. Hence $A \subseteq B$. Conversely, if $x \in B$, then $x \in B_n$ for some n, and therefore $x \in A_n$ for this same n. Thus $x \in A$ and this proves that $B \subseteq A$.

Using Theorems 2.25 and 2.26, we immediately obtain

Theorem 2.27. *If F is a countable collection of countable sets, then the union of all sets in F is also a countable set.*

Example 1. The set $\mathbf{Q}$ of all rational numbers is a countable set.

Proof. Let A_n denote the set of all positive rational numbers having denominator n. The set of all positive rational numbers is equal to $\bigcup_{k=1}^{\infty} A_k$. From this it follows that $\mathbf{Q}$ is countable, since each A_n is countable.

Example 2. The set S of intervals with rational endpoints is a countable set.

Proof. Let $\{x_1, x_2, \ldots\}$ denote the set of rational numbers and let A_n be the set of all intervals whose left endpoint is x_n and whose right endpoint is rational. Then A_n is countable and $S = \bigcup_{k=1}^{\infty} A_k$.

EXERCISES

2.1 Prove Theorem 2.2.   *Hint.* $(a, b) = (c, d)$ means $\{\{a\}, \{a, b\}\} = \{\{c\}, \{c, d\}\}$. Now appeal to the definition of set equality.

2.2 Let S be a relation and let $\mathscr{D}(S)$ be its domain. The relation S is said to be

   i) *reflexive* if $a \in \mathscr{D}(S)$ implies $(a, a) \in S$,

   ii) *symmetric* if $(a, b) \in S$ implies $(b, a) \in S$,

   iii) *transitive* if $(a, b) \in S$ and $(b, c) \in S$ implies $(a, c) \in S$.

A relation which is symmetric, reflexive, and transitive is called an *equivalence relation*. Determine which of these properties is possessed by S, if S is the set of all pairs of real numbers (x, y) such that

  a) $x \leq y$,   b) $x < y$,   c) $x < |y|$,

  d) $x^2 + y^2 = 1$,   e) $x^2 + y^2 < 0$,   f) $x^2 + x = y^2 + y$.

2.3 The following functions F and G are defined for all real x by the equations given. In each case where the composite function $G \circ F$ can be formed, give the domain of $G \circ F$ and a formula (or formulas) for $(G \circ F)(x)$.

  a) $F(x) = 1 - x$, $G(x) = x^2 + 2x$.

  b) $F(x) = x + 5$, $G(x) = |x|/x$, if $x \neq 0$, $G(0) = 1$.

  c) $F(x) = \begin{cases} 2x, & \text{if } 0 \leq x \leq 1, \\ 1, & \text{otherwise,} \end{cases}$ $G(x) = \begin{cases} x^2, & \text{if } 0 \leq x \leq 1, \\ 0, & \text{otherwise.} \end{cases}$

Find $F(x)$ if $G(x)$ and $G[F(x)]$ are given as follows:

    d) $G(x) = x^3$, $G[F(x)] = x^3 - 3x^2 + 3x - 1$.

    e) $G(x) = 3 + x + x^2$, $G[F(x)] = x^2 - 3x + 5$.

2.4 Given three functions F, G, H, what restrictions must be placed on their domains so that the following four composite functions can be defined?

$$G \circ F, \qquad H \circ G, \qquad H \circ (G \circ F), \qquad (H \circ G) \circ F.$$

Assuming that $H \circ (G \circ F)$ and $(H \circ G) \circ F$ can be defined, prove the associative law:

$$H \circ (G \circ F) = (H \circ G) \circ F.$$

2.5 Prove the following set-theoretic identities for union and intersection:

    a) $A \cup (B \cup C) = (A \cup B) \cup C$, $A \cap (B \cap C) = (A \cap B) \cap C$.

    b) $A \cap (B \cup C) = (A \cap B) \cup (A \cap C)$.

    c) $(A \cup B) \cap (A \cup C) = A \cup (B \cap C)$.

    d) $(A \cup B) \cap (B \cup C) \cap (C \cup A) = (A \cap B) \cup (A \cap C) \cup (B \cap C)$.

    e) $A \cap (B - C) = (A \cap B) - (A \cap C)$.

    f) $(A - C) \cap (B - C) = (A \cap B) - C$.

    g) $(A - B) \cup B = A$ if, and only if, $B \subseteq A$.

2.6 Let $f : S \to T$ be a function. If A and B are arbitrary subsets of S, prove that

$$f(A \cup B) = f(A) \cup f(B) \qquad \text{and} \qquad f(A \cap B) \subseteq f(A) \cap f(B).$$

Generalize to arbitrary unions and intersections.

2.7 Let $f : S \to T$ be a function. If $Y \subseteq T$, we denote by $f^{-1}(Y)$ the largest subset of S which f maps into Y. That is,

$$f^{-1}(Y) = \{x : x \in S \text{ and } f(x) \in Y\}.$$

The set $f^{-1}(Y)$ is called the *inverse image* of Y under f. Prove the following for arbitrary subsets X of S and Y of T.

    a) $X \subseteq f^{-1}[f(X)]$,   b) $f[f^{-1}(Y)] \subseteq Y$,

    c) $f^{-1}[Y_1 \cup Y_2] = f^{-1}(Y_1) \cup f^{-1}(Y_2)$,

    d) $f^{-1}(Y_1 \cap Y_2) = f^{-1}(Y_1) \cap f^{-1}(Y_2)$,

    e) $f^{-1}(T - Y) = S - f^{-1}(Y)$.

    f) Generalize (c) and (d) to arbitrary unions and intersections.

2.8 Refer to Exercise 2.7. Prove that $f[f^{-1}(Y)] = Y$ for every subset Y of T if, and only if, $T = f(S)$.

2.9 Let $f : S \to T$ be a function. Prove that the following statements are equivalent.

    a) f is one-to-one on S.

    b) $f(A \cap B) = f(A) \cap f(B)$ for all subsets A, B of S.

    c) $f^{-1}[f(A)] = A$ for every subset A of S.

    d) For all disjoint subsets A and B of S, the images $f(A)$ and $f(B)$ are disjoint.

e) For all subsets A and B of S with $B \subseteq A$, we have

$$f(A - B) = f(A) - f(B).$$

2.10 Prove that if $A \sim B$ and $B \sim C$, then $A \sim C$.

2.11 If $\{1, 2, \ldots, n\} \sim \{1, 2, \ldots, m\}$, prove that $m = n$.

2.12 If S is an infinite set, prove that S contains a countably infinite subset. *Hint.* Choose an element a_1 in S and consider $S - \{a_1\}$.

2.13 Prove that every infinite set S contains a proper subset similar to S.

2.14 If A is a countable set and B an uncountable set, prove that $B - A$ is similar to B.

2.15 A real number is called *algebraic* if it is a root of an algebraic equation $f(x) = 0$, where $f(x) = a_0 + a_1 x + \cdots + a_n x^n$ is a polynomial with integer coefficients. Prove that the set of all polynomials with integer coefficients is countable and deduce that the set of algebraic numbers is also countable.

2.16 Let S be a finite set consisting of n elements and let T be the collection of all subsets of S. Show that T is a finite set and find the number of elements in T.

2.17 Let $\mathbf{R}$ denote the set of real numbers and let S denote the set of all real-valued functions whose domain is $\mathbf{R}$. Show that S and $\mathbf{R}$ are not equinumerous. *Hint.* Assume $S \sim \mathbf{R}$ and let f be a one-to-one function such that $f(\mathbf{R}) = S$. If $a \in \mathbf{R}$, let $g_a = f(a)$ be the real-valued function in S which corresponds to the real number a. Now define h by the equation $h(x) = 1 + g_x(x)$ if $x \in \mathbf{R}$, and show that $h \notin S$.

2.18 Let S be the collection of all sequences whose terms are the integers 0 and 1. Show that S is uncountable.

2.19 Show that the following sets are countable:

    a) the set of circles in the complex plane having rational radii and centers with rational coordinates,

    b) any collection of disjoint intervals of positive length.

2.20 Let f be a real-valued function defined for every x in the interval $0 \le x \le 1$. Suppose there is a positive number M having the following property: for every choice of a finite number of points $x_1, x_2, \ldots, x_n$ in the interval $0 \le x \le 1$, the sum

$$|f(x_1) + \cdots + f(x_n)| \le M.$$

Let S be the set of those x in $0 \le x \le 1$ for which $f(x) \ne 0$. Prove that S is countable.

2.21 Find the fallacy in the following "proof" that the set of all intervals of positive length is countable.

    Let $\{x_1, x_2, \ldots\}$ denote the countable set of rational numbers and let I be any interval of positive length. Then I contains infinitely many rational points x_n, but among these there will be one with *smallest index n*. Define a function F by means of the equation $F(I) = n$, if x_n is the rational number with smallest index in the interval I. This function establishes a one-to-one correspondence between the set of all intervals and a subset of the positive integers. Hence the set of all intervals is countable.

2.22 Let S denote the collection of all subsets of a given set T. Let $f : S \to \mathbf{R}$ be a real-valued function defined on S. The function f is called *additive* if $f(A \cup B) = f(A) + f(B)$ whenever A and B are disjoint subsets of T. If f is additive, prove that for any two subsets

A and *B* we have

$$f(A \cup B) = f(A) + f(B - A) \quad \text{and} \quad f(A \cup B) = f(A) + f(B) - f(A \cap B).$$

2.23 Refer to Exercise 2.22. Assume f is additive and assume also that the following relations hold for two particular subsets A and B of T:

$$f(A \cup B) = f(A') + f(B') - f(A')f(B')$$

$$f(A \cap B) = f(A)f(B), \quad f(A) + f(B) \neq f(T),$$

where $A' = T - A$, $B' = T - B$. Prove that these relations determine $f(T)$, and compute the value of $f(T)$.

SUGGESTED REFERENCES FOR FURTHER STUDY

2.1 Boas, R. P., *A Primer of Real Functions*. Carus Monograph No. 13. Wiley, New York, 1960.

2.2 Fraenkel, A., *Abstract Set Theory*, 3rd ed. North–Holland, Amsterdam, 1965.

2.3 Gleason, A., *Fundamentals of Abstract Analysis*. Addison–Wesley, Reading, 1966.

2.4 Halmos, P. R., *Naive Set Theory*. Van Nostrand, New York, 1960.

2.5 Kamke, E., *Theory of Sets*. F. Bagemihl, translator. Dover, New York, 1950.

2.6 Kaplansky, I., *Set Theory and Metric Spaces*. Allyn and Bacon, Boston, 1972.

2.7 Rotman, B., and Kneebone, G. T., *The Theory of Sets and Transfinite Numbers*. Elsevier, New York, 1968.

CHAPTER 3

ELEMENTS OF
POINT SET TOPOLOGY

3.1 INTRODUCTION

A large part of the previous chapter dealt with "abstract" sets, that is, sets of arbitrary objects. In this chapter we specialize our sets to be sets of real numbers, sets of complex numbers, and more generally, sets in higher-dimensional spaces.

In this area of study it is convenient and helpful to use geometric terminology. Thus, we speak about sets of points on the real line, sets of points in the plane, or sets of points in some higher-dimensional space. Later in this book we will study functions defined on point sets, and it is desirable to become acquainted with certain fundamental types of point sets, such as *open* sets, *closed* sets, and *compact* sets, before beginning the study of functions. The study of these sets is called *point set topology*.

3.2 EUCLIDEAN SPACE R^n

A point in two-dimensional space is an ordered pair of real numbers (x_1, x_2). Similarly, a point in three-dimensional space is an ordered triple of real numbers (x_1, x_2, x_3). It is just as easy to consider an ordered n-tuple of real numbers $(x_1, x_2, \ldots, x_n)$ and to refer to this as a point in n-dimensional space.

Definition 3.1. Let $n > 0$ be an integer. An ordered set of n real numbers $(x_1, x_2, \ldots, x_n)$ is called an n-dimensional point or a vector with n components. Points or vectors will usually be denoted by single bold-face letters; for example,

$$\mathbf{x} = (x_1, x_2, \ldots, x_n) \qquad or \qquad \mathbf{y} = (y_1, y_2, \ldots, y_n).$$

The number x_k is called the kth coordinate of the point $\mathbf{x}$ or the kth component of the vector $\mathbf{x}$. The set of all n-dimensional points is called n-dimensional Euclidean space or simply n-space, and is denoted by $\mathbf{R}^n$.

The reader may wonder whether there is any advantage in discussing spaces of dimension greater than three. Actually, the language of n-space makes many complicated situations much easier to comprehend. The reader is probably familiar enough with three-dimensional vector analysis to realize the advantage of writing the equations of motion of a system having three degrees of freedom as a single vector equation rather than as three scalar equations. There is a similar advantage if the system has n degrees of freedom.

47

Another advantage in studying n-space for a general n is that we are able to deal in one stroke with many properties common to 1-space, 2-space, 3-space, etc., that is, properties independent of the dimensionality of the space.

Higher-dimensional spaces arise quite naturally in such fields as relativity, and statistical and quantum mechanics. In fact, even infinite-dimensional spaces are quite common in quantum mechanics.

Algebraic operations on n-dimensional points are defined as follows:

Definition 3.2. *Let* $\mathbf{x} = (x_1, \ldots, x_n)$ *and* $\mathbf{y} = (y_1, \ldots, y_n)$ *be in* $\mathbf{R}^n$. *We define:*

a) *Equality:*
$$\mathbf{x} = \mathbf{y} \text{ if, and only if, } x_1 = y_1, \ldots, x_n = y_n.$$

b) *Sum:*
$$\mathbf{x} + \mathbf{y} = (x_1 + y_1, \ldots, x_n + y_n).$$

c) *Multiplication by real numbers* (*scalars*):
$$a\mathbf{x} = (ax_1, \ldots, ax_n) \qquad (a \text{ real}).$$

d) *Difference:*
$$\mathbf{x} - \mathbf{y} = \mathbf{x} + (-1)\mathbf{y}.$$

e) *Zero vector or origin:*
$$\mathbf{0} = (0, \ldots, 0).$$

f) *Inner product or dot product:*
$$\mathbf{x} \cdot \mathbf{y} = \sum_{k=1}^{n} x_k y_k.$$

g) *Norm or length:*
$$\|\mathbf{x}\| = (\mathbf{x} \cdot \mathbf{x})^{1/2} = \left(\sum_{k=1}^{n} x_k^2 \right)^{1/2}.$$

The norm $\|\mathbf{x} - \mathbf{y}\|$ is called the *distance* between $\mathbf{x}$ and $\mathbf{y}$.

NOTE. In the terminology of linear algebra, $\mathbf{R}^n$ is an example of a *linear space*.

Theorem 3.3. *Let* $\mathbf{x}$ *and* $\mathbf{y}$ *denote points in* $\mathbf{R}^n$. *Then we have:*

a) $\|\mathbf{x}\| \geq 0$, *and* $\|\mathbf{x}\| = 0$ *if, and only if,* $\mathbf{x} = \mathbf{0}$.

b) $\|a\mathbf{x}\| = |a| \|\mathbf{x}\|$ *for every real* a.

c) $\|\mathbf{x} - \mathbf{y}\| = \|\mathbf{y} - \mathbf{x}\|$.

d) $|\mathbf{x} \cdot \mathbf{y}| \leq \|\mathbf{x}\| \|\mathbf{y}\|$ (*Cauchy–Schwarz inequality*).

e) $\|\mathbf{x} + \mathbf{y}\| \leq \|\mathbf{x}\| + \|\mathbf{y}\|$ (*triangle inequality*).

Proof. Statements (a), (b) and (c) are immediate from the definition, and the Cauchy–Schwarz inequality was proved in Theorem 1.23. Statement (e) follows from (d) because

$$\|x + y\|^2 = \sum_{k=1}^{n} (x_k + y_k)^2 = \sum_{k=1}^{n} (x_k^2 + 2x_k y_k + y_k^2)$$

$$= \|x\|^2 + 2x \cdot y + \|y\|^2 \le \|x\|^2 + 2\|x\| \|y\| + \|y\|^2 = (\|x\| + \|y\|)^2.$$

NOTE. Sometimes the triangle inequality is written in the form

$$\|x - z\| \le \|x - y\| + \|y - z\|.$$

This follows from (e) by replacing x by $x - y$ and y by $y - z$. We also have

$$|\|x\| - \|y\|| \le \|x - y\|.$$

Definition 3.4. *The unit coordinate vector u_k in R^n is the vector whose kth component is 1 and whose remaining components are zero. Thus,*

$$u_1 = (1, 0, \ldots, 0), \qquad u_2 = (0, 1, 0, \ldots, 0), \ldots, u_n = (0, 0, \ldots, 0, 1).$$

If $x = (x_1, \ldots, x_n)$ then $x = x_1 u_1 + \cdots + x_n u_n$ and $x_1 = x \cdot u_1$, $x_2 = x \cdot u_2, \ldots, x_n = x \cdot u_n$. The vectors $u_1, \ldots, u_n$ are also called *basis vectors*.

3.3 OPEN BALLS AND OPEN SETS IN R^n

Let a be a given point in R^n and let r be a given positive number. The set of all points x in R^n such that

$$\|x - a\| < r,$$

is called an open *n-ball* of radius r and center a. We denote this set by $B(a)$ or by $B(a; r)$.

The ball $B(a; r)$ consists of all points whose distance from a is less than r. In R^1 this is simply an open interval with center at a. In R^2 it is a circular disk, and in R^3 it is a spherical solid with center at a and radius r.

3.5 Definition of an interior point. *Let S be a subset of R^n, and assume that $a \in S$. Then a is called an interior point of S if there is an open n-ball with center at a, all of whose points belong to S.*

In other words, every interior point a of S can be surrounded by an n-ball $B(a) \subseteq S$. The set of all interior points of S is called the *interior* of S and is denoted by int S. Any set containing a ball with center a is sometimes called a *neighborhood* of a.

3.6 Definition of an open set. *A set S in R^n is called open if all its points are interior points.*

NOTE. A set S is open if and only if $S = $ int S. (See Exercise 3.9.)

Examples. In $\mathbf{R}^1$ the simplest type of nonempty open set is an open interval. The union of two or more open intervals is also open. A closed interval $[a, b]$ is not an open set because the endpoints a and b are not interior points of the interval.

Examples of open sets in the plane are: the interior of a disk; the cartesian product of two one-dimensional open intervals. The reader should be cautioned that an open interval in $\mathbf{R}^1$ is no longer an open set when it is considered as a subset of the plane. In fact, *no* subset of $\mathbf{R}^1$ (except the empty set) can be open in $\mathbf{R}^2$, because such a set cannot contain a 2-ball.

In $\mathbf{R}^n$ the empty set is open (Why?) as is the whole space $\mathbf{R}^n$. Every open n-ball is an open set in $\mathbf{R}^n$. The cartesian product

$$(a_1, b_1) \times \cdots \times (a_n, b_n)$$

of n one-dimensional open intervals $(a_1, b_1), \ldots, (a_n, b_n)$ is an open set in $\mathbf{R}^n$ called an *n-dimensional open interval*. We denote it by $(\mathbf{a}, \mathbf{b})$, where $\mathbf{a} = (a_1, \ldots, a_n)$ and $\mathbf{b} = (b_1, \ldots, b_n)$.

The next two theorems show how additional open sets in $\mathbf{R}^n$ can be constructed from given open sets.

Theorem 3.7. *The union of any collection of open sets is an open set.*

Proof. Let F be a collection of open sets and let S denote their union, $S = \bigcup_{A \in F} A$. Assume $\mathbf{x} \in S$. Then $\mathbf{x}$ must belong to at least one of the sets in F, say $\mathbf{x} \in A$. Since A is open, there exists an open n-ball $B(\mathbf{x}) \subseteq A$. But $A \subseteq S$, so $B(\mathbf{x}) \subseteq S$ and hence $\mathbf{x}$ is an interior point of S. Since every point of S is an interior point, S is open.

Theorem 3.8. *The intersection of a finite collection of open sets is open.*

Proof. Let $S = \bigcap_{k=1}^{m} A_k$ where each A_k is open. Assume $\mathbf{x} \in S$. (If S is empty, there is nothing to prove.) Then $\mathbf{x} \in A_k$ for every $k = 1, 2, \ldots, m$, and hence there is an open n-ball $B(\mathbf{x}; r_k) \subseteq A_k$. Let r be the smallest of the positive numbers $r_1, r_2, \ldots, r_m$. Then $\mathbf{x} \in B(\mathbf{x}; r) \subseteq S$. That is, $\mathbf{x}$ is an interior point, so S is open.

Thus we see that from given open sets, new open sets can be formed by taking arbitrary unions or finite intersections. Arbitrary intersections, on the other hand, will not always lead to open sets. For example, the intersection of all open intervals of the form $(-1/n, 1/n)$, where $n = 1, 2, 3, \ldots$, is the set consisting of 0 alone.

3.4 THE STRUCTURE OF OPEN SETS IN $\mathbf{R}^1$

In $\mathbf{R}^1$ the union of a countable collection of disjoint open intervals is an open set and, remarkably enough, every nonempty open set in $\mathbf{R}^1$ can be obtained in this way. This section is devoted to a proof of this statement.

First we introduce the concept of a component interval.

3.9 Definition of component interval. *Let S be an open subset of* $\mathbf{R}^1$. *An open interval I (which may be finite or infinite) is called a component interval of S if* $I \subseteq S$ *and if there is no open interval* $J \neq I$ *such that* $I \subseteq J \subseteq S$.

In other words, a component interval of S is not a proper subset of any other open interval contained in S.

Theorem 3.10. *Every point of a nonempty open set S belongs to one and only one component interval of S.*

Proof. Assume $x \in S$. Then x is contained in some open interval I with $I \subseteq S$. There are many such intervals but the "largest" of these will be the desired component interval. We leave it to the reader to verify that this largest interval is $I_x = (a(x), b(x))$, where

$$a(x) = \inf \{a : (a, x) \subseteq S\}, \qquad b(x) = \sup \{b : (x, b) \subseteq S\}.$$

Here $a(x)$ might be $-\infty$ and $b(x)$ might be $+\infty$. Clearly, there is no open interval J such that $I_x \subseteq J \subseteq S$, so I_x is a component interval of S containing x. If J_x is another component interval of S containing x, then the union $I_x \cup J_x$ is an open interval contained in S and containing both I_x and J_x. Hence, by the definition of component interval, it follows that $I_x \cup J_x = I_x$ and $I_x \cup J_x = J_x$, so $I_x = J_x$.

Theorem 3.11 (Representation theorem for open sets on the real line). *Every nonempty open set S in* $\mathbf{R}^1$ *is the union of a countable collection of disjoint component intervals of S.*

Proof. If $x \in S$, let I_x denote the component interval of S containing x. The union of all such intervals I_x is clearly S. If two of them, I_x and I_y, have a point in common, then their union $I_x \cup I_y$ is an open interval contained in S and containing both I_x and I_y. Hence $I_x \cup I_y = I_x$ and $I_x \cup I_y = I_y$ so $I_x = I_y$. Therefore the intervals I_x form a disjoint collection.

It remains to show that they form a countable collection. For this purpose, let $\{x_1, x_2, x_3, \ldots\}$ denote the countable set of rational numbers. In each component interval I_x there will be infinitely many x_n, but among these there will be exactly one with *smallest index n*. We then define a function F by means of the equation $F(I_x) = n$, if x_n is the rational number in I_x with smallest index n. This function F is one-to-one since $F(I_x) = F(I_y) = n$ implies that I_x and I_y have x_n in common and this implies $I_x = I_y$. Therefore F establishes a one-to-one correspondence between the intervals I_x and a subset of the positive integers. This completes the proof.

NOTE. This representation of S is unique. In fact, if S is a union of disjoint open intervals, then these intervals must be the component intervals of S. This is an immediate consequence of Theorem 3.10.

If S is an open interval, then the representation contains only one component interval, namely S itself. Therefore an open interval in $\mathbf{R}^1$ cannot be expressed as the union of two nonempty disjoint open sets. This property is also described by saying that an open interval is *connected*. The concept of connectedness for sets in $\mathbf{R}^n$ will be discussed further in Section 4.16.

3.5 CLOSED SETS

3.12 Definition of a closed set. *A set S in $\mathbf{R}^n$ is called closed if its complement $\mathbf{R}^n - S$ is open.*

Examples. A closed interval $[a, b]$ in $\mathbf{R}^1$ is a closed set. The cartesian product

$$[a_1, b_1] \times \cdots \times [a_n, b_n]$$

of n one-dimensional closed intervals is a closed set in $\mathbf{R}^n$ called an *n-dimensional closed interval* $[\mathbf{a}, \mathbf{b}]$.

The next theorem, a consequence of Theorems 3.7 and 3.8, shows how to construct further closed sets from given ones.

Theorem 3.13. *The union of a finite collection of closed sets is closed, and the intersection of an arbitrary collection of closed sets is closed.*

A further relation between open and closed sets is described by the following theorem.

Theorem 3.14. *If A is open and B is closed, then $A - B$ is open and $B - A$ is closed.*

Proof. We simply note that $A - B = A \cap (\mathbf{R}^n - B)$, the intersection of two open sets, and that $B - A = B \cap (\mathbf{R}^n - A)$, the intersection of two closed sets.

3.6 ADHERENT POINTS. ACCUMULATION POINTS

Closed sets can also be described in terms of adherent points and accumulation points.

3.15 Definition of an adherent point. *Let S be a subset of $\mathbf{R}^n$, and $\mathbf{x}$ a point in $\mathbf{R}^n$, $\mathbf{x}$ not necessarily in S. Then $\mathbf{x}$ is said to be adherent to S if every n-ball $B(\mathbf{x})$ contains at least one point of S.*

Examples
  1. If $\mathbf{x} \in S$, then $\mathbf{x}$ adheres to S for the trivial reason that every n-ball $B(\mathbf{x})$ contains $\mathbf{x}$.
  2. If S is a subset of $\mathbf{R}$ which is bounded above, then sup S is adherent to S.

Some points adhere to S because every ball $B(\mathbf{x})$ contains points of S *distinct* from $\mathbf{x}$. These are called accumulation points.

3.16 Definition of an accumulation point. *If $S \subseteq \mathbf{R}^n$ and $\mathbf{x} \in \mathbf{R}^n$, then $\mathbf{x}$ is called an accumulation point of S if every n-ball $B(\mathbf{x})$ contains at least one point of S distinct from $\mathbf{x}$.*

In other words, $\mathbf{x}$ is an accumulation point of S if, and only if, $\mathbf{x}$ adheres to $S - \{\mathbf{x}\}$. If $\mathbf{x} \in S$ but $\mathbf{x}$ is not an accumulation point of S, then $\mathbf{x}$ is called an *isolated point* of S.

Examples

1. The set of numbers of the form $1/n$, $n = 1, 2, 3, \ldots$, has 0 as an accumulation point.

2. The set of rational numbers has every real number as an accumulation point.

3. Every point of the closed interval $[a, b]$ is an accumulation point of the set of numbers in the open interval (a, b).

Theorem 3.17. *If* $\mathbf{x}$ *is an accumulation point of* S, *then every* n*-ball* $B(\mathbf{x})$ *contains infinitely many points of* S.

Proof. Assume the contrary; that is, suppose an n-ball $B(\mathbf{x})$ exists which contains only a finite number of points of S distinct from $\mathbf{x}$, say $\mathbf{a}_1, \mathbf{a}_2, \ldots, \mathbf{a}_m$. If r denotes the smallest of the positive numbers

$$\|\mathbf{x} - \mathbf{a}_1\|, \qquad \|\mathbf{x} - \mathbf{a}_2\|, \qquad \ldots, \qquad \|\mathbf{x} - \mathbf{a}_m\|,$$

then $B(\mathbf{x}; r/2)$ will be an n-ball about $\mathbf{x}$ which contains no points of S distinct from $\mathbf{x}$. This is a contradiction.

This theorem implies, in particular, that a set cannot have an accumulation point unless it contains infinitely many points to begin with. The converse, however, is not true in general. For example, the set of integers $\{1, 2, 3, \ldots\}$ is an infinite set with no accumulation points. In a later section we will show that infinite sets contained in some n-ball always have an accumulation point. This is an important result known as the Bolzano–Weierstrass theorem.

3.7 CLOSED SETS AND ADHERENT POINTS

A closed set was defined to be the complement of an open set. The next theorem describes closed sets in another way.

Theorem 3.18. *A set* S *in* $\mathbf{R}^n$ *is closed if, and only if, it contains all its adherent points.*

Proof. Assume S is closed and let $\mathbf{x}$ be adherent to S. We wish to prove that $\mathbf{x} \in S$. We assume $\mathbf{x} \notin S$ and obtain a contradiction. If $\mathbf{x} \notin S$ then $\mathbf{x} \in \mathbf{R}^n - S$ and, since $\mathbf{R}^n - S$ is open, some n-ball $B(\mathbf{x})$ lies in $\mathbf{R}^n - S$. Thus $B(\mathbf{x})$ contains no points of S, contradicting the fact that $\mathbf{x}$ adheres to S.

To prove the converse, we assume S contains all its adherent points and show that S is closed. Assume $\mathbf{x} \in \mathbf{R}^n - S$. Then $\mathbf{x} \notin S$, so $\mathbf{x}$ does not adhere to S. Hence some ball $B(\mathbf{x})$ does not intersect S, so $B(\mathbf{x}) \subseteq \mathbf{R}^n - S$. Therefore $\mathbf{R}^n - S$ is open, and hence S is closed.

3.19 Definition of closure. *The set of all adherent points of a set* S *is called the closure of* S *and is denoted by* $\bar{S}$.

For any set we have $S \subseteq \bar{S}$ since every point of S adheres to S. Theorem 3.18 shows that the opposite inclusion $\bar{S} \subseteq S$ holds if and only if S is closed. Therefore we have:

Theorem 3.20. *A set S is closed if and only if $S = \bar{S}$.*

3.21 Definition of derived set. *The set of all accumulation points of a set S is called the derived set of S and is denoted by S'.*

Clearly, we have $\bar{S} = S \cup S'$ for any set S. Hence Theorem 3.20 implies that S is closed if and only if $S' \subseteq S$. In other words, we have:

Theorem 3.22. *A set S in $\mathbf{R}^n$ is closed if, and only if, it contains all its accumulation points.*

3.8 THE BOLZANO–WEIERSTRASS THEOREM

3.23 Definition of a bounded set. *A set S in $\mathbf{R}^n$ is said to be bounded if it lies entirely within an n-ball $B(\mathbf{a}; r)$ for some $r > 0$ and some $\mathbf{a}$ in $\mathbf{R}^n$.*

Theorem 3.24 (Bolzano–Weierstrass). *If a bounded set S in $\mathbf{R}^n$ contains infinitely many points, then there is at least one point in $\mathbf{R}^n$ which is an accumulation point of S.*

Proof. To help fix the ideas we give the proof first for $\mathbf{R}^1$. Since S is bounded, it lies in some interval $[-a, a]$. At least one of the subintervals $[-a, 0]$ or $[0, a]$ contains an infinite subset of S. Call one such subinterval $[a_1, b_1]$. Bisect $[a_1, b_1]$ and obtain a subinterval $[a_2, b_2]$ containing an infinite subset of S, and continue this process. In this way a countable collection of intervals is obtained, the nth interval $[a_n, b_n]$ being of length $b_n - a_n = a/2^{n-1}$. Clearly, the sup of the left endpoints a_n and the inf of the right endpoints b_n must be equal, say to x. [Why are they equal?] The point x will be an accumulation point of S because, if r is any positive number, the interval $[a_n, b_n]$ will be contained in $B(x; r)$ as soon as n is large enough so that $b_n - a_n < r/2$. The interval $B(x; r)$ contains a point of S distinct from x and hence x is an accumulation point of S. This proves the theorem for $\mathbf{R}^1$. (Observe that the accumulation point x may or may not belong to S.)

Next we give a proof for $\mathbf{R}^n$, $n > 1$, by an extension of the ideas used in treating $\mathbf{R}^1$. (The reader may find it helpful to visualize the proof in $\mathbf{R}^2$ by referring to Fig. 3.1.)

Since S is bounded, S lies in some n-ball $B(\mathbf{0}; a)$, $a > 0$, and therefore within the n-dimensional interval J_1 defined by the inequalities

$$-a \le x_k \le a \qquad (k = 1, 2, \ldots, n).$$

Here J_1 denotes the cartesian product

$$J_1 = I_1^{(1)} \times I_2^{(1)} \times \cdots \times I_n^{(1)};$$

that is, the set of points $(x_1, \ldots, x_n)$, where $x_k \in I_k^{(1)}$ and where each $I_k^{(1)}$ is a one-dimensional interval $-a \le x_k \le a$. Each interval $I_k^{(1)}$ can be bisected to

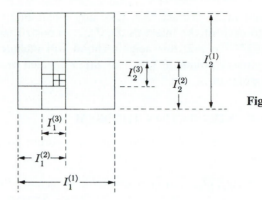

Figure 3.1

form two subintervals $I_{k,1}^{(1)}$ and $I_{k,2}^{(1)}$, defined by the inequalities

$$I_{k,1}^{(1)} : -a \leq x_k \leq 0; \qquad I_{k,2}^{(1)} : 0 \leq x_k \leq a.$$

Next, we consider all possible cartesian products of the form

$$I_{1,k_1}^{(1)} \times I_{2,k_2}^{(1)} \times \cdots \times I_{n,k_n}^{(1)}, \tag{a}$$

where each $k_i = 1$ or 2. There are exactly 2^n such products and, of course, each such product is an n-dimensional interval. The union of these 2^n intervals is the original interval J_1, which contains S; and hence at least one of the 2^n intervals in (a) must contain infinitely many points of S. One of these we denote by J_2, which can then be expressed as

$$J_2 = I_1^{(2)} \times I_2^{(2)} \times \cdots \times I_n^{(2)},$$

where each $I_k^{(2)}$ is one of the subintervals of $I_k^{(1)}$ of length a. We now proceed with J_2 as we did with J_1, bisecting each interval $I_k^{(2)}$ and arriving at an n-dimensional interval J_3 containing an infinite subset of S. If we continue the process, we obtain a countable collection of n-dimensional intervals $J_1, J_2, J_3, \ldots$, where the mth interval J_m has the property that it contains an infinite subset of S and can be expressed in the form

$$J_m = I_1^{(m)} \times I_2^{(m)} \times \cdots \times I_n^{(m)}, \qquad \text{where } I_k^{(m)} \subseteq I_k^{(1)}.$$

Writing

$$I_k^{(m)} = [a_k^{(m)}, b_k^{(m)}],$$

we have

$$b_k^{(m)} - a_k^{(m)} = \frac{a}{2^{m-2}} \qquad (k = 1, 2, \cdots, n).$$

For each fixed k, the sup of all left endpoints $a_k^{(m)}$, $(m = 1, 2, \ldots)$, must therefore be equal to the inf of all right endpoints $b_k^{(m)}$, $(m = 1, 2, \ldots)$, and their common value we denote by t_k. We now assert that the point $\mathbf{t} = (t_1, t_2, \ldots, t_n)$ is an accumulation point of S. To see this, take any n-ball $B(\mathbf{t}; r)$. The point $\mathbf{t}$, of course, belongs to each of the intervals $J_1, J_2, \ldots$ constructed above, and when m is such that $a/2^{m-2} < r/2$, this neighborhood will include J_m. But since J_m contains infinitely many points of S, so will $B(\mathbf{t}; r)$, which proves that $\mathbf{t}$ is indeed an accumulation point of S.

3.9 THE CANTOR INTERSECTION THEOREM

As an application of the Bolzano–Weierstrass theorem we prove the Cantor intersection theorem.

Theorem 3.25. *Let* $\{Q_1, Q_2, \ldots\}$ *be a countable collection of nonempty sets in* $\mathbf{R}^n$ *such that:*

i) $Q_{k+1} \subseteq Q_k$ $(k = 1, 2, 3, \ldots)$.

ii) *Each set* Q_k *is closed and* Q_1 *is bounded.*

Then the intersection $\bigcap_{k=1}^{\infty} Q_k$ *is closed and nonempty.*

Proof. Let $S = \bigcap_{k=1}^{\infty} Q_k$. Then S is closed because of Theorem 3.13. To show that S is nonempty, we exhibit a point $\mathbf{x}$ in S. We can assume that each Q_k contains infinitely many points; otherwise the proof is trivial. Now form a collection of distinct points $A = \{\mathbf{x}_1, \mathbf{x}_2, \ldots\}$, where $\mathbf{x}_k \in Q_k$. Since A is an infinite set contained in the bounded set Q_1, it has an accumulation point, say $\mathbf{x}$. We shall show that $\mathbf{x} \in S$ by verifying that $\mathbf{x} \in Q_k$ for each k. It will suffice to show that $\mathbf{x}$ is an accumulation point of each Q_k, since they are all closed sets. But every neighborhood of $\mathbf{x}$ contains infinitely many points of A, and since all except (possibly) a finite number of the points of A belong to Q_k, this neighborhood also contains infinitely many points of Q_k. Therefore $\mathbf{x}$ is an accumulation point of Q_k and the theorem is proved.

3.10 THE LINDELÖF COVERING THEOREM

In this section we introduce the concept of a *covering* of a set and prove the *Lindelöf covering theorem.* The usefulness of this concept will become apparent in some of the later work.

3.26 Definition of a covering. *A collection F of sets is said to be a covering of a given set S if* $S \subseteq \bigcup_{A \in F} A$. *The collection F is also said to cover S. If F is a collection of open sets, then F is called an open covering of S.*

Examples

 1. The collection of all intervals of the form $1/n < x < 2/n$, $(n = 2, 3, 4, \ldots)$, is an open covering of the interval $0 < x < 1$. This is an example of a countable covering.
 2. The real line $\mathbf{R}^1$ is covered by the collection of all open intervals (a, b). This covering is not countable. However, it contains a countable covering of $\mathbf{R}^1$, namely, all intervals of the form $(n, n + 2)$, where n runs through the integers.

3. Let $S = \{(x, y) : x > 0, y > 0\}$. The collection F of all circular disks with centers at (x, x) and with radius x, where $x > 0$, is a covering of S. This covering is not countable. However, it contains a countable covering of S, namely, all those disks in which x is rational. (See Exercise 3.18.)

The Lindelöf covering theorem states that every open covering of a set S in $\mathbf{R}^n$ contains a countable subcollection which also covers S. The proof makes use of the following preliminary result:

Theorem 3.27 *Let $G = \{A_1, A_2, \ldots\}$ denote the countable collection of all n-balls having rational radii and centers at points with rational coordinates. Assume $\mathbf{x} \in \mathbf{R}^n$ and let S be an open set in $\mathbf{R}^n$ which contains $\mathbf{x}$. Then at least one of the n-balls in G contains $\mathbf{x}$ and is contained in S. That is, we have*

$$\mathbf{x} \in A_k \subseteq S \qquad \text{for some } A_k \text{ in } G.$$

Proof. The collection G is countable because of Theorem 2.27. If $\mathbf{x} \in \mathbf{R}^n$ and if S is an open set containing $\mathbf{x}$, then there is an n-ball $B(\mathbf{x}; r) \subseteq S$. We shall find a point $\mathbf{y}$ in S with rational coordinates that is "near" $\mathbf{x}$ and, using this point as center, will then find a neighborhood in G which lies within $B(\mathbf{x}; r)$ and which contains $\mathbf{x}$. Write

$$\mathbf{x} = (x_1, x_2, \ldots, x_n),$$

and let y_k be a rational number such that $|y_k - x_k| < r/(4n)$ for each $k = 1, 2, \ldots, n$. Then

$$\|\mathbf{y} - \mathbf{x}\| \le |y_1 - x_1| + \cdots + |y_n - x_n| < \frac{r}{4}.$$

Next, let q be a rational number such that $r/4 < q < r/2$. Then $\mathbf{x} \in B(\mathbf{y}; q)$ and $B(\mathbf{y}; q) \subseteq B(\mathbf{x}; r) \subseteq S$. But $B(\mathbf{y}; q) \in G$ and hence the theorem is proved. (See Fig. 3.2 for the situation in $\mathbf{R}^2$.)

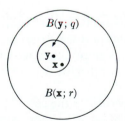

$B(\mathbf{y}; q)$

$\mathbf{y} \bullet$
$\mathbf{x} \bullet$

$B(\mathbf{x}; r)$

Figure 3.2

Theorem 3.28 (Lindelöf covering theorem). *Assume $A \subseteq \mathbf{R}^n$ and let F be an open covering of A. Then there is a countable subcollection of F which also covers A.*

Proof. Let $G = \{A_1, A_2, \ldots\}$ denote the countable collection of all n-balls having rational centers and rational radii. This set G will be used to help us extract a countable subcollection of F which covers A.

Assume $\mathbf{x} \in A$. Then there is an open set S in F such that $\mathbf{x} \in S$. By Theorem 3.27 there is an n-ball A_k in G such that $\mathbf{x} \in A_k \subseteq S$. There are, of course, infinitely many such A_k corresponding to each S, but we choose only one of these, for example, the one of smallest index, say $m = m(\mathbf{x})$. Then we have $\mathbf{x} \in A_{m(\mathbf{x})} \subseteq S$. The set of all n-balls $A_{m(\mathbf{x})}$ obtained as $\mathbf{x}$ varies over all elements of A is a countable collection of open sets which covers A. To get a countable subcollection of F which covers A, we simply correlate to each set $A_{k(\mathbf{x})}$ one of the sets S of F which contained $A_{k(\mathbf{x})}$. This completes the proof.

3.11 THE HEINE–BOREL COVERING THEOREM

The Lindelöf covering theorem states that from any open covering of an arbitrary set A in $\mathbf{R}^n$ we can extract a countable covering. The Heine–Borel theorem tells us that if, in addition, we know that A is closed and bounded, we can reduce the covering to a finite covering. The proof makes use of the Cantor intersection theorem.

Theorem 3.29 (Heine–Borel). Let F be an open covering of a closed and bounded set A in $\mathbf{R}^n$. Then a finite subcollection of F also covers A.

Proof. A countable subcollection of F, say $\{I_1, I_2, \ldots\}$, covers A, by Theorem 3.28. Consider, for $m \geq 1$, the finite union

$$S_m = \bigcup_{k=1}^{m} I_k.$$

This is open, since it is the union of open sets. We shall show that for some value of m the union S_m covers A.

For this purpose we consider the complement $\mathbf{R}^n - S_m$, which is closed. Define a countable collection of sets $\{Q_1, Q_2, \ldots\}$ as follows: $Q_1 = A$, and for $m > 1$,

$$Q_m = A \cap (\mathbf{R}^n - S_m).$$

That is, Q_m consists of those points of A which lie outside of S_m. If we can show that for some value of m the set Q_m is empty, then we will have shown that for this m no point of A lies outside S_m; in other words, we will have shown that some S_m covers A.

Observe the following properties of the sets Q_m: Each set Q_m is closed, since it is the intersection of the closed set A and the closed set $\mathbf{R}^n - S_m$. The sets Q_m are decreasing, since the S_m are increasing; that is, $Q_{m+1} \subseteq Q_m$. The sets Q_m, being subsets of A, are all bounded. Therefore, if no set Q_m is empty, we can apply the Cantor intersection theorem to conclude that the intersection $\bigcap_{k=1}^{\infty} Q_k$ is also not empty. This means that there is some point in A which is in all the sets Q_m, or, what is the same thing, outside all the sets S_m. But this is impossible, since $A \subseteq \bigcup_{k=1}^{\infty} S_k$. Therefore some Q_m must be empty, and this completes the proof.

3.12 COMPACTNESS IN **R**^{n}

We have just seen that if a set S in **R**^{n} is closed and bounded, then any open covering of S can be reduced to a finite covering. It is natural to inquire whether there might be sets other than closed and bounded sets which also have this property. Such sets will be called *compact*.

3.30 Definition of a compact set. *A set S in* **R**^{n} *is said to be compact if, and only if, every open covering of S contains a finite subcover, that is, a finite subcollection which also covers S.*

The Heine–Borel theorem states that every closed and bounded set in **R**^{n} is compact. Now we prove the converse result.

Theorem 3.31. *Let S be a subset of* **R**^{n}. *Then the following three statements are equivalent:*

a) *S is compact.*

b) *S is closed and bounded.*

c) *Every infinite subset of S has an accumulation point in S.*

Proof. As noted above, (b) implies (a). If we prove that (a) implies (b), that (b) implies (c) and that (c) implies (b), this will establish the equivalence of all three statements.

Assume (a) holds. We shall prove first that S is bounded. Choose a point **p** in S. The collection of n-balls $B(\mathbf{p}; k)$, $k = 1, 2, \ldots$, is an open covering of S. By compactness a finite subcollection also covers S and hence S is bounded.

Next we prove that S is closed. Suppose S is not closed. Then there is an accumulation point **y** of S such that $\mathbf{y} \notin S$. If $\mathbf{x} \in S$, let $r_\mathbf{x} = \|\mathbf{x} - \mathbf{y}\|/2$. Each $r_\mathbf{x}$ is positive since $\mathbf{y} \notin S$ and the collection $\{B(\mathbf{x}; r_\mathbf{x}) : \mathbf{x} \in S\}$ is an open covering of S. By compactness, a finite number of these neighborhoods cover S, say

$$S \subseteq \bigcup_{k=1}^{p} B(\mathbf{x}_k; r_k).$$

Let r denote the smallest of the radii $r_1, r_2, \ldots, r_p$. Then it is easy to prove that the ball $B(\mathbf{y}; r)$ has no points in common with any of the balls $B(\mathbf{x}_k; r_k)$. In fact, if $\mathbf{x} \in B(\mathbf{y}; r)$, then $\|\mathbf{x} - \mathbf{y}\| < r \leq r_k$, and by the triangle inequality we have $\|\mathbf{y} - \mathbf{x}_k\| \leq \|\mathbf{y} - \mathbf{x}\| + \|\mathbf{x} - \mathbf{x}_k\|$, so

$$\|\mathbf{x} - \mathbf{x}_k\| \geq \|\mathbf{y} - \mathbf{x}_k\| - \|\mathbf{x} - \mathbf{y}\| = 2r_k - \|\mathbf{x} - \mathbf{y}\| > r_k.$$

Hence $\mathbf{x} \notin B(\mathbf{x}_k; r_k)$. Therefore $B(\mathbf{y}; r) \cap S$ is empty, contradicting the fact that **y** is an accumulation point of S. This contradiction shows that S is closed and hence (a) implies (b).

Assume (b) holds. In this case the proof of (c) is immediate, because if T is an infinite subset of S then T is bounded (since S is bounded), and hence by the Bolzano–Weierstrass theorem T has an accumulation point **x**, say. Now **x** is also an accumulation point of S and hence $\mathbf{x} \in S$, since S is closed. Therefore (b) implies (c).

Assume (c) holds. We shall prove (b). If S is unbounded, then for every $m > 0$ there exists a point $\mathbf{x}_m$ in S with $\|\mathbf{x}_m\| > m$. The collection $T = \{\mathbf{x}_1, \mathbf{x}_2, \ldots\}$ is an infinite subset of S and hence, by (c), T has an accumulation point $\mathbf{y}$ in S. But for $m > 1 + \|\mathbf{y}\|$ we have

$$\|\mathbf{x}_m - \mathbf{y}\| \geq \|\mathbf{x}_m\| - \|\mathbf{y}\| > m - \|\mathbf{y}\| > 1,$$

contradicting the fact that $\mathbf{y}$ is an accumulation point of T. This proves that S is bounded.

To complete the proof we must show that S is closed. Let $\mathbf{x}$ be an accumulation point of S. Since every neighborhood of $\mathbf{x}$ contains infinitely many points of S, we can consider the neighborhoods $B(\mathbf{x}; 1/k)$, where $k = 1, 2, \ldots$, and obtain a countable set of distinct points, say $T = \{\mathbf{x}_1, \mathbf{x}_2, \ldots\}$, contained in S, such that $\mathbf{x}_k \in B(\mathbf{x}; 1/k)$. The point $\mathbf{x}$ is also an accumulation point of T. Since T is an infinite subset of S, part (c) of the theorem tells us that T must have an accumulation point in S. The theorem will then be proved if we show that $\mathbf{x}$ is the only accumulation point of T.

To do this, suppose that $\mathbf{y} \neq \mathbf{x}$. Then by the triangle inequality we have

$$\|\mathbf{y} - \mathbf{x}\| \leq \|\mathbf{y} - \mathbf{x}_k\| + \|\mathbf{x}_k - \mathbf{x}\| < \|\mathbf{y} - \mathbf{x}_k\| + 1/k, \qquad \text{if } \mathbf{x}_k \in T.$$

If k_0 is taken so large that $1/k < \frac{1}{2}\|\mathbf{y} - \mathbf{x}\|$ whenever $k \geq k_0$, the last inequality leads to $\frac{1}{2}\|\mathbf{y} - \mathbf{x}\| < \|\mathbf{y} - \mathbf{x}_k\|$. This shows that $\mathbf{x}_k \notin B(\mathbf{y}; r)$ when $k \geq k_0$, if $r = \frac{1}{2}\|\mathbf{y} - \mathbf{x}\|$. Hence $\mathbf{y}$ cannot be an accumulation point of T. This completes the proof that (c) implies (b).

3.13 METRIC SPACES

The proofs of some of the theorems of this chapter depend only on a few properties of the distance between points and not on the fact that the points are in $\mathbf{R}^n$. When these properties of distance are studied abstractly they lead to the concept of a metric space.

3.32 Definition of a metric space. A metric space is a nonempty set M of objects (called points) together with a function d from $M \times M$ to $\mathbf{R}$ (called the metric of the space) satisfying the following four properties for all points x, y, z in M:

1. $d(x, x) = 0$.
2. $d(x, y) > 0$ *if* $x \neq y$.
3. $d(x, y) = d(y, x)$.
4. $d(x, y) \leq d(x, z) + d(z, y)$.

The nonnegative number $d(x, y)$ is to be thought of as the distance from x to y. In these terms the intuitive meaning of properties 1 through 4 is clear. Property 4 is called the *triangle inequality*.

We sometimes denote a metric space by (M, d) to emphasize that both the set M and the metric d play a role in the definition of a metric space.

Examples

1. $M = \mathbf{R}^n$; $d(\mathbf{x}, \mathbf{y}) = \|\mathbf{x} - \mathbf{y}\|$. This is called the *Euclidean metric*. Whenever we refer to Euclidean space $\mathbf{R}^n$, it will be understood that the metric is the Euclidean metric unless another metric is specifically mentioned.

2. $M = \mathbf{C}$, the complex plane; $d(z_1, z_2) = |z_1 - z_2|$. As a metric space, $\mathbf{C}$ is indistinguishable from Euclidean space $\mathbf{R}^2$ because it has the same points and the same metric.

3. M any nonempty set; $d(x, y) = 0$ if $x = y$, $d(x, y) = 1$ if $x \neq y$. This is called the *discrete metric*, and (M, d) is called a *discrete metric space*.

4. If (M, d) is a metric space and if S is any nonempty subset of M, then (S, d) is also a metric space with the same metric or, more precisely, with the restriction of d to $S \times S$ as metric. This is sometimes called the *relative metric* induced by d on S, and S is called a *metric subspace* of M. For example, the rational numbers $\mathbf{Q}$ with the metric $d(x, y) = |x - y|$ form a metric subspace of $\mathbf{R}$.

5. $M = \mathbf{R}^2$; $d(\mathbf{x}, \mathbf{y}) = \sqrt{(x_1 - y_1)^2 + 4(x_2 - y_2)^2}$, where $\mathbf{x} = (x_1, x_2)$ and $\mathbf{y} = (y_1, y_2)$. The metric space (M, d) is not a metric subspace of Euclidean space $\mathbf{R}^2$ because the metric is different.

6. $M = \{(x_1, x_2) : x_1^2 + x_2^2 = 1\}$, the unit circle in $\mathbf{R}^2$; $d(\mathbf{x}, \mathbf{y}) =$ the length of the smaller arc joining the two points $\mathbf{x}$ and $\mathbf{y}$ on the unit circle.

7. $M = \{(x_1, x_2, x_3) : x_1^2 + x_2^2 + x_3^2 = 1\}$, the unit sphere in $\mathbf{R}^3$; $d(\mathbf{x}, \mathbf{y}) =$ the length of the smaller arc along the great circle joining the two points $\mathbf{x}$ and $\mathbf{y}$.

8. $M = \mathbf{R}^n$; $d(\mathbf{x}, \mathbf{y}) = |x_1 - y_1| + \cdots + |x_n - y_n|$.

9. $M = \mathbf{R}^n$; $d(\mathbf{x}, \mathbf{y}) = \max \{|x_1 - y_1|, \ldots, |x_n - y_n|\}$.

3.14 POINT SET TOPOLOGY IN METRIC SPACES

The basic notions of point set topology can be extended to an arbitrary metric space (M, d).

If $a \in M$, the *ball* $B(a; r)$ with center a and radius $r > 0$ is defined to be the set of all x in M such that

$$d(x, a) < r.$$

Sometimes we denote this ball by $B_M(a; r)$ to emphasize the fact that its points come from M. If S is a metric subspace of M, the ball $B_S(a; r)$ is the intersection of S with the ball $B_M(a; r)$.

Examples. In Euclidean space $\mathbf{R}^1$ the ball $B(0; 1)$ is the open interval $(-1, 1)$. In the metric subspace $S = [0, 1]$ the ball $B_S(0; 1)$ is the half-open interval $[0, 1)$.

NOTE. The geometric appearance of a ball in $\mathbf{R}^n$ need not be "spherical" if the metric is not the Euclidean metric. (See Exercise 3.27.)

If $S \subseteq M$, a point a in S is called an *interior point* of S if some ball $B_M(a; r)$ lies entirely in S. The *interior*, int S, is the set of interior points of S. A set S is called *open* in M if all its points are interior points; it is called *closed* in M if $M - S$ is open in M.

Examples.

1. Every ball $B_M(a; r)$ in a metric space M is open in M.

2. In a discrete metric space M *every* subset S is open. In fact, if $x \in S$, the ball $B(x; \frac{1}{2})$ consists entirely of points of S (since it contains only x), so S is open. Therefore every subset of M is also closed!

3. In the metric subspace $S = [0, 1]$ of Euclidean space $\mathbf{R}^1$, every interval of the form $[0, x)$ or $(x, 1]$, where $0 < x < 1$, is an open set in S. These sets are not open in $\mathbf{R}^1$.

Example 3 shows that if S is a metric subspace of M the open sets in S need not be open in M. The next theorem describes the relation between open sets in M and those in S.

Theorem 3.33. *Let (S, d) be a metric subspace of (M, d), and let X be a subset of S. Then X is open in S if, and only if,*

$$X = A \cap S$$

for some set A which is open in M.

Proof. Assume A is open in M and let $X = A \cap S$. If $x \in X$, then $x \in A$ so $B_M(x; r) \subseteq A$ for some $r > 0$. Hence $B_S(x; r) = B_M(x; r) \cap S \subseteq A \cap S = X$ so X is open in S.

Conversely, assume X is open in S. We will show that $X = A \cap S$ for some open set A in M. For every x in X there is a ball $B_S(x; r_x)$ contained in X. Now $B_S(x; r_x) = B_M(x; r_x) \cap S$, so if we let

$$A = \bigcup_{x \in X} B_M(x; r_x),$$

then A is open in M and it is easy to verify that $A \cap S = X$.

Theorem 3.34. *Let (S, d) be a metric subspace of (M, d) and let Y be a subset of S. Then Y is closed in S if, and only if, $Y = B \cap S$ for some set B which is closed in M.*

Proof. If $Y = B \cap S$, where B is closed in M, then $B = M - A$ where A is open in M so $Y = S \cap B = S \cap (M - A) = S - A$; hence Y is closed in S.

Conversely, if Y is closed in S, let $X = S - Y$. Then X is open in S so $X = A \cap S$, where A is open in M and

$$Y = S - X = S - (A \cap S) = S - A = S \cap (M - A) = S \cap B,$$

where $B = M - A$ is closed in M. This completes the proof.

If $S \subseteq M$, a point x in M is called an *adherent point* of S if every ball $B_M(x; r)$ contains at least one point of S. If x adheres to $S - \{x\}$ then x is called an *accumulation point* of S. The *closure* $\bar{S}$ of S is the set of all adherent points of S, and the *derived set* S' is the set of all accumulation points of S. Thus, $\bar{S} = S \cup S'$.

The following theorems are valid in every metric space (M, d) and are proved exactly as they were for Euclidean space $\mathbf{R}^n$. In the proofs, the Euclidean distance $\|\mathbf{x} - \mathbf{y}\|$ need only be replaced by the metric $d(x, y)$.

Theorem 3.35. a) *The union of any collection of open sets is open, and the intersection of a finite collection of open sets is open.*

b) *The union of a finite collection of closed sets is closed, and the intersection of any collection of closed sets is closed.*

Theorem 3.36. *If A is open and B is closed, then $A - B$ is open and $B - A$ is closed.*

Theorem 3.37. *For any subset S of M the following statements are equivalent:*

a) *S is closed in M.*

b) *S contains all its adherent points.*

c) *S contains all its accumulation points.*

d) *$S = \bar{S}$.*

Example. Let $M = \mathbf{Q}$, the set of rational numbers, with the Euclidean metric of $\mathbf{R}^1$. Let S consist of all rational numbers in the open interval (a, b), where both a and b are irrational. Then S is a closed subset of $\mathbf{Q}$.

Our proofs of the Bolzano–Weierstrass theorem, the Cantor intersection theorem, and the covering theorems of Lindelöf and Heine–Borel used not only the metric properties of Euclidean space $\mathbf{R}^n$ but also special properties of $\mathbf{R}^n$ not generally valid in an arbitrary metric space (M, d). Further restrictions on M are required to extend these theorems to metric spaces. One of these extensions is outlined in Exercise 3.34.

The next section describes compactness in an arbitrary metric space.

3.15 COMPACT SUBSETS OF A METRIC SPACE

Let (M, d) be a metric space and let S be a subset of M. A collection F of open subsets of M is said to be an *open covering* of S if $S \subseteq \bigcup_{A \in F} A$.

A subset S of M is called *compact* if every open covering of S contains a finite subcover. S is called *bounded* if $S \subseteq B(a; r)$ for some $r > 0$ and some a in M.

Theorem 3.38. *Let S be a compact subset of a metric space M. Then:*

i) *S is closed and bounded.*

ii) *Every infinite subset of S has an accumulation point in S.*

Proof. To prove (i) we refer to the proof of Theorem 3.31 and use that part of the argument which showed that (a) implies (b). The only change is that the Euclidean distance $\|\mathbf{x} - \mathbf{y}\|$ is to be replaced throughout by the metric $d(x, y)$.

To prove (ii) we argue by contradiction. Let T be an infinite subset of S and assume that no point of S is an accumulation point of T. Then for each point x in S there is a ball $B(x)$ which contains no point of T (if $x \notin T$) or exactly one point of T (x itself, if $x \in T$). As x runs through S, the union of these balls $B(x)$ is an open covering of S. Since S is compact, a finite subcollection covers S and hence also covers T. But this is a contradiction because T is an infinite set and each ball contains at most one point of T.

NOTE. In Euclidean space $\mathbf{R}^n$, each of properties (i) and (ii) is equivalent to compactness (Theorem 3.31). In a general metric space, property (ii) is equivalent to compactness (for a proof see Reference 3.4), but property (i) is not. Exercise 3.42 gives an example of a metric space M in which certain closed and bounded subsets are not compact.

Theorem 3.39. *Let X be a closed subset of a compact metric space M. Then X is compact.*

Proof. Let F be an open covering of X, say $X \subseteq \bigcup_{A \in F} A$. We will show that a finite number of the sets A cover X. Since X is closed its complement $M - X$ is open, so $F \cup \{(M - X)\}$ is an open covering of M. But M is compact, so this covering contains a finite subcover which we can assume includes $M - X$. Therefore

$$M \subseteq A_1 \cup \cdots \cup A_p \cup (M - X).$$

This subcover also covers X and, since $M - X$ contains no points of X, we can delete the set $M - X$ from the subcover and still cover X. Thus $X \subseteq A_1 \cup \cdots \cup A_p$ so X is compact.

3.16 BOUNDARY OF A SET

Definition 3.40. *Let S be a subset of a metric space M. A point x in M is called a boundary point of S if every ball $B_M(x; r)$ contains at least one point of S and at least one point of $M - S$. The set of all boundary points of S is called the boundary of S and is denoted by ∂S.*

The reader can easily verify that

$$\partial S = \bar{S} \cap \overline{M - S}.$$

This formula shows that ∂S is closed in M.

Example In $\mathbf{R}^n$, the boundary of a ball $B(\mathbf{a}; r)$ is the set of points $\mathbf{x}$ such that $\|\mathbf{x} - \mathbf{a}\| = r$. In $\mathbf{R}^1$, the boundary of the set of rational numbers is all of $\mathbf{R}^1$.

Further properties of metric spaces are developed in the Exercises and also in Chapter 4.

EXERCISES

Open and closed sets in $\mathbf{R}^1$ and $\mathbf{R}^2$

3.1 Prove that an open interval in $\mathbf{R}^1$ is an open set and that a closed interval is a closed set.

3.2 Determine all the accumulation points of the following sets in $\mathbf{R}^1$ and decide whether the sets are open or closed (or neither).

    a) All integers.

    b) The interval $(a, b]$.

    c) All numbers of the form $1/n$, $(n = 1, 2, 3, \ldots)$.

    d) All rational numbers.

    e) All numbers of the form $2^{-n} + 5^{-m}$, $(m, n = 1, 2, \ldots)$.

    f) All numbers of the form $(-1)^n + (1/m)$, $(m, n = 1, 2, \ldots)$.

    g) All numbers of the form $(1/n) + (1/m)$, $(m, n = 1, 2, \ldots)$.

    h) All numbers of the form $(-1)^n/[1 + (1/n)]$, $(n = 1, 2, \ldots)$.

3.3 The same as Exercise 3.2 for the following sets in $\mathbf{R}^2$:

    a) All complex z such that $|z| > 1$.

    b) All complex z such that $|z| \geq 1$.

    c) All complex numbers of the form $(1/n) + (i/m)$, $(m, n = 1, 2, \ldots)$.

    d) All points (x, y) such that $x^2 - y^2 < 1$.

    e) All points (x, y) such that $x > 0$.

    f) All points (x, y) such that $x \geq 0$.

3.4 Prove that every nonempty open set S in $\mathbf{R}^1$ contains both rational and irrational numbers.

3.5 Prove that the only sets in $\mathbf{R}^1$ which are both open and closed are the empty set and $\mathbf{R}^1$ itself. Is a similar statement true for $\mathbf{R}^2$?

3.6 Prove that every closed set in $\mathbf{R}^1$ is the intersection of a countable collection of open sets.

3.7 Prove that a nonempty, bounded closed set S in $\mathbf{R}^1$ is either a closed interval, or that S can be obtained from a closed interval by removing a countable disjoint collection of open intervals whose endpoints belong to S.

Open and closed sets in $\mathbf{R}^n$

3.8 Prove that open n-balls and n-dimensional open intervals are open sets in $\mathbf{R}^n$.

3.9 Prove that the interior of a set in $\mathbf{R}^n$ is open in $\mathbf{R}^n$.

3.10 If $S \subseteq \mathbf{R}^n$, prove that int S is the union of all open subsets of $\mathbf{R}^n$ which are contained in S. This is described by saying that int S is the largest open subset of S.

3.11 If S and T are subsets of $\mathbf{R}^n$, prove that

$$(\text{int } S) \cap (\text{int } T) = \text{int } (S \cap T), \quad \text{and} \quad (\text{int } S) \cup (\text{int } T) \subseteq \text{int } (S \cup T).$$

3.12 Let S' denote the derived set and $\bar{S}$ the closure of a set S in $\mathbf{R}^n$. Prove that:

    a) S' is closed in $\mathbf{R}^n$; that is, $(S')' \subseteq S'$.

    b) If $S \subseteq T$, then $S' \subseteq T'$.   c) $(S \cup T)' = S' \cup T'$.

    d) $(\bar{S})' = S'$.   e) $\bar{S}$ is closed in $\mathbf{R}^n$.

    f) $\bar{S}$ is the intersection of all closed subsets of $\mathbf{R}^n$ containing S. That is, $\bar{S}$ is the smallest closed set containing S.

3.13 Let S and T be subsets of $\mathbf{R}^n$. Prove that $\overline{S \cap T} \subseteq \bar{S} \cap \bar{T}$ and that $S \cap \bar{T} \subseteq \overline{S \cap T}$ if S is open.

NOTE. The statements in Exercises 3.9 through 3.13 are true in any metric space.

3.14 A set S in $\mathbf{R}^n$ is called *convex* if, for every pair of points $\mathbf{x}$ and $\mathbf{y}$ in S and every real θ satisfying $0 < \theta < 1$, we have $\theta\mathbf{x} + (1 - \theta)\mathbf{y} \in S$. Interpret this statement geometrically (in $\mathbf{R}^2$ and $\mathbf{R}^3$) and prove that:

    a) Every n-ball in $\mathbf{R}^n$ is convex.

    b) Every n-dimensional open interval is convex.

    c) The interior of a convex set is convex.

    d) The closure of a convex set is convex.

3.15 Let F be a collection of sets in $\mathbf{R}^n$, and let $S = \bigcup_{A \in F} A$ and $T = \bigcap_{A \in F} A$. For each of the following statements, either give a proof or exhibit a counterexample.

    a) If $\mathbf{x}$ is an accumulation point of T, then $\mathbf{x}$ is an accumulation point of each set A in F.

    b) If $\mathbf{x}$ is an accumulation point of S, then $\mathbf{x}$ is an accumulation point of at least one set A in F.

3.16 Prove that the set S of rational numbers in the interval $(0, 1)$ cannot be expressed as the intersection of a countable collection of open sets. *Hint.* Write $S = \{x_1, x_2, \dots\}$, assume $S = \bigcap_{k=1}^{\infty} S_k$, where each S_k is open, and construct a sequence $\{Q_n\}$ of closed intervals such that $Q_{n+1} \subseteq Q_n \subseteq S_n$ and such that $x_n \notin Q_n$. Then use the Cantor intersection theorem to obtain a contradiction.

Covering theorems in $\mathbf{R}^n$

3.17 If $S \subseteq \mathbf{R}^n$, prove that the collection of isolated points of S is countable.

3.18 Prove that the set of open disks in the xy-plane with center at (x, x) and radius $x > 0$, x rational, is a countable covering of the set $\{(x, y) : x > 0, y > 0\}$.

3.19 The collection F of open intervals of the form $(1/n, 2/n)$, where $n = 2, 3, \dots$, is an open covering of the open interval $(0, 1)$. Prove (without using Theorem 3.31) that no finite subcollection of F covers $(0, 1)$.

3.20 Give an example of a set S which is closed but not bounded and exhibit a countable open covering F such that no finite subset of F covers S.

3.21 Given a set S in $\mathbf{R}^n$ with the property that for every $\mathbf{x}$ in S there is an n-ball $B(\mathbf{x})$ such that $B(\mathbf{x}) \cap S$ is countable. Prove that S is countable.

3.22 Prove that a collection of disjoint open sets in $\mathbf{R}^n$ is necessarily countable. Give an example of a collection of disjoint closed sets which is not countable.

-**3.23** Assume that $S \subseteq \mathbf{R}^n$. A point $\mathbf{x}$ in $\mathbf{R}^n$ is said to be a *condensation point* of S if every n-ball $B(\mathbf{x})$ has the property that $B(\mathbf{x}) \cap S$ is not countable. Prove that if S is not countable, then there exists a point $\mathbf{x}$ *in* S such that $\mathbf{x}$ is a condensation point of S.

3.24 Assume that $S \subseteq \mathbf{R}^n$ and assume that S is not countable. Let T denote the set of condensation points of S. Prove that:

    a) $S - T$ is countable,   b) $S \cap T$ is not countable,

    c) T is a closed set,   d) T contains no isolated points.

Note that Exercise 3.23 is a special case of (b).

3.25 A set in $\mathbf{R}^n$ is called *perfect* if $S = S'$, that is, if S is a closed set which contains no isolated points. Prove that every uncountable closed set F in $\mathbf{R}^n$ can be expressed in the form $F = A \cup B$, where A is perfect and B is countable (*Cantor–Bendixon theorem*).

Hint. Use Exercise 3.24.

Metric spaces

3.26 In any metric space (M, d), prove that the empty set $\emptyset$ and the whole space M are both open and closed.

3.27 Consider the following two metrics in $\mathbf{R}^n$:

$$d_1(\mathbf{x}, \mathbf{y}) = \max_{1 \leq i \leq n} |x_i - y_i|, \qquad d_2(\mathbf{x}, \mathbf{y}) = \sum_{i=1}^{n} |x_i - y_i|.$$

In each of the following metric spaces prove that the ball $B(\mathbf{a}; r)$ has the geometric appearance indicated:

    a) In $(\mathbf{R}^2, d_1)$, a square with sides parallel to the coordinate axes.

    b) In $(\mathbf{R}^2, d_2)$, a square with diagonals parallel to the axes.

    c) A cube in $(\mathbf{R}^3, d_1)$.

    d) An octahedron in $(\mathbf{R}^3, d_2)$.

3.28 Let d_1 and d_2 be the metrics of Exercise 3.27 and let $\|\mathbf{x} - \mathbf{y}\|$ denote the usual Euclidean metric. Prove the following inequalities for all $\mathbf{x}$ and $\mathbf{y}$ in $\mathbf{R}^n$:

    $d_1(\mathbf{x}, \mathbf{y}) \leq \|\mathbf{x} - \mathbf{y}\| \leq d_2(\mathbf{x}, \mathbf{y})$ and $d_2(\mathbf{x}, \mathbf{y}) \leq \sqrt{n}\|\mathbf{x} - \mathbf{y}\| \leq nd_1(\mathbf{x}, \mathbf{y})$.

3.29 If (M, d) is a metric space, define

$$d'(x, y) = \frac{d(x, y)}{1 + d(x, y)}.$$

Prove that d' is also a metric for M. Note that $0 \leq d'(x, y) < 1$ for all x, y in M.

3.30 Prove that every finite subset of a metric space is closed.

3.31 In a metric space (M, d) the *closed ball* of radius $r > 0$ about a point a in M is the set $\bar{B}(a; r) = \{x : d(x, a) \leq r\}$.

    a) Prove that $\bar{B}(a; r)$ is a closed set.

    b) Give an example of a metric space in which $\bar{B}(a; r)$ is not the closure of the open ball $B(a; r)$.

3.32 In a metric space M, if subsets satisfy $A \subseteq S \subseteq \bar{A}$, where $\bar{A}$ is the closure of A, then A is said to be *dense* in S. For example, the set $\mathbf{Q}$ of rational numbers is dense in $\mathbf{R}$. If A is dense in S and if S is dense in T, prove that A is dense in T.

3.33 Refer to Exercise 3.32. A metric space M is said to be *separable* if there is a *countable* subset A which is dense in M. For example, $\mathbf{R}$ is separable because the set $\mathbf{Q}$ of rational numbers is a countable dense subset. Prove that every Euclidean space $\mathbf{R}^k$ is separable.

3.34 Refer to Exercise 3.33. Prove that the Lindelöf covering theorem (Theorem 3.28) is valid in any separable metric space.

3.35 Refer to Exercise 3.32. If A is dense in S and if B is open in S, prove that $B \subseteq \overline{A \cap B}$.
Hint. Exercise 3.13.

3.36 Refer to Exercise 3.32. If each of A and B is dense in S and if B is open in S, prove that $A \cap B$ is dense in S.

3.37 Given two metric spaces (S_1, d_1) and (S_2, d_2), a metric ρ for the Cartesian product $S_1 \times S_2$ can be constructed from d_1 and d_2 in many ways. For example, if $x = (x_1, x_2)$ and $y = (y_1, y_2)$ are in $S_1 \times S_2$, let $\rho(x, y) = d_1(x_1, y_1) + d_2(x_2, y_2)$. Prove that ρ is a metric for $S_1 \times S_2$ and construct further examples.

Compact subsets of a metric space

Prove each of the following statements concerning an arbitrary metric space (M, d) and subsets S, T of M.

3.38 Assume $S \subseteq T \subseteq M$. Then S is compact in (M, d) if, and only if, S is compact in the metric subspace (T, d).

3.39 If S is closed and T is compact, then $S \cap T$ is compact.

3.40 The intersection of an arbitrary collection of compact subsets of M is compact.

3.41 The union of a finite number of compact subsets of M is compact.

3.42 Consider the metric space $\mathbf{Q}$ of rational numbers with the Euclidean metric of $\mathbf{R}$. Let S consist of all rational numbers in the open interval (a, b), where a and b are irrational. Then S is a closed and bounded subset of $\mathbf{Q}$ which is not compact.

Miscellaneous properties of the interior and the boundary

If A and B denote arbitrary subsets of a metric space M, prove that:

3.43 $\operatorname{int} A = M - \overline{M - A}$.

3.44 $\operatorname{int} (M - A) = M - \bar{A}$.

3.45 $\operatorname{int} (\operatorname{int} A) = \operatorname{int} A$.

3.46 a) $\operatorname{int} (\bigcap_{i=1}^{n} A_i) = \bigcap_{i=1}^{n} (\operatorname{int} A_i)$, where each $A_i \subseteq M$.

b) $\operatorname{int} (\bigcap_{A \in F} A) \subseteq \bigcap_{A \in F} (\operatorname{int} A)$, if F is an infinite collection of subsets of M.

c) Give an example where equality does not hold in (b).

3.47 a) $\bigcup_{A \in F} (\operatorname{int} A) \subseteq \operatorname{int} (\bigcup_{A \in F} A)$.

b) Give an example of a finite collection F in which equality does not hold in (a).

3.48 a) $\operatorname{int} (\partial A) = \emptyset$ if A is open or if A is closed in M.

b) Give an example in which $\operatorname{int} (\partial A) = M$.

3.49 If int A = int B = $\emptyset$ and if A is closed in M, then int $(A \cup B)$ = $\emptyset$.

3.50 Give an example in which int A = int B = $\emptyset$ but int $(A \cup B)$ = M.

3.51 $\partial A = \bar{A} \cap \overline{M - A}$ and $\partial A = \partial(M - A)$.

3.52 If $\bar{A} \cap \bar{B}$ = $\emptyset$, then $\partial(A \cup B)$ = $\partial A \cup \partial B$.

SUGGESTED REFERENCES FOR FURTHER STUDY

3.1 Boas, R. P., *A Primer of Real Functions*. Carus Monograph No. 13. Wiley, New York, 1960.

3.2 Gleason, A., *Fundamentals of Abstract Analysis*. Addison-Wesley, Reading, 1966.

3.3 Kaplansky, I., *Set Theory and Metric Spaces*. Allyn and Bacon, Boston, 1972.

3.4 Simmons, G. F., *Introduction to Topology and Modern Analysis*. McGraw-Hill, New York, 1963.

CHAPTER 4

LIMITS AND CONTINUITY

4.1 INTRODUCTION

The reader is already familiar with the limit concept as introduced in elementary calculus where, in fact, several kinds of limits are usually presented. For example, the *limit of a sequence* of real numbers $\{x_n\}$, denoted symbolically by writing

$$\lim_{n \to \infty} x_n = A,$$

means that for every number $\varepsilon > 0$ there is an integer N such that

$$|x_n - A| < \varepsilon \qquad \text{whenever } n \geq N.$$

This limit process conveys the intuitive idea that x_n can be made arbitrarily close to A provided that n is sufficiently large. There is also the *limit of a function*, indicated by notation such as

$$\lim_{x \to p} f(x) = A,$$

which means that for every $\varepsilon > 0$ there is another number $\delta > 0$ such that

$$|f(x) - A| < \varepsilon \qquad \text{whenever } 0 < |x - p| < \delta.$$

This conveys the idea that $f(x)$ can be made arbitrarily close to A by taking x sufficiently close to p.

Applications of calculus to geometrical and physical problems in 3-space and to functions of several variables make it necessary to extend these concepts to $\mathbf{R}^n$. It is just as easy to go one step further and introduce limits in the more general setting of metric spaces. This achieves a simplification in the theory by stripping it of unnecessary restrictions and at the same time covers nearly all the important aspects needed in analysis.

First we discuss limits of sequences of points in a metric space, then we discuss limits of functions and the concept of continuity.

4.2 CONVERGENT SEQUENCES IN A METRIC SPACE

Definition 4.1. *A sequence $\{x_n\}$ of points in a metric space (S, d) is said to converge if there is a point p in S with the following property:*

For every $\varepsilon > 0$ there is an integer N such that

$$d(x_n, p) < \varepsilon \qquad \text{whenever } n \geq N.$$

We also say that $\{x_n\}$ converges to p and we write $x_n \to p$ as $n \to \infty$, or simply $x_n \to p$. If there is no such p in S, the sequence $\{x_n\}$ is said to diverge.

NOTE. The definition of convergence implies that

$$x_n \to p \quad \text{if and only if} \quad d(x_n, p) \to 0.$$

The convergence of the sequence $\{d(x_n, p)\}$ to 0 takes place in the Euclidean metric space $\mathbf{R}^1$.

Examples

1. In Euclidean space $\mathbf{R}^1$, a sequence $\{x_n\}$ is called *increasing* if $x_n \leq x_{n+1}$ for all n. If an increasing sequence is bounded above (that is, if $x_n \leq M$ for some $M > 0$ and all n), then $\{x_n\}$ converges to the supremum of its range, sup $\{x_1, x_2, \ldots\}$. Similarly, $\{x_n\}$ is called *decreasing* if $x_{n+1} \leq x_n$ for all n. Every decreasing sequence which is bounded below converges to the infimum of its range. For example, $\{1/n\}$ converges to 0.

2. If $\{a_n\}$ and $\{b_n\}$ are real sequences converging to 0, then $\{a_n + b_n\}$ also converges to 0. If $0 \leq c_n \leq a_n$ for all n and if $\{a_n\}$ converges to 0, then $\{c_n\}$ also converges to 0. These elementary properties of sequences in $\mathbf{R}^1$ can be used to simplify some of the proofs concerning limits in a general metric space.

3. In the complex plane $\mathbf{C}$, let $z_n = 1 + n^{-2} + (2 - 1/n)i$. Then $\{z_n\}$ converges to $1 + 2i$ because

$$d(z_n, 1 + 2i)^2 = |z_n - (1 + 2i)|^2 = \frac{1}{n^4} + \frac{1}{n^2} \to 0 \text{ as } n \to \infty,$$

so $d(z_n, 1 + 2i) \to 0$.

Theorem 4.2. *A sequence $\{x_n\}$ in a metric space (S, d) can converge to at most one point in S.*

Proof. Assume that $x_n \to p$ and $x_n \to q$. We will prove that $p = q$. By the triangle inequality we have

$$0 \leq d(p, q) \leq d(p, x_n) + d(x_n, q).$$

Since $d(p, x_n) \to 0$ and $d(x_n, q) \to 0$ this implies that $d(p, q) = 0$, so $p = q$.

If a sequence $\{x_n\}$ converges, the unique point to which it converges is called the *limit* of the sequence and is denoted by lim x_n or by $\lim_{n \to \infty} x_n$.

Example. In Euclidean space $\mathbf{R}^1$ we have $\lim_{n \to \infty} 1/n = 0$. The same sequence in the metric subspace $T = (0, 1]$ does *not* converge because the only candidate for the limit is 0 and $0 \notin T$. This example shows that the convergence or divergence of a sequence depends on the underlying space as well as on the metric.

Theorem 4.3. *In a metric space (S, d), assume $x_n \to p$ and let $T = \{x_1, x_2, \ldots\}$ be the range of $\{x_n\}$. Then:*

a) *T is bounded.*

b) *p is an adherent point of T.*

Proof. a) Let N be the integer corresponding to $\varepsilon = 1$ in the definition of convergence. Then every x_n with $n \geq N$ lies in the ball $B(p; 1)$, so every point in T lies in the ball $B(p; r)$, where

$$r = 1 + \max \{d(p, x_1), \ldots, d(p, x_{N-1})\}.$$

Therefore T is bounded.

b) Since every ball $B(p; \varepsilon)$ contains a point of T, p is an adherent point of T.

NOTE. If T is *infinite*, every ball $B(p; \varepsilon)$ contains infinitely many points of T, so p is an accumulation point of T.

The next theorem provides a converse to part (b).

Theorem 4.4. *Given a metric space (S, d) and a subset $T \subseteq S$. If a point p in S is an adherent point of T, then there is a sequence $\{x_n\}$ of points in T which converges to p.*

Proof. For every integer $n \geq 1$ there is a point x_n in T with $d(p, x_n) \leq 1/n$. Hence $d(p, x_n) \to 0$, so $x_n \to p$.

Theorem 4.5. *In a metric space (S, d) a sequence $\{x_n\}$ converges to p if, and only if, every subsequence $\{x_{k(n)}\}$ converges to p.*

Proof. Assume $x_n \to p$ and consider any subsequence $\{x_{k(n)}\}$. For every $\varepsilon > 0$ there is an N such that $n \geq N$ implies $d(x_n, p) < \varepsilon$. Since $\{x_{k(n)}\}$ is a subsequence, there is an integer M such that $k(n) \geq N$ for $n \geq M$. Hence $n \geq M$ implies $d(x_{k(n)}, p) < \varepsilon$, which proves that $x_{k(n)} \to p$. The converse statement holds trivially since $\{x_n\}$ is itself a subsequence.

4.3 CAUCHY SEQUENCES

If a sequence $\{x_n\}$ converges to a limit p, its terms must ultimately become close to p and hence close to each other. This property is stated more formally in the next theorem.

Theorem 4.6. *Assume that $\{x_n\}$ converges in a metric space (S, d). Then for every $\varepsilon > 0$ there is an integer N such that*

$$d(x_n, x_m) < \varepsilon \qquad \text{whenever } n \geq N \text{ and } m \geq N.$$

Proof. Let $p = \lim x_n$. Given $\varepsilon > 0$, let N be such that $d(x_n, p) < \varepsilon/2$ whenever $n \geq N$. Then $d(x_m, p) < \varepsilon/2$ if $m \geq N$. If both $n \geq N$ and $m \geq N$ the triangle inequality gives us

$$d(x_n, x_m) \leq d(x_n, p) + d(p, x_m) < \frac{\varepsilon}{2} + \frac{\varepsilon}{2} = \varepsilon.$$

4.7 Definition of a Cauchy Sequence. *A sequence $\{x_n\}$ in a metric space (S, d) is called a Cauchy sequence if it satisfies the following condition (called the Cauchy condition):*

For every $\varepsilon > 0$ there is an integer N such that

$$d(x_n, x_m) < \varepsilon \qquad \text{whenever } n \geq N \text{ and } m \geq N.$$

Theorem 4.6 states that every convergent sequence is a Cauchy sequence. The converse is not true in a general metric space. For example, the sequence $\{1/n\}$ is a Cauchy sequence in the Euclidean subspace $T = (0, 1]$ of $\mathbf{R}^1$, but this sequence does not converge in T. However, the converse of Theorem 4.6 is true in every Euclidean space $\mathbf{R}^k$.

Theorem 4.8. *In Euclidean space $\mathbf{R}^k$ every Cauchy sequence is convergent.*

Proof. Let $\{x_n\}$ be a Cauchy sequence in $\mathbf{R}^k$ and let $T = \{x_1, x_2, \ldots\}$ be the range of the sequence. If T is finite, then all except a finite number of the terms $\{x_n\}$ are equal and hence $\{x_n\}$ converges to this common value.

Now suppose T is infinite. We use the Bolzano–Weierstrass theorem to show that T has an accumulation point $\mathbf{p}$, and then we show that $\{x_n\}$ converges to $\mathbf{p}$. First we need to know that T is bounded. This follows from the Cauchy condition. In fact, when $\varepsilon = 1$ there is an N such that $n \geq N$ implies $\|x_n - x_N\| < 1$. This means that all points x_n with $n \geq N$ lie inside a ball of radius 1 about x_N as center, so T lies inside a ball of radius $1 + M$ about $\mathbf{0}$, where M is the largest of the numbers $\|x_1\|, \ldots, \|x_N\|$. Therefore, since T is a bounded infinite set it has an accumulation point $\mathbf{p}$ in $\mathbf{R}^k$ (by the Bolzano–Weierstrass theorem). We show next that $\{x_n\}$ converges to $\mathbf{p}$.

Given $\varepsilon > 0$ there is an N such that $\|x_n - x_m\| < \varepsilon/2$ whenever $n \geq N$ and $m \geq N$. The ball $B(\mathbf{p}; \varepsilon/2)$ contains a point x_m with $m \geq N$. Hence if $n \geq N$ we have

$$\|x_n - \mathbf{p}\| \leq \|x_n - x_m\| + \|x_m - \mathbf{p}\| < \frac{\varepsilon}{2} + \frac{\varepsilon}{2} = \varepsilon,$$

so $\lim x_n = \mathbf{p}$. This completes the proof.

Examples

1. Theorem 4.8 is often used for proving the convergence of a sequence when the limit is not known in advance. For example, consider the sequence in $\mathbf{R}^1$ defined by

$$x_n = 1 - \frac{1}{2} + \frac{1}{3} - \frac{1}{4} + \cdots + \frac{(-1)^{n-1}}{n}.$$

If $m > n \geq N$, we find (by taking successive terms in pairs) that

$$|x_m - x_n| = \left| \frac{1}{n+1} - \frac{1}{n+2} + \cdots \pm \frac{1}{m} \right| < \frac{1}{n} \leq \frac{1}{N},$$

so $|x_m - x_n| < \varepsilon$ as soon as $N > 1/\varepsilon$. Therefore $\{x_n\}$ is a Cauchy sequence and hence it converges to some limit. It can be shown (see Exercise 8.18) that this limit is log 2, a fact which is not immediately obvious.

2. Given a real sequence $\{a_n\}$ such that $|a_{n+2} - a_{n+1}| \leq \frac{1}{2}|a_{n+1} - a_n|$ for all $n \geq 1$. We can prove that $\{a_n\}$ converges without knowing its limit. Let $b_n = |a_{n+1} - a_n|$. Then $0 \leq b_{n+1} \leq b_n/2$ so, by induction, $b_{n+1} \leq b_1/2^n$. Hence $b_n \to 0$. Also, if $m > n$ we have

$$a_m - a_n = \sum_{k=n}^{m-1} (a_{k+1} - a_k);$$

hence

$$|a_m - a_n| \leq \sum_{k=n}^{m-1} b_k \leq b_n \left(1 + \frac{1}{2} + \cdots + \frac{1}{2^{m-1-n}} \right) < 2b_n.$$

This implies that $\{a_n\}$ is a Cauchy sequence, so $\{a_n\}$ converges.

4.4 COMPLETE METRIC SPACES

Definition 4.9. *A metric space* (S, d) *is called complete if every Cauchy sequence in S converges in S. A subset T of S is called complete if the metric subspace* (T, d) *is complete.*

Example 1. Every Euclidean space $\mathbf{R}^k$ is complete (Theorem 4.8). In particular, $\mathbf{R}^1$ is complete, but the subspace $T = (0, 1]$ is not complete.

Example 2. The space $\mathbf{R}^n$ with the metric $d(\mathbf{x}, \mathbf{y}) = \max_{1 \leq i \leq n} |x_i - y_i|$ is complete.

The next theorem relates completeness with compactness.

Theorem 4.10. *In any metric space* (S, d) *every compact subset T is complete.*

Proof. Let $\{x_n\}$ be a Cauchy sequence in T and let $A = \{x_1, x_2, \ldots\}$ denote the range of $\{x_n\}$. If A is finite, then $\{x_n\}$ converges to one of the elements of A, hence $\{x_n\}$ converges in T.

If A is infinite, Theorem 3.38 tells us that A has an accumulation point p in T since T is compact. We show next that $x_n \to p$. Given $\varepsilon > 0$, choose N so that $n \geq N$ and $m \geq N$ implies $d(x_n, x_m) < \varepsilon/2$. The ball $B(p; \varepsilon/2)$ contains a point x_m with $m \geq N$. Therefore if $n \geq N$ the triangle inequality gives us

$$d(x_n, p) \leq d(x_n, x_m) + d(x_m, p) < \frac{\varepsilon}{2} + \frac{\varepsilon}{2} = \varepsilon,$$

so $x_n \to p$. Therefore every Cauchy sequence in T has a limit in T, so T is complete.

4.5 LIMIT OF A FUNCTION

In this section we consider two metric spaces (S, d_S) and (T, d_T), where d_S and d_T denote the respective metrics. Let A be a subset of S and let $f : A \to T$ be a function from A to T.

Definition 4.11. *If p is an accumulation point of A and if b $\in$ T, the notation*

$$\lim_{x \to p} f(x) = b, \tag{1}$$

is defined to mean the following:

For every $\varepsilon > 0$ there is a $\delta > 0$ such that

$$d_T(f(x), b) < \varepsilon \qquad \textit{whenever } x \in A, \ x \neq p, \textit{ and } d_S(x,p) < \delta.$$

The symbol in (1) is read "the limit of $f(x)$, as x tends to p, is b," or "$f(x)$ approaches b as x approaches p." We sometimes indicate this by writing $f(x) \to b$ as $x \to p$.

The definition conveys the intuitive idea that $f(x)$ can be made arbitrarily close to b by taking x sufficiently close to p. (See Fig. 4.1.) We require that p be an accumulation point of A to make certain that there will be points x in A sufficiently close to p, with $x \neq p$. However, p need not be in the domain of f, and b need not be in the range of f.

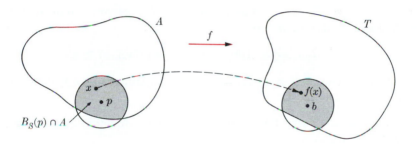

Figure 4.1

NOTE. The definition can also be formulated in terms of balls. Thus, (1) holds if, and only if, for every ball $B_T(b)$, there is a ball $B_S(p)$ such that $B_S(p) \cap A$ is not empty and such that

$$f(x) \in B_T(b) \qquad \textit{whenever } x \in B_S(p) \cap A, \ x \neq p.$$

When formulated this way, the definition is meaningful when p or b (or both) are in the extended real number system **R*** or in the extended complex number system **C***. However, in what follows, it is to be understood that p and b are finite unless it is explicitly stated that they can be infinite.

The next theorem relates limits of functions to limits of convergent sequences.

Theorem 4.12. *Assume p is an accumulation point of A and assume b $\in$ T. Then*

$$\lim_{x \to p} f(x) = b, \tag{2}$$

if, and only if,

$$\lim_{n \to \infty} f(x_n) = b, \tag{3}$$

for every sequence $\{x_n\}$ of points in $A - \{p\}$ which converges to p.

Proof. If (2) holds, then for every $\varepsilon > 0$ there is a $\delta > 0$ such that

$$d_T(f(x), b) < \varepsilon \qquad \text{whenever } x \in A \text{ and } 0 < d_S(x, p) < \delta. \tag{4}$$

Now take any sequence $\{x_n\}$ in $A - \{p\}$ which converges to p. For the δ in (4), there is an integer N such that $n \geq N$ implies $d_S(x_n, p) < \delta$. Therefore (4) implies $d_T(f(x_n), b) < \varepsilon$ for $n \geq N$, and hence $\{f(x_n)\}$ converges to b. Therefore (2) implies (3).

To prove the converse we assume that (3) holds and that (2) is false and arrive at a contradiction. If (2) is false, then for some $\varepsilon > 0$ and every $\delta > 0$ there is a point x in A (where x may depend on δ) such that

$$0 < d_S(x, p) < \delta \qquad \text{but} \qquad d_T(f(x), b) \geq \varepsilon. \tag{5}$$

Taking $\delta = 1/n$, $n = 1, 2, \ldots$, this means there is a corresponding sequence of points $\{x_n\}$ in $A - \{p\}$ such that

$$0 < d_S(x_n, p) < 1/n \qquad \text{but} \qquad d_T(f(x_n), b) \geq \varepsilon.$$

Clearly, this sequence $\{x_n\}$ converges to p but the sequence $\{f(x_n)\}$ does not converge to b, contradicting (3).

NOTE. Theorems 4.12 and 4.2 together show that a function cannot have two different limits as $x \to p$.

4.6 LIMITS OF COMPLEX-VALUED FUNCTIONS

Let (S, d) be a metric space, let A be a subset of S, and consider two complex-valued functions f and g defined on A,

$$f : A \to \mathbf{C}, \qquad g : A \to \mathbf{C}.$$

The *sum* $f + g$ is defined to be the function whose value at each point x of A is the complex number $f(x) + g(x)$. The *difference* $f - g$, the *product* $f \cdot g$, and the *quotient* f/g are similarly defined. It is understood that the quotient is defined only at those points x for which $g(x) \neq 0$.

The usual rules for calculating with limits are given in the next theorem.

Theorem 4.13. *Let f and g be complex-valued functions defined on a subset A of a metric space (S, d). Let p be an accumulation point of A, and assume that*

$$\lim_{x \to p} f(x) = a, \qquad \lim_{x \to p} g(x) = b.$$

Then we also have:

a) $\lim_{x \to p} [f(x) \pm g(x)] = a \pm b$,

b) $\lim_{x \to p} f(x)g(x) = ab$,

c) $\lim_{x \to p} f(x)/g(x) = a/b$ *if* $b \neq 0$.

Proof. We prove (b), leaving the other parts as exercises. Given ε with $0 < \varepsilon < 1$, let ε' be a second number satisfying $0 < \varepsilon' < 1$, which will be made to depend on ε in a way to be described later. There is a $\delta > 0$ such that if $x \in A$ and $d(x, p) < \delta$, then

$$|f(x) - a| < \varepsilon' \qquad \text{and} \qquad |g(x) - b| < \varepsilon'.$$

Then

$$|f(x)| = |a + (f(x) - a)| < |a| + \varepsilon' < |a| + 1.$$

Writing $f(x)g(x) - ab = f(x)g(x) - bf(x) + bf(x) - ab$, we have

$$|f(x)g(x) - ab| \leq |f(x)| \, |g(x) - b| + |b| \, |f(x) - a|$$
$$< (|a| + 1)\varepsilon' + |b|\varepsilon' = \varepsilon'(|a| + |b| + 1).$$

If we choose $\varepsilon' = \varepsilon/(|a| + |b| + 1)$, we see that $|f(x)g(x) - ab| < \varepsilon$ whenever $x \in A$ and $d(x, p) < \delta$, and this proves (b).

4.7 LIMITS OF VECTOR-VALUED FUNCTIONS

Again, let (S, d) be a metric space and let A be a subset of S. Consider two vector-valued functions $\mathbf{f}$ and $\mathbf{g}$ defined on A, each with values in $\mathbf{R}^k$,

$$\mathbf{f} : A \to \mathbf{R}^k, \qquad \mathbf{g} : A \to \mathbf{R}^k.$$

Quotients of vector-valued functions are not defined (if $k > 2$), but we can define the *sum* $\mathbf{f} + \mathbf{g}$, the *product* $\lambda\mathbf{f}$ (if λ is real) and the *inner product* $\mathbf{f} \cdot \mathbf{g}$ by the respective formulas

$$(\mathbf{f} + \mathbf{g})(x) = \mathbf{f}(x) + \mathbf{g}(x), \qquad (\lambda\mathbf{f})(x) = \lambda\mathbf{f}(x), \qquad (\mathbf{f} \cdot \mathbf{g})(x) = \mathbf{f}(x) \cdot \mathbf{g}(x)$$

for each x in A. We then have the following rules for calculating with limits of vector-valued functions.

Theorem 4.14. *Let p be an accumulation point of A and assume that*

$$\lim_{x \to p} \mathbf{f}(x) = \mathbf{a}, \qquad \lim_{x \to p} \mathbf{g}(x) = \mathbf{b}.$$

Then we also have:

a) $\lim_{x \to p} [\mathbf{f}(x) + \mathbf{g}(x)] = \mathbf{a} + \mathbf{b}$,

b) $\lim_{x \to p} \lambda\mathbf{f}(x) = \lambda\mathbf{a}$ *for every scalar* λ,

c) $\lim_{x \to p} \mathbf{f}(x) \cdot \mathbf{g}(x) = \mathbf{a} \cdot \mathbf{b}$,

d) $\lim_{x \to p} \|\mathbf{f}(x)\| = \|\mathbf{a}\|$.

Proof. We prove only parts (c) and (d). To prove (c) we write

$$\mathbf{f}(x)\cdot\mathbf{g}(x) - \mathbf{a}\cdot\mathbf{b} = [\mathbf{f}(x) - \mathbf{a}]\cdot[\mathbf{g}(x) - \mathbf{b}] + \mathbf{a}\cdot[\mathbf{g}(x) - \mathbf{b}] + \mathbf{b}\cdot[\mathbf{f}(x) - \mathbf{a}].$$

The triangle inequality and the Cauchy–Schwarz inequality give us

$$0 \le |\mathbf{f}(x)\cdot\mathbf{g}(x) - \mathbf{a}\cdot\mathbf{b}|$$
$$\le \|\mathbf{f}(x) - \mathbf{a}\| \, \|\mathbf{g}(x) - \mathbf{b}\| + \|\mathbf{a}\| \, \|\mathbf{g}(x) - \mathbf{b}\| + \|\mathbf{b}\| \, \|\mathbf{f}(x) - \mathbf{a}\|.$$

Each term on the right tends to 0 as $x \to p$, so $\mathbf{f}(x)\cdot\mathbf{g}(x) \to \mathbf{a}\cdot\mathbf{b}$. This proves (c). To prove (d) note that $|\,\|\mathbf{f}(x)\| - \|\mathbf{a}\|\,| \le \|\mathbf{f}(x) - \mathbf{a}\|$.

NOTE. Let $f_1, \ldots, f_n$ be n real-valued functions defined on A, and let $\mathbf{f}: A \to \mathbf{R}^n$ be the vector-valued function defined by the equation

$$\mathbf{f}(x) = (f_1(x), f_2(x), \ldots, f_n(x)) \qquad \text{if } x \in A.$$

Then $f_1, \ldots, f_n$ are called the components of $\mathbf{f}$, and we also write $\mathbf{f} = (f_1, \ldots, f_n)$ to denote this relationship.

If $\mathbf{a} = (a_1, \ldots, a_n)$, then for each $r = 1, 2, \ldots, n$ we have

$$|f_r(x) - a_r| \le \|\mathbf{f}(x) - \mathbf{a}\| \le \sum_{r=1}^{n} |f_r(x) - a_r|.$$

These inequalities show that $\lim_{x \to p} \mathbf{f}(x) = \mathbf{a}$ if, and only if, $\lim_{x \to p} f_r(x) = a_r$ for each r.

4.8 CONTINUOUS FUNCTIONS

The definition of continuity presented in elementary calculus can be extended to functions from one metric space to another.

Definition 4.15. *Let (S, d_S) and (T, d_T) be metric spaces and let $f: S \to T$ be a function from S to T. The function f is said to be continuous at a point p in S if for every $\varepsilon > 0$ there is a $\delta > 0$ such that*

$$d_T(f(x), f(p)) < \varepsilon \qquad \text{whenever } d_S(x, p) < \delta.$$

If f is continuous at every point of a subset A of S, we say f is continuous on A.

This definition reflects the intuitive idea that points close to p are mapped by f into points close to $f(p)$. It can also be stated in terms of balls: A function f is continuous at p if and only if, for every $\varepsilon > 0$, there is a $\delta > 0$ such that

$$f(B_S(p; \delta)) \subseteq B_T(f(p); \varepsilon).$$

Here $B_S(p; \delta)$ is a ball in S; its image under f must be contained in the ball $B_T(f(p); \varepsilon)$ in T. (See Fig. 4.2.)

If p is an accumulation point of S, the definition of continuity implies that

$$\lim_{x \to p} f(x) = f(p).$$

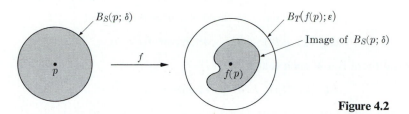

Figure 4.2

If p is an isolated point of S (a point of S which is not an accumulation point of S), then every f defined at p will be continuous at p because for sufficiently small δ there is only one x satisfying $d_S(x, p) < \delta$, namely $x = p$, and $d_T(f(p), f(p)) = 0$.

Theorem 4.16. *Let $f : S \to T$ be a function from one metric space (S, d_S) to another (T, d_T), and assume $p \in S$. Then f is continuous at p if, and only if, for every sequence $\{x_n\}$ in S convergent to p, the sequence $\{f(x_n)\}$ in T converges to $f(p)$; in symbols,*

$$\lim_{n \to \infty} f(x_n) = f\left(\lim_{n \to \infty} x_n\right).$$

The proof of this theorem is similar to that of Theorem 4.12 and is left as an exercise for the reader. (The result can also be deduced from 4.12 but there is a minor complication in the argument due to the fact that some terms of the sequence $\{x_n\}$ could be equal to p.)

The theorem is often described by saying that for continuous functions the limit symbol can be interchanged with the function symbol. Some care is needed in interchanging these symbols because sometimes $\{f(x_n)\}$ converges when $\{x_n\}$ diverges.

Example If $x_n \to x$ and $y_n \to y$ in a metric space (S, d), then $d(x_n, y_n) \to d(x, y)$ (Exercise 4.7). The reader can verify that d is continuous on the metric space $(S \times S, \rho)$, where ρ is the metric of Exercise 3.37 with $S_1 = S_2 = S$.

NOTE. Continuity of a function f at a point p is called a *local property* of f because it depends on the behavior of f only in the immediate vicinity of p. A property of f which concerns the whole domain of f is called a *global property*. Thus, continuity of f on its domain is a global property.

4.9 CONTINUITY OF COMPOSITE FUNCTIONS

Theorem 4.17. *Let (S, d_S), (T, d_T), and (U, d_U) be metric spaces. Let $f : S \to T$ and $g : f(S) \to U$ be functions, and let h be the composite function defined on S by the equation*

$$h(x) = g(f(x)) \quad \text{for } x \text{ in } S.$$

If f is continuous at p and if g is continuous at $f(p)$, then h is continuous at p.

Proof. Let $b = f(p)$. Given $\varepsilon > 0$, there is a $\delta > 0$ such that

$$d_U\big(g(y), g(b)\big) < \varepsilon \qquad \text{whenever } d_T(y, b) < \delta.$$

For this δ there is a δ' such that

$$d_T\big(f(x), f(p)\big) < \delta \qquad \text{whenever } d_S(x, p) < \delta'.$$

Combining these two statements and taking $y = f(x)$, we find that

$$d_U\big(h(x), h(p)\big) < \varepsilon \qquad \text{whenever } d_S(x, p) < \delta',$$

so h is continuous at p.

4.10 CONTINUOUS COMPLEX-VALUED AND VECTOR-VALUED FUNCTIONS

Theorem 4.18. *Let f and g be complex-valued functions continuous at a point p in a metric space (S, d). Then $f + g$, $f - g$, and $f \cdot g$ are each continuous at p. The quotient f/g is also continuous at p if $g(p) \neq 0$.*

Proof. The result is trivial if p is an isolated point of S. If p is an accumulation point of S, we obtain the result from Theorem 4.13.

There is, of course, a corresponding theorem for vector-valued functions, which is proved in the same way, using Theorem 4.14.

Theorem 4.19. *Let $\mathbf{f}$ and $\mathbf{g}$ be functions continuous at a point p in a metric space (S, d), and assume that $\mathbf{f}$ and $\mathbf{g}$ have values in $\mathbf{R}^n$. Then each of the following is continuous at p: the sum $\mathbf{f} + \mathbf{g}$, the product $\lambda\mathbf{f}$ for every real λ, the inner product $\mathbf{f} \cdot \mathbf{g}$, and the norm $\|\mathbf{f}\|$.*

Theorem 4.20. *Let $f_1, \ldots, f_n$ be n real-valued functions defined on a subset A of a metric space (S, d_S), and let $\mathbf{f} = (f_1, \ldots, f_n)$. Then $\mathbf{f}$ is continuous at a point p of A if and only if each of the functions $f_1, \ldots, f_n$ is continuous at p.*

Proof. If p is an isolated point of A there is nothing to prove. If p is an accumulation point, we note that $\mathbf{f}(x) \to \mathbf{f}(p)$ as $x \to p$ if and only if $f_k(x) \to f_k(p)$ for each $k = 1, 2, \ldots, n$.

4.11 EXAMPLES OF CONTINUOUS FUNCTIONS

Let $S = \mathbf{C}$, the complex plane. It is a trivial exercise to show that the following complex-valued functions are continuous on $\mathbf{C}$:

a) constant functions, defined by $f(z) = c$ for every z in $\mathbf{C}$;
b) the identity function defined by $f(z) = z$ for every z in $\mathbf{C}$.

Repeated application of Theorem 4.18 establishes the continuity of every polynomial:

$$f(z) = a_0 + a_1 z + a_2 z^2 + \cdots + a_n z^n,$$

the a_i being complex numbers.

If S is a subset on $\mathbf{C}$ on which the polynomial f does not vanish, then $1/f$ is continuous on S. Therefore a rational function g/f, where g and f are polynomials, is continuous at those points of $\mathbf{C}$ at which the denominator does not vanish.

The familiar real-valued functions of elementary calculus, such as the exponential, trigonometric, and logarithmic functions, are all continuous wherever they are defined. The continuity of these elementary functions justifies the common practice of evaluating certain limits by substituting the limiting value of the "independent variable"; for example,

$$\lim_{x \to 0} e^x = e^0 = 1.$$

The continuity of the complex exponential and trigonometric functions is a consequence of the continuity of the corresponding real-valued functions and Theorem 4.20.

4.12 CONTINUITY AND INVERSE IMAGES OF OPEN OR CLOSED SETS

The concept of inverse image can be used to give two important global descriptions of continuous functions.

4.21 Definition of inverse image. *Let $f : S \to T$ be a function from a set S to a set T. If Y is a subset of T, the inverse image of Y under f, denoted by $f^{-1}(Y)$, is defined to be the largest subset of S which f maps into Y; that is,*

$$f^{-1}(Y) = \{x : x \in S \quad and \quad f(x) \in Y\}.$$

NOTE. If f has an inverse function f^{-1}, the inverse image of Y under f is the same as the image of Y under f^{-1}, and in this case there is no ambiguity in the notation $f^{-1}(Y)$. Note also that $f^{-1}(A) \subseteq f^{-1}(B)$ if $A \subseteq B \subseteq T$.

Theorem 4.22. *Let $f : S \to T$ be a function from S to T. If $X \subseteq S$ and $Y \subseteq T$, then we have:*

a) $X = f^{-1}(Y)$ *implies* $f(X) \subseteq Y$.

b) $Y = f(X)$ *implies* $X \subseteq f^{-1}(Y)$.

The proof of Theorem 4.22 is a direct translation of the definition of the symbols $f^{-1}(Y)$ and $f(X)$, and is left to the reader. It should be observed that; in general, we cannot conclude that $Y = f(X)$ implies $X = f^{-1}(Y)$. (See the example in Fig. 4.3.)

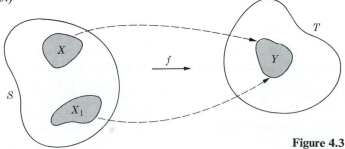

Figure 4.3

Note that the statements in Theorem 4.22 can also be expressed as follows:

$$f[f^{-1}(Y)] \subseteq Y, \qquad X \subseteq f^{-1}[f(X)].$$

Note also that $f^{-1}(A \cup B) = f^{-1}(A) \cup f^{-1}(B)$ for all subsets A and B of T.

Theorem 4.23. *Let $f : S \to T$ be a function from one metric space (S, d_S) to another (T, d_T). Then f is continuous on S if, and only if, for every open set Y in T, the inverse image $f^{-1}(Y)$ is open in S.*

Proof. Let f be continuous on S, let Y be open in T, and let p be any point of $f^{-1}(Y)$. We will prove that p is an interior point of $f^{-1}(Y)$. Let $y = f(p)$. Since Y is open we have $B_T(y; \varepsilon) \subseteq Y$ for some $\varepsilon > 0$. Since f is continuous at p, there is a $\delta > 0$ such that $f(B_S(p; \delta)) \subseteq B_T(y; \varepsilon)$. Hence,

$$B_S(p; \delta) \subseteq f^{-1}[f(B_S(p; \delta))] \subseteq f^{-1}[B_T(y; \varepsilon)] \subseteq f^{-1}(Y),$$

so p is an interior point of $f^{-1}(Y)$.

Conversely, assume that $f^{-1}(Y)$ is open in S for every open subset Y in T. Choose p in S and let $y = f(p)$. We will prove that f is continuous at p. For every $\varepsilon > 0$, the ball $B_T(y; \varepsilon)$ is open in T, so $f^{-1}(B_T(y; \varepsilon))$ is open in S. Now, $p \in f^{-1}(B_T(y; \varepsilon))$ so there is a $\delta > 0$ such that $B_S(p; \delta) \subseteq f^{-1}(B_T(y; \varepsilon))$. Therefore, $f(B_S(p; \delta)) \subseteq B_T(y; \varepsilon)$ so f is continuous at p.

Theorem 4.24. *Let $f : S \to T$ be a function from one metric space (S, d_S) to another (T, d_T). Then f is continuous on S if, and only if, for every closed set Y in T, the inverse image $f^{-1}(Y)$ is closed in S.*

Proof. If Y is closed in T, then $T - Y$ is open in T and

$$f^{-1}(T - Y) = S - f^{-1}(Y).$$

Now apply Theorem 4.23.

Examples. The image of an open set under a continuous mapping is not necessarily open. A simple counterexample is a constant function which maps all of S onto a single point in $\mathbf{R}^1$. Similarly, the image of a closed set under a continuous mapping need not be closed. For example, the real-valued function $f(x) = \arctan x$ maps $\mathbf{R}^1$ onto the open interval $(-\pi/2, \pi/2)$.

4.13 FUNCTIONS CONTINUOUS ON COMPACT SETS

The next theorem shows that the continuous image of a compact set is compact. This is another global property of continuous functions.

Theorem 4.25. *Let $f : S \to T$ be a function from one metric space (S, d_S) to another (T, d_T). If f is continuous on a compact subset X of S, then the image $f(X)$ is a compact subset of T; in particular, $f(X)$ is closed and bounded in T.*

Proof. Let F be an open covering of $f(X)$, so that $f(X) \subseteq \bigcup_{A \in F} A$. We will show that a finite number of the sets A cover $f(X)$. Since f is continuous on the metric subspace (X, d_S) we can apply Theorem 4.23 to conclude that each set $f^{-1}(A)$ is open in (X, d_S). The sets $f^{-1}(A)$ form an open covering of X and, since X is compact, a finite number of them cover X, say $X \subseteq f^{-1}(A_1) \cup \cdots \cup f^{-1}(A_p)$. Hence

$$f(X) \subseteq f[f^{-1}(A_1) \cup \cdots \cup f^{-1}(A_p)] = f[f^{-1}(A_1)] \cup \cdots \cup f[f^{-1}(A_p)]$$
$$\subseteq A_1 \cup \cdots \cup A_p,$$

so $f(X)$ is compact. As a corollary of Theorem 3.38, we see that $f(X)$ is closed and bounded.

Definition 4.26. *A function* $\mathbf{f} : S \to \mathbf{R}^k$ *is called bounded on S if there is a positive number M such that* $\|\mathbf{f}(x)\| \le M$ *for all x in S.*

Since $\mathbf{f}$ is bounded on S if and only if $\mathbf{f}(S)$ is a bounded subset of $\mathbf{R}^k$, we have the following corollary of Theorem 4.25.

Theorem 4.27. *Let* $\mathbf{f} : S \to \mathbf{R}^k$ *be a function from a metric space S to Euclidean space $\mathbf{R}^k$. If $\mathbf{f}$ is continuous on a compact subset X of S, then $\mathbf{f}$ is bounded on X.*

This theorem has important implications for real-valued functions. If f is real-valued and bounded on X, then $f(X)$ is a bounded subset of $\mathbf{R}$, so it has a supremum, $\sup f(X)$, and an infimum, $\inf f(X)$. Moreover,

$$\inf f(X) \le f(x) \le \sup f(X) \qquad \text{for every } x \text{ in } X.$$

The next theorem shows that a continuous f actually takes on the values $\sup f(X)$ and $\inf f(X)$ if X is compact.

Theorem 4.28. *Let $f : S \to \mathbf{R}$ be a real-valued function from a metric space S to Euclidean space $\mathbf{R}$. Assume that f is continuous on a compact subset X of S. Then there exist points p and q in X such that*

$$f(p) = \inf f(X) \qquad \text{and} \qquad f(q) = \sup f(X).$$

NOTE. Since $f(p) \le f(x) \le f(q)$ for all x in X, the numbers $f(p)$ and $f(q)$ are called, respectively, the absolute or global *minimum* and *maximum* values of f on X.

Proof. Theorem 4.25 shows that $f(X)$ is a closed and bounded subset of $\mathbf{R}$. Let $m = \inf f(X)$. Then m is adherent to $f(X)$ and, since $f(X)$ is closed, $m \in f(X)$. Therefore $m = f(p)$ for some p in X. Similarly, $f(q) = \sup f(X)$ for some q in X.

Theorem 4.29. *Let $f : S \to T$ be a function from one metric space (S, d_S) to another (T, d_T). Assume that f is one-to-one on S, so that the inverse function f^{-1} exists. If S is compact and if f is continuous on S, then f^{-1} is continuous on $f(S)$.*

Proof. By Theorem 4.24 (applied to f^{-1}) we need only show that for every closed set X in S the image $f(X)$ is closed in T. (Note that $f(X)$ is the inverse image of

X under f^{-1}.) Since X is closed and S is compact, X is compact (by Theorem 3.39), so $f(X)$ is compact (by Theorem 4.25) and hence $f(X)$ is closed (by Theorem 3.38). This completes the proof.

Example. This example shows that compactness of S is an essential part of Theorem 4.29. Let $S = [0, 1)$ with the usual metric of $\mathbf{R}^1$ and consider the complex-valued function f defined by

$$f(x) = e^{2\pi i x} \qquad \text{for } 0 \le x < 1.$$

This is a one-to-one continuous mapping of the half-open interval $[0, 1)$ onto the unit circle $|z| = 1$ in the complex plane. However, f^{-1} is not continuous at the point $f(0)$. For example, if $x_n = 1 - 1/n$, the sequence $\{f(x_n)\}$ converges to $f(0)$ but $\{x_n\}$ does not converge in S.

4.14 TOPOLOGICAL MAPPINGS (HOMEOMORPHISMS)

Definition 4.30. *Let $f : S \to T$ be a function from one metric space (S, d_S) to another (T, d_T). Assume also that f is one-to-one on S, so that the inverse function f^{-1} exists. If f is continuous on S and if f^{-1} is continuous on $f(S)$, then f is called a topological mapping or a homeomorphism, and the metric spaces (S, d_S) and $(f(S), d_T)$ are said to be homeomorphic.*

If f is a homeomorphism, then so is f^{-1}. Theorem 4.23 shows that a homeomorphism maps open subsets of S onto open subsets of $f(S)$. It also maps closed subsets of S onto closed subsets of $f(S)$.

A property of a set which remains invariant under every topological mapping is called a *topological property*. Thus the properties of being open, closed, or compact are topological properties.

An important example of a homeomorphism is an *isometry*. This is a function $f : S \to T$ which is one-to-one on S and which preserves the metric; that is,

$$d_T(f(x), f(y)) = d_S(x, y)$$

for all points x and y in S. If there is an isometry from (S, d_S) to $(f(S), d_T)$ the two metric spaces are called *isometric*.

Topological mappings are particularly important in the theory of space curves. For example, a *simple arc* is the topological image of an interval, and a *simple closed curve* is the topological image of a circle.

4.15 BOLZANO'S THEOREM

This section is devoted to a famous theorem of Bolzano which concerns a global property of real-valued functions continuous on compact intervals $[a, b]$ in $\mathbf{R}$. If the graph of f lies above the x-axis at a and below the x-axis at b, Bolzano's theorem asserts that the graph must cross the axis somewhere in between. Our proof will be based on a local property of continuous functions known as the *sign-preserving property*.

Theorem 4.31. *Let f be defined on an interval S in* **R***. Assume that f is continuous at a point c in S and that $f(c) \neq 0$. Then there is a 1-ball $B(c; \delta)$ such that $f(x)$ has the same sign as $f(c)$ in $B(c; \delta) \cap S$.*

Proof. Assume $f(c) > 0$. For every $\varepsilon > 0$ there is a $\delta > 0$ such that

$$f(c) - \varepsilon < f(x) < f(c) + \varepsilon \qquad \text{whenever } x \in B(c; \delta) \cap S.$$

Take the δ corresponding to $\varepsilon = f(c)/2$ (this ε is positive). Then we have

$$\tfrac{1}{2}f(c) < f(x) < \tfrac{3}{2}f(c) \qquad \text{whenever } x \in B(c; \delta) \cap S,$$

so $f(x)$ has the same sign as $f(c)$ in $B(c; \delta) \cap S$. The proof is similar if $f(c) < 0$, except that we take $\varepsilon = -\tfrac{1}{2}f(c)$.

Theorem 4.32 (Bolzano). *Let f be real-valued and continuous on a compact interval $[a, b]$ in* **R***, and suppose that $f(a)$ and $f(b)$ have opposite signs; that is, assume $f(a)f(b) < 0$. Then there is at least one point c in the open interval (a, b) such that $f(c) = 0$.*

Proof. For definiteness, assume $f(a) > 0$ and $f(b) < 0$. Let

$$A = \{x : x \in [a, b] \quad \text{and} \quad f(x) \geq 0\}.$$

Then A is nonempty since $a \in A$, and A is bounded above by b. Let $c = \sup A$. Then $a < c < b$. We will prove that $f(c) = 0$.

If $f(c) \neq 0$, there is a 1-ball $B(c; \delta)$ in which f has the same sign as $f(c)$. If $f(c) > 0$, there are points $x > c$ at which $f(x) > 0$, contradicting the definition of c. If $f(c) < 0$, then $c - \delta/2$ is an upper bound for A, again contradicting the definition of c. Therefore we must have $f(c) = 0$.

From Bolzano's theorem we can easily deduce the *intermediate value theorem* for continuous functions.

Theorem 4.33. *Assume f is real-valued and continuous on a compact interval S in* **R***. Suppose there are two points $\alpha < \beta$ in S such that $f(\alpha) \neq f(\beta)$. Then f takes every value between $f(\alpha)$ and $f(\beta)$ in the interval (α, β).*

Proof. Let k be a number between $f(\alpha)$ and $f(\beta)$ and apply Bolzano's theorem to the function g defined on $[\alpha, \beta]$ by the equation $g(x) = f(x) - k$.

The intermediate value theorem, together with Theorem 4.28, implies that the continuous image of a compact interval S under a real-valued function is another compact interval, namely,

$$[\inf f(S), \sup f(S)].$$

(If f is constant on S, this will be a degenerate interval.) The next section extends this property to the more general setting of metric spaces.

4.16 CONNECTEDNESS

This section describes the concept of connectedness and its relation to continuity.

Definition 4.34. *A metric space S is called disconnected if $S = A \cup B$, where A and B are disjoint nonempty open sets in S. We call S connected if it is not disconnected.*

NOTE. A subset X of a metric space S is called connected if, when regarded as a metric subspace of S, it is a connected metric space.

Examples

1. The metric space $S = \mathbf{R} - \{0\}$ with the usual Euclidean metric is disconnected, since it is the union of two disjoint nonempty open sets, the positive real numbers and the negative real numbers.

2. Every open interval in $\mathbf{R}$ is connected. This was proved in Section 3.4 as a consequence of Theorem 3.11.

3. The set $\mathbf{Q}$ of rational numbers, regarded as a metric subspace of Euclidean space $\mathbf{R}^1$, is disconnected. In fact, $\mathbf{Q} = A \cup B$, where A consists of all rational numbers $< \sqrt{2}$ and B of all rational numbers $> \sqrt{2}$. Similarly, every ball in $\mathbf{Q}$ is disconnected.

4. Every metric space S contains nonempty connected subsets. In fact, for each p in S the set $\{p\}$ is connected.

To relate connectedness with continuity we introduce the concept of a two-valued function.

Definition 4.35. *A real-valued function f which is continuous on a metric space S is said to be two-valued on S if $f(S) \subseteq \{0, 1\}$.*

In other words, a two-valued function is a *continuous* function whose only possible values are 0 and 1. This can be regarded as a continuous function from S to the metric space $T = \{0, 1\}$, where T has the discrete metric. We recall that *every* subset of a discrete metric space T is both open and closed in T.

Theorem 4.36 *A metric space S is connected if, and only if, every two-valued function on S is constant.*

Proof. Assume S is connected and let f be a two-valued function on S. We must show that f is constant. Let $A = f^{-1}(\{0\})$ and $B = f^{-1}(\{1\})$ be the inverse images of the subsets $\{0\}$ and $\{1\}$. Since $\{0\}$ and $\{1\}$ are open subsets of the discrete metric space $\{0, 1\}$, both A and B are open in S. Hence, $S = A \cup B$, where A and B are disjoint open sets. But since S is connected, either A is empty and $B = S$, or else B is empty and $A = S$. In either case, f is constant on S.

Conversely, assume that S is disconnected, so that $S = A \cup B$, where A and B are disjoint nonempty open subsets of S. We will exhibit a two-valued function on S which is not constant. Let

$$f(x) = \begin{cases} 0 & \text{if } x \in A, \\ 1 & \text{if } x \in B. \end{cases}$$

Since A and B are nonempty, f takes both values 0 and 1, so f is not constant. Also, f is continuous on S because the inverse image of every open subset of $\{0, 1\}$ is open in S.

Next we show that the continuous image of a connected set is connected.

Theorem 4.37. *Let $f : S \to M$ be a function from a metric space S to another metric space M. Let X be a connected subset of S. If f is continuous on X, then $f(X)$ is a connected subset of M.*

Proof. Let g be a two-valued function on $f(X)$. We will show that g is constant. Consider the composite function h defined on X by the equation $h(x) = g(f(x))$. Then h is continuous on X and can only take the values 0 and 1, so h is two-valued on X. Since X is connected, h is constant on X and this implies that g is constant on $f(X)$. Therefore $f(X)$ is connected.

Example. Since an interval X in $\mathbf{R}^1$ is connected, every continuous image $f(X)$ is connected. If f has real values, the image $f(X)$ is another interval. If f has values in $\mathbf{R}^n$, the image $f(X)$ is called a *curve* in $\mathbf{R}^n$. Thus, every curve in $\mathbf{R}^n$ is connected.

As a corollary of Theorem 4.37 we have the following extension of Bolzano's theorem.

Theorem 4.38 (Intermediate-value theorem for real continuous functions). *Let f be real-valued and continuous on a connected subset S of $\mathbf{R}^n$. If f takes on two different values in S, say a and b, then for each real c between a and b there exists a point $\mathbf{x}$ in S such that $f(\mathbf{x}) = c$.*

Proof. The image $f(S)$ is a connected subset of $\mathbf{R}^1$. Hence, $f(S)$ is an interval containing a and b (see Exercise 4.38). If some value c between a and b were not in $f(S)$, then $f(S)$ would be disconnected.

4.17 COMPONENTS OF A METRIC SPACE

This section shows that every metric space S can be expressed in a unique way as a union of connected "pieces" called components. First we prove the following:

Theorem 4.39. *Let F be a collection of connected subsets of a metric space S such that the intersection $T = \bigcap_{A \in F} A$ is not empty. Then the union $U = \bigcup_{A \in F} A$ is connected.*

Proof. Since $T \neq \emptyset$, there is some t in T. Let f be a two-valued function on U. We will show that f is constant on U by showing that $f(x) = f(t)$ for all x in U. If $x \in U$, then $x \in A$ for some A in F. Since A is connected, f is constant on A and, since $t \in A$, $f(x) = f(t)$.

Every point x in a metric space S belongs to at least one connected subset of S, namely $\{x\}$. By Theorem 4.39, the union of all the connected subsets which contain x is also connected. We call this union a *component* of S, and we denote it by $U(x)$. Thus, $U(x)$ is the maximal connected subset of S which contains x.

Theorem 4.40. *Every point of a metric space S belongs to a uniquely determined component of S. In other words, the components of S form a collection of disjoint sets whose union is S.*

Proof. Two distinct components cannot contain a point x; otherwise (by Theorem 4.39) their union would be a larger connected set containing x.

4.18 ARCWISE CONNECTEDNESS

This section describes a special property, called *arcwise connectedness*, which is possessed by some (but not all) connected sets in Euclidean space $\mathbf{R}^n$.

Definition 4.41. *A set S in $\mathbf{R}^n$ is called arcwise connected if for any two points $\mathbf{a}$ and $\mathbf{b}$ in S there is a continuous function $\mathbf{f} : [0, 1] \to S$ such that*

$$\mathbf{f}(0) = \mathbf{a} \qquad and \qquad \mathbf{f}(1) = \mathbf{b}.$$

NOTE. Such a function is called a *path* from $\mathbf{a}$ to $\mathbf{b}$. If $\mathbf{f}(0) \neq \mathbf{f}(1)$, the image of $[0, 1]$ under $\mathbf{f}$ is called an *arc* joining $\mathbf{a}$ and $\mathbf{b}$. Thus, S is arcwise connected if every pair of distinct points in S can be joined by an arc lying in S. Arcwise connected sets are also called *pathwise connected*. If $\mathbf{f}(t) = t\mathbf{b} + (1 - t)\mathbf{a}$ for $0 \leq t \leq 1$, the curve joining $\mathbf{a}$ and $\mathbf{b}$ is called a *line segment*.

Examples

1. Every convex set in $\mathbf{R}^n$ is arcwise connected, since the line segment joining two points of such a set lies in the set. In particular, every n-ball is arcwise connected.

2. The set in Fig. 4.4 (a union of two tangent closed disks) is arcwise connected.

Figure 4.4

3. The set in Fig. 4.5 consists of those points on the curve described by $y = \sin(1/x)$, $0 < x \leq 1$, along with the points on the horizontal segment $-1 \leq x \leq 0$. This set is connected but not arcwise connected (Exercise 4.46).

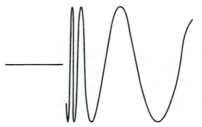

Figure 4.5

The next theorem relates arcwise connectedness with connectedness.

Theorem 4.42. *Every arcwise connected set S in $\mathbf{R}^n$ is connected.*

Proof. Let g be two-valued on S. We will prove that g is constant on S. Choose a point $\mathbf{a}$ in S. If $\mathbf{x} \in S$, join $\mathbf{a}$ to $\mathbf{x}$ by an arc Γ lying in S. Since Γ is connected, g is constant on Γ so $g(\mathbf{x}) = g(\mathbf{a})$. But since $\mathbf{x}$ is an arbitrary point of S, this shows that g is constant on S, so S is connected.

We have already noted that there are connected sets which are not arcwise connected. However, the concepts are equivalent for *open* sets.

Theorem 4.43. *Every open connected set in $\mathbf{R}^n$ is arcwise connected.*

Proof. Let S be an open connected set in $\mathbf{R}^n$ and assume $\mathbf{x} \in S$. We will show that $\mathbf{x}$ can be joined to every point $\mathbf{y}$ in S by an arc lying in S. Let A denote that subset of S which can be so joined to $\mathbf{x}$, and let $B = S - A$. Then $S = A \cup B$, where A and B are disjoint. We will show that A and B are both open in $\mathbf{R}^n$.

Assume that $\mathbf{a} \in A$ and join $\mathbf{a}$ to $\mathbf{x}$ by an arc, say Γ, lying in S. Since $\mathbf{a} \in S$ and S is open, there is an n-ball $B(\mathbf{a}) \subseteq S$. Every $\mathbf{y}$ in $B(\mathbf{a})$ can be joined to $\mathbf{a}$ by a line segment (in S) and thence to $\mathbf{x}$ by Γ. Thus $\mathbf{y} \in A$ if $\mathbf{y} \in B(\mathbf{a})$. That is, $B(\mathbf{a}) \subseteq A$, and hence A is open.

To see that B is also open, assume that $\mathbf{b} \in B$. Then there is an n-ball $B(\mathbf{b}) \subseteq S$, since S is open. But if a point $\mathbf{y}$ in $B(\mathbf{b})$ could be joined to $\mathbf{x}$ by an arc, say Γ', lying in S, the point $\mathbf{b}$ itself could also be so joined by first joining $\mathbf{b}$ to $\mathbf{y}$ (by a line segment in $B(\mathbf{b})$) and then using Γ'. But since $\mathbf{b} \notin A$, no point of $B(\mathbf{b})$ can be in A. That is, $B(\mathbf{b}) \subseteq B$, so B is open.

Therefore we have a decomposition $S = A \cup B$, where A and B are disjoint open sets in $\mathbf{R}^n$. Moreover, A is not empty since $\mathbf{x} \in A$. Since S is connected, it follows that B must be empty, so $S = A$. Now A is clearly arcwise connected, because any two of its points can be suitably joined by first joining each of them to $\mathbf{x}$. Therefore, S is arcwise connected and the proof is complete.

NOTE. A path $\mathbf{f} : [0, 1] \to S$ is said to be *polygonal* if the image of $[0, 1]$ under $\mathbf{f}$ is the union of a finite number of line segments. The same argument used to prove Theorem 4.43 also shows that every open connected set in $\mathbf{R}^n$ is *polygonally connected*. That is, every pair of points in the set can be joined by a polygonal arc lying in the set.

Theorem 4.44. *Every open set S in $\mathbf{R}^n$ can be expressed in one and only one way as a countable disjoint union of open connected sets.*

Proof. By Theorem 4.40, the components of S form a collection of disjoint sets whose union is S. Each component T of S is open, because if $\mathbf{x} \in T$ then there is an n-ball $B(\mathbf{x})$ contained in S. Since $B(\mathbf{x})$ is connected, $B(\mathbf{x}) \subseteq T$, so T is open. By the Lindelöf theorem (Theorem 3.28), the components of S form a countable collection, and by Theorem 4.40 the decomposition into components is unique.

Definition 4.45. *A set in $\mathbf{R}^n$ is called a region if it is the union of an open connected set with some, none, or all its boundary points. If none of the boundary points are* *included, the region is called an open region. If all the boundary points are included, the region is called a closed region.*

NOTE. Some authors use the term *domain* instead of *open region*, especially in the complex plane.

4.19 UNIFORM CONTINUITY

Suppose f is defined on a metric space (S, d_S), with values in another metric space (T, d_T), and assume that f is continuous on a subset A of S. Then, given any point p in A and any $\varepsilon > 0$, there is a $\delta > 0$ (depending on p and on ε) such that, if $x \in A$, then

$$d_T(f(x), f(p)) < \varepsilon \qquad \text{whenever } d_S(x, p) < \delta.$$

In general we cannot expect that for a fixed ε the same value of δ will serve equally well for *every* point p in A. This might happen, however. When it does, the function is called *uniformly continuous* on A.

Definition 4.46. *Let $f : S \to T$ be a function from one metric space (S, d_S) to another (T, d_T). Then f is said to be uniformly continuous on a subset A of S if the following condition holds:*

> *For every $\varepsilon > 0$ there exists a $\delta > 0$ (depending only on ε) such that if $x \in A$ and $p \in A$ then*

$$d_T(f(x), f(p)) < \varepsilon \qquad \text{whenever } d_S(x, p) < \delta. \tag{6}$$

To emphasize the difference between *continuity* on A and *uniform continuity* on A we consider the following examples of real-valued functions.

Examples

1. Let $f(x) = 1/x$ for $x > 0$ and take $A = (0, 1]$. This function is continuous on A but not uniformly continuous on A. To prove this, let $\varepsilon = 10$, and suppose we could find a $\delta, 0 < \delta < 1$, to satisfy the condition of the definition. Taking $x = \delta, p = \delta/11$, we obtain $|x - p| < \delta$ and

$$|f(x) - f(p)| = \frac{11}{\delta} - \frac{1}{\delta} = \frac{10}{\delta} > 10.$$

Hence, for these two points we would always have $|f(x) - f(p)| > 10$, contradicting the definition of uniform continuity.

2. Let $f(x) = x^2$ if $x \in \mathbf{R}^1$ and take $A = (0, 1]$ as above. This function is uniformly continuous on A. To prove this, observe that

$$|f(x) - f(p)| = |x^2 - p^2| = |(x - p)(x + p)| < 2|x - p|.$$

If $|x - p| < \delta$, then $|f(x) - f(p)| < 2\delta$. Hence, if ε is given, we need only take $\delta = \varepsilon/2$ to guarantee that $|f(x) - f(p)| < \varepsilon$ for every pair x, p with $|x - p| < \delta$. This shows that f is uniformly continuous on A.

An instructive exercise is to show that the function in Example 2 is not uniformly continuous on $\mathbf{R}^1$.

4.20 UNIFORM CONTINUITY AND COMPACT SETS

Uniform continuity on a set A implies continuity on A. (The reader should verify this.) The converse is also true if A is compact.

Theorem 4.47 (Heine). *Let $f : S \to T$ be a function from one metric space (S, d_S) to another (T, d_T). Let A be a compact subset of S and assume that f is continuous on A. Then f is uniformly continuous on A.*

Proof. Let $\varepsilon > 0$ be given. Then each point a in A has associated with it a ball $B_S(a; r)$, with r depending on a, such that

$$d_T(f(x), f(a)) < \frac{\varepsilon}{2} \qquad \text{whenever } x \in B_S(a; r) \cap A.$$

Consider the collection of balls $B_S(a; r/2)$ each with radius $r/2$. These cover A and, since A is compact, a finite number of them also cover A, say

$$A \subseteq \bigcup_{k=1}^{m} B_S\left(a_k; \frac{r_k}{2}\right).$$

In any ball of twice the radius, $B(a_k; r_k)$, we have

$$d_T(f(x), f(a_k)) < \frac{\varepsilon}{2} \qquad \text{whenever } x \in B_S(a_k; r_k) \cap A.$$

Let δ be the smallest of the numbers $r_1/2, \ldots, r_m/2$. We shall show that this δ works in the definition of uniform continuity.

For this purpose, consider two points of A, say x and p with $d_S(x, p) < \delta$. By the above discussion there is some ball $B_S(a_k; r_k/2)$ containing x, so

$$d_T(f(x), f(a_k)) < \frac{\varepsilon}{2}.$$

By the triangle inequality we have

$$d_S(p, a_k) \le d_S(p, x) + d_S(x, a_k) < \delta + \frac{r_k}{2} \le \frac{r_k}{2} + \frac{r_k}{2} = r_k.$$

Hence, $p \in B_S(a_k; r_k) \cap S$, so we also have $d_T(f(p), f(a_k)) < \varepsilon/2$. Using the triangle inequality once more we find

$$d_T(f(x), f(p)) \le d_T(f(x), f(a_k)) + d_T(f(a_k), f(p)) < \frac{\varepsilon}{2} + \frac{\varepsilon}{2} = \varepsilon.$$

This completes the proof.

4.21 FIXED-POINT THEOREM FOR CONTRACTIONS

Let $f : S \to S$ be a function from a metric space (S, d) into itself. A point p in S is called a *fixed point* of f if $f(p) = p$. The function f is called a *contraction* of S if there is a positive number $\alpha < 1$ (called a *contraction constant*), such that

$$d(f(x), f(y)) \leq \alpha d(x, y) \qquad \text{for all } x, y \text{ in } S. \tag{7}$$

Clearly, a contraction of any metric space is uniformly continuous on S.

Theorem 4.48 (Fixed-point theorem). A contraction f of a complete metric space S has a unique fixed point p.

Proof. If p and p' are two fixed points, (7) implies $d(p, p') \leq \alpha d(p, p')$, so so $d(p, p') = 0$ and $p = p'$. Hence f has at most one fixed point.

To prove it has one, take any point x in S and consider the sequence of iterates:

$$x, \quad f(x), \quad f(f(x)), \quad \ldots$$

That is, define a sequence $\{p_n\}$ inductively as follows:

$$p_0 = x, \qquad p_{n+1} = f(p_n), \qquad n = 0, 1, 2, \ldots$$

We will prove that $\{p_n\}$ converges to a fixed point of f. First we show that $\{p_n\}$ is a Cauchy sequence. From (7) we have

$$d(p_{n+1}, p_n) = d(f(p_n), f(p_{n-1})) \leq \alpha d(p_n, p_{n-1}),$$

so, by induction, we find

$$d(p_{n+1}, p_n) \leq \alpha^n \, d(p_1, p_0) = c\alpha^n,$$

where $c = d(p_1, p_0)$. Using the triangle inequality we find, for $m > n$,

$$d(p_m, p_n) \leq \sum_{k=n}^{m-1} d(p_{k+1}, p_k) \leq c \sum_{k=n}^{m-1} \alpha^k = c \, \frac{\alpha^n - \alpha^m}{1 - \alpha} < \frac{c}{1 - \alpha} \alpha^n.$$

Since $\alpha^n \to 0$ as $n \to \infty$, this inequality shows that $\{p_n\}$ is a Cauchy sequence. But S is complete so there is a point p in S such that $p_n \to p$. By continuity of f,

$$f(p) = f\left(\lim_{n \to \infty} p_n \right) = \lim_{n \to \infty} f(p_n) = \lim_{n \to \infty} p_{n+1} = p,$$

so p is a fixed point of f. This completes the proof.

Many important existence theorems in analysis are easy consequences of the fixed point theorem. Examples are given in Exercises 7.36 and 7.37. Reference 4.4 gives applications to numerical analysis.

4.22 DISCONTINUITIES OF REAL-VALUED FUNCTIONS

The rest of this chapter is devoted to special properties of real-valued functions defined on subintervals of **R**.

Let f be defined on an interval (a, b). Assume $c \in [a, b)$. If $f(x) \to A$ as $x \to c$ through values greater than c, we say that A is the *righthand limit* of f at c and we indicate this by writing

$$\lim_{x \to c+} f(x) = A.$$

The righthand limit A is also denoted by $f(c+)$. In the ε, δ terminology this means that for every $\varepsilon > 0$ there is a $\delta > 0$ such that

$$|f(x) - f(c+)| < \varepsilon \qquad \text{whenever } c < x < c + \delta < b.$$

Note that f need not be defined at the point c itself. If f is defined at c and if $f(c+) = f(c)$, we say that f is *continuous from the right* at c.

Lefthand limits and continuity from the left at c are similarly defined if $c \in (a, b]$.

If $a < c < b$, then f is continuous at c if, and only if,

$$f(c) = f(c+) = f(c-).$$

We say c is a *discontinuity* of f if f is not continuous at c. In this case one of the following conditions is satisfied:

a) Either $f(c+)$ or $f(c-)$ does not exist.

b) Both $f(c+)$ and $f(c-)$ exist but have different values.

c) Both $f(c+)$ and $f(c-)$ exist and $f(c+) = f(c-) \neq f(c)$.

In case (c), the point c is called a *removable discontinuity*, since the discontinuity could be removed by redefining f at c to have the value $f(c+) = f(c-)$. In cases (a) and (b), we call c an *irremovable discontinuity* because the discontinuity cannot be removed by redefining f at c.

Definition 4.49. *Let f be defined on a closed interval $[a, b]$. If $f(c+)$ and $f(c-)$ both exist at some interior point c, then:*

a) $f(c) - f(c-)$ *is called the lefthand jump of f at c,*

b) $f(c+) - f(c)$ *is called the righthand jump of f at c,*

c) $f(c+) - f(c-)$ *is called the jump of f at c.*

If any one of these three numbers is different from 0, then c is called a jump discontinuity of f.

For the endpoints a and b, only one-sided jumps are considered, the righthand jump at a, $f(a+) - f(a)$, and the lefthand jump at b, $f(b) - f(b-)$.

Examples

1. The function f defined by $f(x) = x/|x|$ if $x \neq 0$, $f(0) = A$, has a jump discontinuity at 0, regardless of the value of A. Here $f(0+) = 1$ and $f(0-) = -1$. (See Fig. 4.6.)

2. The function f defined by $f(x) = 1$ if $x \neq 0$, $f(0) = 0$, has a removable jump discontinuity at 0. In this case $f(0+) = f(0-) = 1$.

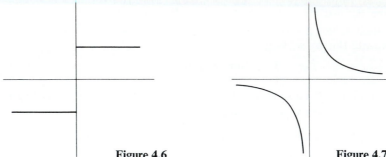

Figure 4.6   Figure 4.7

3. The function f defined by $f(x) = 1/x$ if $x \neq 0$, $f(0) = A$, has an irremovable discontinuity at 0. In this case neither $f(0+)$ nor $f(0-)$ exists. (See Fig. 4.7.)

4. The function f defined by $f(x) = \sin(1/x)$ if $x \neq 0$, $f(0) = A$, has an irremovable discontinuity at 0 since neither $f(0+)$ nor $f(0-)$ exists. (See Fig. 4.8.)

5. The function f defined by $f(x) = x \sin(1/x)$ if $x \neq 0$, $f(0) = 1$, has a removable jump discontinuity at 0, since $f(0+) = f(0-) = 0$. (See Fig. 4.9.)

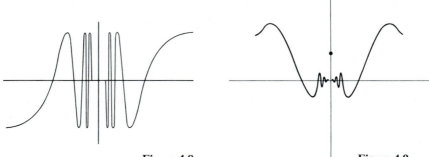

Figure 4.8   Figure 4.9

4.23 MONOTONIC FUNCTIONS

Definition 4.50. *Let f be a real-valued function defined on a subset S of* **R**. *Then f is said to be increasing (or nondecreasing) on S if for every pair of points x and y in S,*

$$x < y \qquad implies \qquad f(x) \le f(y).$$

If $x < y$ implies $f(x) < f(y)$, then f is said to be strictly increasing on S. (Decreasing functions are similarly defined.) A function is called monotonic on S if it is increasing on S or decreasing on S.

If f is an increasing function, then $-f$ is a decreasing function. Because of this simple fact, in many situations involving monotonic functions it suffices to consider only the case of increasing functions.

We shall prove that functions which are monotonic on compact intervals always have finite right- and lefthand limits. Hence their discontinuities (if any) must be jump discontinuities.

Theorem 4.51. *If f is increasing on* $[a, b]$, *then* $f(c+)$ *and* $f(c-)$ *both exist for each c in* (a, b) *and we have*

$$f(c-) \leq f(c) \leq f(c+).$$

At the endpoints we have

$$f(a) \leq f(a+) \quad and \quad f(b-) \leq f(b).$$

Proof. Let $A = \{f(x) : a < x < c\}$. Since f is increasing, this set is bounded above by $f(c)$. Let $\alpha = \sup A$. Then $\alpha \leq f(c)$ and we shall prove that $f(c-)$ exists and equals α.

To do this we must show that for every $\varepsilon > 0$ there is a $\delta > 0$ such that

$$c - \delta < x < c \quad \text{implies} \quad |f(x) - \alpha| < \varepsilon.$$

But since $\alpha = \sup A$, there is an element $f(x_1)$ of A such that $\alpha - \varepsilon < f(x_1) \leq \alpha$. Since f is increasing, for every x in (x_1, c) we also have $\alpha - \varepsilon < f(x) \leq \alpha$, and hence $|f(x) - \alpha| < \varepsilon$. Therefore the number $\delta = c - x_1$ has the required property. (The proof that $f(c+)$ exists and is $\geq f(c)$ is similar, and only trivial modifications are needed for the endpoints.)

There is, of course, a corresponding theorem for decreasing functions which the reader can formulate for himself.

Theorem 4.52. *Let f be strictly increasing on a set S in* **R**. *Then* f^{-1} *exists and is strictly increasing on* $f(S)$.

Proof. Since f is strictly increasing it is one-to-one on S, so f^{-1} exists. To see that f^{-1} is strictly increasing, let $y_1 < y_2$ be two points in $f(S)$ and let $x_1 = f^{-1}(y_1)$, $x_2 = f^{-1}(y_2)$. We cannot have $x_1 \geq x_2$, for then we would also have $y_1 \geq y_2$. The only alternative is

$$x_1 < x_2,$$

and this means that f^{-1} is strictly increasing.

Theorem 4.52, together with Theorem 4.29, now gives us:

Theorem 4.53. *Let f be strictly increasing and continuous on a compact interval* $[a, b]$. *Then* f^{-1} *is continuous and strictly increasing on the interval* $[f(a), f(b)]$.

NOTE. Theorem 4.53 tells us that a continuous, strictly increasing function is a topological mapping. Conversely, every topological mapping of an interval $[a, b]$ onto an interval $[c, d]$ must be a strictly monotonic function. The verification of this fact will be an instructive exercise for the reader (Exercise 4.62).

EXERCISES

Limits of sequences

4.1 Prove each of the following statements about sequences in $\mathbf{C}$.

   a) $z^n \to 0$ if $|z| < 1$; $\{z^n\}$ diverges   if $|z| > 1$.

   b) If $z_n \to 0$ and if $\{c_n\}$ is bounded, then $\{c_n z_n\} \to 0$.

   c) $z^n/n! \to 0$ for every complex z.

   d) If $a_n = \sqrt{n^2 + 2} - n$, then $a_n \to 0$.

4.2 If $a_{n+2} = (a_{n+1} + a_n)/2$ for all $n \geq 1$, show that $a_n \to (a_1 + 2a_2)/3$. *Hint.* $a_{n+2} - a_{n+1} = \frac{1}{2}(a_n - a_{n+1})$.

4.3 If $0 < x_1 < 1$ and if $x_{n+1} = 1 - \sqrt{1 - x_n}$ for all $n \geq 1$, prove that $\{x_n\}$ is a decreasing sequence with limit 0. Prove also that $x_{n+1}/x_n \to \frac{1}{2}$.

4.4 Two sequences of positive integers $\{a_n\}$ and $\{b_n\}$ are defined recursively by taking $a_1 = b_1 = 1$ and equating rational and irrational parts in the equation

$$a_n + b_n\sqrt{2} = (a_{n-1} + b_{n-1}\sqrt{2})^2 \qquad \text{for } n \geq 2.$$

Prove that $a_n^2 - 2b_n^2 = 1$ for $n \geq 2$. Deduce that $a_n/b_n \to \sqrt{2}$ through values $> \sqrt{2}$, and that $2b_n/a_n \to \sqrt{2}$ through values $< \sqrt{2}$.

4.5 A real sequence $\{x_n\}$ satisfies $7x_{n+1} = x_n^3 + 6$ for $n \geq 1$. If $x_1 = \frac{1}{2}$, prove that the sequence increases and find its limit. What happens if $x_1 = \frac{3}{2}$ or if $x_1 = \frac{5}{2}$?

4.6 If $|a_n| < 2$ and $|a_{n+2} - a_{n+1}| \leq \frac{1}{8}|a_{n+1}^2 - a_n^2|$ for all $n \geq 1$, prove that $\{a_n\}$ converges.

4.7 In a metric space (S, d), assume that $x_n \to x$ and $y_n \to y$. Prove that $d(x_n, y_n) \to d(x, y)$.

4.8 Prove that in a compact metric space (S, d), every sequence in S has a subsequence which converges in S. This property also implies that S is compact but you are not required to prove this. (For a proof see either Reference 4.2 or 4.3.)

4.9 Let A be a subset of a metric space S. If A is complete, prove that A is closed. Prove that the converse also holds if S is complete.

Limits of functions

NOTE. In Exercises 4.10 through 4.28, all functions are real-valued.

4.10 Let f be defined on an open interval (a, b) and assume $x \in (a, b)$. Consider the two statements

   a) $\lim\limits_{h \to 0} |f(x + h) - f(x)| = 0$;   b) $\lim\limits_{h \to 0} |f(x + h) - f(x - h)| = 0$.

Prove that (a) always implies (b), and give an example in which (b) holds but (a) does not.

4.11 Let f be defined on $\mathbf{R}^2$. If

$$\lim_{(x,y) \to (a,b)} f(x, y) = L$$

and if the one-dimensional limits $\lim_{x \to a} f(x, y)$ and $\lim_{y \to b} f(x, y)$ both exist, prove that

$$\lim_{x \to a} \left[\lim_{y \to b} f(x, y)\right] = \lim_{y \to b} \left[\lim_{x \to a} f(x, y)\right] = L.$$

Now consider the functions f defined on $\mathbf{R}^2$ as follows:

a) $f(x, y) = \dfrac{x^2 - y^2}{x^2 + y^2}$ $\qquad\qquad$ if $(x, y) \neq (0, 0)$, $f(0, 0) = 0$.

b) $f(x, y) = \dfrac{(xy)^2}{(xy)^2 + (x - y)^2}$ $\qquad$ if $(x, y) \neq (0, 0)$, $f(0, 0) = 0$.

c) $f(x, y) = \dfrac{1}{x} \sin(xy)$ $\qquad\qquad\qquad$ if $x \neq 0$, $f(0, y) = y$.

d) $f(x, y) = \begin{cases} (x + y) \sin(1/x) \sin(1/y) & \text{if } x \neq 0 \text{ and } y \neq 0, \\ 0 & \text{if } x = 0 \text{ or } y = 0. \end{cases}$

e) $f(x, y) = \begin{cases} \dfrac{\sin x - \sin y}{\tan x - \tan y} & \text{if } \tan x \neq \tan y, \\ \cos^3 x & \text{if } \tan x = \tan y. \end{cases}$

In each of the preceding examples, determine whether the following limits exist and evaluate those limits that do exist:

$$\lim_{x \to 0} \left[\lim_{y \to 0} f(x, y) \right] ; \qquad \lim_{y \to 0} \left[\lim_{x \to 0} f(x, y) \right] ; \qquad \lim_{(x,y) \to (0,0)} f(x, y).$$

4.12 If $x \in [0, 1]$ prove that the following limit exists,

$$\lim_{m \to \infty} \left[\lim_{n \to \infty} \cos^{2n}(m!\, \pi x) \right],$$

and that its value is 0 or 1, according to whether x is irrational or rational.

Continuity of real-valued functions

4.13 Let f be continuous on $[a, b]$ and let $f(x) = 0$ when x is rational. Prove that $f(x) = 0$ for every x in $[a, b]$.

4.14 Let f be continuous at the point $\mathbf{a} = (a_1, a_2, \ldots, a_n)$ in $\mathbf{R}^n$. Keep $a_2, a_3, \ldots, a_n$ fixed and define a new function g of one real variable by the equation

$$g(x) = f(x, a_2, \ldots, a_n).$$

Prove that g is continuous at the point $x = a_1$. (This is sometimes stated as follows: *A continuous function of n variables is continuous in each variable separately*.)

4.15 Show by an example that the converse of the statement in Exercise 4.14 is not true in general.

4.16 Let f, g, and h be defined on $[0, 1]$ as follows:

$\qquad\qquad f(x) = g(x) = h(x) = 0,\qquad$ whenever x is irrational;

$\qquad\qquad f(x) = 1$ and $g(x) = x,\qquad$ whenever x is rational;

$\qquad\qquad h(x) = 1/n$, if x is the rational number m/n (in lowest terms);

$\qquad\qquad h(0) = 1$.

Prove that f is not continuous anywhere in $[0, 1]$, that g is continuous only at $x = 0$, and that h is continuous only at the irrational points in $[0, 1]$.

4.17 For each x in $[0, 1]$, let $f(x) = x$ if x is rational, and let $f(x) = 1 - x$ if x is irrational. Prove that:

a) $f(f(x)) = x$ for all x in $[0, 1]$. $\qquad$ b) $f(x) + f(1 - x) = 1$ for all x in $[0, 1]$.

c) f is continuous only at the point $x = \frac{1}{2}$.

d) f assumes every value between 0 and 1.

e) $f(x + y) - f(x) - f(y)$ is rational for all x and y in $[0, 1]$.

4.18 Let f be defined on $\mathbf{R}$ and assume that there exists at least one point x_0 in $\mathbf{R}$ at which f is continuous. Suppose also that, for every x and y in $\mathbf{R}$, f satisfies the equation

$$f(x + y) = f(x) + f(y).$$

Prove that there exists a constant a such that $f(x) = ax$ for all x.

4.19 Let f be continuous on $[a, b]$ and define g as follows: $g(a) = f(a)$ and, for $a < x \leq b$, let $g(x)$ be the maximum value of f in the subinterval $[a, x]$. Show that g is continuous on $[a, b]$.

4.20 Let $f_1, \ldots, f_m$ be m real-valued functions defined on a set S in $\mathbf{R}^n$. Assume that each f_k is continuous at the point $\mathbf{a}$ of S. Define a new function f as follows: For each $\mathbf{x}$ in S, $f(\mathbf{x})$ is the largest of the m numbers $f_1(\mathbf{x}), \ldots, f_m(\mathbf{x})$. Discuss the continuity of f at $\mathbf{a}$.

4.21 Let $f: S \to \mathbf{R}$ be continuous on an open set S in $\mathbf{R}^n$, assume that $\mathbf{p} \in S$, and assume that $f(\mathbf{p}) > 0$. Prove that there is an n-ball $B(\mathbf{p}; r)$ such that $f(\mathbf{x}) > 0$ for every $\mathbf{x}$ in the ball.

4.22 Let f be defined and continuous on a closed set S in $\mathbf{R}$. Let

$$A = \{x : x \in S \text{ and } f(x) = 0\}.$$

Prove that A is a closed subset of $\mathbf{R}$.

4.23 Given a function $f: \mathbf{R} \to \mathbf{R}$, define two sets A and B in $\mathbf{R}^2$ as follows:

$$A = \{(x, y) : y < f(x)\}, \qquad B = \{(x, y) : y > f(x)\}.$$

Prove that f is continuous on $\mathbf{R}$ if, and only if, both A and B are open subsets of $\mathbf{R}^2$.

4.24 Let f be defined and bounded on a compact interval S in $\mathbf{R}$. If $T \subseteq S$, the number

$$\Omega_f(T) = \sup \{f(x) - f(y) : x \in T, y \in T\}$$

is called the *oscillation* (or *span*) of f on T. If $x \in S$, the oscillation of f at x is defined to be the number

$$\omega_f(x) = \lim_{h \to 0+} \Omega_f(B(x; h) \cap S).$$

Prove that this limit always exists and that $\omega_f(x) = 0$ if, and only if, f is continuous at x.

4.25 Let f be continuous on a compact interval $[a, b]$. Suppose that f has a local maximum at x_1 and a local maximum at x_2. Show that there must be a third point between x_1 and x_2 where f has a local minimum.

NOTE. To say that f has a local maximum at x_1 means that there is a 1-ball $B(x_1)$ such that $f(x) \leq f(x_1)$ for all x in $B(x_1) \cap [a, b]$. Local minimum is similarly defined.

4.26 Let f be a real-valued function, continuous on $[0, 1]$, with the following property: For every real y, either there is no x in $[0, 1]$ for which $f(x) = y$ or there is exactly one such x. Prove that f is strictly monotonic on $[0, 1]$.

4.27 Let f be a function defined on $[0, 1]$ with the following property: For every real number y, either there is no x in $[0, 1]$ for which $f(x) = y$ or there are exactly two values of x in $[0, 1]$ for which $f(x) = y$.

a) Prove that f cannot be continuous on $[0, 1]$.

b) Construct a function f which has the above property.

c) Prove that any function with this property has infinitely many discontinuities on $[0, 1]$.

4.28 In each case, give an example of a function f, continuous on S and such that $f(S) = T$, or else explain why there can be no such f:

a) $S = (0, 1)$, $\qquad\qquad$ $T = (0, 1]$.

b) $S = (0, 1)$, $\qquad\qquad$ $T = (0, 1) \cup (1, 2)$.

c) $S = \mathbf{R}^1$, $\qquad\qquad$ $T =$ the set of rational numbers.

d) $S = [0, 1] \cup [2, 3]$, $\qquad$ $T = \{0, 1\}$.

e) $S = [0, 1] \times [0, 1]$, $\qquad$ $T = \mathbf{R}^2$.

f) $S = [0, 1] \times [0, 1]$, $\qquad$ $T = (0, 1) \times (0, 1)$.

g) $S = (0, 1) \times (0, 1)$, $\qquad$ $T = \mathbf{R}^2$.

Continuity in metric spaces

In Exercises 4.29 through 4.33, we assume that $f : S \to T$ is a function from one metric space (S, d_S) to another (T, d_T).

4.29 Prove that f is continuous on S if, and only if,

$$f^{-1}(\text{int } B) \subseteq \text{int } f^{-1}(B) \qquad \text{for every subset } B \text{ of } T.$$

4.30 Prove that f is continuous on S if, and only if,

$$f(\bar{A}) \subseteq \overline{f(A)} \qquad \text{for every subset } A \text{ of } S.$$

4.31 Prove that f is continuous on S if, and only if, f is continuous on every compact subset of S. *Hint.* If $x_n \to p$ in S, the set $\{p, x_1, x_2, \dots\}$ is compact.

4.32 A function $f : S \to T$ is called a *closed mapping* on S if the image $f(A)$ is closed in T for every closed subset A of S. Prove that f is continuous and closed on S if, and only if, $f(\bar{A}) = \overline{f(A)}$ for every subset A of S.

4.33 Give an example of a continuous f and a Cauchy sequence $\{x_n\}$ in some metric space S for which $\{f(x_n)\}$ is not a Cauchy sequence in T.

4.34 Prove that the interval $(-1, 1)$ in $\mathbf{R}^1$ is homeomorphic to $\mathbf{R}^1$. This shows that neither boundedness nor completeness is a topological property.

4.35 Section 9.7 contains an example of a function f, continuous on $[0, 1]$, with $f([0, 1]) = [0, 1] \times [0, 1]$. Prove that no such f can be one-to-one on $[0, 1]$.

Connectedness

4.36 Prove that a metric space S is disconnected if, and only if, there is a nonempty subset A of S, $A \neq S$, which is both open and closed in S.

4.37 Prove that a metric space S is connected if, and only if, the only subsets of S which are both open and closed in S are the empty set and S itself.

4.38 Prove that the only connected subsets of $\mathbf{R}$ are (a) the empty set, (b) sets consisting of a single point, and (c) intervals (open, closed, half-open, or infinite).

4.39 Let X be a connected subset of a metric space S. Let Y be a subset of S such that $X \subseteq Y \subseteq \bar{X}$, where $\bar{X}$ is the closure of X. Prove that Y is also connected. In particular, this shows that $\bar{X}$ is connected.

4.40 If x is a point in a metric space S, let $U(x)$ be the component of S containing x. Prove that $U(x)$ is closed in S.

4.41 Let S be an open subset of $\mathbf{R}$. By Theorem 3.11, S is the union of a countable disjoint collection of open intervals in $\mathbf{R}$. Prove that each of these open intervals is a component of the metric subspace S. Explain why this does not contradict Exercise 4.40.

4.42 Given a compact set S in $\mathbf{R}^m$ with the following property: For every pair of points $\mathbf{a}$ and $\mathbf{b}$ in S and for every $\varepsilon > 0$ there exists a finite set of points $\{\mathbf{x}_0, \mathbf{x}_1, \ldots, \mathbf{x}_n\}$ in S with $\mathbf{x}_0 = \mathbf{a}$ and $\mathbf{x}_n = \mathbf{b}$ such that

$$\|\mathbf{x}_k - \mathbf{x}_{k-1}\| < \varepsilon \qquad \text{for } k = 1, 2, \ldots, n.$$

Prove or disprove: S is connected.

4.43 Prove that a metric space S is connected if, and only if, every nonempty proper subset of S has a nonempty boundary.

4.44 Prove that every convex subset of $\mathbf{R}^n$ is connected.

4.45 Given a function $\mathbf{f}: \mathbf{R}^n \to \mathbf{R}^m$ which is one-to-one and continuous on $\mathbf{R}^n$. If A is open and disconnected in $\mathbf{R}^n$, prove that $\mathbf{f}(A)$ is open and disconnected in $\mathbf{f}(\mathbf{R}^n)$.

4.46 Let $A = \{(x, y) : 0 < x \leq 1, \quad y = \sin 1/x\}$, $B = \{(x, y) : y = 0, \quad -1 \leq x \leq 0\}$, and let $S = A \cup B$. Prove that S is connected but not arcwise connected. (See Fig. 4.5, Section 4.18.)

4.47 Let $F = \{F_1, F_2, \ldots\}$ be a countable collection of connected compact sets in $\mathbf{R}^n$ such that $F_{k+1} \subseteq F_k$ for each $k \geq 1$. Prove that the intersection $\bigcap_{k=1}^{\infty} F_k$ is connected and closed.

4.48 Let S be an open connected set in $\mathbf{R}^n$. Let T be a component of $\mathbf{R}^n - S$. Prove that $\mathbf{R}^n - T$ is connected.

4.49 Let (S, d) be a connected metric space which is not bounded. Prove that for every a in S and every $r > 0$, the set $\{x : d(x, a) = r\}$ is nonempty.

Uniform continuity

4.50 Prove that a function which is uniformly continuous on S is also continuous on S.

4.51 If $f(x) = x^2$ for x in $\mathbf{R}$, prove that f is not uniformly continuous on $\mathbf{R}$.

4.52 Assume that f is uniformly continuous on a bounded set S in $\mathbf{R}^n$. Prove that f must be bounded on S.

4.53 Let $\mathbf{f}$ be a function defined on a set S in $\mathbf{R}^n$ and assume that $\mathbf{f}(S) \subseteq \mathbf{R}^m$. Let $\mathbf{g}$ be defined on $\mathbf{f}(S)$ with value in $\mathbf{R}^k$, and let $\mathbf{h}$ denote the composite function defined by $\mathbf{h}(\mathbf{x}) = \mathbf{g}[\mathbf{f}(\mathbf{x})]$ if $\mathbf{x} \in S$. If $\mathbf{f}$ is uniformly continuous on S and if $\mathbf{g}$ is uniformly continuous on $\mathbf{f}(S)$, show that $\mathbf{h}$ is uniformly continuous on S.

4.54 Assume $f: S \to T$ is uniformly continuous on S, where S and T are metric spaces. If $\{x_n\}$ is any Cauchy sequence in S, prove that $\{f(x_n)\}$ is a Cauchy sequence in T. (Compare with Exercise 4.33.)

4.55 Let $f : S \to T$ be a function from a metric space S to another metric space T. Assume f is uniformly continuous on a subset A of S and that T is complete. Prove that there is a unique extension of f to $\bar{A}$ which is uniformly continuous on $\bar{A}$.

4.56 In a metric space (S, d), let A be a nonempty subset of S. Define a function $f_A : S \to \mathbf{R}$ by the equation

$$f_A(x) = \inf \{d(x, y) : y \in A\}$$

for each x in S. The number $f_A(x)$ is called the distance from x to A.

    a) Prove that f_A is uniformly continuous on S.

    b) Prove that $\bar{A} = \{x : x \in S \text{ and } f_A(x) = 0\}$.

4.57 In a metric space (S, d), let A and B be disjoint closed subsets of S. Prove that there exist disjoint open subsets U and V of S such that $A \subseteq U$ and $B \subseteq V$. *Hint.* Let $g(x) = f_A(x) - f_B(x)$, in the notation of Exercise 4.56, and consider $g^{-1}(-\infty, 0)$ and $g^{-1}(0, +\infty)$.

Discontinuities

4.58 Locate and classify the discontinuities of the functions f defined on $\mathbf{R}^1$ by the following equations:

    a) $f(x) = (\sin x)/x$ if $x \neq 0$, $f(0) = 0$.

    b) $f(x) = e^{1/x}$ if $x \neq 0$, $f(0) = 0$.

    c) $f(x) = e^{1/x} + \sin (1/x)$ if $x \neq 0$, $f(0) = 0$.

    d) $f(x) = 1/(1 - e^{1/x})$ if $x \neq 0$, $f(0) = 0$.

4.59 Locate the points in $\mathbf{R}^2$ at which each of the functions in Exercise 4.11 is not continuous.

Monotonic functions

4.60 Let f be defined in the open interval (a, b) and assume that for each interior point x of (a, b) there exists a 1-ball $B(x)$ in which f is increasing. Prove that f is an increasing function throughout (a, b).

4.61 Let f be continuous on a compact interval $[a, b]$ and assume that f does not have a local maximum or a local minimum at any interior point. (See the NOTE following Exercise 4.25.) Prove that f must be monotonic on $[a, b]$.

4.62 If f is one-to-one and continuous on $[a, b]$, prove that f must be strictly monotonic on $[a, b]$. That is, prove that every topological mapping of $[a, b]$ onto an interval $[c, d]$ must be strictly monotonic.

4.63 Let f be an increasing function defined on $[a, b]$ and let $x_1, \ldots, x_n$ be n points in the interior such that $a < x_1 < x_2 < \cdots < x_n < b$.

    a) Show that $\sum_{k=1}^{n} [f(x_k+) - f(x_k-)] \leq f(b-) - f(a+)$.

    b) Deduce from part (a) that the set of discontinuities of f is countable.

    c) Prove that f has points of continuity in every open subinterval of $[a, b]$.

4.64 Give an example of a function f, defined and strictly increasing on a set S in $\mathbf{R}$, such that f^{-1} is not continuous on $f(S)$.

4.65 Let f be strictly increasing on a subset S of $\mathbf{R}$. Assume that the image $f(S)$ has one of the following properties: (a) $f(S)$ is open; (b) $f(S)$ is connected; (c) $f(S)$ is closed. Prove that f must be continuous on S.

Metric spaces and fixed points

4.66 Let $B(S)$ denote the set of all real-valued functions which are defined and *bounded* on a nonempty set S. If $f \in B(S)$, let

$$\|f\| = \sup_{x \in S} |f(x)|.$$

The number $\|f\|$ is called the "sup norm" of f.

a) Prove that the formula $d(f, g) = \|f - g\|$ defines a metric d on $B(S)$.

b) Prove that the metric space $(B(S), d)$ is complete. *Hint.* If $\{f_n\}$ is a Cauchy sequence in $B(S)$, show that $\{f_n(x)\}$ is a Cauchy sequence of real numbers for each x in S.

4.67 Refer to Exercise 4.66 and let $C(S)$ denote the subset of $B(S)$ consisting of all functions *continuous* and bounded on S, where now S is a metric space.

a) Prove that $C(S)$ is a closed subset of $B(S)$.

b) Prove that the metric subspace $C(S)$ is complete.

4.68 Refer to the proof of the fixed-point theorem (Theorem 4.48) for notation.

a) Prove that $d(p, p_n) \le d(x, f(x))\alpha^n/(1 - \alpha)$.

This inequality, which is useful in numerical work, provides an estimate for the distance from p_n to the fixed point p. An example is given in (b).

b) Take $f(x) = \frac{1}{2}(x + 2/x)$, $S = [1, +\infty)$. Prove that f is a contraction of S with contraction constant $\alpha = \frac{1}{2}$ and fixed point $p = \sqrt{2}$. Form the sequence $\{p_n\}$ starting with $x = p_0 = 1$ and show that $|p_n - \sqrt{2}| \le 2^{-n}$.

4.69 Show by counterexamples that the fixed-point theorem for contractions need not hold if either (a) the underlying metric space is not complete, or (b) the contraction constant $\alpha \ge 1$.

4.70 Let $f: S \to S$ be a function from a complete metric space (S, d) into itself. Assume there is a real sequence $\{\alpha_n\}$ which converges to 0 such that $d(f^n(x), f^n(y)) \le \alpha_n d(x, y)$ for all $n \ge 1$ and all x, y in S, where f^n is the nth iterate of f; that is,

$$f^1(x) = f(x), \qquad f^{n+1}(x) = f(f^n(x)) \qquad \text{for } n \ge 1.$$

Prove that f has a unique fixed point. *Hint.* Apply the fixed-point theorem to f^m for a suitable m.

4.71 Let $f: S \to S$ be a function from a metric space (S, d) into itself such that

$$d(f(x), f(y)) < d(x, y)$$

whenever $x \ne y$.

a) Prove that f has at most one fixed point, and give an example of such an f with no fixed point.

b) If S is compact, prove that f has exactly one fixed point. *Hint.* Show that $g(x) = d(x, f(x))$ attains its minimum on S.

c) Give an example with S compact in which f is not a contraction.

4.72 Assume that f satisfies the condition in Exercise 4.71. If $x \in S$, let $p_0 = x$, $p_{n+1} = f(p_n)$, and $c_n = d(p_n, p_{n+1})$ for $n \geq 0$.

a) Prove that $\{c_n\}$ is a decreasing sequence, and let $c = \lim c_n$.

b) Assume there is a subsequence $\{p_{k(n)}\}$ which converges to a point q in S. Prove that

$$c = d(q, f(q)) = d(f(q), f[f(q)]).$$

Deduce that q is a fixed point of f and that $p_n \to q$.

SUGGESTED REFERENCES FOR FURTHER STUDY

4.1 Boas, R. P., *A Primer of Real Functions*. Carus Monograph No. 13. Wiley, New York, 1960.

4.2 Gleason, A., *Fundamentals of Abstract Analysis*. Addison-Wesley, Reading, 1966.

4.3 Simmons, G. F., *Introduction to Topology and Modern Analysis*. McGraw-Hill, New York, 1963.

4.4 Todd, J., *Survey of Numerical Analysis*. McGraw-Hill, New York, 1962.

CHAPTER 5

DERIVATIVES

5.1 INTRODUCTION

This chapter treats the derivative, the central concept of differential calculus. Two different types of problem—the physical problem of finding the instantaneous velocity of a moving particle, and the geometrical problem of finding the tangent line to a curve at a given point—both lead quite naturally to the notion of derivative. Here, we shall not be concerned with applications to mechanics and geometry, but instead will confine our study to general properties of derivatives.

This chapter deals primarily with derivatives of functions of one real variable, specifically, real-valued functions defined on intervals in **R**. It also discusses briefly derivatives of vector-valued functions of one real variable, and partial derivatives, since these topics involve no new ideas. Much of this material should be familiar to the reader from elementary calculus. A more detailed treatment of derivative theory for functions of several variables involves significant changes and is dealt with in Chapter 12.

The last part of the chapter discusses derivatives of complex-valued functions of a complex variable.

5.2 DEFINITION OF DERIVATIVE

If f is defined on an open interval (a, b), then for two distinct points x and c in (a, b) we can form the difference quotient

$$\frac{f(x) - f(c)}{x - c}.$$

We keep c fixed and study the behavior of this quotient as $x \to c$.

Definition 5.1. *Let f be defined on an open interval (a, b), and assume that $c \in (a, b)$. Then f is said to be differentiable at c whenever the limit*

$$\lim_{x \to c} \frac{f(x) - f(c)}{x - c}$$

exists. The limit, denoted by $f'(c)$, is called the derivative of f at c.

This limit process defines a new function f', whose domain consists of those points in (a, b) at which f is differentiable. The function f' is called the *first derivative* of f. Similarly, the nth derivative of f, denoted by $f^{(n)}$, is defined to be the first derivative of $f^{(n-1)}$, for $n = 2, 3, \ldots$. (By our definition, we do not consider $f^{(n)}$ unless $f^{(n-1)}$ is defined on an open interval.) Other notations with which the reader may be familiar are

$$f'(c) = Df(c) = \frac{df}{dx}(c) = \frac{dy}{dx}\bigg|_{x=c} \qquad [\text{where } y = f(x)],$$

or similar notations. The function f itself is sometimes written $f^{(0)}$. The process which produces f' from f is called *differentiation*.

5.3 DERIVATIVES AND CONTINUITY

The next theorem makes it possible to reduce some of the theorems on derivatives to theorems on continuity.

Theorem 5.2. *If f is defined on (a, b) and differentiable at a point c in (a, b), then there is a function f^* (depending on f and on c) which is continuous at c and which satisfies the equation*

$$f(x) - f(c) = (x - c)f^*(x), \tag{1}$$

for all x in (a, b), with $f^*(c) = f'(c)$. Conversely, if there is a function f^*, continuous at c, which satisfies (1), then f is differentiable at c and $f'(c) = f^*(c)$.

Proof. If $f'(c)$ exists, let f^* be defined on (a, b) as follows:

$$f^*(x) = \frac{f(x) - f(c)}{x - c} \qquad \text{if } x \neq c, \quad f^*(c) = f'(c).$$

Then f^* is continuous at c and (1) holds for all x in (a, b).

Conversely, if (1) holds for some f^* continuous at c, then by dividing by $x - c$ and letting $x \to c$ we see that $f'(c)$ exists and equals $f^*(c)$.

- As an immediate consequence of (1) we obtain:

Theorem 5.3. *If f is differentiable at c, then f is continuous at c.*

Proof. Let $x \to c$ in (1).

NOTE. Equation (1) has a geometric interpretation which helps us gain insight into its meaning. Since f^* is continuous at c, $f^*(x)$ is nearly equal to $f^*(c) = f'(c)$ if x is near c. Replacing $f^*(x)$ by $f'(c)$ in (1) we obtain the equation

$$f(x) = f(c) + f'(c)(x - c),$$

which should be approximately correct when $x - c$ is small. In other words, if f is differentiable at c, then f is approximately a linear function near c. (See Fig. 5.1). Differential calculus continually exploits this geometric property of functions.

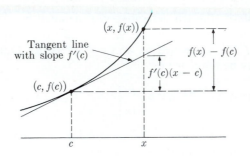

Figure 5.1

5.4 ALGEBRA OF DERIVATIVES

The next theorem describes the usual formulas for differentiating the sum, difference, product and quotient of two functions.

Theorem 5.4. *Assume f and g are defined on (a, b) and differentiable at c. Then
$f + g, f - g,$ and $f \cdot g$ are also differentiable at c. This is also true of f/g if $g(c) \neq 0$.
The derivatives at c are given by the following formulas:*

a) $(f \pm g)'(c) = f'(c) \pm g'(c),$

b) $(f \cdot g)'(c) = f(c)g'(c) + f'(c)g(c),$

c) $(f/g)'(c) = \dfrac{g(c)f'(c) - f(c)g'(c)}{g(c)^2},$ *provided* $g(c) \neq 0.$

Proof. We shall prove (b). Using Theorem 5.2 we write

$$f(x) = f(c) + (x - c)f^*(x), \qquad g(x) = g(c) + (x - c)g^*(x).$$

Thus,

$$f(x)g(x) - f(c)g(c) = (x - c)[f(c)g^*(x) + f^*(x)g(c)] + (x - c)^2 f^*(x)g^*(x).$$

Dividing by $x - c$ and letting $x \to c$ we obtain (b). Proofs of the other statements
are similar.

From the definition we see at once that if f is constant on (a, b) then $f' = 0$
on (a, b). Also, if $f(x) = x$, then $f'(x) = 1$ for all x. Repeated application of
Theorem 5.4 tells us that if $f(x) = x^n$ (n a positive integer), then $f'(x) = nx^{n-1}$
for all x. Applying Theorem 5.4 again, we see that every polynomial has a derivative everywhere in **R** and every rational function has a derivative wherever it is
defined.

5.5 THE CHAIN RULE

A much deeper result is the so-called *chain rule* for differentiating composite functions.

Theorem 5.5 (Chain rule). *Let f be defined on an open interval S, let g be defined on f(S), and consider the composite function g ∘ f defined on S by the equation*

$$(g \circ f)(x) = g[f(x)].$$

Assume there is a point c in S such that f(c) is an interior point of f(S). If f is differentiable at c and if g is differentiable at f(c) then g ∘ f is differentiable at c and we have

$$(g \circ f)'(c) = g'[f(c)]f'(c).$$

Proof. Using Theorem 5.2 we can write

$$f(x) - f(c) = (x - c)f^*(x) \qquad \text{for all } x \text{ in } S,$$

where f^* is continuous at c and $f^*(c) = f'(c)$. Similarly,

$$g(y) - g[f(c)] = [y - f(c)]g^*(y),$$

for all y in some open subinterval T of $f(S)$ containing $f(c)$. Here g^* is continuous at $f(c)$ and $g^*[f(c)] = g'[f(c)]$.

Choosing x in S so that $y = f(x) \in T$, we then have

$$g[f(x)] - g[f(c)] = [f(x) - f(c)]g^*[f(x)] = (x - c)f^*(x)g^*[f(x)]. \quad (2)$$

By the continuity theorem for composite functions,

$$g^*[f(x)] \to g^*[f(c)] = g'[f(c)] \qquad \text{as } x \to c.$$

Therefore, if we divide by $x - c$ in (2) and let $x \to c$, we obtain

$$\lim_{x \to c} \frac{g[f(x)] - g[f(c)]}{x - c} = g'[f(c)]f'(c),$$

as required.

5.6 ONE-SIDED DERIVATIVES AND INFINITE DERIVATIVES

Up to this point, the statement that f has a derivative at c has meant that c was *interior* to an interval in which f was defined and that the limit defining $f'(c)$ was *finite*. It is convenient to extend the scope of our ideas somewhat in order to discuss derivatives at endpoints of intervals. It is also desirable to introduce *infinite* derivatives, so that the usual geometric interpretation of a derivative as the slope of a tangent line will still be valid in case the tangent line happens to be vertical. In such a case we cannot prove that f is continuous at c. Therefore, we explicitly require it to be so.

Definition 5.6. *Let f be defined on a closed interval S and assume that f is continuous at the point c in S. Then f is said to have a righthand derivative at c if the righthand limit*

$$\lim_{x \to c+} \frac{f(x) - f(c)}{x - c}$$

exists as a finite value, or if the limit is $+\infty$ *or* $-\infty$. *This limit will be denoted by* $f'_+(c)$. *Lefthand derivatives, denoted by* $f'_-(c)$, *are similarly defined. In addition, if c is an interior point of S, then we say that f has the derivative* $f'(c) = +\infty$ *if both the right- and lefthand derivatives at c are* $+\infty$. (*The derivative* $f'(c) = -\infty$ *is similarly defined.*)

It is clear that f has a derivative (finite or infinite) at an interior point c if, and only if, $f'_+(c) = f'_-(c)$, in which case $f'_+(c) = f'_-(c) = f'(c)$.

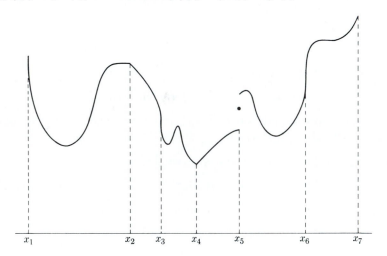

Figure 5.2

Figure 5.2 illustrates some of these concepts. At the point x_1 we have $f'_+(x_1) = -\infty$. At x_2 the lefthand derivative is 0 and the righthand derivative is -1. Also, $f'(x_3) = -\infty$, $f'_-(x_4) = -1$, $f'_+(x_4) = +1$, $f'(x_6) = +\infty$, and $f'_-(x_7) = 2$. There is no derivative (one-sided or otherwise) at x_5, since f is not continuous there.

5.7 FUNCTIONS WITH NONZERO DERIVATIVE

Theorem 5.7. *Let f be defined on an open interval* (a, b) *and assume that for some c in* (a, b) *we have* $f'(c) > 0$ *or* $f'(c) = +\infty$. *Then there is a 1-ball* $B(c) \subseteq (a, b)$ *in which*

$$f(x) > f(c) \quad \text{if } x > c, \qquad \text{and} \qquad f(x) < f(c) \quad \text{if } x < c.$$

Proof. If $f'(c)$ is finite and positive we can write

$$f(x) - f(c) = (x - c)f^*(x),$$

where f^* is continuous at c and $f^*(c) = f'(c) > 0$. By the sign preserving property of continuous functions there is a 1-ball $B(c) \subseteq (a, b)$ in which $f^*(x)$ has the same sign as $f^*(c)$, and this means that $f(x) - f(c)$ has the same sign as $x - c$.

If $f'(c) = +\infty$, there is a 1-ball $B(c)$ in which

$$\frac{f(x) - f(c)}{x - c} > 1 \qquad \text{whenever } x \neq c.$$

In this ball the quotient is again positive and the conclusion follows as before.

A result similar to Theorem 5.7 holds, of course, if $f'(c) < 0$ or if $f'(c) = -\infty$ at some interior point c of (a, b).

5.8 ZERO DERIVATIVES AND LOCAL EXTREMA

Definition 5.8. *Let f be a real-valued function defined on a subset S of a metric space M, and assume a ∈ S. Then f is said to have a local maximum at a if there is a ball B(a) such that*

$$f(x) \leq f(a) \qquad \text{for all } x \text{ in } B(a) \cap S.$$

If $f(x) \geq f(a)$ for all x in $B(a) \cap S$, then f is said to have a local minimum at a.

NOTE. A local maximum at a is the absolute maximum of f on the subset $B(a) \cap S$. If f has an absolute maximum at a, then a is also a local maximum. However, f can have local maxima at several points in S without having an absolute maximum on the whole set S.

The next theorem shows a connection between zero derivatives and local extrema (maxima or minima) at interior points.

Theorem 5.9. *Let f be defined on an open interval (a, b) and assume that f has a local maximum or a local minimum at an interior point c of (a, b). If f has a derivative (finite or infinite) at c, then f'(c) must be 0.*

Proof. If $f'(c)$ is positive or $+\infty$, then f cannot have a local extremum at c because of Theorem 5.7. Similarly, $f'(c)$ cannot be negative or $-\infty$. However, because there is a derivative at c, the only other possibility is $f'(c) = 0$.

The converse of Theorem 5.9 is not true. In general, knowing that $f'(c) = 0$ is not enough to determine whether f has an extremum at c. In fact, it may have neither, as can be verified by the example $f(x) = x^3$ and $c = 0$. In this case, $f'(0) = 0$ but f is increasing in every neighborhood of 0.

Furthermore, it should be emphasized that f can have a local extremum at c without $f'(c)$ being zero. The example $f(x) = |x|$ has a minimum at $x = 0$ but, of course, there is no derivative at 0. Theorem 5.9 assumes that f has a derivative (finite or infinite) at c. The theorem also assumes that c is an interior point of (a, b). In the example $f(x) = x$, where $a \leq x \leq b$, f takes on its maximum and minimum at the endpoints but $f'(x)$ is never zero in $[a, b]$.

5.9 ROLLE'S THEOREM

It is geometrically evident that a sufficiently "smooth" curve which crosses the
x-axis at both endpoints of an interval $[a, b]$ must have a "turning point" somewhere between a and b. The precise statement of this fact is known as Rolle's
theorem.

Theorem 5.10 (Rolle). *Assume f has a derivative (finite or infinite) at each point of
an open interval (a, b), and assume that f is continuous at both endpoints a and b.
If $f(a) = f(b)$ there is at least one interior point c at which $f'(c) = 0$.*

Proof. We assume f' is never 0 in (a, b) and obtain a contradiction. Since f is
continuous on a compact set, it attains its maximum M and its minimum m somewhere in $[a, b]$. Neither extreme value is attained at an interior point (otherwise
f' would vanish there) so both are attained at the endpoints. Since $f(a) = f(b)$,
then $m = M$, and hence f is constant on $[a, b]$. This contradicts the assumption
that f' is never 0 on (a, b). Therefore $f'(c) = 0$ for some c in (a, b).

5.10 THE MEAN-VALUE THEOREM FOR DERIVATIVES

Theorem 5.11 (Mean-Value Theorem). *Assume that f has a derivative (finite or
infinite) at each point of an open interval (a, b), and assume also that f is continuous
at both endpoints a and b. Then there is a point c in (a, b) such that*

$$f(b) - f(a) = f'(c)(b - a).$$

Geometrically, this states that a sufficiently smooth curve joining two points
A and B has a tangent line with the same slope as the chord AB. We will deduce
Theorem 5.11 from a more general version which involves two functions f and g in
a symmetric fashion.

Theorem 5.12 (Generalized Mean-Value Theorem). *Let f and g be two functions,
each having a derivative (finite or infinite) at each point of an open interval (a, b)
and each continuous at the endpoints a and b. Assume also that there is no interior
point x at which both $f'(x)$ and $g'(x)$ are infinite. Then for some interior point c we
have*

$$f'(c)[g(b) - g(a)] = g'(c)[f(b) - f(a)].$$

NOTE. When $g(x) = x$, this gives Theorem 5.11.

Proof. Let $h(x) = f(x)[g(b) - g(a)] - g(x)[f(b) - f(a)]$. Then $h'(x)$ is finite if
both $f'(x)$ and $g'(x)$ are finite, and $h'(x)$ is infinite if exactly one of $f'(x)$ or $g'(x)$ is
infinite. (The hypothesis excludes the case of *both* $f'(x)$ and $g'(x)$ being infinite.)
Also, h is continuous at the endpoints, and $h(a) = h(b) = f(a)g(b) - g(a)f(b)$.
By Rolle's theorem we have $h'(c) = 0$ for some interior point, and this proves the
assertion.

NOTE. The reader should interpret Theorem 5.12 geometrically by referring to the curve in the xy-plane described by the parametric equations $x = g(t)$, $y = f(t)$, $a \le t \le b$.

There is also an extension which does not require continuity at the endpoints.

Theorem 5.13. *Let f and g be two functions, each having a derivative (finite or infinite) at each point of (a, b). At the endpoints assume that the limits $f(a+)$, $g(a+)$, $f(b-)$ and $g(b-)$ exist as finite values. Assume further that there is no interior point x at which both $f'(x)$ and $g'(x)$ are infinite. Then for some interior point c we have*

$$f'(c)[g(b-) - g(a+)] = g'(c)[f(b-) - f(a+)].$$

Proof. Define new functions F and G on $[a, b]$ as follows:

$$F(x) = f(x) \quad \text{and} \quad G(x) = g(x) \qquad \text{if } x \in (a, b);$$

$$F(a) = f(a+), \quad G(a) = g(a+), \quad F(b) = f(b-), \quad G(b) = g(b-).$$

Then F and G are continuous on $[a, b]$ and we can apply Theorem 5.12 to F and G to obtain the desired conclusion.

The next result is an immediate consequence of the Mean-Value Theorem.

Theorem 5.14. *Assume f has a derivative (finite or infinite) at each point of an open interval (a, b) and that f is continuous at the endpoints a and b.*

a) *If f' takes only positive values (finite or infinite) in (a, b), then f is strictly increasing on $[a, b]$.*

b) *If f' takes only negative values (finite or infinite) in (a, b), then f is strictly decreasing on $[a, b]$.*

c) *If f' is zero everywhere in (a, b) then f is constant on $[a, b]$.*

Proof. Choose $x < y$ and apply the Mean-Value Theorem to the subinterval $[x, y]$ of $[a, b]$ to obtain

$$f(y) - f(x) = f'(c)(y - x) \qquad \text{where } c \in (x, y).$$

All the statements of the theorem follow at once from this equation.

By applying Theorem 5.14 (c) to the difference $f - g$ we obtain:

Corollary 5.15. *If f and g are continuous on $[a, b]$ and have equal finite derivatives in (a, b), then $f - g$ is constant on $[a, b]$.*

5.11 INTERMEDIATE-VALUE THEOREM FOR DERIVATIVES

In Theorem 4.33 we proved that a function f which is continuous on a compact interval $[a, b]$ assumes every value between its maximum and its minimum on the interval. In particular, f assumes every value between $f(a)$ and $f(b)$. A similar result will now be proved for functions which are derivatives.

Theorem 5.16 (Intermediate-value theorem for derivatives). *Assume that f is defined on a compact interval $[a, b]$ and that f has a derivative (finite or infinite) at each interior point. Assume also that f has finite one-sided derivatives $f'_+(a)$ and $f'_-(b)$ at the endpoints, with $f'_+(a) \neq f'_-(b)$. Then, if c is a real number between $f'_+(a)$ and $f'_-(b)$, there exists at least one interior point x such that $f'(x) = c$.*

Proof. Define a new function g as follows:

$$g(x) = \frac{f(x) - f(a)}{x - a} \qquad \text{if } x \neq a, \quad g(a) = f'_+(a).$$

Then g is continuous on the closed interval $[a, b]$. By the intermediate-value theorem for continuous functions, g takes on every value between $f'_+(a)$ and $[f(b) - f(a)]/(b - a)$ in the interior (a, b). By the Mean-Value Theorem, we have $g(x) = f'(k)$ for some k in (a, x) whenever $x \in (a, b)$. Therefore f' takes on every value between $f'_+(a)$ and $[f(b) - f(a)]/(b - a)$ in the interior (a, b). A similar argument applied to the function h, defined by

$$h(x) = \frac{f(x) - f(b)}{x - b} \qquad \text{if } x \neq b, \quad h(b) = f'_-(b),$$

shows that f' takes on every value between $[f(b) - f(a)]/(b - a)$ and $f'_-(b)$ in the interior (a, b). Combining these results, we see that f' takes on every value between $f'_+(a)$ and $f'_-(b)$ in the interior (a, b), and this proves the theorem.

NOTE. Theorem 5.16 is still valid if one or both of the one-sided derivatives $f'_+(a)$, $f'_-(b)$, is infinite. The proof in this case can be given by considering the auxiliary function g defined by the equation $g(x) = f(x) - cx$, if $x \in [a, b]$. Details are left to the reader.

The intermediate-value theorem shows that a derivative cannot change sign in an interval without taking the value 0. Therefore, we have the following strengthening of Theorem 5.14(a) and (b).

Theorem 5.17. *Assume f has a derivative (finite or infinite) on (a, b) and is continuous at the endpoints a and b. If $f'(x) \neq 0$ for all x in (a, b) then f is strictly monotonic on $[a, b]$.*

The intermediate-value theorem also shows that monotonic derivatives are necessarily continuous.

Theorem 5.18. *Assume f' exists and is monotonic on an open interval (a, b). Then f' is continuous on (a, b).*

Proof. We assume f' has a discontinuity at some point c in (a, b) and arrive at a contradiction. Choose a closed subinterval $[\alpha, \beta]$ of (a, b) which contains c in its interior. Since f' is monotonic on $[\alpha, \beta]$ the discontinuity at c must be a jump discontinuity (by Theorem 4.51). Hence f' omits some value between $f'(\alpha)$ and $f'(\beta)$, contradicting the intermediate-value theorem.

5.12 TAYLOR'S FORMULA WITH REMAINDER

As noted earlier, if f is differentiable at c, then f is approximately a linear function near c. That is, the equation

$$f(x) = f(c) + f'(c)(x - c),$$

is approximately correct when $x - c$ is small. Taylor's theorem tells us that, more generally, f can be approximated by a polynomial of degree $n - 1$ if f has a derivative of order n. Moreover, Taylor's theorem gives a useful expression for the error made by this approximation.

Theorem 5.19 (Taylor). *Let f be a function having finite nth derivative $f^{(n)}$ everywhere in an open interval (a, b) and assume that $f^{(n-1)}$ is continuous on the closed interval $[a, b]$. Assume that $c \in [a, b]$. Then, for every x in $[a, b]$, $x \neq c$, there exists a point x_1 interior to the interval joining x and c such that*

$$f(x) = f(c) + \sum_{k=1}^{n-1} \frac{f^{(k)}(c)}{k!} (x - c)^k + \frac{f^{(n)}(x_1)}{n!} (x - c)^n.$$

Taylor's theorem will be obtained as a consequence of a more general result that is a direct extension of the generalized Mean-Value Theorem.

Theorem 5.20. *Let f and g be two functions having finite nth derivatives $f^{(n)}$ and $g^{(n)}$ in an open interval (a, b) and continuous $(n - 1)$st derivatives in the closed interval $[a, b]$. Assume that $c \in [a, b]$. Then, for every x in $[a, b]$, $x \neq c$, there exists a point x_1 interior to the interval joining x and c such that*

$$\left[f(x) - \sum_{k=0}^{n-1} \frac{f^{(k)}(c)}{k!} (x - c)^k \right] g^{(n)}(x_1) = f^{(n)}(x_1) \left[g(x) - \sum_{k=0}^{n-1} \frac{g^{(k)}(c)}{k!} (x - c)^k \right].$$

NOTE. For the special case in which $g(x) = (x - c)^n$, we have $g^{(k)}(c) = 0$ for $0 \leq k \leq n - 1$ and $g^{(n)}(x) = n!$. This theorem then reduces to Taylor's theorem.

Proof. For simplicity, assume that $c < b$ and that $x > c$. Keep x fixed and define new functions F and G as follows:

$$F(t) = f(t) + \sum_{k=1}^{n-1} \frac{f^{(k)}(t)}{k!} (x - t)^k,$$

$$G(t) = g(t) + \sum_{k=1}^{n-1} \frac{g^{(k)}(t)}{k!} (x - t)^k,$$

for each t in $[c, x]$. Then F and G are continuous on the closed interval $[c, x]$ and have finite derivatives in the open interval (c, x). Therefore, Theorem 5.12 is applicable and we can write

$$F'(x_1)[G(x) - G(c)] = G'(x_1)[F(x) - F(c)], \qquad \text{where } x_1 \in (c, x).$$

This reduces to the equation

$$F'(x_1)[g(x) - G(c)] = G'(x_1)[f(x) - F(c)], \qquad \text{(a)}$$

since $G(x) = g(x)$ and $F(x) = f(x)$. If, now, we compute the derivative of the sum defining $F(t)$, keeping in mind that each term of the sum is a product, we find that all terms cancel but one, and we are left with

$$F'(t) = \frac{(x - t)^{n-1}}{(n - 1)!} f^{(n)}(t).$$

Similarly, we obtain

$$G'(t) = \frac{(x - t)^{n-1}}{(n - 1)!} g^{(n)}(t).$$

If we put $t = x_1$ and substitute into (a), we obtain the formula of the theorem.

5.13 DERIVATIVES OF VECTOR-VALUED FUNCTIONS

Let $\mathbf{f} : (a, b) \to \mathbf{R}^n$ be a vector-valued function defined on an open interval (a, b) in $\mathbf{R}$. Then $\mathbf{f} = (f_1, \ldots, f_n)$ where each component f_k is a real-valued function defined on (a, b). We say that $\mathbf{f}$ is differentiable at a point c in (a, b) if each component f_k is differentiable at c and we define

$$\mathbf{f}'(c) = (f_1'(c), \ldots, f_n'(c)).$$

In other words, the derivative $\mathbf{f}'(c)$ is obtained by differentiating each component of $\mathbf{f}$ at c. In view of this definition, it is not surprising to find that many of the theorems on differentiation are also valid for vector-valued functions. For example, if $\mathbf{f}$ and $\mathbf{g}$ are vector-valued functions differentiable at c and if λ is a real-valued function differentiable at c, then the sum $\mathbf{f} + \mathbf{g}$, the product $\lambda\mathbf{f}$, and the dot product $\mathbf{f} \cdot \mathbf{g}$ are differentiable at c and we have

$$(\mathbf{f} + \mathbf{g})'(c) = \mathbf{f}'(c) + \mathbf{g}'(c),$$
$$(\lambda\mathbf{f})'(c) = \lambda'(c)\mathbf{f}(c) + \lambda(c)\mathbf{f}'(c),$$
$$(\mathbf{f} \cdot \mathbf{g})'(c) = \mathbf{f}'(c) \cdot \mathbf{g}(c) + \mathbf{f}(c) \cdot \mathbf{g}'(c).$$

The proofs follow easily by considering components. There is also a chain rule for differentiating composite functions which is proved in the same way. If $\mathbf{f}$ is vector-valued and if u is real-valued, then the composite function $\mathbf{g}$ given by $\mathbf{g}(x) = \mathbf{f}[u(x)]$ is vector-valued. The chain rule states that

$$\mathbf{g}'(c) = \mathbf{f}'[u(c)]u'(c),$$

if the domain of $\mathbf{f}$ contains a neighborhood of $u(c)$ and if $u'(c)$ and $\mathbf{f}'[u(c)]$ both exist.

The Mean-Value Theorem, as stated in Theorem 5.11, does not hold for vector-valued functions. For example, if $\mathbf{f}(t) = (\cos t, \sin t)$ for all real t, then

$$\mathbf{f}(2\pi) - \mathbf{f}(0) = \mathbf{0},$$

but $\mathbf{f}'(t)$ is never zero. In fact, $\|\mathbf{f}'(t)\| = 1$ for all t. A modified version of the Mean-Value Theorem for vector-valued functions is given in Chapter 12 (Theorem 12.8).

5.14 PARTIAL DERIVATIVES

Let S be an open set in Euclidean space $\mathbf{R}^n$, and let $f : S \to \mathbf{R}$ be a real-valued function defined on S. If $\mathbf{x} = (x_1, \ldots, x_n)$ and $\mathbf{c} = (c_1, \ldots, c_n)$ are two points of S having corresponding coordinates equal except for the kth, that is, if $x_i = c_i$ for $i \neq k$ and if $x_k \neq c_k$, then we can consider the limit

$$\lim_{x_k \to c_k} \frac{f(\mathbf{x}) - f(\mathbf{c})}{x_k - c_k}.$$

When this limit exists, it is called the *partial derivative* of f with respect to the kth coordinate and is denoted by

$$D_k f(\mathbf{c}), \qquad f_k(\mathbf{c}), \qquad \frac{\partial f}{\partial x_k}(\mathbf{c}),$$

or by a similar expression. We shall adhere to the notation $D_k f(\mathbf{c})$.

This process produces n further functions $D_1 f, D_2 f, \ldots, D_n f$ defined at those points in S where the corresponding limits exist.

Partial differentiation is not really a new concept. We are merely treating $f(x_1, \ldots, x_n)$ as a function of one variable at a time, holding the others fixed. That is, if we introduce a function g defined by

$$g(x_k) = f(c_1, \ldots, c_{k-1}, x_k, c_{k+1}, \ldots, c_n),$$

then the partial derivative $D_k f(\mathbf{c})$ is exactly the same as the ordinary derivative $g'(c_k)$. This is usually described by saying that we differentiate f with respect to the kth variable, holding the others fixed.

In generalizing a concept from $\mathbf{R}^1$ to $\mathbf{R}^n$, we seek to preserve the important properties in the one-dimensional case. For example, in the one-dimensional case, the existence of the derivative at c implies continuity at c. Therefore it seems desirable to have a concept of derivative for functions of several variables which will imply continuity. Partial derivatives do *not* do this. A function of n variables can have partial derivatives at a point with respect to each of the variables and yet not be continuous at the point. We illustrate with the following example of a function of two variables:

$$f(x, y) = \begin{cases} x + y, & \text{if } x = 0 \text{ or } y = 0, \\ 1, & \text{otherwise.} \end{cases}$$

The partial derivatives $D_1 f(0, 0)$ and $D_2 f(0, 0)$ both exist. In fact,

$$D_1 f(0, 0) = \lim_{x \to 0} \frac{f(x, 0) - f(0, 0)}{x - 0} = \lim_{x \to 0} \frac{x}{x} = 1,$$

and, similarly, $D_2 f(0, 0) = 1$. On the other hand, it is clear that this function is not continuous at $(0, 0)$.

The existence of the partial derivatives with respect to each variable separately implies continuity in each variable separately; but, as we have just seen, this does not necessarily imply continuity in all the variables simultaneously. The difficulty with partial derivatives is that by their very definition we are forced to consider only one variable at a time. Partial derivatives give us the rate of change of a function in the direction of each coordinate axis. There is a more general concept of derivative which does not restrict our considerations to the special directions of the coordinate axes. This will be studied in detail in Chapter 12.

The purpose of this section is merely to introduce the notation for partial derivatives, since we shall use them occasionally before we reach Chapter 12.

If f has partial derivatives $D_1 f, \ldots, D_n f$ on an open set S, then we can also consider *their* partial derivatives. These are called *second-order* partial derivatives. We write $D_{r,k} f$ for the partial derivative of $D_k f$ with respect to the rth variable. Thus,

$$D_{r,k} f = D_r(D_k f).$$

Higher-order partial derivatives are similarly defined. Other notations are

$$D_{r,k} f = \frac{\partial^2 f}{\partial x_r \, \partial x_k}, \qquad D_{p,q,r} f = \frac{\partial^3 f}{\partial x_p \, \partial x_q \, \partial x_r}.$$

5.15 DIFFERENTIATION OF FUNCTIONS OF A COMPLEX VARIABLE

In this section we shall discuss briefly derivatives of complex-valued functions defined on subsets of the complex plane. Such functions are, of course, vector-valued functions whose domain and range are subsets of $\mathbf{R}^2$. All the considerations of Chapter 4 concerning limits and continuity of vector-valued functions apply, in particular, to functions of a complex variable. There is, however, one essential difference between the set of complex numbers $\mathbf{C}$ and the set of n-dimensional vectors $\mathbf{R}^n$ (when $n > 2$) that plays an important role here. In the complex number system we have the four algebraic operations of addition, subtraction, multiplication, and division, and these operations satisfy most of the "usual" laws of algebra that hold for the real number system. In particular, they satisfy the first five axioms for real numbers listed in Chapter 1. (Axioms 6 through 10 involve the ordering relation $<$, which cannot exist among the complex numbers.) Any algebraic system which satisfies Axioms 1 through 5 is called a *field*. (For a thorough discussion of fields, see Reference 1.4.) Multiplication and division, it turns out, cannot be introduced in $\mathbf{R}^n$ (for $n > 2$) in such a way that $\mathbf{R}^n$ will become a field† which includes $\mathbf{C}$. Since division is possible in $\mathbf{C}$, however, we can form the fundamental difference quotient $[f(z) - f(c)]/(z - c)$ which was used to define the derivative in $\mathbf{R}$, and it now becomes clear how the derivative should be defined in $\mathbf{C}$.

Definition 5.21. *Let f be a complex-valued function defined on an open set S in* $\mathbf{C}$, *and assume* $c \in S$. *Then f is said to be differentiable at c if the limit*

$$\lim_{z \to c} \frac{f(z) - f(c)}{z - c} = f'(c)$$

exists.

By means of this limit process, a new complex-valued function f' is defined at those points z of S where $f'(z)$ exists. Higher-order derivatives $f'', f''', \ldots$ are, of course, similarly defined.

The following statements can now be proved for complex-valued functions defined on an open set S by exactly the same proofs used in the real case:

a) *f is differentiable at c if, and only if, there is a function* f^*, *continuous at c, such that*

$$f(z) - f(c) = (z - c)f^*(z),$$

for all z in S, with $f^*(c) = f'(c)$.

NOTE. If we let $g(z) = f^*(z) - f'(c)$ the equation in (a) can be put in the form

$$f(z) = f(c) + f'(c)(z - c) + g(z)(z - c),$$

where $g(z) \to 0$ as $z \to c$. This is called a *first-order Taylor formula* for f.

b) *If f is differentiable at c, then f is continuous at c.*

c) *If two functions f and g have derivatives at c, then their sum, difference, product, and quotient also have derivatives at c and are given by the usual formulas (as in Theorem 5.4). In the case of* f/g, *we must assume* $g(c) \neq 0$.

d) *The chain rule is valid; that is to say, we have*

$$(g \circ f)'(c) = g'[f(c)]f'(c),$$

if the domain of g contains a neighborhood of $f(c)$ *and if* $f'(c)$ *and* $g'[f(c)]$ *both exist.*

When $f(z) = z$, we find $f'(z) = 1$ for all z in $\mathbf{C}$. Using (c) repeatedly, we find that $f'(z) = nz^{n-1}$ when $f(z) = z^n$ (n is a positive integer). This also holds when

---

† For example, if it were possible to define multiplication in $\mathbf{R}^3$ so as to make $\mathbf{R}^3$ a field including $\mathbf{C}$, we could argue as follows: For every $\mathbf{x}$ in $\mathbf{R}^3$ the vectors $1, \mathbf{x}, \mathbf{x}^2, \mathbf{x}^3$ would be linearly dependent (see Reference 5.1, p. 558). Hence for each $\mathbf{x}$ in $\mathbf{R}^3$, a relation of the form $a_0 + a_1\mathbf{x} + a_2\mathbf{x}^2 + a_3\mathbf{x}^3 = \mathbf{0}$ would hold, where a_0, a_1, a_2, a_3 are real numbers. But every polynomial of degree three with real coefficients is a product of a linear polynomial and a quadratic polynomial with real coefficients. The only roots such polynomials can have are either real numbers or complex numbers.

n is a negative integer, provided $z \neq 0$. Therefore, we may compute derivatives of complex polynomials and complex rational functions by the same techniques used in elementary differential calculus.

5.16 THE CAUCHY–RIEMANN EQUATIONS

If f is a complex-valued function of a complex variable, we can write each function value in the form

$$f(z) = u(z) + iv(z),$$

where u and v are real-valued functions of a complex variable. We can, of course, also consider u and v to be real-valued functions of two real variables and then we write

$$f(z) = u(x, y) + iv(x, y), \qquad \text{if } z = x + iy.$$

In either case, we write $f = u + iv$ and we refer to u and v as the *real* and *imaginary parts* of f. For example, in the case of the complex exponential function f, defined by

$$f(z) = e^z = e^x \cos y + ie^x \sin y,$$

the real and imaginary parts are given by

$$u(x, y) = e^x \cos y, \qquad v(x, y) = e^x \sin y.$$

Similarly, when $f(z) = z^2 = (x + iy)^2$, we find

$$u(x, y) = x^2 - y^2, \qquad v(x, y) = 2xy.$$

In the next theorem we shall see that the existence of the derivative f' places a rather severe restriction on the real and imaginary parts u and v.

Theorem 5.22. *Let $f = u + iv$ be defined on an open set S in $\mathbf{C}$. If $f'(c)$ exists for some c in S, then the partial derivatives $D_1u(c)$, $D_2u(c)$, $D_1v(c)$ and $D_2v(c)$ also exist and we have*

$$f'(c) = D_1u(c) + i\, D_1v(c), \tag{3}$$

and

$$f'(c) = D_2v(c) - i\, D_2u(c). \tag{4}$$

This implies, in particular, that

$$D_1u(c) = D_2v(c) \qquad and \qquad D_1v(c) = -D_2u(c).$$

NOTE. These last two equations are known as the *Cauchy–Riemann equations*. They are usually seen in the form

$$\frac{\partial u}{\partial x} = \frac{\partial v}{\partial y}, \qquad \frac{\partial v}{\partial x} = -\frac{\partial u}{\partial y}.$$

Proof. Since $f'(c)$ exists there is a function f^* defined on S such that

$$f(z) - f(c) = (z - c)f^*(z), \tag{5}$$

where f^* is continuous at c and $f^*(c) = f'(c)$. Write

$$z = x + iy, \qquad c = a + ib, \qquad \text{and} \qquad f^*(z) = A(z) + iB(z),$$

where $A(z)$ and $B(z)$ are real. Note that $A(z) \to A(c)$ and $B(z) \to B(c)$ as $z \to c$. By considering only those z in S with $y = b$ and taking real and imaginary parts of (5), we find

$$u(x, b) - u(a, b) = (x - a)A(x + ib), \qquad v(x, b) - v(a, b) = (x - a)B(x + ib).$$

Dividing by $x - a$ and letting $x \to a$ we find

$$D_1 u(c) = A(c) \qquad \text{and} \qquad D_1 v(c) = B(c).$$

Since $f'(c) = A(c) + iB(c)$, this proves (3).

Similarly, by considering only those z in S with $x = a$ we find

$$D_2 v(c) = A(c) \qquad \text{and} \qquad D_2 u(c) = -B(c),$$

which proves (4).

Applications of the Cauchy–Riemann equations are given in the next theorem.

Theorem 5.23. *Let $f = u + iv$ be a function with a derivative everywhere in an open disk D centered at (a, b). If any one of u, v, or $|f|$ is constant† on D, then f is constant on D. Also, f is constant if $f'(z) = 0$ for all z in D.*

Proof. Suppose u is constant on D. The Cauchy–Riemann equations show that $D_2 v = D_1 v = 0$ on D. Applying the one-dimensional Mean-Value Theorem twice we find, for some y' between b and y,

$$v(x, y) - v(x, b) = (y - b)D_2 v(x, y') = 0,$$

and, for some x' between a and x,

$$v(x, b) - v(a, b) = (x - a)D_1 v(x', b) = 0.$$

Therefore $v(x, y) = v(a, b)$ for all (x, y) in D, so v is constant on D. A similar argument shows that if v is constant then u is constant.

Now suppose $|f|$ is constant on D. Then $|f|^2 = u^2 + v^2$ is constant on D. Taking partial derivatives we find

$$u D_1 u + v D_1 v = 0, \qquad u D_2 u + v D_2 v = 0.$$

By the Cauchy–Riemann equations the second equation can be written as

$$v D_1 u - u D_1 v = 0.$$

Combining this with the first to eliminate $D_1 v$ we find $(u^2 + v^2)D_1 u = 0$. If $u^2 + v^2 = 0$, then $u = v = 0$, so $f = 0$. If $u^2 + v^2 \neq 0$ then $D_1 u = 0$; hence u is constant, so f is constant.

_____

† Here $|f|$ denotes the function whose value at z is $|f(z)|$.

Finally, if $f' = 0$ on D, both partial derivatives D_1v and D_2v are zero on D. Again, as in the first part of the proof, we find f is constant on D.

Theorem 5.22 tells us that a necessary condition for the function $f = u + iv$ to have a derivative at c is that the four partials D_1u, D_2u, D_1v, D_2v, exist at c and satisfy the Cauchy–Riemann equations. This condition, however, is not sufficient, as we see by considering the following example.

Example. Let u and v be defined as follows:

$$u(x, y) = \frac{x^3 - y^3}{x^2 + y^2} \quad \text{if } (x, y) \neq (0, 0), \quad u(0, 0) = 0,$$

$$v(x, y) = \frac{x^3 + y^3}{x^2 + y^2} \quad \text{if } (x, y) \neq (0, 0), \quad v(0, 0) = 0.$$

It is easily seen that $D_1u(0, 0) = D_1v(0, 0) = 1$ and that $D_2u(0, 0) = -D_2v(0, 0) = -1$, so that the Cauchy–Riemann equations hold at $(0, 0)$. Nevertheless, the function $f = u + iv$ cannot have a derivative at $z = 0$. In fact, for $x = 0$, the difference quotient becomes

$$\frac{f(z) - f(0)}{z - 0} = \frac{-y + iy}{iy} = 1 + i,$$

whereas for $x = y$, it becomes

$$\frac{f(z) - f(0)}{z - 0} = \frac{xi}{x + ix} = \frac{1 + i}{2},$$

and hence $f'(0)$ cannot exist.

In Chapter 12 we shall prove that the Cauchy–Riemann equations *do* suffice to establish existence of the derivative of $f = u + iv$ at c if the partial derivatives of u and v are continuous in some neighborhood of c. To illustrate how this result is used in practice, we shall obtain the derivative of the exponential function. Let $f(z) = e^z = u + iv$. Then

$$u(x, y) = e^x \cos y, \qquad v(x, y) = e^x \sin y,$$

and hence

$$D_1u(x, y) = e^x \cos y = D_2v(x, y), \qquad D_2u(x, y) = -e^x \sin y = -D_1v(x, y).$$

Since these partial derivatives are continuous everywhere in $\mathbf{R}^2$ and satisfy the Cauchy–Riemann equations, the derivative $f'(z)$ exists for all z. To compute it we use Theorem 5.22 to obtain

$$f'(z) = e^x \cos y + ie^x \sin y = f(z).$$

Thus, the exponential function is its own derivative (as in the real case).

EXERCISES

Real-valued functions

In the following exercises assume, where necessary, a knowledge of the formulas for differentiating the elementary trigonometric, exponential, and logarithmic functions.

5.1 A function f is said to satisfy a Lipschitz condition of order α at c if there exists a positive number M (which may depend on c) and a 1-ball $B(c)$ such that

$$|f(x) - f(c)| < M|x - c|^\alpha$$

whenever $x \in B(c)$, $x \neq c$.

 a) Show that a function which satisfies a Lipschitz condition of order α is continuous at c if $\alpha > 0$, and has a derivative at c if $\alpha > 1$.

 b) Give an example of a function satisfying a Lipschitz condition of order 1 at c for which $f'(c)$ does not exist.

5.2 In each of the following cases, determine the intervals in which the function f is increasing or decreasing and find the maxima and minima (if any) in the set where each f is defined.

 a) $f(x) = x^3 + ax + b$, $\qquad\qquad\qquad$ $x \in \mathbf{R}$.

 b) $f(x) = \log(x^2 - 9)$, $\qquad\qquad\qquad$ $|x| > 3$.

 c) $f(x) = x^{2/3}(x - 1)^4$, $\qquad\qquad\qquad$ $0 \leq x \leq 1$.

 d) $f(x) = (\sin x)/x$ if $x \neq 0$, $f(0) = 1$, $\quad$ $0 \leq x \leq \pi/2$.

5.3 Find a polynomial f of lowest possible degree such that

$$f(x_1) = a_1, \qquad f(x_2) = a_2, \qquad f'(x_1) = b_1, \qquad f'(x_2) = b_2,$$

where $x_1 \neq x_2$ and a_1, a_2, b_1, b_2 are given real numbers.

5.4 Define f as follows: $f(x) = e^{-1/x^2}$ if $x \neq 0$, $f(0) = 0$. Show that

 a) f is continuous for all x.

 b) $f^{(n)}$ is continuous for all x, and that $f^{(n)}(0) = 0$, $(n = 1, 2, \dots)$.

5.5 Define f, g, and h as follows: $f(0) = g(0) = h(0) = 0$ and, if $x \neq 0$, $f(x) = \sin(1/x)$, $g(x) = x \sin(1/x)$, $h(x) = x^2 \sin(1/x)$. Show that

 a) $f'(x) = -1/x^2 \cos(1/x)$, if $x \neq 0$; $\qquad\qquad$ $f'(0)$ does not exist.

 b) $g'(x) = \sin(1/x) - 1/x \cos(1/x)$, if $x \neq 0$; $\qquad$ $g'(0)$ does not exist.

 c) $h'(x) = 2x \sin(1/x) - \cos(1/x)$, if $x \neq 0$; $\qquad$ $h'(0) = 0$;

$\qquad\qquad\qquad\qquad\qquad\qquad\qquad\qquad\qquad$ $\lim_{x \to 0} h'(x)$ does not exist.

5.6 Derive Leibnitz's formula for the nth derivative of the product h of two functions f and g:

$$h^{(n)}(x) = \sum_{k=0}^{n} \binom{n}{k} f^{(k)}(x) g^{(n-k)}(x), \qquad \text{where} \quad \binom{n}{k} = \frac{n!}{k!\,(n-k)!}.$$

5.7 Let f and g be two functions defined and having finite third-order derivatives $f'''(x)$ and $g'''(x)$ for all x in $\mathbf{R}$. If $f(x)g(x) = 1$ for all x, show that the relations in (a), (b), (c), and (d) hold at those points where the denominators are not zero:

a) $f'(x)/f(x) + g'(x)/g(x) = 0$.

b) $f''(x)/f'(x) - 2f'(x)/f(x) - g''(x)/g'(x) = 0$.

c) $\dfrac{f'''(x)}{f'(x)} - 3\dfrac{f'(x)g''(x)}{f(x)g'(x)} - 3\dfrac{f''(x)}{f(x)} - \dfrac{g'''(x)}{g'(x)} = 0$.

d) $\dfrac{f'''(x)}{f'(x)} - \dfrac{3}{2}\left(\dfrac{f''(x)}{f'(x)}\right)^2 = \dfrac{g'''(x)}{g'(x)} - \dfrac{3}{2}\left(\dfrac{g''(x)}{g'(x)}\right)^2$.

NOTE. The expression which appears on the left side of (d) is called the *Schwarzian derivative* of f at x.

e) Show that f and g have the same Schwarzian derivative if

$$g(x) = [af(x) + b]/[cf(x) + d], \quad \text{where } ad - bc \neq 0.$$

Hint. If $c \neq 0$, write $(af + b)/(cf + d) = (a/c) + (bc - ad)/[c(cf + d)]$, and apply part (d).

5.8 Let f_1, f_2, g_1, g_2 be four functions having derivatives in (a, b). Define F by means of the determinant

$$F(x) = \begin{vmatrix} f_1(x) & f_2(x) \\ g_1(x) & g_2(x) \end{vmatrix}, \quad \text{if } x \in (a, b).$$

a) Show that $F'(x)$ exists for each x in (a, b) and that

$$F'(x) = \begin{vmatrix} f'_1(x) & f'_2(x) \\ g_1(x) & g_2(x) \end{vmatrix} + \begin{vmatrix} f_1(x) & f_2(x) \\ g'_1(x) & g'_2(x) \end{vmatrix}.$$

b) State and prove a more general result for nth order determinants.

5.9 Given n functions $f_1, \ldots, f_n$, each having nth order derivatives in (a, b). A function W, called the *Wronskian* of $f_1, \ldots, f_n$, is defined as follows: For each x in (a, b), $W(x)$ is the value of the determinant of order n whose element in the kth row and mth column is $f_m^{(k-1)}(x)$, where $k = 1, 2, \ldots, n$ and $m = 1, 2, \ldots, n$. [The expression $f_m^{(0)}(x)$ is written for $f_m(x)$.]

a) Show that $W'(x)$ can be obtained by replacing the last row of the determinant defining $W(x)$ by the nth derivatives $f_1^{(n)}(x), \ldots, f_n^{(n)}(x)$.

b) Assuming the existence of n constants $c_1, \ldots, c_n$, not all zero, such that $c_1 f_1(x) + \cdots + c_n f_n(x) = 0$ for every x in (a, b), show that $W(x) = 0$ for each x in (a, b).

NOTE. A set of functions satisfying such a relation is said to be a *linearly dependent set* on (a, b).

c) The vanishing of the Wronskian throughout (a, b) is necessary, but not sufficient, for linear dependence of $f_1, \ldots, f_n$. Show that in the case of two functions, if the Wronskian vanishes throughout (a, b) and if one of the functions does not vanish in (a, b), then they form a linearly dependent set in (a, b).

Mean-Value Theorem

5.10 Given a function f defined and having a finite derivative in (a, b) and such that $\lim_{x \to b-} f(x) = +\infty$. Prove that $\lim_{x \to b-} f'(x)$ either fails to exist or is infinite.

5.11 Show that the formula in the Mean-Value Theorem can be written as follows:

$$\frac{f(x + h) - f(x)}{h} = f'(x + \theta h),$$

where $0 < \theta < 1$. Determine θ as a function of x and h when

a) $f(x) = x^2$,   b) $f(x) = x^3$,

c) $f(x) = e^x$,   d) $f(x) = \log x$, $x > 0$.

Keep $x \neq 0$ fixed, and find $\lim_{h \to 0} \theta$ in each case.

5.12 Take $f(x) = 3x^4 - 2x^3 - x^2 + 1$ and $g(x) = 4x^3 - 3x^2 - 2x$ in Theorem 5.20. Show that $f'(x)/g'(x)$ is never equal to the quotient $[f(1) - f(0)]/[g(1) - g(0)]$ if $0 < x \leq 1$. How do you reconcile this with the equation

$$\frac{f(b) - f(a)}{g(b) - g(a)} = \frac{f'(x_1)}{g'(x_1)}, \qquad a < x_1 < b,$$

obtainable from Theorem 5.20 when $n = 1$?

5.13 In each of the following special cases of Theorem 5.20, take $n = 1$, $c = a$, $x = b$, and show that $x_1 = (a + b)/2$.

a) $f(x) = \sin x$, $g(x) = \cos x$;   b) $f(x) = e^x$, $g(x) = e^{-x}$.

Can you find a general class of such pairs of functions f and g for which x_1 will always be $(a + b)/2$ and such that both examples (a) and (b) are in this class?

5.14 Given a function f defined and having a finite derivative f' in the half-open interval $0 < x \leq 1$ and such that $|f'(x)| < 1$. Define $a_n = f(1/n)$ for $n = 1, 2, 3, \ldots$, and show that $\lim_{n \to \infty} a_n$ exists. *Hint.* Cauchy condition.

5.15 Assume that f has a finite derivative at each point of the open interval (a, b). Assume also that $\lim_{x \to c} f'(x)$ exists and is finite for some interior point c. Prove that the value of this limit must be $f'(c)$.

5.16 Let f be continuous on (a, b) with a finite derivative f' everywhere in (a, b), except possibly at c. If $\lim_{x \to c} f'(x)$ exists and has the value A, show that $f'(c)$ must also exist and have the value A.

5.17 Let f be continuous on $[0, 1]$, $f(0) = 0$, $f'(x)$ finite for each x in $(0, 1)$. Prove that if f' is an increasing function on $(0, 1)$, then so too is the function g defined by the equation $g(x) = f(x)/x$.

5.18 Assume f has a finite derivative in (a, b) and is continuous on $[a, b]$ with $f(a) = f(b) = 0$. Prove that for every real λ there is some c in (a, b) such that $f'(c) = \lambda f(c)$. *Hint.* Apply Rolle's theorem to $g(x)f(x)$ for a suitable g depending on λ.

5.19 Assume f is continuous on $[a, b]$ and has a finite second derivative f'' in the open interval (a, b). Assume that the line segment joining the points $A = (a, f(a))$ and $B = (b, f(b))$ intersects the graph of f in a third point P different from A and B. Prove that $f''(c) = 0$ for some c in (a, b).

5.20 If f has a finite third derivative f'' in $[a, b]$ and if

$$f(a) = f'(a) = f(b) = f'(b) = 0,$$

prove that $f'''(c) = 0$ for some c in (a, b).

5.21 Assume f is nonnegative and has a finite third derivative f'' in the open interval $(0, 1)$. If $f(x) = 0$ for at least two values of x in $(0, 1)$, prove that $f'''(c) = 0$ for some c in $(0, 1)$.

5.22 Assume f has a finite derivative in some interval $(a, +\infty)$.

a) If $f(x) \to 1$ and $f'(x) \to c$ as $x \to +\infty$, prove that $c = 0$.

b) If $f'(x) \to 1$ as $x \to +\infty$, prove that $f(x)/x \to 1$ as $x \to +\infty$.

c) If $f'(x) \to 0$ as $x \to +\infty$, prove that $f(x)/x \to 0$ as $x \to +\infty$.

5.23 Let h be a fixed positive number. Show that there is no function f satisfying the following three conditions: $f'(x)$ exists for $x \geq 0$, $f'(0) = 0$, $f'(x) \geq h$ for $x > 0$.

5.24 If $h > 0$ and if $f'(x)$ exists (and is finite) for every x in $(a - h, a + h)$, and if f is continuous on $[a - h, a + h]$, show that we have:

a) $\dfrac{f(a + h) - f(a - h)}{h} = f'(a + \theta h) + f'(a - \theta h), \qquad 0 < \theta < 1;$

b) $\dfrac{f(a + h) - 2f(a) + f(a - h)}{h} = f'(a + \lambda h) - f'(a - \lambda h), \qquad 0 < \lambda < 1.$

c) If $f''(a)$ exists, show that

$$f''(a) = \lim_{h \to 0} \frac{f(a + h) - 2f(a) + f(a - h)}{h^2}.$$

d) Give an example where the limit of the quotient in (c) exists but where $f''(a)$ does not exist.

5.25 Let f have a finite derivative in (a, b) and assume that $c \in (a, b)$. Consider the following condition: For every $\varepsilon > 0$ there exists a 1-ball $B(c; \delta)$, whose radius δ depends only on ε and not on c, such that if $x \in B(c; \delta)$, and $x \neq c$, then

$$\left| \frac{f(x) - f(c)}{x - c} - f'(c) \right| < \varepsilon.$$

Show that f' is continuous on (a, b) if this condition holds throughout (a, b).

5.26 Assume f has a finite derivative in (a, b) and is continuous on $[a, b]$, with $a \leq f(x) \leq b$ for all x in $[a, b]$ and $|f'(x)| \leq \alpha < 1$ for all x in (a, b). Prove that f has a unique fixed point in $[a, b]$.

5.27 Give an example of a pair of functions f and g having finite derivatives in $(0, 1)$, such that

$$\lim_{x \to 0} \frac{f(x)}{g(x)} = 0,$$

but such that $\lim_{x \to 0} f'(x)/g'(x)$ does not exist, choosing g so that $g'(x)$ is never zero.

5.28 Prove the following theorem:

Let f and g be two functions having finite nth derivatives in (a, b). For some interior point c in (a, b), assume that $f(c) = f'(c) = \cdots = f^{(n-1)}(c) = 0$, and that $g(c) = g'(c) = \cdots = g^{(n-1)}(c) = 0$, but that $g^{(n)}(x)$ is never zero in (a, b). Show that

$$\lim_{x \to c} \frac{f(x)}{g(x)} = \frac{f^{(n)}(c)}{g^{(n)}(c)}.$$

NOTE. $f^{(n)}$ and $g^{(n)}$ are not assumed to be continuous at c. *Hint.* Let

$$F(x) = f(x) - \frac{(x - c)^{n-1} f^{(n-1)}(c)}{(n - 1)!},$$

define G similarly, and apply Theorem 5.20 to the functions F and G.

5.29 Show that the formula in Taylor's theorem can also be written as follows:

$$f(x) = \sum_{k=0}^{n-1} \frac{f^{(k)}(c)}{k!} (x - c)^k + \frac{(x - c)(x - x_1)^{n-1}}{(n - 1)!} f^{(n)}(x_1),$$

where x_1 is interior to the interval joining x and c. Let $1 - \theta = (x - x_1)/(x - c)$. Show that $0 < \theta < 1$ and deduce the following form of the remainder term (due to Cauchy):

$$\frac{(1 - \theta)^{n-1}(x - c)^n}{(n - 1)!} f^{(n)}[\theta x + (1 - \theta)c].$$

Hint. Take $G(t) = g(t) = t$ in the proof of Theorem 5.20.

Vector-valued functions

5.30 If a vector-valued function **f** is differentiable at c, prove that

$$\mathbf{f}'(c) = \lim_{h \to 0} \frac{1}{h} [\mathbf{f}(c + h) - \mathbf{f}(c)].$$

Conversely, if this limit exists, prove that **f** is differentiable at c.

5.31 A vector-valued function **f** is differentiable at each point of (a, b) and has constant norm $\|\mathbf{f}\|$. Prove that $\mathbf{f}(t) \cdot \mathbf{f}'(t) = 0$ on (a, b).

5.32 A vector-valued function **f** is never zero and has a derivative **f**' which exists and is continuous on **R**. If there is a real function λ such that $\mathbf{f}'(t) = \lambda(t)\mathbf{f}(t)$ for all t, prove that there is a positive real function u and a constant vector **c** such that $\mathbf{f}(t) = u(t)\mathbf{c}$ for all t.

Partial derivatives

5.33 Consider the function f defined on $\mathbf{R}^2$ by the following formulas:

$$f(x, y) = \frac{xy}{x^2 + y^2} \quad \text{if } (x, y) \neq (0, 0) \quad f(0, 0) = 0.$$

Prove that the partial derivatives $D_1 f(x, y)$ and $D_2 f(x, y)$ exist for every (x, y) in $\mathbf{R}^2$ and evaluate these derivatives explicitly in terms of x and y. Also, show that f is not continuous at $(0, 0)$.

5.34 Let f be defined on $\mathbf{R}^2$ as follows:

$$f(x, y) = y\,\frac{x^2 - y^2}{x^2 + y^2} \quad \text{if } (x, y) \neq (0, 0), \quad f(0, 0) = 0.$$

Compute the first- and second-order partial derivatives of f at the origin, when they exist.

Complex-valued functions

5.35 Let S be an open set in $\mathbf{C}$ and let S^* be the set of complex conjugates $\bar{z}$, where $z \in S$. If f is defined on S, define g on S^* as follows: $g(\bar{z}) = \overline{f(z)}$, the complex conjugate of $f(z)$. If f is differentiable at c prove that g is differentiable at $\bar{c}$ and that $g'(\bar{c}) = \overline{f'(c)}$.

5.36 i) In each of the following examples write $f = u + iv$ and find explicit formulas for $u(x, y)$ and $v(x, y)$:

a) $f(z) = \sin z$,   b) $f(z) = \cos z$,

c) $f(z) = |z|$,   d) $f(z) = \bar{z}$,

e) $f(z) = \arg z \ (z \neq 0)$,   f) $f(z) = \text{Log } z \ (z \neq 0)$,

g) $f(z) = e^{z^2}$,   h) $f(z) = z^\alpha \ (\alpha \text{ complex}, z \neq 0)$.

(These functions are to be defined as indicated in Chapter 1.)

ii) Show that u and v satisfy the Cauchy–Riemann equations for the following values of z: All z in (a), (b), (g); no z in (c), (d), (e); all z except real $z \leq 0$ in (f), (h). (In part (h), the Cauchy–Riemann equations hold for all z if α is a nonnegative integer, and they hold for all $z \neq 0$ if α is a negative integer.)

iii) Compute the derivative $f'(z)$ in (a), (b), (f), (g), (h), assuming it exists.

5.37 Write $f = u + iv$ and assume that f has a derivative at each point of an open disk D centered at $(0, 0)$. If $au^2 + bv^2$ is constant on D for some real a and b, not both 0, prove that f is constant on D.

SUGGESTED REFERENCES FOR FURTHER STUDY

5.1 Apostol, T. M., *Calculus*, Vol. 1, 2nd ed. Xerox, Waltham, 1967.

5.2 Chaundy, T. W., *The Differential Calculus*. Clarendon Press, Oxford, 1935.

FUNCTIONS OF
BOUNDED VARIATION AND
RECTIFIABLE CURVES

6.1 INTRODUCTION

Some of the basic properties of monotonic functions were derived in Chapter 4. This brief chapter discusses functions of bounded variation, a class of functions closely related to monotonic functions. We shall find that these functions are intimately connected with curves having finite arc length (rectifiable curves). They also play a role in the theory of Riemann–Stieltjes integration which is developed in the next chapter.

6.2 PROPERTIES OF MONOTONIC FUNCTIONS

Theorem 6.1. *Let f be an increasing function defined on* $[a, b]$ *and let* $x_0, x_1, \ldots, x_n$ *be* $n + 1$ *points such that*

$$a = x_0 < x_1 < x_2 < \cdots < x_n = b.$$

Then we have the inequality

$$\sum_{k=1}^{n-1} [f(x_k+) - f(x_k-)] \le f(b) - f(a).$$

Proof. Assume that $y_k \in (x_k, x_{k+1})$. For $1 \le k \le n - 1$, we have $f(x_k+) \le f(y_k)$ and $f(y_{k-1}) \le f(x_k-)$, so that $f(x_k+) - f(x_k-) \le f(y_k) - f(y_{k-1})$. If we add these inequalities, the sum on the right telescopes to $f(y_{n-1}) - f(y_0)$. Since $f(y_{n-1}) - f(y_0) \le f(b) - f(a)$, this completes the proof.

The difference $f(x_k+) - f(x_k-)$ is, of course, the jump of f at x_k. The foregoing theorem tells us that for every finite collection of points x_k in (a, b), the sum of the jumps at these points is always bounded by $f(b) - f(a)$. This result can be used to prove the following theorem.

Theorem 6.2. *If f is monotonic on* $[a, b]$, *then the set of discontinuities of f is countable.*

Proof. Assume that f is increasing and let S_m be the set of points in (a, b) at which the jump of f exceeds $1/m$, $m > 0$. If $x_1 < x_2 < \cdots < x_{n-1}$ are in S_m, Theorem 6.1 tells us that

$$\frac{n - 1}{m} \le f(b) - f(a).$$

This means that S_m must be a finite set. But the set of discontinuities of f in (a, b) is a subset of the union $\bigcup_{m=1}^{\infty} S_m$ and hence is countable. (If f is decreasing, the argument can be applied to $-f$.)

6.3 FUNCTIONS OF BOUNDED VARIATION

Definition 6.3. *If $[a, b]$ is a compact interval, a set of points*

$$P = \{x_0, x_1, \ldots, x_n\},$$

satisfying the inequalities

$$a = x_0 < x_1 \cdots < x_{n-1} < x_n = b,$$

is called a partition of $[a, b]$. The interval $[x_{k-1}, x_k]$ is called the kth subinterval of P and we write $\Delta x_k = x_k - x_{k-1}$, so that $\sum_{k=1}^{n} \Delta x_k = b - a$. The collection of all possible partitions of $[a, b]$ will be denoted by $\mathscr{P}[a, b]$.

Definition 6.4. *Let f be defined on $[a, b]$. If $P = \{x_0, x_1, \ldots, x_n\}$ is a partition of $[a, b]$, write $\Delta f_k = f(x_k) - f(x_{k-1})$, for $k = 1, 2, \ldots, n$. If there exists a positive number M such that*

$$\sum_{k=1}^{n} |\Delta f_k| \leq M$$

for all partitions of $[a, b]$, then f is said to be of bounded variation on $[a, b]$.

Examples of functions of bounded variation are provided by the next two theorems.

Theorem 6.5. *If f is monotonic on $[a, b]$, then f is of bounded variation on $[a, b]$.*

Proof. Let f be increasing. Then for every partition of $[a, b]$ we have $\Delta f_k \geq 0$ and hence

$$\sum_{k=1}^{n} |\Delta f_k| = \sum_{k=1}^{n} \Delta f_k = \sum_{k=1}^{n} [f(x_k) - f(x_{k-1})] = f(b) - f(a).$$

Theorem 6.6. *If f is continuous on $[a, b]$ and if f' exists and is bounded in the interior, say $|f'(x)| \leq A$ for all x in (a, b), then f is of bounded variation on $[a, b]$.*

Proof. Applying the Mean-Value Theorem, we have

$$\Delta f_k = f(x_k) - f(x_{k-1}) = f'(t_k)(x_k - x_{k-1}), \qquad \text{where } t_k \in (x_{k-1}, x_k).$$

This implies

$$\sum_{k=1}^{n} |\Delta f_k| = \sum_{k=1}^{n} |f'(t_k)| \Delta x_k \leq A \sum_{k=1}^{n} \Delta x_k = A(b - a).$$

Theorem 6.7. *If f is of bounded variation on $[a, b]$, say $\sum |\Delta f_k| \leq M$ for all partitions of $[a, b]$, then f is bounded on $[a, b]$. In fact,*

$$|f(x)| \leq |f(a)| + M \qquad \text{for all } x \text{ in } [a, b].$$

Proof. Assume that $x \in (a, b)$. Using the special partition $P = \{a, x, b\}$, we find

$$|f(x) - f(a)| + |f(b) - f(x)| \leq M.$$

This implies $|f(x) - f(a)| \leq M$, $|f(x)| \leq |f(a)| + M$. The same inequality holds if $x = a$ or $x = b$.

Examples

1. It is easy to construct a continuous function which is not of bounded variation. For example, let $f(x) = x \cos \{\pi/(2x)\}$ if $x \neq 0$, $f(0) = 0$. Then f is continuous on $[0, 1]$, but if we consider the partition into $2n$ subintervals

$$P = \left\{ 0, \frac{1}{2n}, \frac{1}{2n-1}, \ldots, \frac{1}{3}, \frac{1}{2}, 1 \right\},$$

an easy calculation shows that we have

$$\sum_{k=1}^{2n} |\Delta f_k| = \frac{1}{2n} + \frac{1}{2n} + \frac{1}{2n-2} + \frac{1}{2n-2} + \cdots + \frac{1}{2} + \frac{1}{2} = 1 + \frac{1}{2} + \cdots + \frac{1}{n}.$$

This is not bounded for all n, since the series $\sum_{n=1}^{\infty} (1/n)$ diverges. In this example the derivative f' exists in $(0, 1)$ but f' is not bounded on $(0, 1)$. However, f' is bounded on any compact interval not containing the origin and hence f will be of bounded variation on such an interval.

2. An example similar to the first is given by $f(x) = x^2 \cos (1/x)$ if $x \neq 0$, $f(0) = 0$. This f is of bounded variation on $[0, 1]$, since f' is bounded on $[0, 1]$. In fact, $f'(0) = 0$ and, for $x \neq 0$, $f'(x) = \sin (1/x) + 2x \cos (1/x)$, so that $|f'(x)| \leq 3$ for all x in $[0, 1]$.

3. Boundedness of f' is not necessary for f to be of bounded variation. For example, let $f(x) = x^{1/3}$. This function is monotonic (and hence of bounded variation) on every finite interval. However, $f'(x) \rightarrow +\infty$ as $x \rightarrow 0$.

6.4 TOTAL VARIATION

Definition 6.8. *Let f be of bounded variation on $[a, b]$, and let $\sum (P)$ denote the sum $\sum_{k=1}^{n} |\Delta f_k|$ corresponding to the partition $P = \{x_0, x_1, \ldots, x_n\}$ of $[a, b]$. The number*

$$V_f(a, b) = \sup \{\textstyle\sum (P) : P \in \mathscr{P}[a, b]\},$$

is called the total variation of f on the interval $[a, b]$.

NOTE. When there is no danger of misunderstanding, we will write V_f instead of $V_f(a, b)$.

Since f is of bounded variation on $[a, b]$, the number V_f is finite. Also, $V_f \geq 0$, since each sum $\sum (P) \geq 0$. Moreover, $V_f(a, b) = 0$ if, and only if, f is constant on $[a, b]$.

Theorem 6.9. *Assume that f and g are each of bounded variation on* $[a, b]$. *Then so are their sum, difference, and product. Also, we have*

$$V_{f \pm g} \le V_f + V_g \quad and \quad V_{f \cdot g} \le AV_f + BV_g,$$

where

$$A = \sup \{|g(x)| : x \in [a, b]\}, \quad B = \sup \{|f(x)| : x \in [a, b]\}.$$

Proof. Let $h(x) = f(x)g(x)$. For every partition P of $[a, b]$, we have

$$|\Delta h_k| = |f(x_k)g(x_k) - f(x_{k-1})g(x_{k-1})|$$
$$= |[f(x_k)g(x_k) - f(x_{k-1})g(x_k)]$$
$$+ [f(x_{k-1})g(x_k) - f(x_{k-1})g(x_{k-1})]| \le A|\Delta f_k| + B|\Delta g_k|.$$

This implies that h is of bounded variation and that $V_h \le AV_f + BV_g$. The proofs for the sum and difference are simpler and will be omitted.

NOTE. Quotients were not included in the foregoing theorem because the reciprocal of a function of bounded variation need not be of bounded variation. For example, if $f(x) \to 0$ as $x \to x_0$, then $1/f$ will not be bounded on any interval containing x_0 and (by Theorem 6.7) $1/f$ cannot be of bounded variation on such an interval. To extend Theorem 6.9 to quotients, it suffices to exclude functions whose values become arbitrarily close to zero.

Theorem 6.10. *Let f be of bounded variation on* $[a, b]$ *and assume that f is bounded away from zero; that is, suppose that there exists a positive number m such that* $0 < m \le |f(x)|$ *for all x in* $[a, b]$. *Then* $g = 1/f$ *is also of bounded variation on* $[a, b]$, *and* $V_g \le V_f/m^2$.

Proof.

$$|\Delta g_k| = \left| \frac{1}{f(x_k)} - \frac{1}{f(x_{k-1})} \right| = \left| \frac{\Delta f_k}{f(x_k)f(x_{k-1})} \right| \le \frac{|\Delta f_k|}{m^2}.$$

6.5 ADDITIVE PROPERTY OF TOTAL VARIATION

In the last two theorems the interval $[a, b]$ was kept fixed and $V_f(a, b)$ was considered as a function of f. If we keep f fixed and study the total variation as a function of the interval $[a, b]$, we can prove the following *additive* property.

Theorem 6.11. *Let f be of bounded variation on* $[a, b]$, *and assume that* $c \in (a, b)$. *Then f is of bounded variation on* $[a, c]$ *and on* $[c, b]$ *and we have*

$$V_f(a, b) = V_f(a, c) + V_f(c, b).$$

Proof. We first prove that f is of bounded variation on $[a, c]$ and on $[c, b]$. Let P_1 be a partition of $[a, c]$ and let P_2 be a partition of $[c, b]$. Then $P_0 = P_1 \cup P_2$ is a partition of $[a, b]$. If $\sum (P)$ denotes the sum $\sum |\Delta f_k|$ corresponding to the partition P (of the appropriate interval), we can write

$$\sum (P_1) + \sum (P_2) = \sum (P_0) \le V_f(a, b). \tag{1}$$

This shows that each sum $\sum (P_1)$ and $\sum (P_2)$ is bounded by $V_f(a, b)$ and this means that f is of bounded variation on $[a, c]$ and on $[c, b]$. From (1) we also obtain the inequality

$$V_f(a, c) + V_f(c, b) \leq V_f(a, b),$$

because of Theorem 1.15.

To obtain the reverse inequality, let $P = \{x_0, x_1, \ldots, x_n\} \in \mathscr{P}[a, b]$ and let $P_0 = P \cup \{c\}$ be the (possibly new) partition obtained by adjoining the point c. If $c \in [x_{k-1}, x_k]$, then we have

$$|f(x_k) - f(x_{k-1})| \leq |f(x_k) - f(c)| + |f(c) - f(x_{k-1})|,$$

and hence $\sum (P) \leq \sum (P_0)$. Now the points of P_0 in $[a, c]$ determine a partition P_1 of $[a, c]$ and those in $[c, b]$ determine a partition P_2 of $[c, b]$. The corresponding sums for all these partitions are connected by the relation

$$\sum (P) \leq \sum (P_0) = \sum (P_1) + \sum (P_2) \leq V_f(a, c) + V_f(c, b).$$

Therefore, $V_f(a, c) + V_f(c, b)$ is an upper bound for every sum $\sum (P)$. Since this cannot be smaller than the least upper bound, we must have

$$V_f(a, b) \leq V_f(a, c) + V_f(c, b),$$

and this completes the proof.

6.6 TOTAL VARIATION ON [a, x] AS A FUNCTION OF x

Now we keep the function f and the left endpoint of the interval fixed and study the total variation as a function of the right endpoint. The additive property implies important consequences for this function.

Theorem 6.12. *Let f be of bounded variation on $[a, b]$. Let V be defined on $[a, b]$ as follows: $V(x) = V_f(a, x)$ if $a < x \leq b$, $V(a) = 0$. Then:*

i) *V is an increasing function on $[a, b]$.*

ii) *$V-f$ is an increasing function on $[a, b]$.*

Proof. If $a < x < y \leq b$, we can write $V_f(a, y) = V_f(a, x) + V_f(x, y)$. This implies $V(y) - V(x) = V_f(x, y) \geq 0$. Hence $V(x) \leq V(y)$, and (i) holds.

To prove (ii), let $D(x) = V(x) - f(x)$ if $x \in [a, b]$. Then, if $a \leq x < y \leq b$, we have

$$D(y) - D(x) = V(y) - V(x) - [f(y) - f(x)] = V_f(x, y) - [f(y) - f(x)].$$

But from the definition of $V_f(x, y)$ it follows that we have

$$f(y) - f(x) \leq V_f(x, y).$$

This means that $D(y) - D(x) \geq 0$, and (ii) holds.

NOTE. For some functions f, the total variation $V_f(a, x)$ can be expressed as an integral. (See Exercise 7.20.)

6.7 FUNCTIONS OF BOUNDED VARIATION EXPRESSED AS THE DIFFERENCE OF INCREASING FUNCTIONS

The following simple and elegant characterization of functions of bounded variation is a consequence of Theorem 6.12.

Theorem 6.13. *Let f be defined on* $[a, b]$. *Then f is of bounded variation on* $[a, b]$ *if, and only if, f can be expressed as the difference of two increasing functions.*

Proof. If f is of bounded variation on $[a, b]$, we can write $f = V - D$, where V is the function of Theorem 6.12 and $D = V - f$. Both V and D are increasing functions on $[a, b]$.

The converse follows at once from Theorems 6.5 and 6.9.

The representation of a function of bounded variation as a difference of two increasing functions is by no means unique. If $f = f_1 - f_2$, where f_1 and f_2 are increasing, we also have $f = (f_1 + g) - (f_2 + g)$, where g is an arbitrary increasing function, and we get a new representation of f. If g is *strictly* increasing, the same will be true of $f_1 + g$ and $f_2 + g$. Therefore, Theorem 6.13 also holds if "increasing" is replaced by "strictly increasing."

6.8 CONTINUOUS FUNCTIONS OF BOUNDED VARIATION

Theorem 6.14. *Let f be of bounded variation on* $[a, b]$. *If* $x \in (a, b]$, *let* $V(x) = V_f(a, x)$ *and put* $V(a) = 0$. *Then every point of continuity of f is also a point of continuity of V. The converse is also true.*

Proof. Since V is monotonic, the right- and lefthand limits $V(x+)$ and $V(x-)$ exist for each point x in (a, b). Because of Theorem 6.13, the same is true of $f(x+)$ and $f(x-)$.

If $a < x < y \leq b$, then we have [by definition of $V_f(x, y)$]

$$0 \leq |f(y) - f(x)| \leq V(y) - V(x).$$

Letting $y \to x$, we find

$$0 \leq |f(x+) - f(x)| \leq V(x+) - V(x).$$

Similarly, $0 \leq |f(x) - f(x-)| \leq V(x) - V(x-)$. These inequalities imply that a point of continuity of V is also a point of continuity of f.

To prove the converse, let f be continuous at the point c in (a, b). Then, given $\varepsilon > 0$, there exists a $\delta > 0$ such that $0 < |x - c| < \delta$ implies $|f(x) - f(c)| < \varepsilon/2$. For this same ε, there also exists a partition P of $[c, b]$, say

$$P = \{x_0, x_1, \ldots, x_n\}, \qquad x_0 = c, \quad x_n = b,$$

such that

$$V_f(c, b) - \frac{\varepsilon}{2} < \sum_{k=1}^{n} |\Delta f_k|.$$

Adding more points to P can only increase the sum $\sum |\Delta f_k|$ and hence we can assume that $0 < x_1 - x_0 < \delta$. This means that

$$|\Delta f_1| = |f(x_1) - f(c)| < \frac{\varepsilon}{2},$$

and the foregoing inequality now becomes

$$V_f(c, b) - \frac{\varepsilon}{2} < \frac{\varepsilon}{2} + \sum_{k=2}^{n} |\Delta f_k| \le \frac{\varepsilon}{2} + V_f(x_1, b),$$

since $\{x_1, x_2, \ldots, x_n\}$ is a partition of $[x_1, b]$. We therefore have

$$V_f(c, b) - V_f(x_1, b) < \varepsilon.$$

But

$$0 \le V(x_1) - V(c) = V_f(a, x_1) - V_f(a, c)$$
$$= V_f(c, x_1) = V_f(c, b) - V_f(x_1, b) < \varepsilon.$$

Hence we have shown that

$$0 < x_1 - c < \delta \qquad \text{implies} \qquad 0 \le V(x_1) - V(c) < \varepsilon.$$

This proves that $V(c+) = V(c)$. A similar argument yields $V(c-) = V(c)$. The theorem is therefore proved for all interior points of $[a, b]$. (Trivial modifications are needed for the endpoints.)

Combining Theorem 6.14 with 6.13, we can state

Theorem 6.15. *Let f be continuous on $[a, b]$. Then f is of bounded variation on $[a, b]$ if, and only if, f can be expressed as the difference of two increasing continuous functions.*

NOTE. The theorem also holds if "increasing" is replaced by "strictly increasing."

Of course, discontinuities (if any) of a function of bounded variation must be jump discontinuities because of Theorem 6.13. Moreover, Theorem 6.2 tells us that they form a countable set.

6.9 CURVES AND PATHS

Let $\mathbf{f} : [a, b] \to \mathbf{R}^n$ be a vector-valued function, continuous on a compact interval $[a, b]$ in $\mathbf{R}$. As t runs through $[a, b]$, the function values $\mathbf{f}(t)$ trace out a set of points in $\mathbf{R}^n$ called the *graph* of $\mathbf{f}$ or the *curve* described by $\mathbf{f}$. A curve is a compact and connected subset of $\mathbf{R}^n$ since it is the continuous image of a compact interval. The function $\mathbf{f}$ itself is called a *path*.

It is often helpful to imagine a curve as being traced out by a moving particle. The interval $[a, b]$ is thought of as a time interval and the vector $\mathbf{f}(t)$ specifies the position of the particle at time t. In this interpretation, the function $\mathbf{f}$ itself is called a *motion*.

Different paths can trace out the same curve. For example, the two complex-valued functions

$$f(t) = e^{2\pi it}, \qquad g(t) = e^{-2\pi it}, \qquad 0 \le t \le 1,$$

each trace out the unit circle $x^2 + y^2 = 1$, but the points are visited in opposite directions. The same circle is traced out five times by the function $h(t) = e^{10\pi it}$, $0 \le t \le 1$.

6.10 RECTIFIABLE PATHS AND ARC LENGTH

Next we introduce the concept of arc length of a curve. The idea is to approximate the curve by inscribed polygons, a technique learned from ancient geometers. Our intuition tells us that the length of any inscribed polygon should not exceed that of the curve (since a straight line is the shortest path between two points), so the length of a curve should be an upper bound to the lengths of all inscribed polygons. Therefore, it seems natural to define the length of a curve to be the least upper bound of the lengths of all possible inscribed polygons.

For most curves that arise in practice, this gives a useful definition of arc length. However, as we will see presently, there are curves for which there is *no* upper bound to the lengths of the inscribed polygons. Therefore, it becomes necessary to classify curves into two categories: those which have a length, and those which do not. The former are called *rectifiable*, the latter *nonrectifiable*.

We now turn to a formal description of these ideas.

Let $\mathbf{f} : [a, b] \to \mathbf{R}^n$ be a path in $\mathbf{R}^n$. For any partition of $[a, b]$ given by

$$P = \{t_0, t_1, \ldots, t_m\},$$

the points $\mathbf{f}(t_0), \mathbf{f}(t_1), \ldots, \mathbf{f}(t_m)$ are the vertices of an inscribed polygon. (An example is shown in Fig. 6.1.) The length of this polygon is denoted by $\Lambda_{\mathbf{f}}(P)$ and is defined to be the sum

$$\Lambda_{\mathbf{f}}(P) = \sum_{k=1}^{m} \|\mathbf{f}(t_k) - \mathbf{f}(t_{k-1})\|.$$

Definition 6.16. *If the set of numbers $\Lambda_{\mathbf{f}}(P)$ is bounded for all partitions P of $[a, b]$, then the path $\mathbf{f}$ is said to be rectifiable and its arc length, denoted by $\Lambda_{\mathbf{f}}(a, b)$, is*

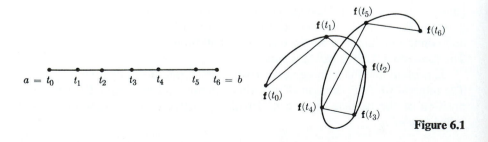

Figure 6.1

defined by the equation

$$\Lambda_f(a, b) = \sup \{\Lambda_f(P) : P \in \mathscr{P}[a, b]\}.$$

*If the set of numbers $\Lambda_f(P)$ is unbounded, **f** is called nonrectifiable.*

It is an easy matter to characterize all rectifiable curves.

Theorem 6.17. *Consider a path* $\mathbf{f} : [a, b] \to \mathbf{R}^n$ *with components* $\mathbf{f} = (f_1, \ldots, f_n)$. *Then* **f** *is rectifiable if, and only if, each component* f_k *is of bounded variation on* $[a, b]$. *If* **f** *is rectifiable, we have the inequalities*

$$V_k(a, b) \le \Lambda_f(a, b) \le V_1(a, b) + \cdots + V_n(a, b), \qquad (k = 1, 2, \ldots, n), \qquad (2)$$

where $V_k(a, b)$ *denotes the total variation of* f_k *on* $[a, b]$.

Proof. If $P = \{t_0, t_1, \ldots, t_m\}$ is a partition of $[a, b]$ we have

$$\sum_{i=1}^{m} |f_k(t_i) - f_k(t_{i-1})| \le \Lambda_f(P) \le \sum_{i=1}^{m} \sum_{j=1}^{n} |f_j(t_i) - f_j(t_{i-1})|, \qquad (3)$$

for each k. All assertions of the theorem follow easily from (3).

Examples

1. As noted earlier, the function given by $f(x) = x \cos \{\pi/(2x)\}$ for $x \ne 0$, $f(0) = 0$, is continuous but not of bounded variation on $[0, 1]$. Therefore its graph is a non-rectifiable curve.
2. It can be shown (Exercise 7.21) that if $\mathbf{f}'$ is continuous on $[a, b]$, then **f** is rectifiable and its arc length can be expressed as an integral,

$$\Lambda_f(a, b) = \int_a^b \|\mathbf{f}'(t)\| \, dt.$$

6.11 ADDITIVE AND CONTINUITY PROPERTIES OF ARC LENGTH

Let $\mathbf{f} = (f_1, \ldots, f_n)$ be a rectifiable path defined on $[a, b]$. Then each component f_k is of bounded variation on every subinterval $[x, y]$ of $[a, b]$. In this section we keep **f** fixed and study the arc length $\Lambda_f(x, y)$ as a function of the interval $[x, y]$. First we prove an additive property.

Theorem 6.18. *If* $c \in (a, b)$ *we have*

$$\Lambda_f(a, b) = \Lambda_f(a, c) + \Lambda_f(c, b).$$

Proof. Adjoining the point c to a partition P of $[a, b]$, we get a partition P_1 of $[a, c]$ and a partition P_2 of $[c, b]$ such that

$$\Lambda_f(P) \le \Lambda_f(P_1) + \Lambda_f(P_2) \le \Lambda_f(a, c) + \Lambda_f(c, b).$$

This implies $\Lambda_f(a, b) \le \Lambda_f(a, c) + \Lambda_f(c, b)$. To obtain the reverse inequality, let P_1 and P_2 be arbitrary partitions of $[a, c]$ and $[c, b]$, respectively. Then

$$P = P_1 \cup P_2,$$

is a partition of $[a, b]$ for which we have

$$\Lambda_f(P_1) + \Lambda_f(P_2) = \Lambda_f(P) \leq \Lambda_f(a, b).$$

Since the supremum of all sums $\Lambda_f(P_1) + \Lambda_f(P_2)$ is the sum $\Lambda_f(a, c) + \Lambda_f(c, b)$ (see Theorem 1.15), the theorem follows.

Theorem 6.19. *Consider a rectifiable path* $\mathbf{f}$ *defined on* $[a, b]$. *If* $x \in (a, b]$, *let* $s(x) = \Lambda_f(a, x)$ *and let* $s(a) = 0$. *Then we have:*

  i) *The function s so defined is increasing and continuous on* $[a, b]$.
 ii) *If there is no subinterval of* $[a, b]$ *on which* $\mathbf{f}$ *is constant, then s is strictly increasing on* $[a, b]$.

Proof. If $a \leq x < y \leq b$, Theorem 6.18 implies $s(y) - s(x) = \Lambda_f(x, y) \geq 0$. This proves that s is increasing on $[a, b]$. Furthermore, we have $s(y) - s(x) > 0$ unless $\Lambda_f(x, y) = 0$. But, by inequality (2), $\Lambda_f(x, y) = 0$ implies $V_k(x, y) = 0$ for each k and this, in turn, implies that $\mathbf{f}$ is constant on $[x, y]$. Hence (ii) holds.

To prove that s is continuous, we use inequality (2) again to write

$$0 \leq s(y) - s(x) = \Lambda_f(x, y) \leq \sum_{k=1}^{n} V_k(x, y).$$

If we let $y \to x$, we find each term $V_k(x, y) \to 0$ and hence $s(x) = s(x+)$. Similarly, $s(x) = s(x-)$ and the proof is complete.

6.12 EQUIVALENCE OF PATHS. CHANGE OF PARAMETER

This section describes a class of paths having the same graph. Let $\mathbf{f} : [a, b] \to \mathbf{R}^n$ be a path in $\mathbf{R}^n$. Let $u : [c, d] \to [a, b]$ be a real-valued function, continuous and strictly monotonic on $[c, d]$ with range $[a, b]$. Then the composite function $\mathbf{g} = \mathbf{f} \circ u$ given by

$$\mathbf{g}(t) = \mathbf{f}[u(t)] \qquad \text{for } c \leq t \leq d,$$

is a path having the same graph as $\mathbf{f}$. Two paths $\mathbf{f}$ and $\mathbf{g}$ so related are called *equivalent*. They are said to provide different parametric representations of the same curve. The function u is said to define a *change of parameter*.

Let C denote the common graph of two equivalent paths $\mathbf{f}$ and $\mathbf{g}$. If u is strictly increasing, we say that $\mathbf{f}$ and $\mathbf{g}$ trace out C in the *same direction*. If u is strictly decreasing, we say that $\mathbf{f}$ and $\mathbf{g}$ trace out C in *opposite directions*. In the first case, u is said to be *orientation-preserving*; in the second case, *orientation-reversing*.

Theorem 6.20. *Let* $\mathbf{f} : [a, b] \to \mathbf{R}^n$ *and* $\mathbf{g} : [c, d] \to \mathbf{R}^n$ *be two paths in* $\mathbf{R}^n$, *each of which is one-to-one on its domain. Then* $\mathbf{f}$ *and* $\mathbf{g}$ *are equivalent if, and only if, they have the same graph.*

Proof. Equivalent paths necessarily have the same graph. To prove the converse, assume that $\mathbf{f}$ and $\mathbf{g}$ have the same graph. Since $\mathbf{f}$ is one-to-one and continuous on the compact set $[a, b]$, Theorem 4.29 tells us that $\mathbf{f}^{-1}$ exists and is continuous on its graph. Define $u(t) = \mathbf{f}^{-1}[\mathbf{g}(t)]$ if $t \in [c, d]$. Then u is continuous on $[c, d]$ and $\mathbf{g}(t) = \mathbf{f}[u(t)]$. The reader can verify that u is strictly monotonic, and hence $\mathbf{f}$ and $\mathbf{g}$ are equivalent paths.

EXERCISES

Functions of bounded variation

6.1 Determine which of the following functions are of bounded variation on $[0, 1]$.

a) $f(x) = x^2 \sin(1/x)$ if $x \neq 0, f(0) = 0$.

b) $f(x) = \sqrt{x} \sin(1/x)$ if $x \neq 0, f(0) = 0$.

6.2 A function f, defined on $[a, b]$, is said to satisfy a uniform Lipschitz condition of order $\alpha > 0$ on $[a, b]$ if there exists a constant $M > 0$ such that $|f(x) - f(y)| < M|x - y|^\alpha$ for all x and y in $[a, b]$. (Compare with Exercise 5.1.)

a) If f is such a function, show that $\alpha > 1$ implies f is constant on $[a, b]$, whereas $\alpha = 1$ implies f is of bounded variation $[a, b]$.

b) Give an example of a function f satisfying a uniform Lipschitz condition of order $\alpha < 1$ on $[a, b]$ such that f is not of bounded variation on $[a, b]$.

c) Give an example of a function f which is of bounded variation on $[a, b]$ but which satisfies no uniform Lipschitz condition on $[a, b]$.

6.3 Show that a polynomial f is of bounded variation on every compact interval $[a, b]$. Describe a method for finding the total variation of f on $[a, b]$ if the zeros of the derivative f' are known.

6.4 A nonempty set S of real-valued functions defined on an interval $[a, b]$ is called a *linear space of functions* if it has the following two properties:

a) If $f \in S$, then $cf \in S$ for every real number c.

b) If $f \in S$ and $g \in S$, then $f + g \in S$.

Theorem 6.9 shows that the set V of all functions of bounded variation on $[a, b]$ is a linear space. If S is any linear space which contains all monotonic functions on $[a, b]$, prove that $V \subseteq S$. This can be described by saying that the functions of bounded variation form the smallest linear space containing all monotonic functions.

6.5 Let f be a real-valued function defined on $[0, 1]$ such that $f(0) > 0, f(x) \neq x$ for all x, and $f(x) \leq f(y)$ whenever $x \leq y$. Let $A = \{x : f(x) > x\}$. Prove that $\sup A \in A$ and that $f(1) > 1$.

6.6 If f is defined everywhere in $\mathbf{R}^1$, then f is said to be of bounded variation on $(-\infty, +\infty)$ if f is of bounded variation on every finite interval and if there exists a positive number M such that $V_f(a, b) < M$ for all compact intervals $[a, b]$. The total variation of f on $(-\infty, +\infty)$ is then defined to be the sup of all numbers $V_f(a, b), -\infty < a < b < +\infty$, and is denoted by $V_f(-\infty, +\infty)$. Similar definitions apply to half-open infinite intervals $[a, +\infty)$ and $(-\infty, b]$.

a) State and prove theorems for the infinite interval $(-\infty, +\infty)$ analogous to Theorems 6.7, 6.9, 6.10, 6.11, and 6.12.

b) Show that Theorem 6.5 is true for $(-\infty, +\infty)$ if "monotonic" is replaced by "bounded and monotonic." State and prove a similar modification of Theorem 6.13.

6.7 Assume that f is of bounded variation on $[a, b]$ and let

$$P = \{x_0, x_1, \ldots, x_n\} \in \mathscr{P}[a, b].$$

As usual, write $\Delta f_k = f(x_k) - f(x_{k-1})$, $k = 1, 2, \ldots, n$. Define

$$A(P) = \{k : \Delta f_k > 0\}, \qquad B(P) = \{k : \Delta f_k < 0\}.$$

The numbers

$$p_f(a, b) = \sup\left\{\sum_{k \in A(P)} \Delta f_k : P \in \mathscr{P}[a, b]\right\}$$

and

$$n_f(a, b) = \sup\left\{\sum_{k \in B(P)} |\Delta f_k| : P \in \mathscr{P}[a, b]\right\}$$

are called, respectively, the positive and negative variations of f on $[a, b]$. For each x in $(a, b]$, let $V(x) = V_f(a, x)$, $p(x) = p_f(a, x)$, $n(x) = n_f(a, x)$, and let $V(a) = p(a) = n(a) = 0$. Show that we have:

a) $V(x) = p(x) + n(x)$.

b) $0 \le p(x) \le V(x)$ and $0 \le n(x) \le V(x)$.

c) p and n are increasing on $[a, b]$.

d) $f(x) = f(a) + p(x) - n(x)$. Part (d) gives an alternative proof of Theorem 6.13.

e) $2p(x) = V(x) + f(x) - f(a)$, $\qquad 2n(x) = V(x) - f(x) + f(a)$.

f) Every point of continuity of f is also a point of continuity of p and of n.

Curves

6.8 Let f and g be complex-valued functions defined as follows:

$$f(t) = e^{2\pi it} \quad \text{if } t \in [0, 1], \qquad g(t) = e^{2\pi it} \quad \text{if } t \in [0, 2].$$

a) Prove that f and g have the same graph but are not equivalent according to the definition in Section 6.12.

b) Prove that the length of g is twice that of f.

6.9 Let $\mathbf{f}$ be a rectifiable path of length L defined on $[a, b]$, and assume that $\mathbf{f}$ is not constant on any subinterval of $[a, b]$. Let s denote the arc-length function given by $s(x) = \Lambda_{\mathbf{f}}(a, x)$ if $a < x \le b$, $s(a) = 0$.

a) Prove that s^{-1} exists and is continuous on $[0, L]$.

b) Define $\mathbf{g}(t) = \mathbf{f}[s^{-1}(t)]$ if $t \in [0, L]$ and show that $\mathbf{g}$ is equivalent to $\mathbf{f}$. Since $\mathbf{f}(t) = \mathbf{g}[s(t)]$, the function $\mathbf{g}$ is said to provide a representation of the graph of $\mathbf{f}$ with arc length as parameter.

6.10 Let f and g be two real-valued continuous functions of bounded variation defined on $[a, b]$, with $0 < f(x) < g(x)$ for each x in (a, b), $f(a) = g(a)$, $f(b) = g(b)$. Let h be the complex-valued function defined on the interval $[a, 2b - a]$ as follows:

$$h(t) = t + if(t), \qquad\qquad \text{if } a \le t \le b,$$
$$h(t) = 2b - t + ig(2b - t), \qquad \text{if } b \le t \le 2b - a.$$

a) Show that h describes a rectifiable curve Γ.

b) Explain, by means of a sketch, the geometric relationship between f, g, and h.

c) Show that the set of points

$$S = \{(x, y) : a \leq x \leq b, \quad f(x) \leq y \leq g(x)\}$$

is a region in $\mathbf{R}^2$ whose boundary is the curve Γ.

d) Let H be the complex-valued function defined on $[a, 2b - a]$ as follows:

$$H(t) = t - \tfrac{1}{2}i\,[g(t) - f(t)], \qquad\qquad \text{if } a \leq t \leq b,$$
$$H(t) = t + \tfrac{1}{2}i\,[g(2b - t) - f(2b - t)], \qquad \text{if } b \leq t \leq 2b - a.$$

Show that H describes a rectifiable curve Γ_0 which is the boundary of the region

$$S_0 = \{(x, y) : a \leq x \leq b, \quad f(x) - g(x) \leq 2y \leq g(x) - f(x)\}.$$

e) Show that S_0 has the x-axis as a line of symmetry. (The region S_0 is called the *symmetrization* of S with respect to the x-axis.)

f) Show that the length of Γ_0 does not exceed the length of Γ.

Absolutely continuous functions

A real-valued function f defined on $[a, b]$ is said to be *absolutely continuous* on $[a, b]$ if for every $\varepsilon > 0$ there is a $\delta > 0$ such that

$$\sum_{k=1}^{n} |f(b_k) - f(a_k)| < \varepsilon$$

for every n disjoint open subintervals (a_k, b_k) of $[a, b]$, $n = 1, 2, \ldots$, the sum of whose lengths $\sum_{k=1}^{n} (b_k - a_k)$ is less than δ.

Absolutely continuous functions occur in the Lebesgue theory of integration and differentiation. The following exercises give some of their elementary properties.

6.11 Prove that every absolutely continuous function on $[a, b]$ is continuous and of bounded variation on $[a, b]$.

NOTE. There exist functions which are continuous and of bounded variation but not absolutely continuous.

6.12 Prove that f is absolutely continuous if it satisfies a uniform Lipschitz condition of order 1 on $[a, b]$. (See Exercise 6.2.)

6.13 If f and g are absolutely continuous on $[a, b]$, prove that each of the following is also: $|f|$, cf (c constant), $f + g$, $f \cdot g$; also f/g if g is bounded away from zero.

SUGGESTED REFERENCES FOR FURTHER STUDY

6.1 Apostol, T. M., *Calculus*, Vol. 1, 2nd ed. Xerox, Waltham, 1967.

6.2 Natanson, I. P., *Theory of Functions of a Real Variable*, Vol. 1, rev. ed. Leo F. Boron, translator. Ungar, New York, 1961.

CHAPTER 7

THE RIEMANN–STIELTJES INTEGRAL

7.1 INTRODUCTION

Calculus deals principally with two geometric problems: finding the tangent line to a curve, and finding the area of a region under a curve. The first is studied by a limit process known as *differentiation*; the second by another limit process—*integration*—to which we turn now.

The reader will recall from elementary calculus that to find the area of the region under the graph of a positive function f defined on $[a, b]$, we subdivide the interval $[a, b]$ into a finite number of subintervals, say n, the kth subinterval having length Δx_k, and we consider sums of the form $\sum_{k=1}^{n} f(t_k) \, \Delta x_k$, where t_k is some point in the kth subinterval. Such a sum is an approximation to the area by means of rectangles. If f is sufficiently well behaved in $[a, b]$—continuous, for example—then there is some hope that these sums will tend to a limit as we let $n \to \infty$, making the successive subdivisions finer and finer. This, roughly speaking, is what is involved in Riemann's definition of the definite integral $\int_a^b f(x) \, dx$. (A precise definition is given below.)

The two concepts, derivative and integral, arise in entirely different ways and it is a remarkable fact indeed that the two are intimately connected. If we consider the definite integral of a continuous function f as a function of its upper limit, say we write

$$F(x) = \int_a^x f(t) \, dt,$$

then F has a derivative and $F'(x) = f(x)$. This important result shows that differentiation and integration are, in a sense, inverse operations.

In this chapter we study the process of integration in some detail. Actually we consider a more general concept than that of Riemann: this is the *Riemann–Stieltjes integral*, which involves two functions f and α. The symbol for such an integral is $\int_a^b f(x) \, d\alpha(x)$, or something similar, and the usual Riemann integral occurs as the special case in which $\alpha(x) = x$. When α has a continuous derivative, the definition is such that the Stieltjes integral $\int_a^b f(x) \, d\alpha(x)$ becomes the Riemann integral $\int_a^b f(x) \, \alpha'(x) \, dx$. However, the Stieltjes integral still makes sense when α is not differentiable or even when α is discontinuous. In fact, it is in dealing with *discontinuous* α that the importance of the Stieltjes integral becomes apparent. By a suitable choice of a discontinuous α, any finite or infinite sum can be expressed as a Stieltjes integral, and summation and ordinary Riemann integration then become special cases of this more general process. Problems in physics which involve mass distributions that are partly discrete and partly continuous can also be treated by using Stieltjes integrals. In the mathematical theory of probability this integral is a very useful tool that makes possible the simultaneous treatment of continuous and discrete random variables.

In Chapter 10 we discuss another generalization of the Riemann integral known as the *Lebesgue integral*.

7.2 NOTATION

For brevity we make certain stipulations concerning notation and terminology to be used in this chapter. We shall be working with a compact interval $[a, b]$ and, unless otherwise stated, all functions denoted by f, g, α, β, etc., will be assumed to be real-valued functions defined and *bounded* on $[a, b]$. Complex-valued functions are dealt with in Section 7.27, and extensions to unbounded functions and infinite intervals will be discussed in Chapter 10.

As in Chapter 6, a partition P of $[a, b]$ is a finite set of points, say

$$P = \{x_0, x_1, \ldots, x_n\},$$

such that $a = x_0 < x_1 < \cdots < x_{n-1} < x_n = b$. A partition P' of $[a, b]$ is said to be *finer* than P (or a *refinement* of P) if $P \subseteq P'$, which we also write $P' \supseteq P$. The symbol $\Delta \alpha_k$ denotes the difference $\Delta \alpha_k = \alpha(x_k) - \alpha(x_{k-1})$, so that

$$\sum_{k=1}^{n} \Delta \alpha_k = \alpha(b) - \alpha(a).$$

The set of all possible partitions of $[a, b]$ is denoted by $\mathscr{P}[a, b]$.

The norm of a partition P is the length of the largest subinterval of P and is denoted by $\|P\|$. Note that

$$P' \supseteq P \qquad \text{implies} \qquad \|P'\| \leq \|P\|.$$

That is, refinement of a partition decreases its norm, but the converse does not necessarily hold.

7.3 THE DEFINITION OF THE RIEMANN–STIELTJES INTEGRAL

Definition 7.1. *Let* $P = \{x_0, x_1, \ldots, x_n\}$ *be a partition of* $[a, b]$ *and let* t_k *be a point in the subinterval* $[x_{k-1}, x_k]$. *A sum of the form*

$$S(P, f, \alpha) = \sum_{k=1}^{n} f(t_k)\, \Delta \alpha_k$$

is called a Riemann–Stieltjes sum of f *with respect to* α. *We say* f *is Riemann-integrable with respect to* α *on* $[a, b]$, *and we write "$f \in R(\alpha)$ on* $[a, b]$" *if there exists a number* A *having the following property: For every* $\varepsilon > 0$, *there exists a partition* P_ε *of* $[a, b]$ *such that for every partition* P *finer than* P_ε *and for every choice of the points* t_k *in* $[x_{k-1}, x_k]$, *we have* $|S(P, f, \alpha) - A| < \varepsilon$.

When such a number A exists, it is uniquely determined and is denoted by $\int_a^b f \, d\alpha$ or by $\int_a^b f(x) \, d\alpha(x)$. We also say that the Riemann–Stieltjes integral $\int_a^b f \, d\alpha$ exists. The functions f and α are referred to as the *integrand* and the *integrator*, respectively. In the special case when $\alpha(x) = x$, we write $S(P, f)$ instead of $S(P, f, \alpha)$, and $f \in R$ instead of $f \in R(\alpha)$. The integral is then called a Riemann integral and is denoted by $\int_a^b f \, dx$ or by $\int_a^b f(x) \, dx$. The numerical value of $\int_a^b f(x) \, d\alpha(x)$ depends only on f, α, a, and b, and does not depend on the symbol x. The letter x is a "dummy variable" and may be replaced by any other convenient symbol.

NOTE. This is one of several accepted definitions of the Riemann–Stieltjes integral. An alternative (but not equivalent) definition is stated in Exercise 7.3.

7.4 LINEAR PROPERTIES

It is an easy matter to prove that the integral operates in a linear fashion on both the integrand and the integrator. This is the context of the next two theorems.

Theorem 7.2. *If $f \in R(\alpha)$ and if $g \in R(\alpha)$ on $[a, b]$, then $c_1 f + c_2 g \in R(\alpha)$ on $[a, b]$ (for any two constants c_1 and c_2) and we have*

$$\int_a^b (c_1 f + c_2 g) \, d\alpha = c_1 \int_a^b f \, d\alpha + c_2 \int_a^b g \, d\alpha.$$

Proof. Let $h = c_1 f + c_2 g$. Given a partition P of $[a, b]$, we can write

$$S(P, h, \alpha) = \sum_{k=1}^n h(t_k) \, \Delta\alpha_k = c_1 \sum_{k=1}^n f(t_k) \, \Delta\alpha_k + c_2 \sum_{k=1}^n g(t_k) \, \Delta\alpha_k$$

$$= c_1 S(P, f, \alpha) + c_2 S(P, g, \alpha).$$

Given $\varepsilon > 0$, choose P'_ε so that $P \supseteq P'_\varepsilon$ implies $|S(P, f, \alpha) - \int_a^b f \, d\alpha| < \varepsilon$, and choose P''_ε so that $P \supseteq P''_\varepsilon$ implies $|S(P, g, \alpha) - \int_a^b g \, d\alpha| < \varepsilon$. If we take $P_\varepsilon = P'_\varepsilon \cup P''_\varepsilon$, then, for P finer than P_ε, we have

$$\left| S(P, h, \alpha) - c_1 \int_a^b f \, d\alpha - c_2 \int_a^b g \, d\alpha \right| \leq |c_1|\varepsilon + |c_2|\varepsilon,$$

and this proves the theorem.

Theorem 7.3. *If $f \in R(\alpha)$ and $f \in R(\beta)$ on $[a, b]$, then $f \in R(c_1\alpha + c_2\beta)$ on $[a, b]$ (for any two constants c_1 and c_2) and we have*

$$\int_a^b f \, d(c_1\alpha + c_2\beta) = c_1 \int_a^b f \, d\alpha + c_2 \int_a^b f \, d\beta.$$

The proof is similar to that of Theorem 7.2 and is left as an exercise.

A result somewhat analogous to the previous two theorems tells us that the integral is also additive with respect to the interval of integration.

Theorem 7.4. *Assume that $c \in (a, b)$. If two of the three integrals in* (1) *exist, then the third also exists and we have*

$$\int_a^c f \, d\alpha + \int_c^b f \, d\alpha = \int_a^b f \, d\alpha. \tag{1}$$

Proof. If P is a partition of $[a, b]$ such that $c \in P$, let

$$P' = P \cap [a, c] \qquad \text{and} \qquad P'' = P \cap [c, b],$$

denote the corresponding partitions of $[a, c]$ and $[c, b]$, respectively. The Riemann–Stieltjes sums for these partitions are connected by the equation

$$S(P, f, \alpha) = S(P', f, \alpha) + S(P'', f, \alpha).$$

Assume that $\int_a^c f \, d\alpha$ and $\int_c^b f \, d\alpha$ exist. Then, given $\varepsilon > 0$, there is a partition P'_ε of $[a, c]$ such that

$$\left| S(P', f, \alpha) - \int_a^c f \, d\alpha \right| < \frac{\varepsilon}{2} \qquad \text{whenever } P' \text{ is finer than } P'_\varepsilon,$$

and a partition P''_ε of $[c, b]$ such that

$$\left| S(P'', f, \alpha) - \int_c^b f \, d\alpha \right| < \frac{\varepsilon}{2} \qquad \text{whenever } P'' \text{ is finer than } P''_\varepsilon.$$

Then $P_\varepsilon = P'_\varepsilon \cup P''_\varepsilon$ is a partition of $[a, b]$ such that P finer than P_ε implies $P' \supseteq P'_\varepsilon$ and $P'' \supseteq P''_\varepsilon$. Hence, if P is finer than P_ε, we can combine the foregoing results to obtain the inequality

$$\left| S(P, f, \alpha) - \int_a^c f \, d\alpha - \int_c^b f \, d\alpha \right| < \varepsilon.$$

This proves that $\int_a^b f \, d\alpha$ exists and equals $\int_a^c f \, d\alpha + \int_c^b f \, d\alpha$. The reader can easily verify that a similar argument proves the theorem in the remaining cases.

Using mathematical induction, we can prove a similar result for a decomposition of $[a, b]$ into a finite number of subintervals.

NOTE. The preceding type of argument cannot be used to prove that the integral $\int_a^c f \, d\alpha$ exists whenever $\int_a^b f \, d\alpha$ exists. The conclusion is correct, however. For integrators α of bounded variation, this fact will later be proved in Theorem 7.25.

Definition 7.5. *If $a < b$, we define $\int_b^a f \, d\alpha = -\int_a^b f \, d\alpha$ whenever $\int_a^b f \, d\alpha$ exists. We also define $\int_a^a f \, d\alpha = 0$.*

The equation in Theorem 7.4 can now be written as follows:

$$\int_a^b f \, d\alpha + \int_b^c f \, d\alpha + \int_c^a f \, d\alpha = 0.$$

7.5 INTEGRATION BY PARTS

A remarkable connection exists between the integrand and the integrator in a Riemann–Stieltjes integral. The existence of $\int_a^b f\,d\alpha$ implies the existence of $\int_a^b \alpha\,df$, and the converse is also true. Moreover, a very simple relation holds between the two integrals.

Theorem 7.6. *If $f \in R(\alpha)$ on $[a, b]$, then $\alpha \in R(f)$ on $[a, b]$ and we have*

$$\int_a^b f(x)\,d\alpha(x) + \int_a^b \alpha(x)\,df(x) = f(b)\alpha(b) - f(a)\alpha(a).$$

NOTE. This equation, which provides a kind of reciprocity law for the integral, is known as the *formula for integration by parts.*

Proof. Let $\varepsilon > 0$ be given. Since $\int_a^b f\,d\alpha$ exists, there is a partition P_ε of $[a, b]$ such that for every P' finer than P_ε, we have

$$\left| S(P', f, \alpha) - \int_a^b f\,d\alpha \right| < \varepsilon. \tag{2}$$

Consider an arbitrary Riemann–Stieltjes sum for the integral $\int_a^b \alpha\,df$, say

$$S(P, \alpha, f) = \sum_{k=1}^n \alpha(t_k)\,\Delta f_k = \sum_{k=1}^n \alpha(t_k)f(x_k) - \sum_{k=1}^n \alpha(t_k)f(x_{k-1}),$$

where P is finer than P_ε. Writing $A = f(b)\alpha(b) - f(a)\alpha(a)$, we have the identity

$$A = \sum_{k=1}^n f(x_k)\alpha(x_k) - \sum_{k=1}^n f(x_{k-1})\alpha(x_{k-1}).$$

Subtracting the last two displayed equations, we find

$$A - S(P, \alpha, f) = \sum_{k=1}^n f(x_k)[\alpha(x_k) - \alpha(t_k)] + \sum_{k=1}^n f(x_{k-1})[\alpha(t_k) - \alpha(x_{k-1})].$$

The two sums on the right can be combined into a single sum of the form $S(P', f, \alpha)$, where P' is that partition of $[a, b]$ obtained by taking the points x_k and t_k together. Then P' is finer than P and hence finer than P_ε. Therefore the inequality (2) is valid and this means that we have

$$\left| A - S(P, \alpha, f) - \int_a^b f\,d\alpha \right| < \varepsilon,$$

whenever P is finer than P_ε. But this is exactly the statement that $\int_a^b \alpha\,df$ exists and equals $A - \int_a^b f\,d\alpha$.

7.6 CHANGE OF VARIABLE IN A RIEMANN–STIELTJES INTEGRAL

Theorem 7.7. *Let $f \in R(\alpha)$ on $[a, b]$ and let g be a strictly monotonic continuous function defined on an interval S having endpoints c and d. Assume that $a = g(c)$,*

$b = g(d)$. *Let h and β be the composite functions defined as follows:*

$$h(x) = f[g(x)], \qquad \beta(x) = \alpha[g(x)], \qquad if \ x \in S.$$

Then $h \in R(\beta)$ on S and we have $\int_a^b f \, d\alpha = \int_c^d h \, d\beta$. That is,

$$\int_{g(c)}^{g(d)} f(t) \, d\alpha(t) = \int_c^d f[g(x)] \, d\{\alpha[g(x)]\}.$$

Proof. For definiteness, assume that g is strictly increasing on S. (This implies $c < d$.) Then g is one-to-one and has a strictly increasing, continuous inverse g^{-1} defined on $[a, b]$. Therefore, for every partition $P = \{y_0, \ldots, y_n\}$ of $[c, d]$, there corresponds one and only one partition $P' = \{x_0, \ldots, x_n\}$ of $[a, b]$ with $x_k = g(y_k)$. In fact, we can write

$$P' = g(P) \qquad \text{and} \qquad P = g^{-1}(P').$$

Furthermore, a refinement of P produces a corresponding refinement of P', and the converse also holds.

If $\varepsilon > 0$ is given, there is a partition P'_ε of $[a, b]$ such that P' finer than P'_ε implies $|S(P', f, \alpha) - \int_a^b f \, d\alpha| < \varepsilon$. Let $P_\varepsilon = g^{-1}(P'_\varepsilon)$ be the corresponding partition of $[c, d]$, and let $P = \{y_0, \ldots, y_n\}$ be a partition of $[c, d]$ finer than P_ε. Form a Riemann–Stieltjes sum

$$S(P, h, \beta) = \sum_{k=1}^n h(u_k) \, \Delta\beta_k,$$

where $u_k \in [y_{k-1}, y_k]$ and $\Delta\beta_k = \beta(y_k) - \beta(y_{k-1})$. If we put $t_k = g(u_k)$ and $x_k = g(y_k)$, then $P' = \{x_0, \ldots, x_n\}$ is a partition of $[a, b]$ finer than P'_ε. Moreover, we then have

$$S(P, h, \beta) = \sum_{k=1}^n f[g(u_k)]\{\alpha[g(y_k)] - \alpha[g(y_{k-1})]\}$$

$$= \sum_{k=1}^n f(t_k)\{\alpha(x_k) - \alpha(x_{k-1})\} = S(P', f, \alpha),$$

since $t_k \in [x_{k-1}, x_k]$. Therefore, $|S(P, h, \beta) - \int_a^b f \, d\alpha| < \varepsilon$ and the theorem is proved.

NOTE. This theorem applies, in particular, to Riemann integrals, that is, when $\alpha(x) = x$. Another theorem of this type, in which g is not required to be monotonic, will later be proved for Riemann integrals. (See Theorem 7.36.)

7.7 REDUCTION TO A RIEMANN INTEGRAL

The next theorem tells us that we are permitted to replace the symbol $d\alpha(x)$ by $\alpha'(x) \, dx$ in the integral $\int_a^b f(x) \, d\alpha(x)$ whenever α has a continuous derivative α'.

Theorem 7.8. *Assume $f \in R(\alpha)$ on $[a, b]$ and assume that α has a continuous derivative α' on $[a, b]$. Then the Riemann integral $\int_a^b f(x)\alpha'(x)\,dx$ exists and we have*

$$\int_a^b f(x)\,d\alpha(x) = \int_a^b f(x)\alpha'(x)\,dx.$$

Proof. Let $g(x) = f(x)\alpha'(x)$ and consider a Riemann sum

$$S(P, g) = \sum_{k=1}^n g(t_k)\,\Delta x_k = \sum_{k=1}^n f(t_k)\alpha'(t_k)\,\Delta x_k.$$

The same partition P and the same choice of the t_k can be used to form the Riemann–Stieltjes sum

$$S(P, f, \alpha) = \sum_{k=1}^n f(t_k)\,\Delta \alpha_k.$$

Applying the Mean-Value Theorem, we can write

$$\Delta \alpha_k = \alpha'(v_k)\,\Delta x_k, \qquad \text{where } v_k \in (x_{k-1}, x_k),$$

and hence

$$S(P, f, \alpha) - S(P, g) = \sum_{k=1}^n f(t_k)[\alpha'(v_k) - \alpha'(t_k)]\,\Delta x_k.$$

Since f is bounded, we have $|f(x)| \leq M$ for all x in $[a, b]$, where $M > 0$. Continuity of α' on $[a, b]$ implies uniform continuity on $[a, b]$. Hence, if $\varepsilon > 0$ is given, there exists a $\delta > 0$ (depending only on ε) such that

$$0 \leq |x - y| < \delta \qquad \text{implies} \qquad |\alpha'(x) - \alpha'(y)| < \frac{\varepsilon}{2M(b-a)}.$$

If we take a partition P'_ε with norm $\|P'_\varepsilon\| < \delta$, then for any finer partition P we will have $|\alpha'(v_k) - \alpha'(t_k)| < \varepsilon/[2M(b-a)]$ in the preceding equation. For such P we therefore have

$$|S(P, f, \alpha) - S(P, g)| < \frac{\varepsilon}{2}.$$

On the other hand, since $f \in R(\alpha)$ on $[a, b]$, there exists a partition P''_ε such that P finer than P''_ε implies

$$\left| S(P, f, \alpha) - \int_a^b f\,d\alpha \right| < \frac{\varepsilon}{2}.$$

Combining the last two inequalities, we see that when P is finer than $P_\varepsilon = P'_\varepsilon \cup P''_\varepsilon$, we will have $|S(P, g) - \int_a^b f\,d\alpha| < \varepsilon$, and this proves the theorem.

NOTE. A stronger result not requiring continuity of α' is proved in Theorem 7.35.

7.8 STEP FUNCTIONS AS INTEGRATORS

If α is constant throughout $[a, b]$, the integral $\int_a^b f \, d\alpha$ exists and has the value 0, since each sum $S(P, f, \alpha) = 0$. However, if α is constant except for a jump discontinuity at one point, the integral $\int_a^b f \, d\alpha$ need not exist and, if it does exist, its value need not be zero. The situation is described more fully in the following theorem:

Theorem 7.9. *Given $a < c < b$. Define α on $[a, b]$ as follows: The values $\alpha(a)$, $\alpha(c)$, $\alpha(b)$ are arbitrary;*

$$\alpha(x) = \alpha(a) \qquad \text{if } a \leq x < c,$$

and

$$\alpha(x) = \alpha(b) \qquad \text{if } c < x \leq b.$$

Let f be defined on $[a, b]$ in such a way that at least one of the functions f or α is continuous from the left at c and at least one is continuous from the right at c. Then $f \in R(\alpha)$ on $[a, b]$ and we have

$$\int_a^b f \, d\alpha = f(c)[\alpha(c+) - \alpha(c-)].$$

NOTE. The result also holds if $c = a$, provided that we write $\alpha(c)$ for $\alpha(c-)$, and it holds for $c = b$ if we write $\alpha(c)$ for $\alpha(c+)$. We will prove later (Theorem 7.29) that the integral does not exist if both f and α are discontinuous from the right or from the left at c.

Proof. If $c \in P$, every term in the sum $S(P, f, \alpha)$ is zero except the two terms arising from the subinterval separated by c, say

$$S(P, f, \alpha) = f(t_{k-1})[\alpha(c) - \alpha(c-)] + f(t_k)[\alpha(c+) - \alpha(c)],$$

where $t_{k-1} \leq c \leq t_k$. This equation can also be written as follows:

$$\Delta = [f(t_{k-1}) - f(c)][\alpha(c) - \alpha(c-)] + [f(t_k) - f(c)][\alpha(c+) - \alpha(c)],$$

where $\Delta = S(P, f, \alpha) - f(c)[\alpha(c+) - \alpha(c-)]$. Hence we have

$$|\Delta| \leq |f(t_{k-1}) - f(c)| \, |\alpha(c) - \alpha(c-)| + |f(t_k) - f(c)| \, |\alpha(c+) - \alpha(c)|.$$

If f is continuous at c, for every $\varepsilon > 0$ there is a $\delta > 0$ such that $\|P\| < \delta$ implies

$$|f(t_{k-1}) - f(c)| < \varepsilon \qquad \text{and} \qquad |f(t_k) - f(c)| < \varepsilon.$$

In this case, we obtain the inequality

$$|\Delta| \leq \varepsilon |\alpha(c) - \alpha(c-)| + \varepsilon |\alpha(c+) - \alpha(c)|.$$

But this inequality holds whether or not f is continuous at c. For example, if f is discontinuous both from the right and from the left at c, then $\alpha(c) = \alpha(c-)$ and $\alpha(c) = \alpha(c+)$ and we get $\Delta = 0$. On the other hand, if f is continuous from the left and discontinuous from the right at c, we must have $\alpha(c) = \alpha(c+)$ and we get

$|\Delta| \le \varepsilon |\alpha(c) - \alpha(c-)|$. Similarly, if f is continuous from the right and discontinuous from the left at c, we have $\alpha(c) = \alpha(c-)$ and $|\Delta| \le \varepsilon |\alpha(c+) - \alpha(c)|$. Hence the last displayed inequality holds in every case. This proves the theorem.

Example. Theorem 7.9 tells us that the value of a Riemann–Stieltjes integral can be altered by changing the value of f at a single point. The following example shows that the *existence* of the integral can also be affected by such a change. Let

$$\alpha(x) = 0, \quad \text{if } x \ne 0, \quad \alpha(0) = -1,$$

$$f(x) = 1, \quad \text{if } -1 \le x \le +1.$$

In this case Theorem 7.9 implies $\int_{-1}^{1} f \, d\alpha = 0$. But if we re-define f so that $f(0) = 2$ and $f(x) = 1$ if $x \ne 0$, we can easily see that $\int_{-1}^{1} f \, d\alpha$ will not exist. In fact, when P is a partition which includes 0 as a point of subdivision, we find

$$S(P, f, \alpha) = f(t_k)[\alpha(x_k) - \alpha(0)] + f(t_{k-1})[\alpha(0) - \alpha(x_{k-2})]$$
$$= f(t_k) - f(t_{k-1}),$$

where $x_{k-2} \le t_{k-1} \le 0 \le t_k \le x_k$. The value of this sum is 0, 1, or -1, depending on the choice of t_k and t_{k-1}. Hence, $\int_{-1}^{1} f \, d\alpha$ does not exist in this case. However, in a Riemann integral $\int_a^b f(x) \, dx$, the values of f can be changed at a finite number of points without affecting either the existence or the value of the integral. To prove this, it suffices to consider the case where $f(x) = 0$ for all x in $[a, b]$ except for one point, say $x = c$. But for such a function it is obvious that $|S(P, f)| \le |f(c)| \|P\|$. Since $\|P\|$ can be made arbitrarily small, it follows that $\int_a^b f(x) \, dx = 0$.

7.9 REDUCTION OF A RIEMANN–STIELTJES INTEGRAL TO A FINITE SUM

The integrator α in Theorem 7.9 is a special case of an important class of functions known as *step functions*. These are functions which are constant throughout an interval except for a finite number of jump discontinuities.

Definition 7.10 (Step function). *A function α defined on $[a, b]$ is called a step function if there is a partition*

$$a = x_1 < x_2 < \cdots < x_n = b$$

such that α is constant on each open subinterval (x_{k-1}, x_k). The number $\alpha(x_k+) - \alpha(x_k-)$ is called the jump at x_k if $1 < k < n$. The jump at x_1 is $\alpha(x_1+) - \alpha(x_1)$, and the jump at x_n is $\alpha(x_n) - \alpha(x_n-)$.

Step functions provide the connecting link between Riemann–Stieltjes integrals and finite sums:

Theorem 7.11 (Reduction of a Riemann–Stieltjes integral to a finite sum). *Let α be a step function defined on $[a, b]$ with jump α_k at x_k, where $x_1, \ldots, x_n$ are as described in Definition 7.10. Let f be defined on $[a, b]$ in such a way that not both f and α are* *discontinuous from the right or from the left at each x_k. Then $\int_a^b f\,d\alpha$ exists and we have*

$$\int_a^b f(x)\,d\alpha(x) = \sum_{k=1}^{n} f(x_k)\alpha_k.$$

Proof. By Theorem 7.4, $\int_a^b f\,d\alpha$ can be written as a sum of integrals of the type considered in Theorem 7.9.

One of the simplest step functions is the *greatest-integer function*. Its value at x is the greatest integer which is less than or equal to x and is denoted by $[x]$. Thus, $[x]$ is the unique integer satisfying the inequalities $[x] \le x < [x] + 1$.

Theorem 7.12. *Every finite sum can be written as a Riemann–Stieltjes integral. In fact, given a sum $\sum_{k=1}^{n} a_k$, define f on $[0, n]$ as follows:*

$$f(x) = a_k \quad \text{if } k - 1 < x \le k \quad (k = 1, 2, \ldots, n), \quad f(0) = 0.$$

Then

$$\sum_{k=1}^{n} a_k = \sum_{k=1}^{n} f(k) = \int_0^n f(x)\,d[x],$$

where $[x]$ is the greatest integer $\le x$.

Proof. The greatest-integer function is a step function, continuous from the right and having jump 1 at each integer. The function f is continuous from the left at $1, 2, \ldots, n$. Now apply Theorem 7.11.

7.10 EULER'S SUMMATION FORMULA

We shall illustrate the use of Riemann–Stieltjes integrals by deriving a remarkable formula known as *Euler's summation formula*, which relates the integral of a function over an interval $[a, b]$ with the sum of the function values at the *integers* in $[a, b]$. It can sometimes be used to approximate integrals by sums or, conversely, to estimate the values of certain sums by means of integrals.

Theorem 7.13 (Euler's summation formula). *If f has a continuous derivative f' on $[a, b]$, then we have*

$$\sum_{a<n\le b} f(n) = \int_a^b f(x)\,dx + \int_a^b f'(x)((x))\,dx + f(a)((a)) - f(b)((b)),$$

where $((x)) = x - [x]$. When a and b are integers, this becomes

$$\sum_{n=a}^{b} f(n) = \int_a^b f(x)\,dx + \int_a^b f'(x)\left(x - [x] - \frac{1}{2}\right)dx + \frac{f(a) + f(b)}{2}.$$

NOTE. $\sum_{a<n\le b}$ means the sum from $n = [a] + 1$ to $n = [b]$.

Proof. Applying Theorem 7.6 (integration by parts), we have

$$\int_a^b f(x)\, d(x - [x]) + \int_a^b (x - [x])\, df(x) = f(b)(b - [b]) - f(a)(a - [a]).$$

Since the greatest-integer function has unit jumps at the integers $[a] + 1$, $[a] + 2, \ldots, [b]$, we can write

$$\int_a^b f(x)\, d[x] = \sum_{a < n \leq b} f(n).$$

If we combine this with the previous equation, the theorem follows at once.

7.11 MONOTONICALLY INCREASING INTEGRATORS. UPPER AND LOWER INTEGRALS

The further theory of Riemann–Stieltjes integration will now be developed for monotonically increasing integrators, and we shall see later (in Theorem 7.24) that for many purposes this is just as general as studying the theory for integrators which are of bounded variation.

When α is increasing, the differences $\Delta \alpha_k$ which appear in the Riemann–Stieltjes sums are all nonnegative. This simple fact plays a vital role in the development of the theory. For brevity, we shall use the abbreviation "$\alpha \nearrow$ on $[a, b]$" to mean that "α is increasing on $[a, b]$."

As stated earlier, to find the area of the region under the graph of a function f we consider Riemann sums $\sum f(t_k)\, \Delta x_k$ as approximations to the area by means of rectangles. Such sums also arise quite naturally in certain physical problems requiring the use of integration for their solution. Another approach to these problems is by means of *upper* and *lower* Riemann sums. For example, in the case of areas, we can consider approximations from "above" and from "below" by means of the sums $\sum M_k\, \Delta x_k$ and $\sum m_k\, \Delta x_k$, where M_k and m_k denote, respectively, the sup and inf of the function values in the kth subinterval. Our geometric intuition tells us that the upper sums are at least as big as the area we seek, whereas the lower sums cannot exceed this area. (See Fig. 7.1.) Therefore it seems natural

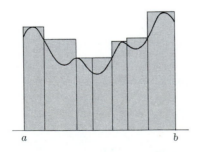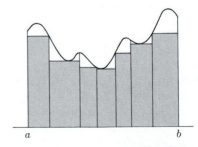

Figure 7.1

to ask: What is the smallest possible value of the upper sums? This leads us to consider the inf of all upper sums, a number called the *upper integral of f*. The *lower integral* is similarly defined to be the sup of all lower sums. For reasonable functions (for example, continuous functions) both these integrals will be equal to $\int_a^b f(x)\, dx$. However, in general, these integrals will be different and it becomes an important problem to find conditions on the function which will ensure that the upper and lower integrals will be the same. We now discuss this type of problem for Riemann–Stieltjes integrals.

Definition 7.14. *Let P be a partition of* $[a, b]$ *and let*

$$M_k(f) = \sup\,\{f(x) : x \in [x_{k-1}, x_k]\},$$

$$m_k(f) = \inf\,\{f(x) : x \in [x_{k-1}, x_k]\}.$$

The numbers

$$U(P, f, \alpha) = \sum_{k=1}^{n} M_k(f)\,\Delta\alpha_k \qquad and \qquad L(P, f, \alpha) = \sum_{k=1}^{n} m_k(f)\,\Delta\alpha_k,$$

are called, respectively, the upper and lower Stieltjes sums of f with respect to α *for the partition P.*

NOTE. We always have $m_k(f) \leq M_k(f)$. If $\alpha \nearrow$ on $[a, b]$, then $\Delta\alpha_k \geq 0$ and we can also write $m_k(f)\,\Delta\alpha_k \leq M_k(f)\,\Delta\alpha_k$, from which it follows that the lower sums do not exceed the upper sums. Furthermore, if $t_k \in [x_{k-1}, x_k]$, then

$$m_k(f) \leq f(t_k) \leq M_k(f).$$

Therefore, when $\alpha \nearrow$, we have the inequalities

$$L(P, f, \alpha) \leq S(P, f, \alpha) \leq U(P, f, \alpha)$$

relating the upper and lower sums to the Riemann–Stieltjes sums. These inequalities, which are frequently used in the material that follows, do not necessarily hold when α is not an increasing function.

The next theorem shows that, for increasing α, refinement of the partition increases the lower sums and decreases the upper sums.

Theorem 7.15. *Assume that* $\alpha \nearrow$ *on* $[a, b]$. *Then:*

i) *If P' is finer than P, we have*

$$U(P', f, \alpha) \leq U(P, f, \alpha) \qquad and \qquad L(P', f, \alpha) \geq L(P, f, \alpha).$$

ii) *For any two partitions* P_1 *and* P_2, *we have*

$$L(P_1, f, \alpha) \leq U(P_2, f, \alpha).$$

Proof. It suffices to prove (i) when P' contains exactly one more point than P, say the point c. If c is in the ith subinterval of P, we can write

$$U(P', f, \alpha) = \sum_{\substack{k=1 \\ k \neq i}}^{n} M_k(f)\,\Delta\alpha_k + M'[\alpha(c) - \alpha(x_{i-1})] + M''[\alpha(x_i) - \alpha(c)],$$

where M' and M'' denote the sup of f in $[x_{i-1}, c]$ and $[c, x_i]$. But, since

$$M' \leq M_i(f) \qquad \text{and} \qquad M'' \leq M_i(f),$$

we have $U(P', f, \alpha) \leq U(P, f, \alpha)$. (The inequality for lower sums is proved in a similar fashion.)

To prove (ii), let $P = P_1 \cup P_2$. Then we have

$$L(P_1, f, \alpha) \leq L(P, f, \alpha) \leq U(P, f, \alpha) \leq U(P_2, f, \alpha).$$

NOTE. It follows from this theorem that we also have (for increasing α)

$$m[\alpha(b) - \alpha(a)] \leq L(P_1, f, \alpha) \leq U(P_2, f, \alpha) \leq M[\alpha(b) - \alpha(a)],$$

where M and m denote the sup and inf of f on $[a, b]$.

Definition 7.16. *Assume that $\alpha \nearrow$ on $[a, b]$. The upper Stieltjes integral of f with respect to α is defined as follows:*

$$\overline{\int_a^b} f\, d\alpha = \inf\{U(P, f, \alpha) : P \in \mathscr{P}[a, b]\}.$$

The lower Stieltjes integral is similarly defined:

$$\underline{\int_a^b} f\, d\alpha = \sup\{L(P, f, \alpha) : P \in \mathscr{P}[a, b]\}.$$

NOTE. We sometimes write $\bar{I}(f, \alpha)$ and $\underline{I}(f, \alpha)$ for the upper and lower integrals. In the special case where $\alpha(x) = x$, the upper and lower sums are denoted by $U(P, f)$ and $L(P, f)$ and are called upper and lower Riemann sums. The corresponding integrals, denoted by $\overline{\int_a^b} f(x)\, dx$ and by $\underline{\int_a^b} f(x)\, dx$, are called upper and lower Riemann integrals. They were first introduced by J. G. Darboux (1875).

Theorem 7.17. *Assume that $\alpha \nearrow$ on $[a, b]$. Then $\underline{I}(f, \alpha) \leq \bar{I}(f, \alpha)$.*

Proof. If $\varepsilon > 0$ is given, there exists a partition P_1 such that

$$U(P_1, f, \alpha) < \bar{I}(f, \alpha) + \varepsilon.$$

By Theorem 7.15, it follows that $\bar{I}(f, \alpha) + \varepsilon$ is an upper bound to all lower sums $L(P, f, \alpha)$. Hence, $\underline{I}(f, \alpha) \leq \bar{I}(f, \alpha) + \varepsilon$, and, since ε is arbitrary, this implies $\underline{I}(f, \alpha) \leq \bar{I}(f, \alpha)$.

Example. It is easy to give an example in which $\underline{I}(f, \alpha) < \bar{I}(f, \alpha)$. Let $\alpha(x) = x$ and define f on $[0, 1]$ as follows:

$$f(x) = 1, \quad \text{if } x \text{ is rational}, \qquad f(x) = 0, \quad \text{if } x \text{ is irrational}.$$

Then for every partition P of $[0, 1]$, we have $M_k(f) = 1$ and $m_k(f) = 0$, since every subinterval contains both rational and irrational numbers. Therefore, $U(P, f) = 1$ and $L(P, f) = 0$ for all P. It follows that we have, for $[a, b] = [0, 1]$,

$$\overline{\int_a^b} f \, dx = 1 \qquad \text{and} \qquad \underline{\int_a^b} f \, dx = 0.$$

Observe that the same result holds if $f(x) = 0$ when x is rational, and $f(x) = 1$ when x is irrational.

7.12 ADDITIVE AND LINEARITY PROPERTIES OF UPPER AND LOWER INTEGRALS

Upper and lower integrals share many of the properties of the integral. For example, we have

$$\overline{\int_a^b} f \, d\alpha = \overline{\int_a^c} f \, d\alpha + \overline{\int_c^b} f \, d\alpha,$$

if $a < c < b$, and the same equation holds for lower integrals. However, certain equations which hold for integrals must be replaced by inequalities when they are stated for upper and lower integrals. For example, we have

$$\overline{\int_a^b} (f + g) \, d\alpha \leq \overline{\int_a^b} f \, d\alpha + \overline{\int_a^b} g \, d\alpha,$$

and

$$\underline{\int_a^b} (f + g) \, d\alpha \geq \underline{\int_a^b} f \, d\alpha + \underline{\int_a^b} g \, d\alpha.$$

These remarks can be easily verified by the reader. (See Exercise 7.11.)

7.13 RIEMANN'S CONDITION

If we are to expect equality of the upper and lower integrals, then we must also expect the upper sums to become arbitrarily close to the lower sums. Hence it seems reasonable to seek those functions f for which the difference $U(P, f, \alpha) - L(P, f, \alpha)$ can be made arbitrarily small.

Definition 7.18. *We say that f satisfies Riemann's condition with respect to α on $[a, b]$ if, for every $\varepsilon > 0$, there exists a partition P_ε such that P finer than P_ε implies*

$$0 \leq U(P, f, \alpha) - L(P, f, \alpha) < \varepsilon.$$

Theorem 7.19. *Assume that $\alpha \nearrow$ on $[a, b]$. Then the following three statements are equivalent:*

i) $f \in R(\alpha)$ on $[a, b]$.

ii) *f satisfies Riemann's condition with respect to α on* $[a, b]$.

iii) $I(f, \alpha) = \bar{I}(f, \alpha)$.

Proof. We will prove that part (i) implies (ii), part (ii) implies (iii), and part (iii) implies (i). Assume that (i) holds. If $\alpha(b) = \alpha(a)$, then (ii) holds trivially, so we can assume that $\alpha(a) < \alpha(b)$. Given $\varepsilon > 0$, choose P_ε so that for any finer P and all choices of t_k and t'_k in $[x_{k-1}, x_k]$, we have

$$\left| \sum_{k=1}^{n} f(t_k) \, \Delta\alpha_k - A \right| < \frac{\varepsilon}{3} \quad \text{and} \quad \left| \sum_{k=1}^{n} f(t'_k) \, \Delta\alpha_k - A \right| < \frac{\varepsilon}{3},$$

where $A = \int_a^b f \, d\alpha$. Combining these inequalities, we find

$$\left| \sum_{k=1}^{n} [f(t_k) - f(t'_k)] \, \Delta\alpha_k \right| < \frac{2}{3} \varepsilon.$$

Since $M_k(f) - m_k(f) = \sup \{f(x) - f(x') : x, x' \text{ in } [x_{k-1}, x_k]\}$, it follows that for every $h > 0$ we can choose t_k and t'_k so that

$$f(t_k) - f(t'_k) > M_k(f) - m_k(f) - h.$$

Making a choice corresponding to $h = \frac{1}{3}\varepsilon/[\alpha(b) - \alpha(a)]$, we can write

$$U(P, f, \alpha) - L(P, f, \alpha) = \sum_{k=1}^{n} [M_k(f) - m_k(f)] \, \Delta\alpha_k$$

$$< \sum_{k=1}^{n} [f(t_k) - f(t'_k)] \, \Delta\alpha_k + h \sum_{k=1}^{n} \Delta\alpha_k < \varepsilon.$$

Hence, (i) implies (ii).

Next, assume that (ii) holds. If $\varepsilon > 0$ is given, there exists a partition P_ε such that P finer than P_ε implies $U(P, f, \alpha) < L(P, f, \alpha) + \varepsilon$. Hence, for such P we have

$$\bar{I}(f, \alpha) \leq U(P, f, \alpha) < L(P, f, \alpha) + \varepsilon \leq I(f, \alpha) + \varepsilon.$$

That is, $\bar{I}(f, \alpha) \leq I(f, \alpha) + \varepsilon$ for every $\varepsilon > 0$. Therefore, $\bar{I}(f, \alpha) \leq I(f, \alpha)$. But, by Theorem 7.17, we also have the opposite inequality. Hence (ii) implies (iii).

Finally, assume that $\bar{I}(f, \alpha) = I(f, \alpha)$ and let A denote their common value. We will prove that $\int_a^b f \, d\alpha$ exists and equals A. Given $\varepsilon > 0$, choose P'_ε so that $U(P, f, \alpha) < \bar{I}(f, \alpha) + \varepsilon$ for all P finer than P'_ε. Also choose P''_ε such that

$$L(P, f, \alpha) > I(f, \alpha) - \varepsilon$$

for all P finer than P''_ε. If $P_\varepsilon = P'_\varepsilon \cup P''_\varepsilon$, we can write

$$I(f, \alpha) - \varepsilon < L(P, f, \alpha) \leq S(P, f, \alpha) \leq U(P, f, \alpha) < \bar{I}(f, \alpha) + \varepsilon$$

for every P finer than P_ε. But, since $I(f, \alpha) = \bar{I}(f, \alpha) = A$, this means that $|S(P, f, \alpha) - A| < \varepsilon$ whenever P is finer than P_ε. This proves that $\int_a^b f \, d\alpha$ exists and equals A, and the proof of the theorem is now complete.

7.14 COMPARISON THEOREMS

Theorem 7.20. *Assume that* $\alpha \nearrow$ *on* $[a, b]$. *If* $f \in R(\alpha)$ *and* $g \in R(\alpha)$ *on* $[a, b]$ *and if* $f(x) \le g(x)$ *for all* x *in* $[a, b]$, *then we have*

$$\int_a^b f(x) \, d\alpha(x) \le \int_a^b g(x) \, d\alpha(x).$$

Proof. For every partition P, the corresponding Riemann–Stieltjes sums satisfy

$$S(P, f, \alpha) = \sum_{k=1}^n f(t_k) \, \Delta\alpha_k \le \sum_{k=1}^n g(t_k) \, \Delta a_k = S(P, g, \alpha),$$

since $\alpha \nearrow$ on $[a, b]$. From this the theorem follows easily.

In particular, this theorem implies that $\int_a^b g(x) \, d\alpha(x) \ge 0$ whenever $g(x) \ge 0$ and $\alpha \nearrow$ on $[a, b]$.

Theorem 7.21. *Assume that* $\alpha \nearrow$ *on* $[a, b]$. *If* $f \in R(\alpha)$ *on* $[a, b]$, *then* $|f| \in R(\alpha)$ *on* $[a, b]$ *and we have the inequality*

$$\left| \int_a^b f(x) \, d\alpha(x) \right| \le \int_a^b |f(x)| \, d\alpha(x).$$

Proof. Using the notation of Definition 7.14, we can write

$$M_k(f) - m_k(f) = \sup \{f(x) - f(y) : x, y \text{ in } [x_{k-1}, x_k]\}.$$

Since the inequality $\left| |f(x)| - |f(y)| \right| \le |f(x) - f(y)|$ always holds, it follows that we have

$$M_k(|f|) - m_k(|f|) \le M_k(f) - m_k(f).$$

Multiplying by $\Delta\alpha_k$ and summing on k, we obtain

$$U(P, |f|, \alpha) - L(P, |f|, \alpha) \le U(P, f, \alpha) - L(P, f, \alpha),$$

for every partition P of $[a, b]$. By applying Riemann's condition, we find that $|f| \in R(\alpha)$ on $[a, b]$. The inequality in the theorem follows by taking $g = |f|$ in Theorem 7.20.

NOTE. The converse of Theorem 7.21 is not true. (See Exercise 7.12.)

Theorem 7.22. *Assume that* $\alpha \nearrow$ *on* $[a, b]$. *If* $f \in R(\alpha)$ *on* $[a, b]$, *then* $f^2 \in R(\alpha)$ *on* $[a, b]$.

Proof. Using the notation of Definition 7.14, we have

$$M_k(f^2) = [M_k(|f|)]^2 \qquad \text{and} \qquad m_k(f^2) = [m_k(|f|)]^2.$$

Hence we can write

$$M_k(f^2) - m_k(f^2) = [M_k(|f|) + m_k(|f|)][M_k(|f|) - m_k(|f|)]$$
$$\le 2M[M_k(|f|) - m_k(|f|)],$$

where M is an upper bound for $|f|$ on $[a, b]$. By applying Riemann's condition, the conclusion follows.

Theorem 7.23. *Assume that $\alpha \nearrow$ on $[a, b]$. If $f \in R(\alpha)$ and $g \in R(\alpha)$ on $[a, b]$, then the product $f \cdot g \in R(\alpha)$ on $[a, b]$.*

Proof. We use Theorem 7.22 along with the identity

$$2f(x)g(x) = [f(x) + g(x)]^2 - [f(x)]^2 - [g(x)]^2.$$

7.15 INTEGRATORS OF BOUNDED VARIATION

In Theorem 6.13 we found that every function α of bounded variation on $[a, b]$ can be expressed as the difference of two increasing functions. If $\alpha = \alpha_1 - \alpha_2$ is such a decomposition and if $f \in R(\alpha_1)$ and $f \in R(\alpha_2)$ on $[a, b]$, it follows by linearity that $f \in R(\alpha)$ on $[a, b]$. However, the converse is not always true. If $f \in R(\alpha)$ on $[a, b]$, it is quite possible to choose increasing functions α_1 and α_2 such that $\alpha = \alpha_1 - \alpha_2$, but such that neither integral $\int_a^b f\,d\alpha_1, \int_a^b f\,d\alpha_2$ exists. The difficulty, of course, is due to the nonuniqueness of the decomposition $\alpha = \alpha_1 - \alpha_2$. However, we can prove that there is at least one decomposition for which the converse *is* true, namely, when α_1 is the total variation of α and $\alpha_2 = \alpha_1 - \alpha$. (Recall Definition 6.8.)

Theorem 7.24. *Assume that α is of bounded variation on $[a, b]$. Let $V(x)$ denote the total variation of α on $[a, x]$ if $a < x \le b$, and let $V(a) = 0$. Let f be defined and bounded on $[a, b]$. If $f \in R(\alpha)$ on $[a, b]$, then $f \in R(V)$ on $[a, b]$.*

Proof. If $V(b) = 0$, then V is contant and the result is trivial. Suppose therefore, that $V(b) > 0$. Suppose also that $|f(x)| \le M$ if $x \in [a, b]$. Since V is increasing, we need only verify that f satisfies Riemann's condition with respect to V on $[a, b]$.

Given $\varepsilon > 0$, choose P_ε so that for any finer P and all choices of points t_k and t_k' in $[x_{k-1}, x_k]$ we have

$$\left| \sum_{k=1}^n [f(t_k) - f(t_k')] \Delta\alpha_k \right| < \frac{\varepsilon}{4} \quad \text{and} \quad V(b) < \sum_{k=1}^n |\Delta\alpha_k| + \frac{\varepsilon}{4M}.$$

For P finer than P_ε we will establish the two inequalities

$$\sum_{k=1}^n [M_k(f) - m_k(f)](\Delta V_k - |\Delta\alpha_k|) < \frac{\varepsilon}{2},$$

and

$$\sum_{k=1}^n [M_k(f) - m_k(f)] |\Delta\alpha_k| < \frac{\varepsilon}{2},$$

which, by addition, yield $U(P, f, V) - L(P, f, V) < \varepsilon$.

To prove the first inequality, we note that $\Delta V_k - |\Delta\alpha_k| \geq 0$ and hence

$$\sum_{k=1}^{n} [M_k(f) - m_k(f)](\Delta V_k - |\Delta\alpha_k|) \leq 2M \sum_{k=1}^{n} (\Delta V_k - |\Delta\alpha_k|)$$

$$= 2M \left(V(b) - \sum_{k=1}^{n} |\Delta\alpha_k|\right) < \frac{\varepsilon}{2}.$$

To prove the second inequality, let

$$A(P) = \{k : \Delta\alpha_k \geq 0\}, \qquad B(P) = \{k : \Delta\alpha_k < 0\},$$

and let $h = \frac{1}{4}\varepsilon/V(b)$. If $k \in A(P)$, choose t_k and t_k' so that

$$f(t_k) - f(t_k') > M_k(f) - m_k(f) - h;$$

but, if $k \in B(P)$, choose t_k and t_k' so that $f(t_k') - f(t_k) > M_k(f) - m_k(f) - h$. Then

$$\sum_{k=1}^{n} [M_k(f) - m_k(f)] |\Delta\alpha_k| < \sum_{k\in A(P)} [f(t_k) - f(t_k')] |\Delta\alpha_k|$$

$$+ \sum_{k\in B(P)} [f(t_k') - f(t_k)] |\Delta\alpha_k| + h \sum_{k=1}^{n} |\Delta\alpha_k|$$

$$= \sum_{k=1}^{n} [f(t_k) - f(t_k')] \Delta\alpha_k + h \sum_{k=1}^{n} |\Delta\alpha_k|$$

$$< \frac{\varepsilon}{4} + hV(b) = \frac{\varepsilon}{4} + \frac{\varepsilon}{4} = \frac{\varepsilon}{2}.$$

It follows that $f \in R(V)$ on $[a, b]$.

NOTE. This theorem (together with Theorem 6.12) enables us to reduce the theory of Riemann–Stieltjes integration for integrators of bounded variation to the case of *increasing integrators*. Riemann's condition then becomes available and it turns out to be a particularly useful tool in this work. As a first application we shall obtain a result which is closely related to Theorem 7.4.

Theorem 7.25. *Let α be of bounded variation on $[a, b]$ and assume that $f \in R(\alpha)$ on $[a, b]$. Then $f \in R(\alpha)$ on every subinterval $[c, d]$ of $[a, b]$.*

Proof. Let $V(x)$ denote the total variation of α on $[a, x]$, with $V(a) = 0$. Then $\alpha = V - (V - \alpha)$, where both V and $V - \alpha$ are increasing on $[a, b]$ (Theorem 6.12). By Theorem 7.24, $f \in R(V)$, and hence $f \in R(V - \alpha)$ on $[a, b]$. Therefore, if the theorem is true for increasing integrators, it follows that $f \in R(V)$ on $[c, d]$ and $f \in R(V - \alpha)$ on $[c, d]$, so $f \in R(\alpha)$ on $[c, d]$.

Hence, it suffices to prove the theorem when $\alpha \nearrow$ on $[a, b]$. By Theorem 7.4 it suffices to prove that each integral $\int_a^c f \, d\alpha$ and $\int_a^d f \, d\alpha$ exists. Assume that $a < c < b$. If P is a partition of $[a, x]$, let $\Delta(P, x)$ denote the difference

$$\Delta(P, x) = U(P, f, \alpha) - L(P, f, \alpha),$$

of the upper and lower sums associated with the interval $[a, x]$. Since $f \in R(\alpha)$ on $[a, b]$, Riemann's condition holds. Hence, if $\varepsilon > 0$ is given, there exists a partition P_ε of $[a, b]$ such that $\Delta(P, b) < \varepsilon$ if P is finer than P_ε. We can assume that $c \in P_\varepsilon$. The points of P_ε in $[a, c]$ form a partition P_ε' of $[a, c]$. If P' is a partition of $[a, c]$ finer than P_ε', then $P = P' \cup P_\varepsilon$ is a partition of $[a, b]$ composed of the points of P' along with those points of P_ε in $[c, b]$. Now the sum defining $\Delta(P', c)$ contains only part of the terms in the sum defining $\Delta(P, b)$. Since each term is ≥ 0 and since P is finer than P_ε, we have

$$\Delta(P', c) \leq \Delta(P, b) < \varepsilon.$$

That is, P' finer than P_ε' implies $\Delta(P', c) < \varepsilon$. Hence, f satisfies Riemann's condition on $[a, c]$ and $\int_a^c f \, d\alpha$ exists. The same argument, of course, shows that $\int_a^d f \, d\alpha$ exists, and by Theorem 7.4 it follows that $\int_c^d f \, d\alpha$ exists.

The next theorem is an application of Theorems 7.23, 7.21, and 7.25.

Theorem 7.26. *Assume* $f \in R(\alpha)$ *and* $g \in R(\alpha)$ *on* $[a, b]$, *where* $\alpha \nearrow$ *on* $[a, b]$. *Define*

$$F(x) = \int_a^x f(t) \, d\alpha(t)$$

and

$$G(x) = \int_a^x g(t) \, d\alpha(t) \quad \text{if } x \in [a, b].$$

Then $f \in R(G)$, $g \in R(F)$, *and the product* $f \cdot g \in R(\alpha)$ *on* $[a, b]$, *and we have*

$$\int_a^b f(x)g(x) \, d\alpha(x) = \int_a^b f(x) \, dG(x)$$

$$= \int_a^b g(x) \, dF(x).$$

Proof. The integral $\int_a^b f \cdot g \, d\alpha$ exists by Theorem 7.23. For every partition P of $[a, b]$ we have

$$S(P, f, G) = \sum_{k=1}^n f(t_k) \int_{x_{k-1}}^{x_k} g(t) \, d\alpha(t) = \sum_{k=1}^n \int_{x_{k-1}}^{x_k} f(t_k)g(t) \, d\alpha(t),$$

and

$$\int_a^b f(x)g(x) \, d\alpha(x) = \sum_{k=1}^n \int_{x_{k-1}}^{x_k} f(t)g(t) \, d\alpha(t).$$

Therefore, if $M_g = \sup\{|g(x)| : x \in [a, b]\}$, we have

$$\left| S(P, f, G) - \int_a^b f \cdot g \, d\alpha \right| = \left| \sum_{k=1}^n \int_{x_{k-1}}^{x_k} \{f(t_k) - f(t)\} g(t) \, d\alpha(t) \right|$$

$$\leq M_g \sum_{k=1}^n \int_{x_{k-1}}^{x_k} |f(t_k) - f(t)| \, d\alpha(t) \leq M_g \sum_{k=1}^n \int_{x_{k-1}}^{x_k} [M_k(f) - m_k(f)] \, d\alpha(t)$$

$$= M_g\{U(P, f, \alpha) - L(P, f, \alpha)\}.$$

Since $f \in R(\alpha)$, for every $\varepsilon > 0$ there is a partition P_ε such that P finer than P_ε implies $U(P, f, \alpha) - L(P, f, \alpha) < \varepsilon$. This proves that $f \in R(G)$ on $[a, b]$ and that $\int_a^b f \cdot g \, d\alpha = \int_a^b f \, dG$. A similar argument shows that $g \in R(F)$ on $[a, b]$ and that $\int_a^b f \cdot g \, d\alpha = \int_a^b g \, dF$.

NOTE. Theorem 7.26 is also valid if α is of bounded variation on $[a, b]$.

7.16 SUFFICIENT CONDITIONS FOR EXISTENCE OF RIEMANN–STIELTJES INTEGRALS

In most of the previous theorems we have assumed that certain integrals existed and then studied their properties. It is quite natural to ask: When does the integral exist? Two useful sufficient conditions will be obtained.

Theorem 7.27. *If f is continuous on $[a, b]$ and if α is of bounded variation on $[a, b]$, then $f \in R(\alpha)$ on $[a, b]$.*

NOTE. By Theorem 7.6, a second sufficient condition can be obtained by interchanging f and α in the hypothesis.

Proof. It suffices to prove the theorem when $\alpha \nearrow$ with $\alpha(a) < \alpha(b)$. Continuity of f on $[a, b]$ implies uniform continuity, so that if $\varepsilon > 0$ is given, we can find $\delta > 0$ (depending only on ε) such that $|x - y| < \delta$ implies $|f(x) - f(y)| < \varepsilon/A$, where $A = 2[\alpha(b) - \alpha(a)]$. If P_ε is a partition with norm $\|P_\varepsilon\| < \delta$, then for P finer than P_ε we must have

$$M_k(f) - m_k(f) \leq \varepsilon/A,$$

since $M_k(f) - m_k(f) = \sup\{f(x) - f(y) : x, y \text{ in } [x_{k-1}, x_k]\}$. Multiplying the inequality by $\Delta\alpha_k$ and summing, we find

$$U(P, f, \alpha) - L(P, f, \alpha) \leq \frac{\varepsilon}{A} \sum_{k=1}^n \Delta\alpha_k = \frac{\varepsilon}{2} < \varepsilon,$$

and we see that Riemann's condition holds. Hence, $f \in R(\alpha)$ on $[a, b]$.

For the special case in which $\alpha(x) = x$, Theorems 7.27 and 7.6 give the following corollary:

Theorem 7.28. *Each of the following conditions is sufficient for the existence of the Riemann integral $\int_a^b f(x) \, dx$:*

a) *f is continuous on $[a, b]$.*   b) *f is of bounded variation on $[a, b]$.*

7.17 NECESSARY CONDITIONS FOR EXISTENCE OF RIEMANN–STIELTJES INTEGRALS

When α is of bounded variation on $[a, b]$, continuity of f is sufficient for the existence of $\int_a^b f \, d\alpha$. Continuity of f throughout $[a, b]$ is by no means necessary, however. For example, in Theorem 7.9 we found that when α is a step function, then f can be defined quite arbitrarily in $[a, b]$ provided only that f is continuous at the discontinuities of α. The next theorem tells us that common discontinuities from the right or from the left must be avoided if the integral is to exist.

Theorem 7.29. *Assume that $\alpha \nearrow$ on $[a, b]$ and let $a < c < b$. Assume further that both α and f are discontinuous from the right at $x = c$; that is, assume that there exists an $\varepsilon > 0$ such that for every $\delta > 0$ there are values of x and y in the interval $(c, c + \delta)$ for which*

$$|f(x) - f(c)| \geq \varepsilon \qquad and \qquad |\alpha(y) - \alpha(c)| \geq \varepsilon.$$

Then the integral $\int_a^b f(x) \, d\alpha(x)$ cannot exist. The integral also fails to exist if α and f are discontinuous from the left at c.

Proof. Let P be a partition of $[a, b]$ containing c as a point of subdivision and form the difference

$$U(P, f, \alpha) - L(P, f, \alpha) = \sum_{k=1}^{n} [M_k(f) - m_k(f)] \, \Delta\alpha_k.$$

If the ith subinterval has c as its left endpoint, then

$$U(P, f, \alpha) - L(P, f, \alpha) \geq [M_i(f) - m_i(f)][\alpha(x_i) - \alpha(c)],$$

since each term of the sum is ≥ 0. If c is a common discontinuity from the right, we can assume that the point x_i is chosen so that $a(x_i) - \alpha(c) \geq \varepsilon$. Furthermore, the hypothesis of the theorem implies $M_i(f) - m_i(f) \geq \varepsilon$. Hence,

$$U(P, f, \alpha) - L(P, f, \alpha) \geq \varepsilon^2,$$

and Riemann's condition cannot be satisfied. (If c is a common discontinuity from the left, the argument is similar.)

7.18 MEAN-VALUE THEOREMS FOR RIEMANN–STIELTJES INTEGRALS

Although integrals occur in a wide variety of problems, there are relatively few cases in which the explicit value of the integral can be obtained. However, it often suffices to have an *estimate* for the integral rather than its exact value. The *Mean Value Theorems* of this section are especially useful in making such estimates.

Theorem 7.30 (First Mean-Value Theorem for Riemann–Stieltjes integrals). *Assume that $\alpha \nearrow$ and let $f \in R(\alpha)$ on $[a, b]$. Let M and m denote, respectively, the sup and inf of the set $\{f(x) : x \in [a, b]\}$. Then there exists a real number c satisfying*

$m \leq c \leq M$ such that

$$\int_a^b f(x)\, d\alpha(x) = c \int_a^b d\alpha(x) = c[\alpha(b) - \alpha(a)].$$

In particular, if f is continuous on $[a, b]$, then $c = f(x_0)$ for some x_0 in $[a, b]$.

Proof. If $\alpha(a) = \alpha(b)$, the theorem holds trivially, both sides being 0. Hence we can assume that $\alpha(a) < \alpha(b)$. Since all upper and lower sums satisfy

$$m[\alpha(b) - \alpha(a)] \leq L(P, f, \alpha) \leq U(P, f, \alpha) \leq M[\alpha(b) - \alpha(a)],$$

the integral $\int_a^b f\, d\alpha$ must lie between the same bounds. Therefore, the quotient $c = (\int_a^b f\, d\alpha)/(\int_a^b d\alpha)$ lies between m and M. When f is continuous on $[a, b]$, the intermediate value theorem yields $c = f(x_0)$ for some x_0 in $[a, b]$.

A second theorem of this type can be obtained from the first by using integration by parts.

Theorem 7.31 (Second Mean-Value Theorem for Riemann–Stieltjes integrals). *Assume that α is continuous and that $f \nearrow$ on $[a, b]$. Then there exists a point x_0 in $[a, b]$ such that*

$$\int_a^b f(x)\, d\alpha(x) = f(a) \int_a^{x_0} d\alpha(x) + f(b) \int_{x_0}^b d\alpha(x).$$

Proof. By Theorem 7.6, we have

$$\int_a^b f(x)\, d\alpha(x) = f(b)\alpha(b) - f(a)\alpha(a) - \int_a^b \alpha(x)\, df(x).$$

Applying Theorem 7.30 to the integral on the right, we find

$$\int_a^b f(x)\, d\alpha(x) = f(a)[\alpha(x_0) - \alpha(a)] + f(b)[\alpha(b) - \alpha(x_0)],$$

where $x_0 \in [a, b]$, which is the statement we set out to prove.

7.19 THE INTEGRAL AS A FUNCTION OF THE INTERVAL

If $f \in R(\alpha)$ on $[a, b]$ and if α is of bounded variation, then (by Theorem 7.25) the integral $\int_a^x f\, d\alpha$ exists for each x in $[a, b]$ and can be studied as a function of x. Some properties of this function will now be obtained.

Theorem 7.32. *Let α be of bounded variation on $[a, b]$ and assume that $f \in R(\alpha)$ on $[a, b]$. Define F by the equation*

$$F(x) = \int_a^x f\, d\alpha, \qquad if\ x \in [a, b].$$

Then we have:

i) *F is of bounded variation on* $[a, b]$.

ii) *Every point of continuity of* α *is also a point of continuity of F.*

iii) *If* $\alpha \nearrow$ *on* $[a, b]$, *the derivative* $F'(x)$ *exists at each point x in* (a, b) *where* $\alpha'(x)$ *exists and where f is continuous. For such x, we have*

$$F'(x) = f(x)\alpha'(x).$$

Proof. It suffices to assume that $\alpha \nearrow$ on $[a, b]$. If $x \neq y$, Theorem 7.30 implies that

$$F(y) - F(x) = \int_x^y f \, d\alpha = c[\alpha(y) - \alpha(x)],$$

where $m \leq c \leq M$ (in the notation of Theorem 7.30). Statements (i) and (ii) follow at once from this equation. To prove (iii), we divide by $y - x$ and observe that $c \to f(x)$ as $y \to x$.

When Theorem 7.32 is used in conjunction with Theorem 7.26, we obtain the following theorem which converts a Riemann integral of a product $f \cdot g$ into a Riemann–Stieltjes integral $\int_a^b f \, dG$ with a continuous integrator of bounded variation.

Theorem 7.33. *If* $f \in R$ *and* $g \in R$ *on* $[a, b]$, *let*

$$F(x) = \int_a^x f(t) \, dt, \qquad G(x) = \int_a^x g(t) \, dt \quad \text{if } x \in [a, b].$$

Then F and G are continuous functions of bounded variation on $[a, b]$. *Also,* $f \in R(G)$ *and* $g \in R(F)$ *on* $[a, b]$, *and we have*

$$\int_a^b f(x)g(x) \, dx = \int_a^b f(x) \, dG(x) = \int_a^b g(x) \, dF(x).$$

Proof. Parts (i) and (ii) of Theorem 7.32 show that F and G are continuous functions of bounded variation on $[a, b]$. The existence of the integrals and the two formulas for $\int_a^b f(x)g(x) \, dx$ follow by taking $\alpha(x) = x$ in Theorem 7.26.

NOTE. When $\alpha(x) = x$, part (iii) of Theorem 7.32 is sometimes called the *first fundamental theorem of integral calculus*. It states that $F'(x) = f(x)$ at each point of continuity of f. A companion result, called the *second fundamental theorem*, is given in the next section.

7.20 SECOND FUNDAMENTAL THEOREM OF INTEGRAL CALCULUS

The next theorem tells how to integrate a derivative.

Theorem 7.34 (Second fundamental theorem of integral calculus). *Assume that* $f \in R$ *on* $[a, b]$. *Let g be a function defined on* $[a, b]$ *such that the derivative* g' *exists in* *(a, b) and has the value*

$$g'(x) = f(x) \qquad \text{for every } x \text{ in } (a, b).$$

At the endpoints assume that $g(a+)$ and $g(b-)$ exist and satisfy

$$g(a) - g(a+) = g(b) - g(b-).$$

Then we have

$$\int_a^b f(x)\, dx = \int_a^b g'(x)\, dx = g(b) - g(a).$$

Proof. For every partition of $[a, b]$ we can write

$$g(b) - g(a) = \sum_{k=1}^n [g(x_k) - g(x_{k-1})] = \sum_{k=1}^n g'(t_k)\, \Delta x_k = \sum_{k=1}^n f(t_k)\, \Delta x_k,$$

where t_k is a point in (x_{k-1}, x_k) determined by the Mean-Value Theorem of differential calculus. But, for a given $\varepsilon > 0$, the partition can be taken so fine that

$$\left| g(b) - g(a) - \int_a^b f(x)\, dx \right| = \left| \sum_{k=1}^n f(t_k)\, \Delta x_k - \int_a^b f(x)\, dx \right| < \varepsilon,$$

and this proves the theorem.

The second fundamental theorem can be combined with Theorem 7.33 to give the following strengthening of Theorem 7.8.

Theorem 7.35. *Assume $f \in R$ on $[a, b]$. Let α be a function which is continuous on $[a, b]$ and whose derivative α' is Riemann integrable on $[a, b]$. Then the following integrals exist and are equal:*

$$\int_a^b f(x)\, d\alpha(x) = \int_a^b f(x)\alpha'(x)\, dx.$$

Proof. By the second fundamental theorem we have, for each x in $[a, b]$,

$$\alpha(x) - \alpha(a) = \int_a^x \alpha'(t)\, dt.$$

Taking $g = \alpha'$ in Theorem 7.33 we obtain Theorem 7.35.

NOTE. A related result is described in Exercise 7.34.

7.21 CHANGE OF VARIABLE IN A RIEMANN INTEGRAL

The formula $\int_a^b f\, d\alpha = \int_c^d h\, d\beta$ of Theorem 7.7 for changing the variable in an integral assumes the form

$$\int_{g(c)}^{g(d)} f(x)\, dx = \int_c^d f[g(t)]g'(t)\, dt,$$

when $\alpha(x) = x$ and when g is a strictly monotonic function with a continuous derivative g'. It is valid if $f \in R$ on $[a, b]$. When f is continuous, we can use Theorem 7.32 to remove the restriction that g be monotonic. In fact, we have the following theorem:

Theorem 7.36 (Change of variable in a Riemann integral). *Assume that g has a continuous derivative g' on an interval $[c, d]$. Let f be continuous on $g([c, d])$ and define F by the equation*

$$F(x) = \int_{g(c)}^{x} f(t) \, dt \qquad \text{if } x \in g([c, d]).$$

Then, for each x in $[c, d]$ the integral $\int_{c}^{x} f[g(t)]g'(t) \, dt$ exists and has the value $F[g(x)]$. In particular, we have

$$\int_{g(c)}^{g(d)} f(x) \, dx = \int_{c}^{d} f[g(t)]g'(t) \, dt.$$

Proof. Since both g' and the composite function $f \circ g$ are continuous on $[c, d]$ the integral in question exists. Define G on $[c, d]$ as follows:

$$G(x) = \int_{c}^{x} f[g(t)]g'(t) \, dt.$$

We are to show that $G(x) = F[g(x)]$. By Theorem 7.32, we have

$$G'(x) = f[g(x)]g'(x),$$

and, by the chain rule, the derivative of $F[g(x)]$ is also $f[g(x)]g'(x)$, since $F'(x) = f(x)$. Hence, $G(x) - F[g(x)]$ is constant. But, when $x = c$, we get $G(c) = 0$ and $F[g(c)] = 0$, so this constant must be 0. Hence, $G(x) = F[g(x)]$ for all x in $[c, d]$. In particular, when $x = d$, we get $G(d) = F[g(d)]$ and this is the last equation in the theorem.

NOTE. Some texts prove the preceding theorem under the added hypothesis that g' is never zero on $[c, d]$, which, of course, implies monotonicity of g. The above proof shows that this is not needed. It should be noted that g is continuous on $[c, d]$, so $g([c, d])$ is an interval which contains the interval joining $g(c)$ and $g(d)$.

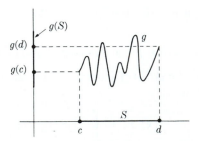

Figure 7.2

In particular, the result is valid if $g(c) = g(d)$. This makes the theorem especially useful in the applications. (See Fig. 7.2 for a permissible g.)

Actually, there is a more general version of Theorem 7.36 which does not require continuity of f or of g', but the proof is considerably more difficult. Assume that $h \in R$ on $[c, d]$ and, if $x \in [c, d]$, let $g(x) = \int_a^x h(t) \, dt$, where a is a fixed point in $[c, d]$. Then if $f \in R$ on $g([c, d])$ the integral $\int_c^d f[g(t)] \, h(t) \, dt$ exists and we have

$$\int_{g(c)}^{g(d)} f(x) \, dx = \int_c^d f[g(t)] h(t) \, dt.$$

This appears to be the most general theorem on change of variable in a Riemann integral. (For a proof, see the article by H. Kestelman, *Mathematical Gazette*, 45 (1961), pp. 17–23.) Theorem 7.36 is the special case in which h is continuous on $[c, d]$ and f is continuous on $g([c, d])$.

7.22 SECOND MEAN-VALUE THEOREM FOR RIEMANN INTEGRALS

Theorem 7.37. *Let g be continuous and assume that $f \nearrow$ on $[a, b]$. Let A and B be two real numbers satisfying the inequalities*

$$A \le f(a+) \qquad and \qquad B \ge f(b-).$$

Then there exists a point x_0 in $[a, b]$ such that

i) $$\int_a^b f(x)g(x) \, dx = A \int_a^{x_0} g(x) \, dx + B \int_{x_0}^b g(x) \, dx.$$

In particular, if $f(x) \ge 0$ for all x in $[a, b]$, we have

ii) $$\int_a^b f(x)g(x) \, dx = B \int_{x_0}^b g(x) \, dx, \qquad where \; x_0 \in [a, b].$$

NOTE. Part (ii) is known as *Bonnet's theorem*.

Proof. If $\alpha(x) = \int_a^x g(t) \, dt$, then $\alpha' = g$, Theorem 7.31 is applicable, and we get

$$\int_a^b f(x)g(x) \, dx = f(a) \int_a^{x_0} g(x) \, dx + f(b) \int_{x_0}^b g(x) \, dx.$$

This proves (i) whenever $A = f(a)$ and $B = f(b)$. Now if A and B are any two real numbers satisfying $A \le f(a+)$ and $B \ge f(b-)$, we can redefine f at the endpoints a and b to have the values $f(a) = A$ and $f(b) = B$. The modified f is still increasing on $[a, b]$ and, as we have remarked before, changing the value of f at a finite number of points does not affect the value of a Riemann integral. (Of course, the point x_0 in (i) will depend on the choice of A and B.) By taking $A = 0$, part (ii) follows from part (i).

7.23 RIEMANN–STIELTJES INTEGRALS DEPENDING ON A PARAMETER

Theorem 7.38. *Let f be continuous at each point (x, y) of a rectangle*

$$Q = \{(x, y) : a \le x \le b, \; c \le y \le d\}.$$

Assume that α is of bounded variation on $[a, b]$ and let F be the function defined on $[c, d]$ by the equation

$$F(y) = \int_a^b f(x, y) \, d\alpha(x).$$

Then F is continuous on $[c, d]$. In other words, if $y_0 \in [c, d]$, we have

$$\lim_{y \to y_0} \int_a^b f(x, y) \, d\alpha(x) = \int_a^b \lim_{y \to y_0} f(x, y) \, d\alpha(x)$$

$$= \int_a^b f(x, y_0) \, d\alpha(x).$$

Proof. Assume that $\alpha \nearrow$ on $[a, b]$. Since Q is a compact set, f is uniformly continuous on Q. Hence, given $\varepsilon > 0$, there exists a $\delta > 0$ (depending only on ε) such that for every pair of points $z = (x, y)$ and $z' = (x', y')$ in Q with $|z - z'| < \delta$, we have $|f(x, y) - f(x', y')| < \varepsilon$. If $|y - y'| < \delta$, we have

$$|F(y) - F(y')| \le \int_a^b |f(x, y) - f(x, y')| \, d\alpha(x) \le \varepsilon[\alpha(b) - \alpha(a)].$$

This establishes the continuity of F on $[c, d]$.

Of course, when $\alpha(x) = x$, this becomes a continuity theorem for Riemann integrals involving a parameter. However, we can derive a much more useful result for Riemann integrals than that obtained by simply setting $\alpha(x) = x$ if we employ Theorem 7.26.

Theorem 7.39. *If f is continuous on the rectangle $[a, b] \times [c, d]$, and if $g \in R$ on $[a, b]$, then the function F defined by the equation*

$$F(y) = \int_a^b g(x)f(x, y) \, dx,$$

is continuous on $[c, d]$. That is, if $y_0 \in [c, d]$, we have

$$\lim_{y \to y_0} \int_a^b g(x)f(x, y) \, dx = \int_a^b g(x)f(x, y_0) \, dx.$$

Proof. If $G(x) = \int_a^x g(t) \, dt$, Theorem 7.26 shows that $F(y) = \int_a^b f(x, y) \, dG(x)$. Now apply Theorem 7.38.

7.24 DIFFERENTIATION UNDER THE INTEGRAL SIGN

Theorem 7.40. *Let* $Q = \{(x, y) : a \le x \le b, c \le y \le d\}$. *Assume that* α *is of bounded variation on* $[a, b]$ *and, for each fixed y in* $[c, d]$, *assume that the integral*

$$F(y) = \int_a^b f(x, y)\, d\alpha(x),$$

exists. If the partial derivative $D_2 f$ *is continuous on Q, the derivative* $F'(y)$ *exists for each y in* (c, d) *and is given by*

$$F'(y) = \int_a^b D_2 f(x, y)\, d\alpha(x).$$

NOTE. In particular, when $g \in R$ on $[a, b]$ and $\alpha(x) = \int_a^x g(t)\, dt$, we get

$$F(y) = \int_a^b g(x) f(x, y)\, dx \qquad \text{and} \qquad F'(y) = \int_a^b g(x)\, D_2 f(x, y)\, dx.$$

Proof. If $y_0 \in (c, d)$ and $y \ne y_0$, we have

$$\frac{F(y) - F(y_0)}{y - y_0} = \int_a^b \frac{f(x, y) - f(x, y_0)}{y - y_0}\, d\alpha(x) = \int_a^b D_2 f(x, \bar{y})\, d\alpha(x),$$

where $\bar{y}$ is between y and y_0. Since $D_2 f$ is continuous on Q, we obtain the conclusion by arguing as in the proof of Theorem 7.38.

7.25 INTERCHANGING THE ORDER OF INTEGRATION

Theorem 7.41. *Let* $Q = \{(x, y) : a \le x \le b, c \le y \le d\}$. *Assume that* α *is of bounded variation on* $[a, b]$, β *is of bounded variation on* $[c, d]$, *and f is continuous on Q. If* $(x, y) \in Q$, *define*

$$F(y) = \int_a^b f(x, y)\, d\alpha(x), \qquad G(x) = \int_c^d f(x, y)\, d\beta(y).$$

Then $F \in R(\beta)$ *on* $[c, d]$, $G \in R(\alpha)$ *on* $[a, b]$, *and we have*

$$\int_c^d F(y)\, d\beta(y) = \int_a^b G(x)\, d\alpha(x).$$

In other words, we may interchange the order of integration as follows:

$$\int_a^b \left[\int_c^d f(x, y)\, d\beta(y) \right] d\alpha(x) = \int_c^d \left[\int_a^b f(x, y)\, d\alpha(x) \right] d\beta(y).$$

Proof. By Theorem 7.38, F is continuous on $[c, d]$ and hence $F \in R(\beta)$ on $[c, d]$. Similarly, $G \in R(\alpha)$ on $[a, b]$. To prove the equality of the two integrals, it suffices to consider the case in which $\alpha \nearrow$ on $[a, b]$ and $\beta \nearrow$ on $[c, d]$.

By uniform continuity, given $\varepsilon > 0$ there is a $\delta > 0$ such that for every pair of points $z = (x, y)$ and $z' = (x', y')$ in Q, with $|z - z'| < \delta$, we have

$$|f(x, y) - f(x', y')| < \varepsilon.$$

Let us now subdivide Q into n^2 equal rectangles by subdividing $[a, b]$ and $[c, d]$ each into n equal parts, where n is chosen so that

$$\frac{(b - a)}{n} < \frac{\delta}{\sqrt{2}} \quad \text{and} \quad \frac{(d - c)}{n} < \frac{\delta}{\sqrt{2}}.$$

Writing

$$x_k = a + \frac{k(b - a)}{n} \quad \text{and} \quad y_k = c + \frac{k(d - c)}{n},$$

for $k = 0, 1, 2, \ldots, n$, we have

$$\int_a^b \left(\int_c^d f(x, y) \, d\beta(y) \right) d\alpha(x) = \sum_{k=0}^{n-1} \sum_{j=0}^{n-1} \int_{x_k}^{x_{k+1}} \left(\int_{y_j}^{y_{j+1}} f(x, y) \, d\beta(y) \right) d\alpha(x).$$

We apply Theorem 7.30 twice on the right. The double sum becomes

$$\sum_{k=0}^{n-1} \sum_{j=0}^{n-1} f(x_k', y_j')[\beta(y_{j+1}) - \beta(y_j)][\alpha(x_{k+1}) - \alpha(x_k)],$$

where (x_k', y_j') is in the rectangle $Q_{k,j}$ having (x_k, y_j) and (x_{k+1}, y_{j+1}) as opposite vertices. Similarly, we find

$$\int_c^d \left(\int_a^b f(x, y) \, d\alpha(x) \right) d\beta(y)$$

$$= \sum_{k=0}^{n-1} \sum_{j=0}^{n-1} f(x_k'', y_j'')[\beta(y_{j+1}) - \beta(y_j)][\alpha(x_{k+1}) - \alpha(x_k)],$$

where $(x_k'', y_j'') \in Q_{k,j}$. But $|f(x_k', y_j') - f(x_k'', y_j'')| < \varepsilon$ and hence

$$\left| \int_a^b G(x) \, d\alpha(x) - \int_c^d F(y) \, d\beta(y) \right|$$

$$< \varepsilon \sum_{j=0}^{n-1} [\beta(y_{j+1}) - \beta(y_j)] \sum_{k=0}^{n-1} [\alpha(x_{k+1}) - \alpha(x_k)]$$

$$= \varepsilon[\beta(d) - \beta(c)][\alpha(b) - \alpha(a)].$$

Since ε is arbitrary, this implies equality of the two integrals.

Theorem 7.41 together with Theorem 7.26 gives the following result for Riemann integrals.

Theorem 7.42. *Let f be continuous on the rectangle $[a, b] \times [c, d]$. If $g \in R$ on $[a, b]$ and if $h \in R$ on $[c, d]$, then we have*

$$\int_a^b \left[\int_c^d g(x)h(y)f(x, y) \, dy \right] dx = \int_c^d \left[\int_a^b g(x)h(y)f(x, y) \, dx \right] dy.$$

Proof. Let $\alpha(x) = \int_a^x g(u) \, du$ and let $\beta(y) = \int_c^y h(v) \, dv$, and apply Theorems 7.26 and 7.41.

7.26 LEBESGUE'S CRITERION FOR EXISTENCE OF RIEMANN INTEGRALS

Every continuous function is Riemann integrable. However, continuity is certainly not necessary, for we have seen that $f \in R$ when f is of bounded variation on $[a, b]$. In particular, f can be a monotonic function with a countable set of discontinuities and yet the integral $\int_a^b f(x) \, dx$ will exist. Actually, there are Riemann-integrable functions whose discontinuities form a noncountable set. (See Exercise 7.32.) Therefore, it is natural to ask "how many" discontinuities a function can have and still be Riemann integrable. The definitive theorem on this question was discovered by Lebesgue and is proved in this section. The idea behind Lebesgue's theorem is revealed by examining Riemann's condition to see the kind of restriction it puts on the set of discontinuities of f.

The difference between the upper and lower Riemann sums is given by

$$\sum_{k=1}^{n} [M_k(f) - m_k(f)] \, \Delta x_k,$$

and, roughly speaking, f will be integrable if, and only if, this sum can be made arbitrarily small. Split this sum into two parts, say $S_1 + S_2$, where S_1 comes from subintervals containing only points of continuity of f, and S_2 contains the remaining terms. In S_1, each difference $M_k(f) - m_k(f)$ is small because of continuity and hence a large number of such terms can occur and still keep S_1 small. In S_2, however, the differences $M_k(f) - m_k(f)$ need not be small; but because they are bounded (say by M), we have $|S_2| \leq M \sum \Delta x_k$, so that S_2 will be small if the sum of the lengths of the subintervals corresponding to S_2 is small. Hence we may expect that the set of discontinuities of an integrable function can be covered by intervals whose total length is small.

This is the central idea in Lebesgue's theorem. To formulate it more precisely we introduce sets of measure zero.

Definition 7.43. *A set S of real numbers is said to have measure zero if, for every $\varepsilon > 0$, there is a countable covering of S by open intervals, the sum of whose lengths is less than ε.*

If the intervals are denoted by (a_k, b_k), the definition requires that

$$S \subseteq \bigcup_k (a_k, b_k) \qquad \text{and} \qquad \sum_k (b_k - a_k) < \varepsilon. \tag{3}$$

If the collection of intervals is finite, the index k in (3) runs over a finite set. If the collection is countably infinite, then k goes from 1 to ∞, and the sum of the lengths is the sum of an infinite series given by

$$\sum_{k=1}^{\infty} (b_k - a_k) = \lim_{N \to \infty} \sum_{k=1}^{N} (b_k - a_k).$$

Besides the definition, we need one more result about sets of measure zero.

Theorem 7.44. *Let F be a countable collection of sets in $\mathbf{R}$, say*

$$F = \{F_1, F_2, \ldots\},$$

each of which has measure zero. Then their union

$$S = \bigcup_{k=1}^{\infty} F_k,$$

also has a measure zero.

Proof. Given $\varepsilon > 0$, there is a countable covering of F_k by open intervals, the sum of whose lengths is less than $\varepsilon/2^k$. The union of all these coverings is itself a countable covering of S by open intervals and the sum of the lengths of all the intervals is less than

$$\sum_{k=1}^{\infty} \frac{\varepsilon}{2^k} = \varepsilon.$$

Examples. Since a set consisting of just one point has measure zero, it follows that every countable subset of $\mathbf{R}$ has measure zero. In particular, the set of rational numbers has measure zero. However, there are *uncountable* sets which have measure zero. (See Exercise 7.32.)

Next we introduce the concept of oscillation.

Definition 7.45. *Let f be defined and bounded on an interval S. If $T \subseteq S$, the number*

$$\Omega_f(T) = \sup \{f(x) - f(y) : x \in T, \quad y \in T\},$$

is called the oscillation of f on T. The oscillation of f at x is defined to be the number

$$\omega_f(x) = \lim_{h \to 0+} \Omega_f(B(x; h) \cap S).$$

NOTE. This limit always exists, since $\Omega_f(B(x; h) \cap S)$ is a decreasing function of h. In fact, $T_1 \subseteq T_2$ implies $\Omega_f(T_1) \leq \Omega_f(T_2)$. Also, $\omega_f(x) = 0$ if, and only if, f is continuous at x (Exercise 4.24).

The next theorem tells us that if $\omega_f(x) < \varepsilon$ at each point of a compact interval $[a, b]$, then $\Omega_f(T) < \varepsilon$ for all sufficiently small subintervals T.

Theorem 7.46. *Let f be defined and bounded on $[a, b]$, and let $\varepsilon > 0$ be given. Assume that $\omega_f(x) < \varepsilon$ for every x in $[a, b]$. Then there exists a $\delta > 0$ (depending* *only on* ε) *such that for every closed subinterval* $T \subseteq [a, b]$, *we have* $\Omega_f(T) < \varepsilon$ *whenever the length of* T *is less than* δ.

Proof. For each x in $[a, b]$ there exists a 1-ball $B_x = B(x; \delta_x)$ such that

$$\Omega_f(B_x \cap [a, b]) < \omega_f(x) + [\varepsilon - \omega_f(x)] = \varepsilon.$$

The set of all halfsize balls $B(x; \delta_x/2)$ forms an open covering of $[a, b]$. By compactness, a finite number (say k) of these cover $[a, b]$. Let their radii be $\delta_1/2, \ldots, \delta_k/2$ and let δ be the smallest of these k numbers. When the interval T has length $< \delta$, then T is partly covered by at least one of these balls, say by $B(x_p; \delta_p/2)$. However, the ball $B(x_p; \delta_p)$ completely covers T (since $\delta_p \geq 2\delta$). Moreover, in $B(x_p; \delta_p) \cap [a, b]$ the oscillation of f is less than ε. This implies that $\Omega_f(T) < \varepsilon$ and the theorem is proved.

Theorem 7.47. *Let f be defined and bounded on $[a, b]$. For each $\varepsilon > 0$ define the set J_ε as follows:*

$$J_\varepsilon = \{x : x \in [a, b], \quad \omega_f(x) \geq \varepsilon\}.$$

Then J_ε is a closed set.

Proof. Let x be an accumulation point of J_ε. If $x \notin J_\varepsilon$, we have $\omega_f(x) < \varepsilon$. Hence there is a 1-ball $B(x)$ such that

$$\Omega_f(B(x) \cap [a, b]) < \varepsilon.$$

Thus no points of $B(x)$ can belong to J_ε, contradicting the statement that x is an accumulation point of J_ε. Therefore, $x \in J_\varepsilon$ and J_ε is closed.

Theorem 7.48 (Lebesgue's criterion for Riemann-integrability). *Let f be defined and bounded on $[a, b]$ and let D denote the set of discontinuities of f in $[a, b]$. Then $f \in R$ on $[a, b]$ if, and only if, D has measure zero.*

Proof. (Necessity). First we assume that D does not have measure zero and show that f is not integrable. We can write D as a countable union of sets

$$D = \bigcup_{r=1}^{\infty} D_r,$$

where

$$D_r = \left\{x : \omega_f(x) \geq \frac{1}{r}\right\}.$$

If $x \in D$, then $\omega_f(x) > 0$, so D is the union of the sets D_r for $r = 1, 2, \ldots$

Now if D does not have measure zero, then some set D_r does not (by Theorem 7.44). Therefore, there is some $\varepsilon > 0$ such that every countable collection of open intervals covering D_r has a sum of lengths $\geq \varepsilon$. For any partition P of $[a, b]$ we have

$$U(P, f) - L(P, f) = \sum_{k=1}^{n} [M_k(f) - m_k(f)] \Delta x_k = S_1 + S_2 \geq S_1,$$

where S_1 contains those terms coming from subintervals containing points of D in their interior, and S_2 contains the remaining terms. The open intervals from S_1 cover D_r except possibly for a finite subset of D_r, which has measure 0, so the sum of their lengths is at least ε. Moreover, in these intervals we have

$$M_k(f) - m_k(f) \geq \frac{1}{r} \quad \text{and hence } S_1 \geq \frac{\varepsilon}{r}.$$

This means that

$$U(P, f) - L(P, f) \geq \frac{\varepsilon}{r},$$

for every partition P, so Riemann's condition cannot be satisfied. Therefore f is not integrable. In other words, if $f \in R$, then D has measure zero.

(*Sufficiency*). Now we assume that D has measure zero and show that the Riemann condition is satisfied. Again we write $D = \bigcup_{r=1}^{\infty} D_r$, where D_r is the set of points x at which $\omega_f(x) \geq 1/r$. Since $D_r \subseteq D$, each D_r has measure 0, so D_r can be covered by open intervals, the sum of whose lengths is $< 1/r$. Since D_r is compact (Theorem 7.47), a finite number of these intervals cover D_r. The union of these intervals is an open set which we denote by A_r. The complement $B_r = [a, b] - A_r$ is the union of a finite number of closed subintervals of $[a, b]$. Let I be a typical subinterval of B_r. If $x \in I$, then $\omega_f(x) < 1/r$ so, by Theorem 7.46, there is a $\delta > 0$ (depending only on r) such that I can be further subdivided into a finite number of subintervals T of length $< \delta$ in which $\Omega_f(T) < 1/r$. The endpoints of all these subintervals determine a partition P_r of $[a, b]$. If P is finer than P_r we can write

$$U(P, f) - L(P, f) = \sum_{k=1}^{n} [M_k(f) - m_k(f)] \, \Delta x_k = S_1 + S_2,$$

where S_1 contains those terms coming from subintervals containing points of D_r, and S_2 contains the remaining terms. In the kth term of S_2 we have

$$M_k(f) - m_k(f) < \frac{1}{r} \quad \text{and hence } S_2 < \frac{b - a}{r}.$$

Since A_r covers all the intervals contributing to S_1, we have

$$S_1 \leq \frac{M - m}{r},$$

where M and m are the sup and inf of f on $[a, b]$. Therefore

$$U(P, f) - L(P, f) < \frac{M - m + b - a}{r}.$$

Since this holds for every $r \geq 1$, we see that Riemann's condition holds, so $f \in R$ on $[a, b]$.

NOTE. A property is said to hold *almost everywhere* on a subset S of $\mathbf{R}^1$ if it holds everywhere on S except for a set of measure 0. Thus, Lebesgue's theorem states that a bounded function f on a compact interval $[a, b]$ is Riemann integrable on $[a, b]$ if, and only if, f is continuous almost everywhere on $[a, b]$.

The following statements (some of which were proved earlier in the chapter) are immediate consequences of Lebesgue's theorem.

Theorem 7.49. a) *If f is of bounded variation on $[a, b]$, then $f \in R$ on $[a, b]$.*

b) *If $f \in R$ on $[a, b]$, then $f \in R$ on $[c, d]$ for every subinterval $[c, d] \subset [a, b]$, $|f| \in R$ and $f^2 \in R$ on $[a, b]$. Also, $f \cdot g \in R$ on $[a, b]$ whenever $g \in R$ on $[a, b]$.*

c) *If $f \in R$ and $g \in R$ on $[a, b]$, then $f/g \in R$ on $[a, b]$ whenever g is bounded away from 0.*

d) *If f and g are bounded functions having the same discontinuities on $[a, b]$, then $f \in R$ on $[a, b]$ if, and only if, $g \in R$ on $[a, b]$.*

e) *Let $g \in R$ on $[a, b]$ and assume that $m \le g(x) \le M$ for all x in $[a, b]$. If f is continuous on $[m, M]$, the composite function h defined by $h(x) = f[g(x)]$ is Riemann-integrable on $[a, b]$.*

NOTE. Statement (e) need not hold if we assume only that $f \in R$ on $[m, M]$. (See Exercise 7.29.)

7.27 COMPLEX-VALUED RIEMANN–STIELTJES INTEGRALS

Riemann–Stieltjes integrals of the form $\int_a^b f\, d\alpha$, in which f and α are *complex-valued* functions defined and bounded on an interval $[a, b]$, are of fundamental importance in the theory of functions of a complex variable. They can be introduced by exactly the same definition we have used in the real case. In fact, Definition 7.1 is meaningful when f and α are complex-valued. The sums of the products $f(t_k)[\alpha(x_k) - \alpha(x_{k-1})]$ which are used to form Riemann–Stieltjes sums need only be interpreted as sums of products of complex numbers. Since complex numbers satisfy the commutative, associative, and distributive laws which hold for real numbers, it is not surprising that complex-valued integrals share many of the properties of real-valued integrals. In particular, Theorems 7.2, 7.3, 7.4, 7.6, and 7.7 (as well as their proofs) are all valid (word for word) when f and α are complex-valued functions. (In Theorems 7.2 and 7.3, the constants c_1 and c_2 may now be complex numbers.) In addition, we have the following theorem which, in effect, reduces the theory of complex Stieltjes integrals to the real case.

Theorem 7.50. *Let $f = f_1 + if_2$ and $\alpha = \alpha_1 + i\alpha_2$ be complex-valued functions defined on an interval $[a, b]$. Then we have*

$$\int_a^b f\, d\alpha = \left(\int_a^b f_1\, d\alpha_1 - \int_a^b f_2\, d\alpha_2 \right) + i \left(\int_a^b f_2\, d\alpha_1 + \int_a^b f_1\, d\alpha_2 \right),$$

whenever all four integrals on the right exist.

The proof of Theorem 7.50 is immediate from the definition and is left to the reader.

The use of this theorem permits us to extend most of the important properties of real integrals to the complex case. For example, the connection between differentiation and integration established in Theorem 7.32 remains valid for complex integrals if we simply define such notions as continuity, differentiability and bounded variation by components, as with vector-valued functions. Thus, we say that the complex-valued function $\alpha = \alpha_1 + i\alpha_2$ is of bounded variation on $[a, b]$ if each component α_1 and α_2 is of bounded variation on $[a, b]$. Similarly, the derivative $\alpha'(t)$ is defined by the equation $\alpha'(t) = \alpha_1'(t) + i\alpha_2'(t)$ whenever the derivatives $\alpha_1'(t)$ and $\alpha_2'(t)$ exist. (One-sided derivatives are defined in the same way.) With this understanding, Theorems 7.32 and 7.34 (the fundamental theorems of integral calculus) both remain valid when f and α are complex-valued. The proofs follow from the real case by using Theorem 7.50 in a straightforward manner.

We shall return to complex-valued integrals in Chapter 16, when we study functions of a complex variable in more detail.

EXERCISES

Riemann–Stieltjes integrals

7.1 Prove that $\int_a^b d\alpha(x) = \alpha(b) - \alpha(a)$, directly from Definition 7.1.

7.2 If $f \in R(\alpha)$ on $[a, b]$ and if $\int_a^b f\, d\alpha = 0$ for every f which is monotonic on $[a, b]$, prove that α must be constant on $[a, b]$.

7.3 The following definition of a Riemann–Stieltjes integral is often used in the literature: We say f is integrable with respect to α if there exists a real number A having the property that for every $\varepsilon > 0$, there exists a $\delta > 0$ such that for every partition P of $[a, b]$ with norm $\|P\| < \delta$ and for every choice of t_k in $[x_{k-1}, x_k]$, we have $|S(P, f, \alpha) - A| < \varepsilon$.

a) Show that if $\int_a^b f\, d\alpha$ exists according to this definition, then it also exists according to Definition 7.1 and the two integrals are equal.

b) Let $f(x) = \alpha(x) = 0$ for $a \le x < c$, $f(x) = \alpha(x) = 1$ for $c < x \le b$, $f(c) = 0$, $\alpha(c) = 1$. Show that $\int_a^b f\, d\alpha$ exists according to Definition 7.1 but does not exist by this second definition.

7.4 If $f \in R$ according to Definition 7.1, prove that $\int_a^b f(x)\, dx$ also exists according to the definition of Exercise 7.3. [Contrast with Exercise 7.3(b).] *Hint.* Let $I = \int_a^b f(x)\, dx$, $M = \sup\{|f(x)| : x \in [a, b]\}$. Given $\varepsilon > 0$, choose P_ε so that $U(P_\varepsilon, f) < I + \varepsilon/2$ (notation of Section 7.11). Let N be the number of subdivision points in P_ε and let $\delta = \varepsilon/(2MN)$. If $\|P\| < \delta$, write

$$U(P, f) = \sum M_k(f)\, \Delta x_k = S_1 + S_2,$$

where S_1 is the sum of terms arising from those subintervals of P containing no points of P_ε and S_2 is the sum of the remaining terms. Then

$$S_1 \le U(P_\varepsilon, f) < I + \varepsilon/2 \quad\text{and}\quad S_2 \le NM\|P\| < NM\delta = \varepsilon/2,$$

and hence $U(P, f) < I + \varepsilon$. Similarly,

$$L(P, f) > I - \varepsilon \text{ if } \|P\| < \delta' \qquad \text{for some } \delta'.$$

Hence $|S(P, f) - I| < \varepsilon$ if $\|P\| < \min (\delta, \delta')$.

7.5 Let $\{a_n\}$ be a sequence of real numbers. For $x \geq 0$, define

$$A(x) = \sum_{n \leq x} a_n = \sum_{n=1}^{[x]} a_n,$$

where $[x]$ is the greatest integer in x and empty sums are interpreted as zero. Let f have a continuous derivative in the interval $1 \leq x \leq a$. Use Stieltjes integrals to derive the following formula:

$$\sum_{n \leq a} a_n f(n) = - \int_1^a A(x) f'(x) \, dx + A(a) f(a).$$

7.6 Use Euler's summation formula, or integration by parts in a Stieltjes integral, to derive the following identities:

a) $\displaystyle\sum_{k=1}^n \frac{1}{k^s} = \frac{1}{n^{s-1}} + s \int_1^n \frac{[x]}{x^{s+1}} \, dx \qquad$ if $s \neq 1$.

b) $\displaystyle\sum_{k=1}^n \frac{1}{k} = \log n - \int_1^n \frac{x - [x]}{x^2} \, dx + 1$.

7.7 Assume f' is continuous on $[1, 2n]$ and use Euler's summation formula or integration by parts to prove that

$$\sum_{k=1}^{2n} (-1)^k f(k) = \int_1^{2n} f'(x)([x] - 2[x/2]) \, dx.$$

7.8 Let $\varphi_1(x) = x - [x] - \frac{1}{2}$ if $x \neq$ integer, and let $\varphi_1(x) = 0$ if $x =$ integer. Also, let $\varphi_2(x) = \int_0^x \varphi_1(t) \, dt$. If f'' is continuous on $[1, n]$ prove that Euler's summation formula implies that

$$\sum_{k=1}^n f(k) = \int_1^n f(x) \, dx - \int_1^n \varphi_2(x) f''(x) \, dx + \frac{f(1) + f(n)}{2}.$$

7.9 Take $f(x) = \log x$ in Exercise 7.8 and prove that

$$\log n! = (n + \tfrac{1}{2}) \log n - n + 1 + \int_1^n \frac{\varphi_2(t)}{t^2} \, dt.$$

7.10 If $x \geq 1$, let $\pi(x)$ denote the number of primes $\leq x$, that is,

$$\pi(x) = \sum_{p \leq x} 1,$$

where the sum is extended over all primes $p \leq x$. The *prime number theorem* states that

$$\lim_{x \to \infty} \pi(x) \frac{\log x}{x} = 1.$$

This is usually proved by studying a related function ϑ given by

$$\vartheta(x) = \sum_{p \leq x} \log p,$$

where again the sum is extended over all primes $p \leq x$. Both functions π and ϑ are step functions with jumps at the primes. This exercise shows how the Riemann–Stieltjes integral can be used to relate these two functions.

    a) If $x \geq 2$, prove that $\pi(x)$ and $\vartheta(x)$ can be expressed as the following Riemann–Stieltjes integrals:

$$\vartheta(x) = \int_{3/2}^{x} \log t \, d\pi(t), \qquad \pi(x) = \int_{3/2}^{x} \frac{1}{\log t} \, d\vartheta(t).$$

NOTE. The lower limit can be replaced by any number in the open interval $(1, 2)$.

    b) If $x \geq 2$, use integration by parts to show that

$$\vartheta(x) = \pi(x) \log x - \int_{2}^{x} \frac{\pi(t)}{t} \, dt,$$

$$\pi(x) = \frac{\vartheta(x)}{\log x} + \int_{2}^{x} \frac{\vartheta(t)}{t \log^2 t} \, dt.$$

    These equations can be used to prove that the prime number theorem is equivalent to the relation $\lim_{x \to \infty} \vartheta(x)/x = 1$.

7.11 If $\alpha \nearrow$ on $[a, b]$, prove that we have

    a) $\displaystyle \overline{\int_{a}^{b}} f \, d\alpha = \overline{\int_{a}^{c}} f \, d\alpha + \overline{\int_{c}^{b}} f \, d\alpha, \qquad (a < c < b),$

    b) $\displaystyle \overline{\int_{a}^{b}} (f + g) \, d\alpha \leq \overline{\int_{a}^{b}} f \, d\alpha + \overline{\int_{a}^{b}} g \, d\alpha,$

    c) $\displaystyle \underline{\int_{a}^{b}} (f + g) \, d\alpha \geq \underline{\int_{a}^{b}} f \, d\alpha + \underline{\int_{a}^{b}} g \, d\alpha.$

7.12 Give an example of a bounded function f and an increasing function α defined on $[a, b]$ such that $|f| \in R(\alpha)$ but for which $\int_{a}^{b} f \, d\alpha$ does not exist.

7.13 Let α be a continuous function of bounded variation on $[a, b]$. Assume $g \in R(\alpha)$ on $[a, b]$ and define $\beta(x) = \int_{a}^{x} g(t) \, d\alpha(t)$ if $x \in [a, b]$. Show that:

    a) If $f \nearrow$ on $[a, b]$, there exists a point x_0 in $[a, b]$ such that

$$\int_{a}^{b} f \, d\beta = f(a) \int_{a}^{x_0} g \, d\alpha + f(b) \int_{x_0}^{b} g \, d\alpha.$$

    b) If, in addition, f is continuous on $[a, b]$, we also have

$$\int_{a}^{b} f(x)g(x) \, d\alpha(x) = f(a) \int_{a}^{x_0} g \, d\alpha + f(b) \int_{x_0}^{b} g \, d\alpha.$$

7.14 Assume $f \in R(\alpha)$ on $[a, b]$, where α is of bounded variation on $[a, b]$. Let $V(x)$ denote the total variation of α on $[a, x]$ for each x in $(a, b]$, and let $V(a) = 0$. Show that

$$\left| \int_a^b f \, d\alpha \right| \le \int_a^b |f| \, dV \le MV(b),$$

where M is an upper bound for $|f|$ on $[a, b]$. In particular, when $\alpha(x) = x$, the inequality becomes

$$\left| \int_a^b f(x) \, dx \right| \le M(b - a).$$

7.15 Let $\{\alpha_n\}$ be a sequence of functions of bounded variation on $[a, b]$. Suppose there exists a function α defined on $[a, b]$ such that the total variation of $\alpha - \alpha_n$ on $[a, b]$ tends to 0 as $n \to \infty$. Assume also that $\alpha(a) = \alpha_n(a) = 0$ for each $n = 1, 2, \ldots$ If f is continuous on $[a, b]$, prove that

$$\lim_{n \to \infty} \int_a^b f(x) \, d\alpha_n(x) = \int_a^b f(x) \, d\alpha(x).$$

7.16 If $f \in R(\alpha)$, $f^2 \in R(\alpha)$, $g \in R(\alpha)$, and $g^2 \in R(\alpha)$ on $[a, b]$, prove that

$$\frac{1}{2} \int_a^b \left[\int_a^b \left| \begin{matrix} f(x) & g(x) \\ f(y) & g(y) \end{matrix} \right|^2 d\alpha(y) \right] d\alpha(x)$$

$$= \left(\int_a^b f(x)^2 \, d\alpha(x) \right)\left(\int_a^b g(x)^2 \, d\alpha(x) \right) - \left(\int_a^b f(x)g(x) \, d\alpha(x) \right)^2.$$

When $\alpha \nearrow$ on $[a, b]$, deduce the Cauchy–Schwarz inequality

$$\left(\int_a^b f(x)g(x) \, d\alpha(x) \right)^2 \le \left(\int_a^b f(x)^2 \, d\alpha(x) \right)\left(\int_a^b g(x)^2 \, d\alpha(x) \right).$$

(Compare with Exercise 1.23.)

7.17 Assume that $f \in R(\alpha)$, $g \in R(\alpha)$, and $f \cdot g \in R(\alpha)$ on $[a, b]$. Show that

$$\frac{1}{2} \int_a^b \left[\int_a^b (f(y) - f(x))(g(y) - g(x)) \, d\alpha(y) \right] d\alpha(x)$$

$$= (\alpha(b) - \alpha(a)) \int_a^b f(x)g(x) \, d\alpha(x) - \left(\int_a^b f(x) \, d\alpha(x) \right)\left(\int_a^b g(x) \, d\alpha(x) \right).$$

If $\alpha \nearrow$ on $[a, b]$, deduce the inequality

$$\left(\int_a^b f(x) \, d\alpha(x) \right)\left(\int_a^b g(x) \, d\alpha(x) \right) \le (\alpha(b) - \alpha(a)) \int_a^b f(x)g(x) \, d\alpha(x)$$

when both f and g are increasing (or both are decreasing) on $[a, b]$. Show that the reverse inequality holds if f increases and g decreases on $[a, b]$.

Riemann integrals

7.18 Assume $f \in R$ on $[a, b]$. Use Exercise 7.4 to prove that the limit

$$\lim_{n \to \infty} \frac{b - a}{n} \sum_{k=1}^{n} f\left(a + k \frac{b - a}{n}\right)$$

exists and has the value $\int_a^b f(x) \, dx$. Deduce that

$$\lim_{n \to \infty} \sum_{k=1}^{n} \frac{n}{k^2 + n^2} = \frac{\pi}{4}, \qquad \lim_{n \to \infty} \sum_{k=1}^{n} (n^2 + k^2)^{-1/2} = \log (1 + \sqrt{2}).$$

7.19 Define

$$f(x) = \left(\int_0^x e^{-t^2} \, dt\right)^2, \qquad g(x) = \int_0^1 \frac{e^{-x^2(t^2+1)}}{t^2 + 1} \, dt.$$

a) Show that $g'(x) + f'(x) = 0$ for all x and deduce that $g(x) + f(x) = \pi/4$.

b) Use (a) to prove that

$$\lim_{x \to \infty} \int_0^x e^{-t^2} \, dt = \frac{1}{2} \sqrt{\pi}.$$

7.20 Assume $g \in R$ on $[a, b]$ and define $f(x) = \int_a^x g(t) \, dt$ if $x \in [a, b]$. Prove that the integral $\int_a^x |g(t)| \, dt$ gives the total variation of f on $[a, x]$.

7.21 Let $\mathbf{f} = (f_1, \ldots, f_n)$ be a vector-valued function with a continuous derivative $\mathbf{f}'$ on $[a, b]$. Prove that the curve described by $\mathbf{f}$ has length

$$\Lambda_{\mathbf{f}}(a, b) = \int_a^b \|\mathbf{f}'(t)\| \, dt.$$

7.22 If $f^{(n+1)}$ is continuous on $[a, x]$, define

$$I_n(x) = \frac{1}{n!} \int_a^x (x - t)^n f^{(n+1)}(t) \, dt.$$

a) Show that

$$I_{k-1}(x) - I_k(x) = \frac{f^{(k)}(a)(x - a)^k}{k!}, \qquad k = 1, 2, \ldots, n.$$

b) Use (a) to express the remainder in Taylor's formula (Theorem 5.19) as an integral.

7.23 Let f be continuous on $[0, a]$. If $x \in [0, a]$, define $f_0(x) = f(x)$ and let

$$f_{n+1}(x) = \frac{1}{n!} \int_0^x (x - t)^n f(t) \, dt, \qquad n = 0, 1, 2, \ldots$$

a) Show that the nth derivative of f_n exists and equals f.

b) Prove the following theorem of M. Fekete: The number of changes in sign of f in $[0, a]$ is not less than the number of changes in sign in the ordered set of numbers

$$f(a), f_1(a), \ldots, f_n(a).$$

Hint. Use mathematical induction.

c) Use (b) to prove the following theorem of L. Fejér: The number of changes in sign of f in $[0, a]$ is not less than the number of changes in sign in the ordered set

$$f(0), \qquad \int_0^a f(t)\, dt, \qquad \int_0^a tf(t)\, dt, \qquad \ldots, \qquad \int_0^a t^n f(t)\, dt.$$

7.24 Let f be a positive continuous function in $[a, b]$. Let M denote the maximum value of f on $[a, b]$. Show that

$$\lim_{n \to \infty} \left(\int_a^b f(x)^n\, dx \right)^{1/n} = M.$$

7.25 A function f of two real variables is defined for each point (x, y) in the unit square $0 \le x \le 1, 0 \le y \le 1$ as follows:

$$f(x, y) = \begin{cases} 1, & \text{if } x \text{ is rational,} \\ 2y, & \text{if } x \text{ is irrational.} \end{cases}$$

a) Compute $\overline{\int}_0^1 f(x, y)\, dx$ and $\underline{\int}_0^1 f(x, y)\, dx$ in terms of y.

b) Show that $\int_0^1 f(x, y)\, dy$ exists for each fixed x and compute $\int_0^t f(x, y)\, dy$ in terms of x and t for $0 \le x \le 1, 0 \le t \le 1$.

c) Let $F(x) = \int_0^1 f(x, y)\, dy$. Show that $\int_0^1 F(x)\, dx$ exists and find its value.

7.26 Let f be defined on $[0, 1]$ as follows: $f(0) = 0$; if $2^{-n-1} < x \le 2^{-n}$, then $f(x) = 2^{-n}$, for $n = 0, 1, 2, \ldots$

a) Give two reasons why $\int_0^1 f(x)\, dx$ exists.

b) Let $F(x) = \int_0^x f(t)\, dt$. Show that for $0 < x \le 1$ we have

$$F(x) = xA(x) - \tfrac{1}{3}A(x)^2,$$

where $A(x) = 2^{-[-\log x/\log 2]}$ and where $[y]$ is the greatest integer in y.

7.27 Assume f has a derivative which is monotonic decreasing and satisfies $f'(x) \ge m > 0$ for all x in $[a, b]$. Prove that

$$\left| \int_a^b \cos f(x)\, dx \right| \le \frac{2}{m}.$$

Hint. Multiply and divide the integrand by $f'(x)$ and use Theorem 7.37(ii).

7.28 Given a decreasing sequence of real numbers $\{G(n)\}$ such that $G(n) \to 0$ as $n \to \infty$. Define a function f on $[0, 1]$ in terms of $\{G(n)\}$ as follows: $f(0) = 1$; if x is irrational, then $f(x) = 0$; if x is the rational m/n (in lowest terms), then $f(m/n) = G(n)$. Compute the oscillation $\omega_f(x)$ at each x in $[0, 1]$ and show that $f \in R$ on $[0, 1]$.

7.29 Let f be defined as in Exercise 7.28 with $G(n) = 1/n$. Let $g(x) = 1$ if $0 < x \le 1$, $g(0) = 0$. Show that the composite function h defined by $h(x) = g[f(x)]$ is not Riemann-integrable on $[0, 1]$, although both $f \in R$ and $g \in R$ on $[0, 1]$.

7.30 Use Lebesgue's theorem to prove Theorem 7.49.

7.31 Use Lebesgue's theorem to prove that if $f \in R$ and $g \in R$ on $[a, b]$ and if $f(x) \ge m > 0$ for all x in $[a, b]$, then the function h defined by

$$h(x) = f(x)^{g(x)}$$

is Riemann-integrable on $[a, b]$.

7.32 Let $I = [0, 1]$ and let $A_1 = I - (\frac{1}{3}, \frac{2}{3})$ be that subset of I obtained by removing those points which lie in the open middle third of I; that is, $A_1 = [0, \frac{1}{3}] \cup [\frac{2}{3}, 1]$. Let A_2 be that subset of A_1 obtained by removing the open middle third of $[0, \frac{1}{3}]$ and of $[\frac{2}{3}, 1]$. Continue this process and define $A_3, A_4, \ldots$ The set $C = \bigcap_{n=1}^{\infty} A_n$ is called the *Cantor set.* Prove that:

    a) C is a compact set having measure zero.

    b) $x \in C$ if, and only if, $x = \sum_{n=1}^{\infty} a_n 3^{-n}$, where each a_n is either 0 or 2.

    c) C is uncountable.

    d) Let $f(x) = 1$ if $x \in C$, $f(x) = 0$ if $x \notin C$. Prove that $f \in R$ on $[0, 1]$.

7.33 This exercise outlines a proof (due to Ivan Niven) that π^2 is irrational. Let $f(x) = x^n(1 - x)^n/n!$. Prove that:

    a) $0 < f(x) < 1/n!$ if $0 < x < 1$.

    b) Each kth derivative $f^{(k)}(0)$ and $f^{(k)}(1)$ is an integer.

Now assume that $\pi^2 = a/b$, where a and b are positive integers, and let

$$F(x) = b^n \sum_{k=0}^{n} (-1)^k f^{(2k)}(x) \, \pi^{2n-2k}.$$

Prove that:

    c) $F(0)$ and $F(1)$ are integers.

    d) $\pi^2 a^n f(x) \sin \pi x = \dfrac{d}{dx} \{F'(x) \sin \pi x - \pi F(x) \cos \pi x\}$.

    e) $F(1) + F(0) = \pi a^n \displaystyle\int_0^1 f(x) \sin \pi x \, dx$.

    f) Use (a) in (e) to deduce that $0 < F(1) + F(0) < 1$ if n is sufficiently large. This contradicts (c) and shows that π^2 (and hence π) is irrational.

7.34 Given a real-valued function α, continuous on the interval $[a, b]$ and having a finite bounded derivative α' on (a, b). Let f be defined and bounded on $[a, b]$ and assume that both integrals

$$\int_a^b f(x) \, d\alpha(x) \qquad \text{and} \qquad \int_a^b f(x) \, \alpha'(x) \, dx$$

exist. Prove that these integrals are equal. (It is not assumed that α' is continuous.)

7.35 Prove the following theorem, which implies that a function with a positive integral must itself be positive on some interval. Assume that $f \in R$ on $[a, b]$ and that $0 \le f(x) \le M$ on $[a, b]$, where $M > 0$. Let $I = \int_a^b f(x) \, dx$, let $h = \frac{1}{2}I/(M + b - a)$, and assume that $I > 0$. Then the set $T = \{x : f(x) \ge h\}$ contains a finite number of intervals, the sum of whose lengths is at least h. *Hint.* Let P be a partition of $[a, b]$ such that every Riemann sum $S(P, f) = \sum_{k=1}^{n} f(t_k) \Delta x_k$ satisfies $S(P, f) > I/2$. Split $S(P, f)$ into two parts, $S(P, f) = \sum_{k \in A} + \sum_{k \in B}$, where

$$A = \{k : [x_{k-1}, x_k] \subseteq T\}, \qquad \text{and} \qquad B = \{k : k \notin A\}.$$

If $k \in A$, use the inequality $f(t_k) \le M$; if $k \in B$, choose t_k so that $f(t_k) < h$. Deduce that $\sum_{k \in A} \Delta x_k > h$.

Existence theorems for integral and differential equations

The following exercises illustrate how the fixed-point theorem for contractions (Theorem 4.48) is used to prove existence theorems for solutions of certain integral and differential equations. We denote by $C[a, b]$ the metric space of all real continuous functions on $[a, b]$ with the metric

$$d(f, g) = \|f - g\| = \max_{a \le x \le b} |f(x) - g(x)|,$$

and recall that $C[a, b]$ is a complete metric space (Exercise 4.67).

7.36 Given a function g in $C[a, b]$, and a function K continuous on the rectangle $Q = [a, b] \times [a, b]$, consider the function T defined on $C[a, b]$ by the equation

$$T(\varphi)(x) = g(x) + \lambda \int_a^b K(x, t)\varphi(t)\, dt,$$

where λ is a given constant.

a) Prove that T maps $C[a, b]$ into itself.

b) If $|K(x, y)| \le M$ on Q, where $M > 0$, and if $|\lambda| < M^{-1}(b - a)^{-1}$, prove that T is a contraction of $C[a, b]$ and hence has a fixed point φ which is a solution of the integral equation $\varphi(x) = g(x) + \lambda \int_a^b K(x, t)\varphi(t)\, dt$.

7.37 Assume f is continuous on a rectangle $Q = [a - h, a + h] \times [b - k, b + k]$, where $h > 0, k > 0$.

a) Let φ be a function, continuous on $[a - h, a + h]$, such that $(x, \varphi(x)) \in Q$ for all x in $[a - h, a + h]$. If $0 < c \le h$, prove that φ satisfies the differential equation $y' = f(x, y)$ on $(a - c, a + c)$ and the initial condition $\varphi(a) = b$ if, and only if, φ satisfies the integral equation

$$\varphi(x) = b + \int_a^x f(t, \varphi(t))\, dt \qquad \text{on} \qquad (a - c, a + c).$$

b) Assume that $|f(x, y)| \le M$ on Q, where $M > 0$, and let $c = \min\{h, k/M\}$. Let S denote the metric subspace of $C[a - c, a + c]$ consisting of all φ such that $|\varphi(x) - b| \le Mc$ on $[a - c, a + c]$. Prove that S is a closed subspace of $C[a - c, a + c]$ and hence that S is itself a complete metric space.

c) Prove that the function T defined on S by the equation

$$T(\varphi)(x) = b + \int_a^x f(t, \varphi(t))\, dt$$

maps S into itself.

d) Now assume that f satisfies a Lipschitz condition of the form

$$|f(x, y) - f(x, z)| \le A|y - z|$$

for every pair of points (x, y) and (x, z) in Q, where $A > 0$. Prove that T is a contraction of S if $h < 1/A$. Deduce that for $h < 1/A$ the differential equation $y' = f(x, y)$ has exactly one solution $y = \varphi(x)$ on $(a - c, a + c)$ such that $\varphi(a) = b$.

SUGGESTED REFERENCES FOR FURTHER STUDY

7.1 Hildebrandt, T. H., *Introduction to the Theory of Integration.* Academic Press, New York, 1963.

7.2 Kestelman, H., *Modern Theories of Integration.* Oxford University Press, Oxford, 1937.

7.3 Rankin, R. A., *An Introduction to Mathematical Analysis.* Pergamon Press, Oxford, 1963.

7.4 Rogosinski, W. W., *Volume and Integral.* Wiley, New York, 1952.

7.5 Shilov, G. E., and Gurevich, B. L., *Integral, Measure and Derivative: A Unified Approach.* R. Silverman, translator. Prentice-Hall, Englewood Cliffs, 1966.

CHAPTER 8

INFINITE SERIES
AND INFINITE PRODUCTS

8.1 INTRODUCTION

This chapter gives a brief development of the theory of infinite series and infinite products. These are merely special infinite sequences whose terms are real or complex numbers. Convergent sequences were discussed in Chapter 4 in the setting of general metric spaces. We recall some of the concepts of Chapter 4 as they apply to sequences in $\mathbf{C}$ with the usual Euclidean metric.

8.2 CONVERGENT AND DIVERGENT SEQUENCES OF COMPLEX NUMBERS

Definition 8.1. *A sequence $\{a_n\}$ of points in $\mathbf{C}$ is said to converge if there is a point p in $\mathbf{C}$ with the following property:*

For every $\varepsilon > 0$ there is an integer N (depending on ε) such that

$$|a_n - p| < \varepsilon \qquad \text{whenever } n \geq N.$$

If $\{a_n\}$ converges to p, we write $\lim_{n \to \infty} a_n = p$ and call p the *limit* of the sequence. A sequence is called *divergent* if it is not convergent.

A sequence in $\mathbf{C}$ is called a *Cauchy sequence* if it satisfies the *Cauchy condition*; that is, for every $\varepsilon > 0$ there is an integer N such that

$$|a_n - a_m| < \varepsilon \qquad \text{whenever } n \geq N \text{ and } m \geq N.$$

Since $\mathbf{C}$ is a complete metric space, we know from Chapter 4 that a sequence in $\mathbf{C}$ is convergent if, and only if, it is a Cauchy sequence.

The Cauchy condition is particularly useful in establishing convergence when we do not know the actual value to which the sequence converges.

Every convergent sequence is bounded (Theorem 4.3) and hence an unbounded sequence necessarily diverges.

If a sequence $\{a_n\}$ converges to p, then every subsequence $\{a_{k_n}\}$ also converges to p (Theorem 4.5).

A sequence $\{a_n\}$ whose terms are real numbers is said to diverge to $+\infty$ if, for every $M > 0$, there is an integer N (depending on M) such that

$$a_n > M \qquad \text{whenever } n \geq N.$$

In this case we write $\lim_{n \to \infty} a_n = +\infty$.

If $\lim_{n \to \infty} (-a_n) = +\infty$, we write $\lim_{n \to \infty} a_n = -\infty$ and say that $\{a_n\}$ diverges to $-\infty$. Of course, there are divergent real-valued sequences which do not diverge to $+\infty$ or to $-\infty$. For example, the sequence $\{(-1)^n(1 + 1/n)\}$ diverges but does not diverge to $+\infty$ or to $-\infty$.

8.3 LIMIT SUPERIOR AND LIMIT INFERIOR OF A REAL-VALUED SEQUENCE

Definition 8.2. *Let* $\{a_n\}$ *be a sequence of real numbers. Suppose there is a real number U satisfying the following two conditions:*

i) *For every* $\varepsilon > 0$ *there exists an integer N such that* $n > N$ *implies*

$$a_n < U + \varepsilon.$$

ii) *Given* $\varepsilon > 0$ *and given* $m > 0$*, there exists an integer* $n > m$ *such that*

$$a_n > U - \varepsilon.$$

Then U is called the limit superior (or upper limit) of $\{a_n\}$ *and we write*

$$U = \limsup_{n \to \infty} a_n.$$

Statement (i) *implies that the set* $\{a_1, a_2, \ldots\}$ *is bounded above. If this set is not bounded above, we define*

$$\limsup_{n \to \infty} a_n = +\infty.$$

If the set is bounded above but not bounded below and if $\{a_n\}$ *has no finite limit superior, then we say* $\limsup_{n \to \infty} a_n = -\infty$*. The limit inferior (or lower limit) of* $\{a_n\}$ *is defined as follows:*

$$\liminf_{n \to \infty} a_n = -\limsup_{n \to \infty} b_n, \qquad \text{where } b_n = -a_n \qquad \text{for } n = 1, 2, \ldots$$

NOTE. Statement (i) means that ultimately *all* terms of the sequence lie to the left of $U + \varepsilon$. Statement (ii) means that *infinitely many* terms lie to the right of $U - \varepsilon$. It is clear that there cannot be more than one U which satisfies both (i) and (ii). Every real sequence has a limit superior and a limit inferior in the extended real number system $\mathbf{R}^*$. (See Exercise 8.1.)

The reader should supply the proofs of the following theorems:

Theorem 8.3. *Let* $\{a_n\}$ *be a sequence of real numbers. Then we have:*

a) $\liminf_{n \to \infty} a_n \leq \limsup_{n \to \infty} a_n$.

b) *The sequence converges if, and only if,* $\limsup_{n \to \infty} a_n$ *and* $\liminf_{n \to \infty} a_n$ *are both finite and equal, in which case* $\lim_{n \to \infty} a_n = \liminf_{n \to \infty} a_n = \limsup_{n \to \infty} a_n$.

c) *The sequence diverges to* $+\infty$ *if, and only if,* $\liminf_{n \to \infty} a_n = \limsup_{n \to \infty} a_n = +\infty$.

d) *The sequence diverges to* $-\infty$ *if, and only if,* $\liminf_{n \to \infty} a_n = \limsup_{n \to \infty} a_n = -\infty$.

NOTE. A sequence for which $\liminf_{n\to\infty} a_n \neq \limsup_{n\to\infty} a_n$ is said to oscillate.

Theorem 8.4. *Assume that $a_n \leq b_n$ for each $n = 1, 2, \ldots$ Then we have:*

$$\liminf_{n\to\infty} a_n \leq \liminf_{n\to\infty} b_n \quad and \quad \limsup_{n\to\infty} a_n \leq \limsup_{n\to\infty} b_n.$$

Examples

1. $a_n = (-1)^n(1 + 1/n)$, $\liminf_{n\to\infty} a_n = -1$, $\limsup a_n = +1$.
2. $a_n = (-1)^n$, $\liminf_{n\to\infty} a_n = -1$, $\limsup_{n\to\infty} a_n = +1$.
3. $a_n = (-1)^n\, n$, $\liminf_{n\to\infty} a_n = -\infty$, $\limsup_{n\to\infty} a_n = +\infty$.
4. $a_n = n^2 \sin^2 (\tfrac{1}{2}n\pi)$, $\liminf_{n\to\infty} a_n = 0$, $\limsup a_n = +\infty$.

8.4 MONOTONIC SEQUENCES OF REAL NUMBERS

Definition 8.5. *Let $\{a_n\}$ be a sequence of real numbers. We say the sequence is increasing and we write $a_n \nearrow$ if $a_n \leq a_{n+1}$ for $n = 1, 2, \ldots$ If $a_n \geq a_{n+1}$ for all n, we say the sequence is decreasing and we write $a_n \searrow$. A sequence is called monotonic if it is increasing or if it is decreasing.*

The convergence or divergence of a monotonic sequence is particularly easy to determine. In fact, we have

Theorem 8.6. *A monotonic sequence converges if, and only if, it is bounded.*

Proof. If $a_n \nearrow$, $\lim_{n\to\infty} a_n = \sup\{a_n : n = 1, 2, \ldots\}$. If $a_n \searrow$, $\lim_{n\to\infty} a_n = \inf\{a_n : n = 1, 2, \ldots\}$.

8.5 INFINITE SERIES

Let $\{a_n\}$ be a given sequence of real or complex numbers, and form a new sequence $\{s_n\}$ as follows:

$$s_n = a_1 + \cdots + a_n = \sum_{k=1}^{n} a_k \quad (n = 1, 2, \ldots). \tag{1}$$

Definition 8.7. *The ordered pair of sequences $(\{a_n\}, \{s_n\})$ is called an infinite series. The number s_n is called the nth partial sum of the series. The series is said to converge or to diverge according as $\{s_n\}$ is convergent or divergent. The following symbols are used to denote the series defined by (1):*

$$a_1 + a_2 + \cdots + a_n + \cdots, \qquad a_1 + a_2 + a_3 + \cdots, \qquad \sum_{k=1}^{\infty} a_k.$$

NOTE. The letter k used in $\sum_{k=1}^{\infty} a_k$ is a "dummy variable" and may be replaced by any other convenient symbol. If p is an integer ≥ 0, a symbol of the form $\sum_{n=p}^{\infty} b_n$ is interpreted to mean $\sum_{n=1}^{\infty} a_n$ where $a_n = b_{n+p-1}$. When there is no danger of misunderstanding, we write Σb_n instead of $\sum_{n=p}^{\infty} b_n$.

If the sequence $\{s_n\}$ defined by (1) converges to s, the number s is called the *sum* of the series and we write

$$s = \sum_{k=1}^{\infty} a_k.$$

Thus, for convergent series the symbol $\sum a_k$ is used to denote both the series and its sum.

Example. If x has the infinite decimal expansion $x = a_0.a_1a_2 \cdots$ (see Section 1.17), then the series $\sum_{k=0}^{\infty} a_k 10^{-k}$ converges to x.

Theorem 8.8. *Let $a = \sum a_n$ and $b = \sum b_n$ be convergent series. Then, for every pair of constants α and β, the series $\sum(\alpha a_n + \beta b_n)$ converges to the sum $\alpha a + \beta b$. That is,*

$$\sum_{n=1}^{\infty} (\alpha a_n + \beta b_n) = \alpha \sum_{n=1}^{\infty} a_n + \beta \sum_{n=1}^{\infty} b_n.$$

Proof. $\sum_{k=1}^{n} (\alpha a_k + \beta b_k) = \alpha \sum_{k=1}^{n} a_k + \beta \sum_{k=1}^{n} b_k.$

Theorem 8.9. *Assume that $a_n \geq 0$ for each $n = 1, 2, \ldots$ Then $\sum a_n$ converges if, and only if, the sequence of partial sums is bounded above.*

Proof. Let $s_n = a_1 + \cdots + a$. Then $s_n \nearrow$ and we can apply Theorem 8.6.

Theorem 8.10 (Telescoping series). *Let $\{a_n\}$ and $\{b_n\}$ be two sequences such that $a_n = b_{n+1} - b_n$ for $n = 1, 2, \ldots$ Then $\sum a_n$ converges if, and only if, $\lim_{n\to\infty} b_n$ exists, in which case we have*

$$\sum_{n=1}^{\infty} a_n = \lim_{n\to\infty} b_n - b_1.$$

Proof. $\sum_{k=1}^{n} a_k = \sum_{k=1}^{n} (b_{k+1} - b_k) = b_{n+1} - b_1.$

Theorem 8.11 (Cauchy condition for series). *The series $\sum a_n$ converges if, and only if, for every $\varepsilon > 0$ there exists an integer N such that $n > N$ implies*

$$|a_{n+1} + \cdots + a_{n+p}| < \varepsilon \qquad \text{for each } p = 1, 2, \ldots \tag{2}$$

Proof. Let $s_n = \sum_{k=1}^{n} a_k$, write $s_{n+p} - s_n = a_{n+1} + \cdots + a_{n+p}$, and apply Theorem 4.8 and Theorem 4.6.

Taking $p = 1$ in (2), we find that $\lim_{n\to\infty} a_n = 0$ is a *necessary* condition for convergence of $\sum a_n$. That this condition is *not sufficient* is seen by considering the example in which $a_n = 1/n$. When $n = 2^m$ and $p = 2^m$ in (2), we find

$$a_{n+1} + \cdots + a_{n+p} = \frac{1}{2^m + 1} + \cdots + \frac{1}{2^m + 2^m} \geq \frac{2^m}{2^m + 2^m} = \frac{1}{2},$$

and hence the Cauchy condition cannot be satisfied when $\varepsilon \leq \frac{1}{2}$. Therefore the series $\sum_{n=1}^{\infty} 1/n$ diverges. This series is called the *harmonic series*.

8.6 INSERTING AND REMOVING PARENTHESES

Definition 8.12. *Let p be a function whose domain is the set of positive integers and whose range is a subset of the positive integers such that*

i)
$$p(n) < p(m), \quad if \ n < m.$$

Let $\sum a_n$ and $\sum b_n$ be two series related as follows:

$$b_1 = a_1 + a_2 + \cdots + a_{p(1)},$$

ii) $$b_{n+1} = a_{p(n)+1} + a_{p(n)+2} + \cdots + a_{p(n+1)}, \quad if \ n = 1, 2, \ldots$$

Then we say that $\sum b_n$ is obtained from $\sum a_n$ by inserting parentheses, and that $\sum a_n$ is obtained from $\sum b_n$ by removing parentheses.

Theorem 8.13. *If $\sum a_n$ converges to s, every series $\sum b_n$ obtained from $\sum a_n$ by inserting parentheses also converges to s.*

Proof. Let $\sum a_n$ and $\sum b_n$ be related by (ii) and write $s_n = \sum_{k=1}^{n} a_k$, $t_n = \sum_{k=1}^{n} b_k$. Then $\{t_n\}$ is a subsequence of $\{s_n\}$. In fact, $t_n = s_{p(n)}$. Therefore, convergence of $\{s_n\}$ to s implies convergence of $\{t_n\}$ to s.

Removing parentheses may destroy convergence. To see this, consider the series $\sum b_n$ in which each term is 0 (obviously convergent). Let $p(n) = 2n$ and let $a_n = (-1)^n$. Then (i) and (ii) hold but $\sum a_n$ diverges.

Parentheses *can* be removed if we further restrict $\sum a_n$ and p.

Theorem 8.14. *Let $\sum a_n$, $\sum b_n$ be related as in Definition 8.12. Assume that there exists a constant $M > 0$ such that $p(n + 1) - p(n) < M$ for all n, and assume that $\lim_{n \to \infty} a_n = 0$. Then $\sum a_n$ converges if, and only if, $\sum b_n$ converges, in which case they have the same sum.*

Proof. If $\sum a_n$ converges, the result follows from Theorem 8.13. The whole difficulty lies in the converse deduction. Let

$$s_n = a_1 + \cdots + a_n, \quad t_n = b_1 + \cdots + b_n, \quad t = \lim_{n \to \infty} t_n.$$

Let $\varepsilon > 0$ be given and choose N so that $n > N$ implies

$$|t_n - t| < \frac{\varepsilon}{2} \quad and \quad |a_n| < \frac{\varepsilon}{2M}.$$

If $n > p(N)$, we can find $m \geq N$ so that $N \leq p(m) \leq n < p(m + 1)$. [Why?] For such n, we have

$$s_n = a_1 + \cdots + a_{p(m+1)} - (a_{n+1} + a_{n+2} + \cdots + a_{p(m+1)})$$

$$= t_{m+1} - (a_{n+1} + a_{n+2} + \cdots + a_{p(m+1)}),$$

and hence

$$|s_n - t| \leq |t_{m+1} - t| + |a_{n+1} + a_{n+2} + \cdots + a_{p(m+1)}|$$

$$\leq |t_{m+1} - t| + |a_{p(m)+1}| + |a_{p(m)+2}| + \cdots + |a_{p(m+1)}|$$

$$< \frac{\varepsilon}{2} + (p(m+1) - p(m)) \frac{\varepsilon}{2M} < \frac{\varepsilon}{2} + \frac{\varepsilon}{2} = \varepsilon.$$

This proves that $\lim_{n \to \infty} s_n = t$.

8.7 ALTERNATING SERIES

Definition 8.15. *If $a_n > 0$ for each n, the series $\sum_{n=1}^{\infty} (-1)^{n+1} a_n$ is called an alternating series.*

Theorem 8.16. *If $\{a_n\}$ is a decreasing sequence converging to 0, the alternating series $\sum (-1)^{n+1} a_n$ converges. If s denotes its sum and s_n its nth partial sum, we have the inequality*

$$0 < (-1)^n(s - s_n) < a_{n+1}, \qquad \text{for } n = 1, 2, \ldots \tag{3}$$

NOTE. Inequality (3) tells us that when we "approximate" s by s_n, the error made has the same sign as the first neglected term and is less than the absolute value of this term.

Proof. We insert parentheses in $\sum (-1)^{n+1} a_n$, grouping together two terms at a time. That is, we take $p(n) = 2n$ and form a new series $\sum b_n$ according to Definition 8.12, with

$$b_1 = a_1 - a_2, \quad b_2 = a_3 - a_4, \quad \ldots, \quad b_n = a_{2n-1} - a_{2n}.$$

Since $a_n \to 0$ and $p(n + 1) - p(n) = 2$, Theorem 8.14 tells us that $\sum (-1)^{n+1} a_n$ converges if $\sum b_n$ converges. But $\sum b_n$ is a series of nonnegative terms (since $a_n \searrow$), and its partial sums are bounded above, since

$$\sum_{k=1}^{n} b_k = a_1 - (a_2 - a_3) - \cdots - (a_{2n-2} - a_{2n-1}) - a_{2n} < a_1.$$

Therefore $\sum b_n$ converges, so $\sum (-1)^{n+1} a_n$ also converges.
   Inequality (3) is a consequence of the following relations:

$$(-1)^n(s - s_n) = \sum_{k=1}^{\infty} (-1)^{k+1} a_{n+k} = \sum_{k=1}^{\infty} (a_{n+2k-1} - a_{n+2k}) > 0,$$

and

$$(-1)^n(s - s_n) = a_{n+1} - \sum_{k=1}^{\infty} (a_{n+2k} - a_{n+2k+1}) < a_{n+1}.$$

8.8 ABSOLUTE AND CONDITIONAL CONVERGENCE

Definition 8.17. *A series $\sum a_n$ is called absolutely convergent if $\sum |a_n|$ converges. It is called conditionally convergent if $\sum a_n$ converges but $\sum |a_n|$ diverges.*

Theorem 8.18. *Absolute convergence of $\sum a_n$ implies convergence.*

Proof. Apply the Cauchy condition to the inequality

$$|a_{n+1} + \cdots + a_{n+p}| \le |a_{n+1}| + \cdots + |a_{n+p}|.$$

To see that the converse is not true, consider the example

$$\sum_{n=1}^{\infty} \frac{(-1)^{n+1}}{n}.$$

This alternating series converges, by Theorem 8.16, but it does not converge absolutely.

Theorem 8.19. *Let $\sum a_n$ be a given series with real-valued terms and define*

$$p_n = \frac{|a_n| + a_n}{2}, \qquad q_n = \frac{|a_n| - a_n}{2} \qquad (n = 1, 2, \dots). \tag{4}$$

Then:

  i) *If $\sum a_n$ is conditionally convergent, both $\sum p_n$ and $\sum q_n$ diverge.*

 ii) *If $\sum |a_n|$ converges, both $\sum p_n$ and $\sum q_n$ converge and we have*

$$\sum_{n=1}^{\infty} a_n = \sum_{n=1}^{\infty} p_n - \sum_{n=1}^{\infty} q_n.$$

NOTE. $p_n = a_n$ and $q_n = 0$ if $a_n \ge 0$, whereas $q_n = -a_n$ and $p_n = 0$ if $a_n \le 0$.

Proof. We have $a_n = p_n - q_n$, $|a_n| = p_n + q_n$. To prove (i), assume that $\sum a_n$ converges and $\sum |a_n|$ diverges. If $\sum q_n$ converges, then $\sum p_n$ also converges (by Theorem 8.8), since $p_n = a_n + q_n$. Similarly, if $\sum p_n$ converges, then $\sum q_n$ also converges. Hence, if either $\sum p_n$ or $\sum q_n$ converges, *both* must converge and we deduce that $\sum |a_n|$ converges, since $|a_n| = p_n + q_n$. This contradiction proves (i).

To prove (ii), we simply use (4) in conjunction with Theorem 8.8.

8.9 REAL AND IMAGINARY PARTS OF A COMPLEX SERIES

Let $\sum c_n$ be a series with complex terms and write $c_n = a_n + ib_n$, where a_n and b_n are real. The series $\sum a_n$ and $\sum b_n$ are called, respectively, the real and imaginary parts of $\sum c_n$. In situations involving complex series, it is often convenient to treat the real and imaginary parts separately. Of course, convergence of both $\sum a_n$ and $\sum b_n$ implies convergence of $\sum c_n$. Conversely, convergence of $\sum c_n$ implies convergence of both $\sum a_n$ and $\sum b_n$. The same remarks hold for *absolute* convergence.

However, when Σc_n is *conditionally* convergent, one (but not both) of Σa_n and Σb_n might be absolutely convergent. (See Exercise 8.19.)

If Σc_n converges absolutely, we can apply part (ii) of Theorem 8.19 to the real and imaginary parts separately, to obtain the decomposition.

$$\Sigma c_n = \Sigma(p_n + iu_n) - \Sigma(q_n + iv_n),$$

where $\Sigma p_n, \Sigma q_n, \Sigma u_n, \Sigma v_n$ are convergent series of nonnegative terms.

8.10 TESTS FOR CONVERGENCE OF SERIES WITH POSITIVE TERMS

Theorem 8.20 (Comparison test). If $a_n > 0$ and $b_n > 0$ for $n = 1, 2, \ldots$, and if there exist positive constants c and N such that

$$a_n < cb_n \quad \text{for } n \geq N,$$

then convergence of Σb_n implies convergence of Σa_n.

Proof. The partial sums of Σa_n are bounded if the partial sums of Σb_n are bounded. By Theorem 8.9, this completes the proof.

Theorem 8.21 (Limit comparison test). Assume that $a_n > 0$ and $b_n > 0$ for $n = 1, 2, \ldots$, and suppose that

$$\lim_{n \to \infty} \frac{a_n}{b_n} = 1.$$

Then Σa_n converges if, and only if, Σb_n converges.

Proof. There exists N such that $n \geq N$ implies $\frac{1}{2} < a_n/b_n < \frac{3}{2}$. The theorem follows by applying Theorem 8.20 twice.

NOTE. Theorem 8.21 also holds if $\lim_{n \to \infty} a_n/b_n = c$, provided that $c \neq 0$. If $\lim_{n \to \infty} a_n/b_n = 0$, we can only conclude that convergence of Σb_n implies convergence of Σa_n.

8.11 THE GEOMETRIC SERIES

To use comparison tests effectively, we must have at our disposal some examples of series of known behavior. One of the most important series for comparison purposes is the *geometric series*.

Theorem 8.22. If $|x| < 1$, the series $1 + x + x^2 + \cdots$ converges and has sum $1/(1 - x)$. If $|x| \geq 1$, the series diverges.

Proof. $(1 - x) \sum_{k=0}^{n} x^k = \sum_{k=0}^{n} (x^k - x^{k+1}) = 1 - x^{n+1}$. When $|x| < 1$, we find $\lim_{n \to \infty} x^{n+1} = 0$. If $|x| \geq 1$, the general term does not tend to zero and the series cannot converge.

8.12 THE INTEGRAL TEST

Further examples of series of known behavior can be obtained very simply by applying the *integral test*.

Theorem 8.23 (Integral test). *Let f be a positive decreasing function defined on $[1, +\infty)$ such that $\lim_{x \to +\infty} f(x) = 0$. For $n = 1, 2, \ldots$, define*

$$s_n = \sum_{k=1}^{n} f(k), \qquad t_n = \int_1^n f(x)\, dx, \qquad d_n = s_n - t_n.$$

Then we have:

i) $0 < f(n + 1) \le d_{n+1} \le d_n \le f(1), \qquad$ *for* $n = 1, 2, \ldots$

ii) $\lim_{n \to \infty} d_n$ *exists.*

iii) $\sum_{n=1}^{\infty} f(n)$ *converges if, and only if, the sequence* $\{t_n\}$ *converges.*

iv) $0 \le d_k - \lim_{n \to \infty} d_n \le f(k), \qquad$ *for* $k = 1, 2, \ldots$

Proof. To prove (i), write

$$t_{n+1} = \int_1^{n+1} f(x)\, dx = \sum_{k=1}^{n} \int_k^{k+1} f(x)\, dx \le \sum_{k=1}^{n} \int_k^{k+1} f(k)\, dx$$

$$= \sum_{k=1}^{n} f(k) = s_n.$$

This implies that $f(n + 1) = s_{n+1} - s_n \le s_{n+1} - t_{n+1} = d_{n+1}$, and we obtain

$$0 < f(n + 1) \le d_{n+1}.$$

But we also have

$$d_n - d_{n+1} = t_{n+1} - t_n - (s_{n+1} - s_n) = \int_n^{n+1} f(x)\, dx - f(n + 1) \qquad (5)$$

$$\ge \int_n^{n+1} f(n + 1)\, dx - f(n + 1) = 0,$$

and hence $d_{n+1} \le d_n \le d_1 = f(1)$. This proves (i). But now it is clear that (i) implies (ii) and that (ii) implies (iii).

To prove part (iv), we use (5) again to write

$$0 \le d_n - d_{n+1} \le \int_n^{n+1} f(n)\, dx - f(n + 1) = f(n) - f(n + 1).$$

Summing on n, we get

$$0 \le \sum_{n=k}^{\infty} (d_n - d_{n+1}) \le \sum_{n=k}^{\infty} (f(n) - f(n + 1)), \qquad \text{if } k \ge 1.$$

When we evaluate the sums of these telescoping series, we get (iv).

NOTE. Let $D = \lim_{n \to \infty} d_n$. Then (i) implies $0 \le D \le f(1)$, whereas (iv) gives us

$$0 \le \sum_{k=1}^{n} f(k) - \int_{1}^{n} f(x)\, dx - D \le f(n). \tag{6}$$

This inequality is extremely useful for approximating certain finite sums by integrals.

8.13 THE BIG OH AND LITTLE OH NOTATION

Definition 8.24. *Given two sequences $\{a_n\}$ and $\{b_n\}$ such that $b_n \ge 0$ for all n. We write*

$$a_n = O(b_n) \qquad (\textit{read: ``a_n is big oh of } b_n\textit{''}),$$

if there exists a constant $M > 0$ such that $|a_n| \le M b_n$ for all n. We write

$$a_n = o(b_n) \qquad \textit{as } n \to \infty \qquad (\textit{read: ``a_n is little oh of } b_n\textit{''}),$$

if $\lim_{n \to \infty} a_n/b_n = 0$.

NOTE. An equation of the form $a_n = c_n + O(b_n)$ means $a_n - c_n = O(b_n)$. Similarly, $a_n = c_n + o(b_n)$ means $a_n - c_n = o(b_n)$. The advantage of this notation is that it allows us to replace certain inequalities by equations. For example, (6) implies

$$\sum_{k=1}^{n} f(k) = \int_{1}^{n} f(x)\, dx + D + O(f(n)). \tag{7}$$

Example 1. Let $f(x) = 1/x$ in Theorem 8.23. We find $t_n = \log n$ and hence $\sum 1/n$ diverges. However, (ii) establishes the existence of the limit

$$\lim_{n \to \infty} \left(\sum_{k=1}^{n} \frac{1}{k} - \log n \right),$$

a famous number known as *Euler's constant*, usually denoted by C (or by γ). Equation (7) becomes

$$\sum_{k=1}^{n} \frac{1}{k} = \log n + C + O\!\left(\frac{1}{n}\right). \tag{8}$$

Example 2. Let $f(x) = x^{-s}$, $s \ne 1$, in Theorem 8.23. We find that $\sum n^{-s}$ converges if $s > 1$ and diverges if $s < 1$. For $s > 1$, this series defines an important function known as the *Riemann zeta function*:

$$\zeta(s) = \sum_{n=1}^{\infty} \frac{1}{n^s} \qquad (s > 1).$$

For $s > 0$, $s \ne 1$, we can apply (7) to write

$$\sum_{k=1}^{n} \frac{1}{k^s} = \frac{n^{1-s} - 1}{1 - s} + C(s) + O\!\left(\frac{1}{n^s}\right),$$

where $C(s) = \lim_{n \to \infty} \left(\sum_{k=1}^{n} k^{-s} - (n^{1-s} - 1)/(1 - s) \right)$.

8.14 THE RATIO TEST AND ROOT TEST

Theorem 8.25 (Ratio test). *Given a series $\sum a_n$ of nonzero complex terms, let*

$$r = \liminf_{n \to \infty} \left| \frac{a_{n+1}}{a_n} \right| , \qquad R = \limsup_{n \to \infty} \left| \frac{a_{n+1}}{a_n} \right| .$$

a) *The series $\sum a_n$ converges absolutely if $R < 1$.*

b) *The series $\sum a_n$ diverges if $r > 1$.*

c) *The test is inconclusive if $r \leq 1 \leq R$.*

Proof. Assume that $R < 1$ and choose x so that $R < x < 1$. The definition of R implies the existence of N such that $|a_{n+1}/a_n| < x$ if $n \geq N$. Since $x = x^{n+1}/x^n$, this means that

$$\frac{|a_{n+1}|}{x^{n+1}} < \frac{|a_n|}{x^n} \leq \frac{|a_N|}{x^N} , \qquad \text{if } n \geq N,$$

and hence $|a_n| \leq cx^n$ if $n \geq N$, where $c = |a_N| x^{-N}$. Statement (a) now follows by applying the comparison test.

To prove (b), we simply observe that $r > 1$ implies $|a_{n+1}| > |a_n|$ for all $n \geq N$ for some N and hence we cannot have $\lim_{n \to \infty} a_n = 0$.

To prove (c), consider the two examples $\sum n^{-1}$ and $\sum n^{-2}$. In both cases, $r = R = 1$ but $\sum n^{-1}$ diverges, whereas $\sum n^{-2}$ converges.

Theorem 8.26 (Root test). *Given a series $\sum a_n$ of complex terms, let*

$$\rho = \limsup_{n \to \infty} \sqrt[n]{|a_n|}.$$

a) *The series $\sum a_n$ converges absolutely if $\rho < 1$.*

b) *The series $\sum a_n$ diverges if $\rho > 1$.*

c) *The test is inconclusive if $\rho = 1$.*

Proof. Assume that $\rho < 1$ and choose x so that $\rho < x < 1$. The definition of ρ implies the existence of N such that $|a_n| < x^n$ for $n \geq N$. Hence, $\sum |a_n|$ converges by the comparison test. This proves (a).

To prove (b), we observe that $\rho > 1$ implies $|a_n| > 1$ infinitely often and hence we cannot have $\lim_{n \to \infty} a_n = 0$.

Finally, (c) is proved by using the same examples as in Theorem 8.25.

NOTE. The root test is more "powerful" than the ratio test. That is, whenever the root test is inconclusive, so is the ratio test. But there are examples where the ratio test fails and the root test *is* conclusive. (See Exercise 8.4.)

8.15 DIRICHLET'S TEST AND ABEL'S TEST.

All the tests in the previous section help us to determine *absolute* convergence of a series with complex terms. It is also important to have tests for determining convergence when the series might not converge absolutely. The tests in this section are particularly useful for this purpose. They all depend on the *partial summation formula* of Abel (equation (9) in the next theorem).

Theorem 8.27. *If $\{a_n\}$ and $\{b_n\}$ are two sequences of complex numbers, define*

$$A_n = a_1 + \cdots + a_n.$$

Then we have the identity

$$\sum_{k=1}^{n} a_k b_k = A_n b_{n+1} - \sum_{k=1}^{n} A_k(b_{k+1} - b_k). \qquad (9)$$

Therefore, $\sum_{k=1}^{\infty} a_k b_k$ converges if both the series $\sum_{k=1}^{\infty} A_k(b_{k+1} - b_k)$ and the sequence $\{A_n b_{n+1}\}$ converge.

Proof. Writing $A_0 = 0$, we have

$$\sum_{k=1}^{n} a_k b_k = \sum_{k=1}^{n} (A_k - A_{k-1})b_k = \sum_{k=1}^{n} A_k b_k - \sum_{k=1}^{n} A_k b_{k+1} + A_n b_{n+1}.$$

The second assertion follows at once from this identity.

NOTE. Formula (9) is analogous to the formula for integration by parts in a Riemann–Stieltjes integral.

Theorem 8.28 (Dirichlet's test). *Let $\sum a_n$ be a series of complex terms whose partial sums form a bounded sequence. Let $\{b_n\}$ be a decreasing sequence which converges to 0. Then $\sum a_n b_n$ converges.*

Proof. Let $A_n = a_1 + \cdots + a_n$ and assume that $|A_n| \leq M$ for all n. Then

$$\lim_{n \to \infty} A_n b_{n+1} = 0.$$

Therefore, to establish convergence of $\sum a_n b_n$ we need only show that $\sum A_k(b_{k+1} - b_k)$ is convergent. Since $b_n \searrow$, we have

$$|A_k(b_{k+1} - b_k)| \leq M(b_k - b_{k+1}).$$

But the series $\sum(b_{k+1} - b_k)$ is a convergent telescoping series. Hence the comparison test implies *absolute* convergence of $\sum A_k(b_{k+1} - b_k)$.

Theorem 8.29 (Abel's test). *The series $\sum a_n b_n$ converges if $\sum a_n$ converges and if $\{b_n\}$ is a monotonic convergent sequence.*

Proof. Convergence of $\sum a_n$ and of $\{b_n\}$ establishes the existence of the limit $\lim_{n \to \infty} A_n b_{n+1}$, where $A_n = a_1 + \cdots + a_n$. Also, $\{A_n\}$ is a bounded sequence. The remainder of the proof is similar to that of Theorem 8.28. (Two further tests, similar to the above, are given in Exercise 8.27.)

8.16 PARTIAL SUMS OF THE GEOMETRIC SERIES $\sum z^n$ ON THE UNIT CIRCLE $|z| = 1$

To use Dirichlet's test effectively, we must be acquainted with a few series having bounded partial sums. Of course, all *convergent* series have this property. The next theorem gives an example of a divergent series whose partial sums are bounded. This is the geometric series $\sum z^n$ with $|z| = 1$, that is, with $z = e^{ix}$ where x is real. The formula for the partial sums of this series is of fundamental importance in the theory of Fourier series.

Theorem 8.30. *For every real* $x \neq 2m\pi$ *(m is an integer), we have*

$$\sum_{k=1}^{n} e^{ikx} = e^{ix} \frac{1 - e^{inx}}{1 - e^{ix}} = \frac{\sin(nx/2)}{\sin(x/2)} e^{i(n+1)x/2}. \tag{10}$$

NOTE. This identity yields the following estimate:

$$\left| \sum_{k=1}^{n} e^{ikx} \right| \leq \frac{1}{|\sin(x/2)|}. \tag{11}$$

Proof. $(1 - e^{ix}) \sum_{k=1}^{n} e^{ikx} = \sum_{k=1}^{n} (e^{ikx} - e^{i(k+1)x}) = e^{ix} - e^{i(n+1)x}$. This establishes the first equality in (10). The second follows from the identity

$$e^{ix} \frac{1 - e^{inx}}{1 - e^{ix}} = \frac{e^{inx/2} - e^{-inx/2}}{e^{ix/2} - e^{-ix/2}} e^{i(n+1)x/2}.$$

NOTE. By considering the real and imaginary parts of (10), we obtain

$$\sum_{k=1}^{n} \cos kx = \sin \frac{nx}{2} \cos (n+1) \frac{x}{2} \Big/ \sin \frac{x}{2}$$

$$= -\frac{1}{2} + \frac{1}{2} \sin (2n+1) \frac{x}{2} \Big/ \sin \frac{x}{2}, \tag{12}$$

$$\sum_{k=1}^{n} \sin kx = \sin \frac{nx}{2} \sin (n+1) \frac{x}{2} \Big/ \sin \frac{x}{2}. \tag{13}$$

Using (10), we can also write

$$\sum_{k=1}^{n} e^{i(2k-1)x} = e^{-ix} \sum_{k=1}^{n} e^{ik(2x)} = \frac{\sin nx}{\sin x} e^{inx}, \tag{14}$$

an identity valid for every $x \neq m\pi$ (m is an integer). Taking real and imaginary parts of (14) we obtain

$$\sum_{k=1}^{n} \cos (2k-1)x = \frac{\sin 2nx}{2 \sin x}, \qquad (15)$$

$$\sum_{k=1}^{n} \sin (2k-1)x = \frac{\sin^2 nx}{\sin x}. \qquad (16)$$

Formulas (12) and (16) occur in the theory of Fourier series.

8.17 REARRANGEMENTS OF SERIES

We recall that $\mathbf{Z}^+$ denotes the set of positive integers, $\mathbf{Z}^+ = \{1, 2, 3, \dots\}$.

Definition 8.31. *Let f be a function whose domain is $\mathbf{Z}^+$ and whose range is $\mathbf{Z}^+$, and assume that f is one-to-one on $\mathbf{Z}^+$. Let $\sum a_n$ and $\sum b_n$ be two series such that*

$$b_n = a_{f(n)} \qquad for\ n = 1, 2, \dots \qquad (17)$$

Then $\sum b_n$ is said to be a rearrangement of $\sum a_n$.

NOTE. Equation (17) implies $a_n = b_{f^{-1}(n)}$ and hence $\sum a_n$ is also a rearrangement of $\sum b_n$.

Theorem 8.32. *Let $\sum a_n$ be an absolutely convergent series having sum s. Then every rearrangement of $\sum a_n$ also converges absolutely and has sum s.*

Proof. Let $\{b_n\}$ be defined by (17). Then

$$|b_1| + \cdots + |b_n| = |a_{f(1)}| + \cdots + |a_{f(n)}| \le \sum_{k=1}^{\infty} |a_k|,$$

so $\sum |b_n|$ has bounded partial sums. Hence $\sum b_n$ converges absolutely.

To show that $\sum b_n = s$, let $t_n = b_1 + \cdots + b_n$, $s_n = a_1 + \cdots + a_n$. Given $\varepsilon > 0$, choose N so that $|s_N - s| < \varepsilon/2$ and so that $\sum_{k=1}^{\infty} |a_{N+k}| \le \varepsilon/2$. Then

$$|t_n - s| \le |t_n - s_N| + |s_N - s| < |t_n - s_N| + \frac{\varepsilon}{2}.$$

Choose M so that $\{1, 2, \dots, N\} \subseteq \{f(1), f(2), \dots, f(M)\}$. Then $n > M$ implies $f(n) > N$, and for such n we have

$$|t_n - s_N| = |b_1 + \cdots + b_n - (a_1 + \cdots + a_N)|$$
$$= |a_{f(1)} + \cdots + a_{f(n)} - (a_1 + \cdots + a_N)|$$
$$\le |a_{N+1}| + |a_{N+2}| + \cdots \le \frac{\varepsilon}{2},$$

since all the terms $a_1, \dots, a_N$ cancel out in the subtraction. Hence, $n > M$ implies $|t_n - s| < \varepsilon$ and this means $\sum b_n = s$.

8.18 RIEMANN'S THEOREM ON CONDITIONALLY CONVERGENT SERIES

The hypothesis of absolute convergence is essential in Theorem 8.32. Riemann discovered that any *conditionally* convergent series of real terms can be rearranged to yield a series which converges to any prescribed sum. This remarkable fact is a consequence of the following theorem:

Theorem 8.33. *Let $\sum a_n$ be a conditionally convergent series with real-valued terms. Let x and y be given numbers in the closed interval $[-\infty, +\infty]$, with $x \le y$. Then there exists a rearrangement $\sum b_n$ of $\sum a_n$ such that*

$$\liminf_{n \to \infty} t_n = x \quad and \quad \limsup_{n \to \infty} t_n = y,$$

where $t_n = b_1 + \cdots + b_n$.

Proof. Discarding those terms of a series which are zero does not affect its convergence or divergence. Hence we might as well assume that no terms of $\sum a_n$ are zero. Let p_n denote the nth positive term of $\sum a_n$ and let $-q_n$ denote its nth negative term. Then $\sum p_n$ and $\sum q_n$ are both divergent series of positive terms. [Why?] Next, construct two sequences of real numbers, say $\{x_n\}$ and $\{y_n\}$, such that

$$\lim_{n \to \infty} x_n = x, \quad \lim_{n \to \infty} y_n = y, \quad \text{with } x_n < y_n, \quad y_1 > 0.$$

The idea of the proof is now quite simple. We take just enough (say k_1) positive terms so that

$$p_1 + \cdots + p_{k_1} > y_1,$$

followed by just enough (say r_1) negative terms so that

$$p_1 + \cdots + p_{k_1} - q_1 - \cdots - q_{r_1} < x_1.$$

Next, we take just enough *further* positive terms so that

$$p_1 + \cdots + p_{k_1} - q_1 - \cdots - q_{r_1} + p_{k_1+1} + \cdots + p_{k_2} > y_2,$$

followed by just enough further negative terms to satisfy the inequality

$$p_1 + \cdots + p_{k_1} - q_1 - \cdots - q_{r_1} + p_{k_1+1} + \cdots$$
$$+ p_{k_2} - q_{r_1+1} - \cdots - q_{r_2} < x_2.$$

These steps are possible since $\sum p_n$ and $\sum q_n$ are both divergent series of positive terms. If the process is continued in this way, we obviously obtain a rearrangement of $\sum a_n$. We leave it to the reader to show that the partial sums of this rearrangement have limit superior y and limit inferior x.

8.19 SUBSERIES

Definition 8.34. *Let f be a function whose domain is $\mathbf{Z}^+$ and whose range is an infinite subset of $\mathbf{Z}^+$, and assume that f is one-to-one on $\mathbf{Z}^+$. Let $\sum a_n$ and $\sum b_n$ be* *two series such that*

$$b_n = a_{f(n)}, \qquad if \ n \in \mathbf{Z}^+.$$

Then Σb_n is said to be a subseries of Σa_n.

Theorem 8.35. *If Σa_n converges absolutely, every subseries Σb_n also converges absolutely. Moreover, we have*

$$\left|\sum_{n=1}^{\infty} b_n\right| \le \sum_{n=1}^{\infty} |b_n| \le \sum_{n=1}^{\infty} |a_n|.$$

Proof. Given n, let N be the largest integer in the set $\{f(1), \ldots, f(n)\}$. Then

$$\left|\sum_{k=1}^{n} b_k\right| \le \sum_{k=1}^{n} |b_k| \le \sum_{k=1}^{N} |a_k| \le \sum_{k=1}^{\infty} |a_k|.$$

The inequality $\sum_{k=1}^{n} |b_k| \le \sum_{k=1}^{\infty} |a_k|$ implies absolute convergence of Σb_n.

Theorem 8.36. *Let $\{f_1, f_2, \ldots\}$ be a countable collection of functions, each defined on $\mathbf{Z}^+$, having the following properties:*

a) *Each f_n is one-to-one on $\mathbf{Z}^+$.*
b) *The range $f_n(\mathbf{Z}^+)$ is a subset Q_n of $\mathbf{Z}^+$.*
c) *$\{Q_1, Q_2, \ldots\}$ is a collection of disjoint sets whose union is $\mathbf{Z}^+$.*

Let Σa_n be an absolutely convergent series and define

$$b_k(n) = a_{f_k(n)}, \qquad if \ n \in \mathbf{Z}^+, \ \ k \in \mathbf{Z}^+.$$

Then:

i) *For each k, $\sum_{n=1}^{\infty} b_k(n)$ is an absolutely convergent subseries of Σa_n.*
ii) *If $s_k = \sum_{n=1}^{\infty} b_k(n)$, the series $\sum_{k=1}^{\infty} s_k$ converges absolutely and has the same sum as $\sum_{k=1}^{\infty} a_k$.*

Proof. Theorem 8.35 implies (i). To prove (ii), let $t_k = |s_1| + \cdots + |s_k|$. Then

$$t_k \le \sum_{n=1}^{\infty} |b_1(n)| + \cdots + \sum_{n=1}^{\infty} |b_k(n)| = \sum_{n=1}^{\infty} (|b_1(n)| + \cdots + |b_k(n)|)$$

$$= \sum_{n=1}^{\infty} (|a_{f_1(n)}| + \cdots + |a_{f_k(n)}|).$$

But $\sum_{n=1}^{\infty} (|a_{f_1(n)}| + \cdots + |a_{f_k(n)}|) \le \sum_{n=1}^{\infty} |a_n|$. This proves that $\Sigma|s_k|$ has bounded partial sums and hence Σs_k converges absolutely.

To find the sum of Σs_k, we proceed as follows: Let $\varepsilon > 0$ be given and choose N so that $n \ge N$ implies

$$\sum_{k=1}^{\infty} |a_k| - \sum_{k=1}^{n} |a_k| < \frac{\varepsilon}{2}. \tag{18}$$

Choose enough functions $f_1, \ldots, f_r$ so that each term $a_1, a_2, \ldots, a_N$ will appear somewhere in the sum

$$\sum_{n=1}^{\infty} a_{f_1(n)} + \cdots + \sum_{n=1}^{\infty} a_{f_r(n)}.$$

The number r depends on N and hence on ε. If $n > r$ and $n > N$, we have

$$\left| s_1 + s_2 + \cdots + s_n - \sum_{k=1}^{n} a_k \right| \leq |a_{N+1}| + |a_{N+2}| + \cdots < \frac{\varepsilon}{2}, \qquad (19)$$

because the terms $a_1, a_2, \ldots, a_N$ cancel in the subtraction. Now (18) implies

$$\left| \sum_{k=1}^{\infty} a_k - \sum_{k=1}^{n} a_k \right| < \frac{\varepsilon}{2}.$$

When this is combined with (19) we find

$$\left| s_1 + \cdots + s_n - \sum_{k=1}^{\infty} a_k \right| < \varepsilon,$$

if $n > r, n > N$. This completes the proof of (ii).

8.20 DOUBLE SEQUENCES

Definition 8.37. *A function f whose domain is $\mathbf{Z}^+ \times \mathbf{Z}^+$ is called a double sequence.*

NOTE. We shall be interested only in real- or complex-valued double sequences.

Definition 8.38. *If $a \in \mathbf{C}$, we write $\lim_{p,q \to \infty} f(p, q) = a$ and we say that the double sequence f converges to a, provided that the following condition is satisfied: For every $\varepsilon > 0$, there exists an N such that $|f(p, q) - a| < \varepsilon$ whenever both $p > N$ and $q > N$.*

Theorem 8.39. *Assume that $\lim_{p,q \to \infty} f(p, q) = a$. For each fixed p, assume that the limit $\lim_{q \to \infty} f(p, q)$ exists. Then the limit $\lim_{p \to \infty} (\lim_{q \to \infty} f(p, q))$ also exists and has the value a.*

NOTE. To distinguish $\lim_{p,q \to \infty} f(p, q)$ from $\lim_{p \to \infty} (\lim_{q \to \infty} f(p, q))$, the first is called a *double limit*, the second an *iterated limit*.

Proof. Let $F(p) = \lim_{q \to \infty} f(p, q)$. Given $\varepsilon > 0$, choose N_1 so that

$$|f(p, q) - a| < \frac{\varepsilon}{2}, \qquad \text{if } p > N_1 \text{ and } q > N_1. \qquad (20)$$

For each p we can choose N_2 so that

$$|F(p) - f(p, q)| < \frac{\varepsilon}{2}, \qquad \text{if } q > N_2. \qquad (21)$$

(Note that N_2 depends on p as well as on ε.) For each $p > N_1$ choose N_2, and then choose a fixed q greater than both N_1 and N_2. Then both (20) and (21) hold and hence

$$|F(p) - a| < \varepsilon, \qquad \text{if } p > N_1.$$

Therefore, $\lim_{p \to \infty} F(p) = a$.

NOTE. A similar result holds if we interchange the roles of p and q.

Thus the existence of the double limit $\lim_{p,q \to \infty} f(p, q)$ and of $\lim_{q \to \infty} f(p, q)$ implies the existence of the iterated limit

$$\lim_{p \to \infty} \left(\lim_{q \to \infty} f(p, q) \right).$$

The following example shows that the converse is not true.

Example. Let

$$f(p, q) = \frac{pq}{p^2 + q^2}, \qquad (p = 1, 2, \ldots, \quad q = 1, 2, \ldots).$$

Then $\lim_{q \to \infty} f(p, q) = 0$ and hence $\lim_{p \to \infty} (\lim_{q \to \infty} f(p, q)) = 0$. But $f(p, q) = \frac{1}{2}$ when $p = q$ and $f(p, q) = \frac{2}{5}$ when $p = 2q$, and hence it is clear that the double limit cannot exist in this case.

A suitable converse of Theorem 8.39 can be established by introducing the notion of *uniform convergence*. (This is done in the next chapter in Theorem 9.16.)

Further examples illustrating the behavior of double sequences are given in Exercise 8.28.

8.21 DOUBLE SERIES

Definition 8.40. *Let f be a double sequence and let s be the double sequence defined by the equation*

$$s(p, q) = \sum_{m=1}^{p} \sum_{n=1}^{q} f(m, n).$$

The pair (f, s) is called a double series and is denoted by the symbol $\sum_{m,n} f(m, n)$ or, more briefly, by $\sum f(m, n)$. The double series is said to converge to the sum a if

$$\lim_{p,q \to \infty} s(p, q) = a.$$

Each number $f(m, n)$ is called a *term* of the double series and each $s(p, q)$ is a partial sum. If $\sum f(m, n)$ has only positive terms, it is easy to show that it converges if, and only if, the set of partial sums is bounded. (See Exercise 8.29.) We say $\sum f(m, n)$ converges *absolutely* if $\sum |f(m, n)|$ converges. Theorem 8.18 is valid for double series. (See Exercise 8.29.)

8.22 REARRANGEMENT THEOREM FOR DOUBLE SERIES

Definition 8.41. *Let f be a double sequence and let g be a one-to-one function defined on* $\mathbf{Z}^+$ *with range* $\mathbf{Z}^+ \times \mathbf{Z}^+$. *Let G be the sequence defined by*

$$G(n) = f[g(n)] \qquad if \quad n \in \mathbf{Z}^+.$$

Then g is said to be an arrangement of the double sequence f into the sequence G.

Theorem 8.42. *Let* $\sum f(m, n)$ *be a given double series and let g be an arrangement of the double sequence f into a sequence G. Then*

a) $\sum G(n)$ *converges absolutely if, and only if,* $\sum f(m, n)$ *converges absolutely.*

Assuming that $\sum f(m, n)$ *does converge absolutely, with sum S, we have further:*

b) $\sum_{n=1}^{\infty} G(n) = S.$

c) $\sum_{n=1}^{\infty} f(m, n)$ *and* $\sum_{m=1}^{\infty} f(m, n)$ *both converge absolutely.*

d) *If* $A_m = \sum_{n=1}^{\infty} f(m, n)$ *and* $B_n = \sum_{m=1}^{\infty} f(m, n)$, *both series* $\sum A_m$ *and* $\sum B_n$ *converge absolutely and both have sum S. That is,*

$$\sum_{m=1}^{\infty} \sum_{n=1}^{\infty} f(m, n) = \sum_{n=1}^{\infty} \sum_{m=1}^{\infty} f(m, n) = S.$$

Proof. Let $T_k = |G(1)| + \cdots + |G(k)|$ and let

$$S(p, q) = \sum_{m=1}^{p} \sum_{n=1}^{q} |f(m, n)|.$$

Then, for each k, there exists a pair (p, q) such that $T_k \leq S(p, q)$ and, conversely, for each pair (p, q) there exists an integer r such that $S(p, q) \leq T_r$. These inequalities tell us that $\sum |G(n)|$ has bounded partial sums if, and only if, $\sum |f(m, n)|$ has bounded partial sums. This proves (a).

Now assume that $\sum |f(m, n)|$ converges. Before we prove (b), we will show that the sum of the series $\sum G(n)$ is independent of the function g used to construct G from f. To see this, let h be another arrangement of the double sequence f into a sequence H. Then we have

$$G(n) = f[g(n)] \qquad and \qquad H(n) = f[h(n)].$$

But this means that $G(n) = H[k(n)]$, where $k(n) = h^{-1}[g(n)]$. Since k is a one-to-one mapping of $\mathbf{Z}^+$ onto $\mathbf{Z}^+$, the series $\sum H(n)$ is a rearrangement of $\sum G(n)$, and hence has the same sum. Let us denote this common sum by S'. We will show later that $S' = S$.

Now observe that each series in (c) is a subseries of $\sum G(n)$. Hence (c) follows from (a). Applying Theorem 8.36, we conclude that $\sum A_m$ converges absolutely and has sum S'. The same thing is true of $\sum B_n$. It remains to prove that $S' = S$.

For this purpose let $T = \lim_{p,q \to \infty} S(p, q)$. Given $\varepsilon > 0$, choose N so that $0 \le T - S(p, q) < \varepsilon/2$ whenever $p > N$ and $q > N$. Now write

$$t_k = \sum_{n=1}^{k} G(n), \qquad s(p, q) = \sum_{m=1}^{p} \sum_{n=1}^{q} f(m, n).$$

Choose M so that t_M includes all terms $f(m, n)$ with

$$1 \le m \le N + 1, \qquad 1 \le n \le N + 1.$$

Then $t_M - s(N + 1, N + 1)$ is a sum of terms $f(m, n)$ with either $m > N$ or $n > N$. Therefore, if $n \ge M$, we have

$$|t_n - s(N + 1, N + 1)| \le T - S(N + 1, N + 1) < \frac{\varepsilon}{2}.$$

Similarly,

$$|S - s(N + 1, N + 1)| \le T - S(N + 1, N + 1) < \frac{\varepsilon}{2}.$$

Thus, given $\varepsilon > 0$, we can always find M so that $|t_n - S| < \varepsilon$ whenever $n \ge M$. Since $\lim_{n \to \infty} t_n = S'$, it follows that $S' = S$.

NOTE. The series $\sum_{m=1}^{\infty} \sum_{n=1}^{\infty} f(m, n)$ and $\sum_{n=1}^{\infty} \sum_{m=1}^{\infty} f(m, n)$ are called "iterated series". Convergence of both iterated series does not imply their equality. For example, suppose

$$f(m, n) = \begin{cases} 1, & \text{if } m = n + 1, \ n = 1, 2, \ldots, \\ -1, & \text{if } m = n - 1, \ n = 1, 2, \ldots, \\ 0, & \text{otherwise.} \end{cases}$$

Then

$$\sum_{m=1}^{\infty} \sum_{n=1}^{\infty} f(m, n) = -1, \qquad \text{but} \quad \sum_{n=1}^{\infty} \sum_{m=1}^{\infty} f(m, n) = 1.$$

8.23 A SUFFICIENT CONDITION FOR EQUALITY OF ITERATED SERIES

Theorem 8.43. *Let f be a complex-valued double sequence. Assume that $\sum_{n=1}^{\infty} f(m, n)$ converges absolutely for each fixed m and that*

$$\sum_{m=1}^{\infty} \sum_{n=1}^{\infty} |f(m, n)|,$$

converges. Then:

a) *The double series $\sum_{m,n} f(m, n)$ converges absolutely.*
b) *The series $\sum_{m=1}^{\infty} f(m, n)$ converges absolutely for each n.*

c) *Both iterated series* $\sum_{n=1}^{\infty} \sum_{m=1}^{\infty} f(m, n)$ *and* $\sum_{m=1}^{\infty} \sum_{n=1}^{\infty} f(m, n)$ *converge absolutely and we have*

$$\sum_{m=1}^{\infty} \sum_{n=1}^{\infty} f(m, n) = \sum_{n=1}^{\infty} \sum_{m=1}^{\infty} f(m, n) = \sum_{m,n} f(m, n).$$

Proof. Let g be an arrangement of the double sequence f into a sequence G. Then $\sum G(n)$ is absolutely convergent since all the partial sums of $\sum |G(n)|$ are bounded by $\sum_{m=1}^{\infty} \sum_{n=1}^{\infty} |f(m, n)|$. By Theorem 8.42(a), the double series $\sum_{m,n} f(m, n)$ converges absolutely, and statements (b) and (c) also follow from Theorem 8.42.

As an application of Theorem 8.43 we prove the following theorem concerning double series $\sum_{m,n} f(m, n)$ whose terms can be factored into a function of m times a function of n.

Theorem 8.44. *Let* $\sum a_m$ *and* $\sum b_n$ *be two absolutely convergent series with sums A and B, respectively. Let f be the double sequence defined by the equation*

$$f(m, n) = a_m b_n, \qquad if\ (m, n) \in \mathbf{Z}^+ \times \mathbf{Z}^+.$$

Then $\sum_{m,n} f(m, n)$ *converges absolutely and has the sum AB.*

Proof. We have

$$\sum_{m=1}^{\infty} |a_m| \sum_{n=1}^{\infty} |b_n| = \sum_{m=1}^{\infty} \left(|a_m| \sum_{n=1}^{\infty} |b_n| \right) = \sum_{m=1}^{\infty} \sum_{n=1}^{\infty} |a_m| |b_n|.$$

Therefore, by Theorem 8.43, the double series $\sum_{m,n} a_m b_n$ converges absolutely and has sum AB.

8.24 MULTIPLICATION OF SERIES

Given two series $\sum a_n$ and $\sum b_n$, we can always form the double series $\sum f(m, n)$, where $f(m, n) = a_m b_n$. For every arrangement g of f into a sequence G, we are led to a further series $\sum G(n)$. By analogy with finite sums, it seems natural to refer to $\sum f(m, n)$ or to $\sum G(n)$ as the "product" of $\sum a_n$ and $\sum b_n$, and Theorem 8.44 justifies this terminology when the two given series $\sum a_n$ and $\sum b_n$ are absolutely convergent. However, if either $\sum a_n$ or $\sum b_n$ is *conditionally* convergent, we have no guarantee that either $\sum f(m, n)$ or $\sum G(n)$ will converge. Moreover, if one of them does converge, its sum need not be AB. The convergence and the sum will depend on the arrangement g. Different choices of g may yield different values of the product. There is one very important case in which the terms $f(m, n)$ are arranged "diagonally" to produce $\sum G(n)$, and then parentheses are inserted by grouping together those terms $a_m b_n$ for which $m + n$ has a fixed value. This product is called the *Cauchy product* and is defined as follows:

Definition 8.45. *Given two series $\sum_{n=0}^{\infty} a_n$ and $\sum_{n=0}^{\infty} b_n$, define*

$$c_n = \sum_{k=0}^{n} a_k b_{n-k}, \qquad \text{if } n = 0, 1, 2, \ldots \tag{22}$$

The series $\sum_{n=0}^{\infty} c_n$ is called the Cauchy product of $\sum a_n$ and $\sum b_n$.

NOTE. The Cauchy product arises in a natural way when we multiply two power series. (See Exercise 8.33.)

Because of Theorems 8.44 and 8.13, absolute convergence of *both* $\sum a_n$ and $\sum b_n$ implies convergence of the Cauchy product to the value

$$\sum_{n=0}^{\infty} c_n = \left(\sum_{n=0}^{\infty} a_n\right)\left(\sum_{n=0}^{\infty} b_n\right). \tag{23}$$

This equation may fail to hold if both $\sum a_n$ and $\sum b_n$ are *conditionally* convergent. (See Exercise 8.32.) However, we can prove that (23) is valid if at least *one* of $\sum a_n$, $\sum b_n$ is absolutely convergent.

Theorem 8.46 (Mertens). *Assume that $\sum_{n=0}^{\infty} a_n$ converges absolutely and has sum A, and suppose $\sum_{n=0}^{\infty} b_n$ converges with sum B. Then the Cauchy product of these two series converges and has sum AB.*

Proof. Define $A_n = \sum_{k=0}^{n} a_k$, $B_n = \sum_{k=0}^{n} b_k$, $C_n = \sum_{k=0}^{n} c_k$, where c_k is given by (22). Let $d_n = B - B_n$ and $e_n = \sum_{k=0}^{n} a_k d_{n-k}$. Then

$$C_p = \sum_{n=0}^{p} \sum_{k=0}^{n} a_k b_{n-k} = \sum_{n=0}^{p} \sum_{k=0}^{p} f_n(k), \tag{24}$$

where

$$f_n(k) = \begin{cases} a_k b_{n-k}, & \text{if } n \geq k, \\ 0, & \text{if } n < k. \end{cases}$$

Then (24) becomes

$$C_p = \sum_{k=0}^{p} \sum_{n=0}^{p} f_n(k) = \sum_{k=0}^{p} \sum_{n=k}^{p} a_k b_{n-k} = \sum_{k=0}^{p} a_k \sum_{m=0}^{p-k} b_m = \sum_{k=0}^{p} a_k B_{p-k}$$

$$= \sum_{k=0}^{p} a_k(B - d_{p-k}) = A_p B - e_p.$$

To complete the proof, it suffices to show that $e_p \to 0$ as $p \to \infty$. The sequence $\{d_n\}$ converges to 0, since $B = \sum b_n$. Choose $M > 0$ so that $|d_n| \leq M$ for all n, and let $K = \sum_{n=0}^{\infty} |a_n|$. Given $\varepsilon > 0$, choose N so that $n > N$ implies $|d_n| < \varepsilon/(2K)$ and also so that

$$\sum_{n=N+1}^{\infty} |a_n| < \frac{\varepsilon}{2M}.$$

Then, for $p > 2N$, we can write

$$|e_p| \le \sum_{k=0}^{N} |a_k d_{p-k}| + \sum_{k=N+1}^{p} |a_k d_{p-k}| \le \frac{\varepsilon}{2K} \sum_{k=0}^{N} |a_k| + M \sum_{k=N+1}^{p} |a_k|$$

$$\le \frac{\varepsilon}{2K} \sum_{k=0}^{\infty} |a_k| + M \sum_{k=N+1}^{\infty} |a_k| < \frac{\varepsilon}{2} + \frac{\varepsilon}{2} = \varepsilon.$$

This proves that $e_p \to 0$ as $p \to \infty$, and hence $C_p \to AB$ as $p \to \infty$.

A related theorem (due to Abel), in which no absolute convergence is assumed, will be proved in the next chapter. (See Theorem 9.32.)

Another product, known as the *Dirichlet product*, is of particular importance in the Theory of Numbers. We take $a_0 = b_0 = 0$ and, instead of defining c_n by (22), we use the formula

$$c_n = \sum_{d|n} a_d b_{n/d}, \qquad (n = 1, 2, \ldots), \tag{25}$$

where $\sum_{d|n}$ means a sum extended over all *positive divisors* of n (including 1 and n). For example, $c_6 = a_1 b_6 + a_2 b_3 + a_3 b_2 + a_6 b_1$, and $c_7 = a_1 b_7 + a_7 b_1$. The analog of Mertens' theorem holds also for this product. The Dirichlet product arises in a natural way when we multiply Dirichlet series. (See Exercise 8.34.)

8.25 CESÀRO SUMMABILITY

Definition 8.47. *Let s_n denote the nth partial sum of the series $\sum a_n$, and let $\{\sigma_n\}$ be the sequence of arithmetic means defined by*

$$\sigma_n = \frac{s_1 + \cdots + s_n}{n}, \qquad \text{if } n = 1, 2, \ldots \tag{26}$$

The series $\sum a_n$ is said to be Cesàro summable (or (C, 1) summable) if $\{\sigma_n\}$ converges. If $\lim_{n \to \infty} \sigma_n = S$, then S is called the Cesàro sum (or (C, 1) sum) of $\sum a_n$, and we write

$$\sum a_n = S \qquad (C, 1).$$

Example 1. Let $a_n = z^n$, $|z| = 1$, $z \ne 1$. Then

$$s_n = \frac{1}{1-z} - \frac{z^n}{1-z} \quad \text{and} \quad \sigma_n = \frac{1}{1-z} - \frac{1}{n} \frac{z(1-z^n)}{(1-z)^2}.$$

Therefore,

$$\sum_{n=1}^{\infty} z^{n-1} = \frac{1}{1-z} \qquad (C, 1).$$

In particular,

$$\sum_{n=1}^{\infty} (-1)^{n-1} = \tfrac{1}{2} \qquad (C, 1).$$

Example 2. Let $a_n = (-1)^{n+1}n$. In this case,

$$\limsup_{n \to \infty} \sigma_n = \tfrac{1}{2}, \qquad \liminf_{n \to \infty} \sigma_n = 0,$$

and hence $\sum(-1)^{n+1}n$ is not (C, 1) summable.

Theorem 8.48. *If a series is convergent with sum S, then it is also* (C, 1) *summable with Cesàro sum S.*

Proof. Let s_n denote the nth partial sum of the series, define σ_n by (26), and introduce $t_n = s_n - S$, $\tau_n = \sigma_n - S$. Then we have

$$\tau_n = \frac{t_1 + \cdots + t_n}{n}, \tag{27}$$

and we must prove that $\tau_n \to 0$ as $n \to \infty$. Choose $A > 0$ so that each $|t_n| \le A$. Given $\varepsilon > 0$, choose N so that $n > N$ implies $|t_n| < \varepsilon$. Taking $n > N$ in (27), we obtain

$$|\tau_n| \le \frac{|t_1| + \cdots + |t_N|}{n} + \frac{|t_{N+1}| + \cdots + |t_n|}{n} < \frac{NA}{n} + \varepsilon.$$

Hence, $\limsup_{n \to \infty} |\tau_n| \le \varepsilon$. Since ε is arbitrary, it follows that $\lim_{n \to \infty} |\tau_n| = 0$.

NOTE. We have really proved that if a sequence $\{s_n\}$ converges, then the sequence $\{\sigma_n\}$ of arithmetic means also converges and, in fact, to the same limit.

Cesàro summability is just one of a large class of "summability methods" which can be used to assign a "sum" to an infinite series. Theorem 8.48 and Example 1 (following Definition 8.47) show us that Cesàro's method has a wider scope than ordinary convergence. The theory of summability methods is an important and fascinating subject, but one which we cannot enter into here. For an excellent account of the subject the reader is referred to Hardy's *Divergent Series* (Reference 8.1). We shall see later that (C, 1) summability plays an important role in the theory of Fourier series. (See Theorem 11.15.)

8.26 INFINITE PRODUCTS

This section gives a brief introduction to the theory of infinite products.

Definition 8.49. *Given a sequence $\{u_n\}$ of real or complex numbers, let*

$$p_1 = u_1, \qquad p_2 = u_1 u_2, \qquad p_n = u_1 u_2 \cdots u_n = \prod_{k=1}^{n} u_k. \tag{28}$$

The ordered pair of sequences $(\{u_n\}, \{p_n\})$ *is called an infinite product (or simply, a product). The number p_n is called the nth partial product and u_n is called the nth* *factor of the product. The following symbols are used to denote the product defined by* (28):

$$u_1 u_2 \cdots u_n \cdots, \qquad \prod_{n=1}^{\infty} u_n. \tag{29}$$

NOTE. The symbol $\prod_{n=N+1}^{\infty} u_n$ means $\prod_{n=1}^{\infty} u_{N+n}$. We also write $\prod u_n$ when there is no danger of misunderstanding.

By analogy with infinite series, it would seem natural to call the product (29) convergent if $\{p_n\}$ converges. However, this definition would be inconvenient since every product having one factor equal to zero would converge, regardless of the behavior of the remaining factors. The following definition turns out to be more useful:

Definition 8.50. *Given an infinite product* $\prod_{n=1}^{\infty} u_n$, *let* $p_n = \prod_{k=1}^{n} u_k$.

a) *If infinitely many factors* u_n *are zero, we say the product diverges to zero.*

b) *If no factor* u_n *is zero, we say the product converges if there exists a number* $p \neq 0$ *such that* $\{p_n\}$ *converges to* p. *In this case,* p *is called the value of the product and we write* $p = \prod_{n=1}^{\infty} u_n$. *If* $\{p_n\}$ *converges to zero, we say the product diverges to zero.*

c) *If there exists an* N *such that* $n > N$ *implies* $u_n \neq 0$, *we say* $\prod_{n=1}^{\infty} u_n$ *converges, provided that* $\prod_{n=N+1}^{\infty} u_n$ *converges as described in* (b). *In this case, the value of the product* $\prod_{n=1}^{\infty} u_n$ *is*

$$u_1 u_2 \cdots u_N \prod_{n=N+1}^{\infty} u_n.$$

d) $\prod_{n=1}^{\infty} u_n$ *is called divergent if it does not converge as described in* (b) *or* (c).

Note that the value of a convergent infinite product can be zero. But this happens if, and only if, a finite number of factors are zero. The convergence of an infinite product is not affected by inserting or removing a finite number of factors, zero or not. It is this fact which makes Definition 8.50 very convenient.

Example. $\prod_{n=1}^{\infty} (1 + 1/n)$ and $\prod_{n=2}^{\infty} (1 - 1/n)$ are both divergent. In the first case, $p_n = n + 1$, and in the second case, $p_n = 1/n$.

Theorem 8.51 (Cauchy condition for products). *The infinite product* $\prod u_n$ *converges if, and only if, for every* $\varepsilon > 0$ *there exists an* N *such that* $n > N$ *implies*

$$|u_{n+1} u_{n+2} \cdots u_{n+k} - 1| < \varepsilon, \qquad \text{for } k = 1, 2, 3, \ldots \tag{30}$$

Proof. Assume that the product $\prod u_n$ converges. We can assume that no u_n is zero (discarding a few terms if necessary). Let $p_n = u_1 \cdots u_n$ and $p = \lim_{n \to \infty} p_n$. Then $p \neq 0$ and hence there exists an $M > 0$ such that $|p_n| > M$. Now $\{p_n\}$ satisfies the Cauchy condition for sequences. Hence, given $\varepsilon > 0$, there is an N such that $n > N$ implies $|p_{n+k} - p_n| < \varepsilon M$ for $k = 1, 2, \ldots$ Dividing by $|p_n|$, we obtain (30).

Now assume that condition (30) holds. Then $n > N$ implies $u_n \neq 0$. [Why?] Take $\varepsilon = \frac{1}{2}$ in (30), let N_0 be the corresponding N, and let $q_n = u_{N_0+1} u_{N_0+2} \cdots u_n$ if $n > N_0$. Then (30) implies $\frac{1}{2} < |q_n| < \frac{3}{2}$. Therefore, if $\{q_n\}$ converges, it cannot converge to zero. To show that $\{q_n\}$ *does* converge, let $\varepsilon > 0$ be arbitrary and write (30) as follows:

$$\left| \frac{q_{n+k}}{q_n} - 1 \right| < \varepsilon.$$

This gives us $|q_{n+k} - q_n| < \varepsilon |q_n| < \frac{3}{2}\varepsilon$. Therefore, $\{q_n\}$ satisfies the Cauchy condition for sequences and hence is convergent. This means that the product $\prod u_n$ converges.

NOTE. Taking $k = 1$ in (30), we find that convergence of $\prod u_n$ implies $\lim_{n \to \infty} u_n = 1$. For this reason, the factors of a product are written as $u_n = 1 + a_n$. Thus convergence of $\prod(1 + a_n)$ implies $\lim_{n \to \infty} a_n = 0$.

Theorem 8.52. *Assume that each $a_n > 0$. Then the product $\prod(1 + a_n)$ converges if, and only if, the series $\sum a_n$ converges.*

Proof. Part of the proof is based on the following inequality:

$$1 + x \le e^x. \tag{31}$$

Although (31) holds for all real x, we need it only for $x \ge 0$. When $x > 0$, (31) is a simple consequence of the Mean-Value Theorem, which gives us

$$e^x - 1 = x e^{x_0}, \qquad \text{where } 0 < x_0 < x.$$

Since $e^{x_0} \ge 1$, (31) follows at once from this equation.

Now let $s_n = a_1 + a_2 + \cdots + a_n, p_n = (1 + a_1)(1 + a_2) \cdots (1 + a_n)$. Both sequences $\{s_n\}$ and $\{p_n\}$ are increasing, and hence to prove the theorem we need only show that $\{s_n\}$ is bounded if, and only if, $\{p_n\}$ is bounded.

First, the inequality $p_n > s_n$ is obvious. Next, taking $x = a_k$ in (31), where $k = 1, 2, \ldots, n$, and multiplying, we find $p_n < e^{s_n}$. Hence, $\{s_n\}$ is bounded if, and only if, $\{p_n\}$ is bounded. Note that $\{p_n\}$ cannot converge to zero since each $p_n \ge 1$. Note also that

$$p_n \to +\infty \qquad \text{if } s_n \to +\infty.$$

Definition 8.53. *The product $\prod(1 + a_n)$ is said to converge absolutely if $\prod(1 + |a_n|)$ converges.*

Theorem 8.54. *Absolute convergence of $\prod(1 + a_n)$ implies convergence.*

Proof. Use the Cauchy condition along with the inequality

$$|(1 + a_{n+1})(1 + a_{n+2}) \cdots (1 + a_{n+k}) - 1|$$
$$\le (1 + |a_{n+1}|)(1 + |a_{n+2}|) \cdots (1 + |a_{n+k}|) - 1.$$

NOTE. Theorem 8.52 tells us that $\prod(1 + a_n)$ converges absolutely if, and only if, $\sum a_n$ converges absolutely. In Exercise 8.43 we give an example in which $\prod(1 + a_n)$ converges but $\sum a_n$ diverges.

A result analogous to Theorem 8.52 is the following:

Theorem 8.55. *Assume that each* $a_n \geq 0$. *Then the product* $\prod(1 - a_n)$ *converges if, and only if, the series* $\sum a_n$ *converges.*

Proof. Convergence of $\sum a_n$ implies absolute convergence (and hence convergence) of $\prod(1 - a_n)$.

To prove the converse, assume that $\sum a_n$ diverges. If $\{a_n\}$ does not converge to zero, then $\prod(1 - a_n)$ also diverges. Therefore we can assume that $a_n \to 0$ as $n \to \infty$. Discarding a few terms if necessary, we can assume that each $a_n \leq \frac{1}{2}$. Then each factor $1 - a_n \geq \frac{1}{2}$ (and hence $\neq 0$). Let

$$p_n = (1 - a_1)(1 - a_2) \cdots (1 - a_n), \qquad q_n = (1 + a_1)(1 + a_2) \cdots (1 + a_n).$$

Since we have

$$(1 - a_k)(1 + a_k) = 1 - a_k^2 \leq 1,$$

we can write $p_n \leq 1/q_n$. But in the proof of Theorem 8.52 we observed that $q_n \to +\infty$ if $\sum a_n$ diverges. Therefore, $p_n \to 0$ as $n \to \infty$ and, by part (b) of Definition 8.50, it follows that $\prod(1 - a_n)$ diverges to 0.

8.27 EULER'S PRODUCT FOR THE RIEMANN ZETA FUNCTION

We conclude this chapter with a theorem of Euler which expresses the Riemann zeta function $\zeta(s) = \sum_{n=1}^{\infty} n^{-s}$ as an infinite product extended over all primes.

Theorem 8.56. *Let* p_k *denote the* kth *prime number. Then if* $s > 1$ *we have*

$$\zeta(s) = \sum_{n=1}^{\infty} \frac{1}{n^s} = \prod_{k=1}^{\infty} \frac{1}{1 - p_k^{-s}}.$$

The product converges absolutely.

Proof. We consider the partial product $P_m = \prod_{k=1}^{m} (1 - p_k^{-s})^{-1}$ and show that $P_m \to \zeta(s)$ as $m \to \infty$. Writing each factor as a geometric series we have

$$P_m = \prod_{k=1}^{m} \left(1 + \frac{1}{p_k^s} + \frac{1}{p_k^{2s}} + \cdots \right),$$

a product of a finite number of absolutely convergent series. When we multiply these series together and rearrange the terms according to increasing denominators, we get another absolutely convergent series, a typical term of which is

$$\frac{1}{p_1^{a_1 s} p_2^{a_2 s} \cdots p_m^{a_m s}} = \frac{1}{n^s}, \qquad \text{where } n = p_1^{a_1} \cdots p_m^{a_m},$$

and each $a_i \geq 0$. Therefore we have

$$P_m = \sum_1 \frac{1}{n^s},$$

where $\sum_1$ is summed over those n having all their prime factors $\leq p_m$. By the unique factorization theorem (Theorem 1.9), each such n occurs once and only once in $\sum_1$. Subtracting P_m from $\zeta(s)$ we get

$$\zeta(s) - P_m = \sum_{n=1}^{\infty} \frac{1}{n^s} - \sum_1 \frac{1}{n^s} = \sum_2 \frac{1}{n^s},$$

where $\sum_2$ is summed over those n having at least one prime factor $> p_m$. Since these n occur among the integers $> p_m$, we have

$$|\zeta(s) - P_m| \leq \sum_{n > p_m} \frac{1}{n^s}.$$

As $m \to \infty$ the last sum tends to 0 because $\sum n^{-s}$ converges, so $P_m \to \zeta(s)$.

To prove that the product converges absolutely we use Theorem 8.52. The product has the form $\prod(1 + a_k)$, where

$$a_k = \frac{1}{p_k^s} + \frac{1}{p_k^{2s}} + \cdots$$

The series $\sum a_k$ converges absolutely since it is dominated by $\sum n^{-s}$. Therefore $\prod(1 + a_k)$ also converges absolutely.

EXERCISES

Sequences

8.1 a) Given a real-valued sequence $\{a_n\}$ bounded above, let $u_n = \sup \{a_k : k \geq n\}$. Then $u_n \searrow$ and hence $U = \lim_{n \to \infty} u_n$ is either finite or $-\infty$. Prove that

$$U = \limsup_{n \to \infty} a_n = \lim_{n \to \infty} (\sup \{a_k : k \geq n\}).$$

b) Similarly, if $\{a_n\}$ is bounded below, prove that

$$V = \liminf_{n \to \infty} a_n = \lim_{n \to \infty} (\inf \{a : k \geq n\}).$$

If U and V are finite, show that:

c) There exists a subsequence of $\{a_n\}$ which converges to U and a subsequence which converges to V.

d) If $U = V$, every subsequence of $\{a_n\}$ converges to U.

8.2 Given two real-valued sequences $\{a_n\}$ and $\{b_n\}$ bounded below. Prove that

a) $\limsup_{n \to \infty} (a_n + b_n) \leq \limsup_{n \to \infty} a_n + \limsup_{n \to \infty} b_n$.

b) $\lim \sup_{n \to \infty} (a_n b_n) \le (\lim \sup_{n \to \infty} a_n)(\lim \sup_{n \to \infty} b_n)$ if $a_n > 0$, $b_n > 0$ for all n, and if both $\lim \sup_{n \to \infty} a_n$ and $\lim \sup_{n \to \infty} b_n$ are finite or both are infinite.

8.3 Prove Theorems 8.3 and 8.4.

8.4 If each $a_n > 0$, prove that

$$\lim_{n \to \infty} \inf \frac{a_{n+1}}{a_n} \le \lim_{n \to \infty} \inf \sqrt[n]{a_n} \le \lim_{n \to \infty} \sup \sqrt[n]{a_n} \le \lim_{n \to \infty} \sup \frac{a_{n+1}}{a_n}.$$

8.5 Let $a_n = n^n/n!$. Show that $\lim_{n \to \infty} a_{n+1}/a_n = e$ and use Exercise 8.4 to deduce that

$$\lim_{n \to \infty} \frac{n}{(n!)^{1/n}} = e.$$

8.6 Let $\{a_n\}$ be a real-valued sequence and let $\sigma_n = (a_1 + \cdots + a_n)/n$. Show that

$$\lim_{n \to \infty} \inf a_n \le \lim_{n \to \infty} \inf \sigma_n \le \lim_{n \to \infty} \sup \sigma_n \le \lim_{n \to \infty} \sup a_n.$$

8.7 Find $\lim \sup_{n \to \infty} a_n$ and $\lim \inf_{n \to \infty} a_n$ if a_n is given by

a) $\cos n$,

b) $\left(1 + \dfrac{1}{n}\right) \cos n\pi$,

c) $n \sin \dfrac{n\pi}{3}$,

d) $\sin \dfrac{n\pi}{2} \cos \dfrac{n\pi}{2}$,

e) $(-1)^n n/(1 + n)^n$,

f) $\dfrac{n}{3} - \left[\dfrac{n}{3}\right]$.

NOTE. In (f), $[x]$ denotes the greatest integer $\le x$.

8.8 Let $a_n = 2\sqrt{n} - \sum_{k=1}^n 1/\sqrt{k}$. Prove that the sequence $\{a_n\}$ converges to a limit p in the interval $1 < p < 2$.

In each of Exercises 8.9 through 8.14, show that the real-valued sequence $\{a_n\}$ is convergent. The given conditions are assumed to hold for all $n \ge 1$. In Exercises 8.10 through 8.14, show that $\{a_n\}$ has the limit L indicated.

8.9 $|a_n| \le 2$, $|a_{n+2} - a_{n+1}| \le \frac{1}{8}|a_{n+1}^2 - a_n^2|$.

8.10 $a_1 \ge 0$, $a_2 \ge 0$, $a_{n+2} = (a_n a_{n+1})^{1/2}$, $L = (a_1 a_2^2)^{1/3}$.

8.11 $a_1 = 2$, $a_2 = 8$, $a_{2n+1} = \frac{1}{2}(a_{2n} + a_{2n+1})$, $a_{2n+2} = \dfrac{a_{2n} a_{2n-1}}{a_{2n+1}}$, $L = 4$.

8.12 $a_1 = -\frac{3}{2}$, $3a_{n+1} = 2 + a_n^3$, $L = 1$. Modify a_1 to make $L = -2$.

8.13 $a_1 = 3$, $a_{n+1} = \dfrac{3(1 + a_n)}{3 + a_n}$, $L = \sqrt{3}$.

8.14 $a_n = \dfrac{b_{n+1}}{b_n}$, where $b_1 = b_2 = 1$, $b_{n+2} = b_n + b_{n+1}$, $L = \dfrac{1 + \sqrt{5}}{2}$.

Hint. Show that $b_{n+2} b_n - b_{n+1}^2 = (-1)^{n+1}$ and deduce that $|a_n - a_{n+1}| < n^{-2}$, if $n > 4$.

Series

8.15 Test for convergence (p and q denote fixed real numbers).

a) $\displaystyle\sum_{n=1}^{\infty} n^3 e^{-n}$,

b) $\displaystyle\sum_{n=2}^{\infty} (\log n)^p$,

c) $\displaystyle\sum_{n=1}^{\infty} p^n n^p \quad (p > 0)$,

d) $\displaystyle\sum_{n=2}^{\infty} \frac{1}{n^p - n^q} \quad (0 < q < p)$,

e) $\displaystyle\sum_{n=1}^{\infty} n^{-1-1/n}$,

f) $\displaystyle\sum_{n=1}^{\infty} \frac{1}{p^n - q^n} \quad (0 < q < p)$,

g) $\displaystyle\sum_{n=1}^{\infty} \frac{1}{n \log (1 + 1/n)}$,

h) $\displaystyle\sum_{n=2}^{\infty} \frac{1}{(\log n)^{\log n}}$,

i) $\displaystyle\sum_{n=3}^{\infty} \frac{1}{n \log n (\log \log n)^p}$,

j) $\displaystyle\sum_{n=3}^{\infty} \left(\frac{1}{\log \log n}\right)^{\log \log n}$,

k) $\displaystyle\sum_{n=1}^{\infty} (\sqrt{1 + n^2} - n)$,

l) $\displaystyle\sum_{n=2}^{\infty} n^p \left(\frac{1}{\sqrt{n-1}} - \frac{1}{\sqrt{n}}\right)$,

m) $\displaystyle\sum_{n=1}^{\infty} (\sqrt[n]{n} - 1)^n$,

n) $\displaystyle\sum_{n=1}^{\infty} n^p(\sqrt{n+1} - 2\sqrt{n} + \sqrt{n-1})$.

8.16 Let $S = \{n_1, n_2, \dots\}$ denote the collection of those positive integers that do not involve the digit 0 in their decimal representation. (For example, $7 \in S$ but $101 \notin S$.) Show that $\sum_{k=1}^{\infty} 1/n_k$ converges and has a sum less than 90.

8.17 Given integers $a_1, a_2, \dots$ such that $1 \le a_n \le n - 1$, $n = 2, 3, \dots$ Show that the sum of the series $\sum_{n=1}^{\infty} a_n/n!$ is rational if, and only if, there exists an integer N such that $a_n = n - 1$ for all $n \ge N$. *Hint.* For sufficiency, show that $\sum_{n=2}^{\infty} (n - 1)/n!$ is a telescoping series with sum 1.

8.18 Let p and q be fixed integers, $p \ge q \ge 1$, and let

$$x_n = \sum_{k=qn+1}^{pn} \frac{1}{k}, \qquad s_n = \sum_{k=1}^{n} \frac{(-1)^{k+1}}{k}.$$

a) Use formula (8) to prove that $\lim_{n \to \infty} x_n = \log (p/q)$.

b) When $q = 1$, $p = 2$, show that $s_{2n} = x_n$ and deduce that

$$\sum_{n=1}^{\infty} \frac{(-1)^{n+1}}{n} = \log 2.$$

c) Rearrange the series in (b), writing alternately p positive terms followed by q negative terms and use (a) to show that this rearrangement has sum

$$\log 2 + \tfrac{1}{2} \log (p/q).$$

d) Find the sum of $\sum_{n=1}^{\infty} (-1)^{n+1}(1/(3n - 2) - 1/(3n - 1))$.

8.19 Let $c_n = a_n + ib_n$, where $a_n = (-1)^n/\sqrt{n}$, $b_n = 1/n^2$. Show that $\sum c_n$ is conditionally convergent.

8.20 Use Theorem 8.23 to derive the following formulas:

a) $\displaystyle\sum_{k=1}^{n} \frac{\log k}{k} = \frac{1}{2} \log^2 n + A + O\left(\frac{\log n}{n}\right)$ (A is constant).

b) $\displaystyle\sum_{k=2}^{n} \frac{1}{k \log k} = \log(\log n) + B + O\left(\frac{1}{n \log n}\right)$ (B is constant).

8.21 If $0 < a \le 1$, $s > 1$, define $\zeta(s, a) = \sum_{n=0}^{\infty} (n + a)^{-s}$.

a) Show that this series converges absolutely for $s > 1$ and prove that

$$\sum_{h=1}^{k} \zeta\left(s, \frac{h}{k}\right) = k^s \zeta(s) \qquad \text{if } k = 1, 2, \ldots,$$

where $\zeta(s) = \zeta(s, 1)$ is the Riemann zeta function.

b) Prove that $\sum_{n=1}^{\infty} (-1)^{n-1}/n^s = (1 - 2^{1-s})\zeta(s)$ if $s > 1$.

8.22 Given a convergent series $\sum a_n$, where each $a_n \ge 0$. Prove that $\sum \sqrt{a_n} n^{-p}$ converges if $p > \frac{1}{2}$. Give a counterexample for $p = \frac{1}{2}$.

8.23 Given that $\sum a_n$ diverges. Prove that $\sum n a_n$ also diverges.

8.24 Given that $\sum a_n$ converges, where each $a_n > 0$. Prove that

$$\sum (a_n a_{n+1})^{1/2}$$

also converges. Show that the converse is also true if $\{a_n\}$ is monotonic.

8.25 Given that $\sum a_n$ converges absolutely. Show that each of the following series also converges absolutely:

a) $\displaystyle\sum a_n^2$,

b) $\displaystyle\sum \frac{a_n}{1 + a_n}$ (if no $a_n = -1$),

c) $\displaystyle\sum \frac{a_n^2}{1 + a_n^2}$.

8.26 Determine all real values of x for which the following series converges:

$$\sum_{n=1}^{\infty} \left(1 + \frac{1}{2} + \cdots + \frac{1}{n}\right) \frac{\sin nx}{n}.$$

8.27 Prove the following statements:

a) $\sum a_n b_n$ converges if $\sum a_n$ converges and if $\sum (b_n - b_{n+1})$ converges absolutely.

b) $\sum a_n b_n$ converges if $\sum a_n$ has bounded partial sums and if $\sum (b_n - b_{n+1})$ converges absolutely, provided that $b_n \to 0$ as $n \to \infty$.

Double sequences and double series

8.28 Investigate the existence of the two iterated limits and the double limit of the double sequence f defined by

a) $\displaystyle f(p, q) = \frac{1}{p + q}$,

b) $\displaystyle f(p, q) = \frac{p}{p + q}$,

c) $f(p, q) = \dfrac{(-1)^p p}{p + q}$,

d) $f(p, q) = (-1)^{p+q} \left(\dfrac{1}{p} + \dfrac{1}{q} \right)$,

e) $f(p, q) = \dfrac{(-1)^p}{q}$,

f) $f(p, q) = (-1)^{p+q}$,

g) $f(p, q) = \dfrac{\cos p}{q}$,

h) $f(p, q) = \dfrac{p}{q^2} \sum_{n=1}^{q} \sin \dfrac{n}{p}$.

Answer. Double limit exists in (a), (d), (e), (g). Both iterated limits exist in (a), (b), (h). Only one iterated limit exists in (c), (e). Neither iterated limit exists in (d), (f).

8.29 Prove the following statements:

a) A double series of positive terms converges if, and only if, the set of partial sums is bounded.

b) A double series converges if it converges absolutely.

c) $\sum_{m,n} e^{-(m^2 + n^2)}$ converges.

8.30 Assume that the double series $\sum_{m,n} a(n) x^{mn}$ converges absolutely for $|x| < 1$. Call its sum $S(x)$. Show that each of the following series also converges absolutely for $|x| < 1$ and has sum $S(x)$:

$$\sum_{n=1}^{\infty} a(n) \frac{x^n}{1 - x^n}, \qquad \sum_{n=1}^{\infty} A(n) x^n, \qquad \text{where } A(n) = \sum_{d|n} a(d).$$

8.31 If α is real, show that the double series $\sum_{m,n} (m + in)^{-\alpha}$ converges absolutely if, and only if, $\alpha > 2$. *Hint.* Let $s(p, q) = \sum_{m=1}^{p} \sum_{n=1}^{q} |m + in|^{-\alpha}$. The set

$$\{m + in : m = 1, 2, \ldots, p, \ n = 1, 2, \ldots, p\}$$

consists of p^2 complex numbers of which one has absolute value $\sqrt{2}$, three satisfy $|1 + 2i| \leq |m + in| \leq 2\sqrt{2}$, five satisfy $|1 + 3i| \leq |m + in| \leq 3\sqrt{2}$, etc. Verify this geometrically and deduce the inequality

$$2^{-\alpha/2} \sum_{n=1}^{p} \frac{2n - 1}{n^\alpha} \leq s(p, p) \leq \sum_{n=1}^{p} \frac{2n - 1}{(n^2 + 1)^{\alpha/2}}.$$

8.32 a) Show that the Cauchy product of $\sum_{n=0}^{\infty} (-1)^{n+1}/\sqrt{n + 1}$ with itself is a divergent series.

b) Show that the Cauchy product of $\sum_{n=0}^{\infty} (-1)^{n+1}/(n + 1)$ with itself is the series

$$2 \sum_{n=1}^{\infty} \frac{(-1)^{n+1}}{n + 1} \left(1 + \frac{1}{2} + \cdots + \frac{1}{n} \right).$$

Does this converge? Why?

8.33 Given two absolutely convergent power series, say $\sum_{n=0}^{\infty} a_n x^n$ and $\sum_{n=0}^{\infty} b_n x^n$, having sums $A(x)$ and $B(x)$, respectively, show that $\sum_{n=0}^{\infty} c_n x^n = A(x)B(x)$ where

$$c_n = \sum_{k=0}^{n} a_k b_{n-k}.$$

8.34 A series of the form $\sum_{n=1}^{\infty} a_n/n^s$ is called a *Dirichlet series*. Given two absolutely convergent Dirichlet series, say $\sum_{n=1}^{\infty} a_n/n^s$ and $\sum_{n=1}^{\infty} b_n/n^s$, having sums $A(s)$ and $B(s)$, respectively, show that $\sum_{n=1}^{\infty} c_n/n^s = A(s)B(s)$ where $c_n = \sum_{d|n} a_d b_{n/d}$.

8.35 If $\zeta(s) = \sum_{n=1}^{\infty} 1/n^s$, $s > 1$, show that $\zeta^2(s) = \sum_{n=1}^{\infty} d(n)/n^s$, where $d(n)$ is the number of positive divisors of n (including 1 and n).

Cesàro summability

8.36 Show that each of the following series has (C, 1) sum 0:

 a) $1 - 1 - 1 + 1 + 1 - 1 - 1 + 1 + 1 - - + + \cdots$.

 b) $\frac{1}{2} - 1 + \frac{1}{2} + \frac{1}{2} - 1 + \frac{1}{2} + \frac{1}{2} - 1 + + - \cdots$.

 c) $\cos x + \cos 3x + \cos 5x + \cdots (x$ real, $x \neq m\pi)$.

8.37 Given a series $\sum a_n$, let

$$s_n = \sum_{k=1}^{n} a_k, \qquad t_n = \sum_{k=1}^{n} k a_k, \qquad \sigma_n = \frac{1}{n} \sum_{k=1}^{n} s_k.$$

Prove that

 a) $t_n = (n + 1)s_n - n\sigma_n$.

 b) If $\sum a_n$ is (C, 1) summable, then $\sum a_n$ converges if, and only if, $t_n = o(n)$ as $n \to \infty$.

 c) $\sum a_n$ is (C, 1) summable if, and only if, $\sum_{n=1}^{\infty} t_n/n(n + 1)$ converges.

8.38 Given a monotonic sequence $\{a_n\}$ of positive terms, such that $\lim_{n \to \infty} a_n = 0$. Let

$$s_n = \sum_{k=1}^{n} a_k, \qquad u_n = \sum_{k=1}^{n} (-1)^k a_k, \qquad v_n = \sum_{k=1}^{n} (-1)^k s_k.$$

Prove that:

 a) $v_n = \frac{1}{2}u_n + (-1)^n s_n/2$.

 b) $\sum_{n=1}^{\infty} (-1)^n s_n$ is (C, 1) summable and has Cesàro sum $\frac{1}{2}\sum_{n=1}^{\infty} (-1)^n a_n$.

 c) $\sum_{n=1}^{\infty} (-1)^n (1 + \frac{1}{2} + \cdots + 1/n) = -\log \sqrt{2}$ (C, 1).

Infinite products

8.39 Determine whether or not the following infinite products converge. Find the value of each convergent product.

 a) $\displaystyle\prod_{n=2}^{\infty} \left(1 - \frac{2}{n(n + 1)}\right)$,

 b) $\displaystyle\prod_{n=2}^{\infty} (1 - n^{-2})$,

 c) $\displaystyle\prod_{n=2}^{\infty} \frac{n^3 - 1}{n^3 + 1}$,

 d) $\displaystyle\prod_{n=0}^{\infty} (1 + z^{2^n})$ if $|z| < 1$.

8.40 If each partial sum s_n of the convergent series $\sum a_n$ is not zero and if the sum itself is not zero, show that the infinite product $a_1 \prod_{n=2}^{\infty} (1 + a_n/s_{n-1})$ converges and has the value $\sum_{n=1}^{\infty} a_n$.

8.41 Find the values of the following products by establishing the following identities and summing the series:

a) $\displaystyle\prod_{n=2}^{\infty} \left(1 + \frac{1}{2^n - 2}\right) = 2 \sum_{n=1}^{\infty} 2^{-n}.$ b) $\displaystyle\prod_{n=2}^{\infty} \left(1 + \frac{1}{n^2 - 1}\right) = 2 \sum_{n=1}^{\infty} \frac{1}{n(n+1)}.$

8.42 Determine all real x for which the product $\prod_{n=1}^{\infty} \cos(x/2^n)$ converges and find the value of the product when it does converge.

8.43 a) Let $a_n = (-1)^n/\sqrt{n}$ for $n = 1, 2, \ldots$ Show that $\prod(1 + a_n)$ diverges but that $\sum a_n$ converges.

b) Let $a_{2n-1} = -1/\sqrt{n}$, $a_{2n} = 1/\sqrt{n} + 1/n$ for $n = 1, 2, \ldots$ Show that $\prod(1 + a_n)$ converges but that $\sum a_n$ diverges.

8.44 Assume that $a_n \geq 0$ for each $n = 1, 2, \ldots$ Assume further that

$$a_{2n+2} < a_{2n+1} < \frac{a_{2n}}{1 + a_{2n}} \qquad \text{for } n = 1, 2, \ldots$$

Show that $\prod_{k=1}^{\infty} (1 + (-1)^k a_k)$ converges if, and only if, $\sum_{k=1}^{\infty} (-1)^k a_k$ converges.

8.45 A complex-valued sequence $\{f(n)\}$ is called *multiplicative* if $f(1) = 1$ and if $f(mn) = f(m)f(n)$ whenever m and n are relatively prime. (See Section 1.7.) It is called *completely multiplicative* if

$$f(1) = 1 \qquad \text{and} \qquad f(mn) = f(m)f(n) \qquad \text{for all } m \text{ and } n.$$

a) If $\{f(n)\}$ is multiplicative and if the series $\sum f(n)$ converges absolutely, prove that

$$\sum_{n=1}^{\infty} f(n) = \prod_{k=1}^{\infty} \{1 + f(p_k) + f(p_k^2) + \cdots\},$$

where p_k denotes the kth prime, the product being absolutely convergent.

b) If, in addition, $\{f(n)\}$ is completely multiplicative, prove that the formula in (a) becomes

$$\sum_{n=1}^{\infty} f(n) = \prod_{k=1}^{\infty} \frac{1}{1 - f(p_k)}.$$

Note that Euler's product for $\zeta(s)$ (Theorem 8.56) is the special case in which $f(n) = n^{-s}$.

8.46 This exercise outlines a simple proof of the formula $\zeta(2) = \pi^2/6$. Start with the inequality $\sin x < x < \tan x$, valid for $0 < x < \pi/2$, take reciprocals, and square each member to obtain

$$\cot^2 x < \frac{1}{x^2} < 1 + \cot^2 x.$$

Now put $x = k\pi/(2m + 1)$, where k and m are integers, with $1 \leq k \leq m$, and sum on k to obtain

$$\sum_{k=1}^{m} \cot^2 \frac{k\pi}{2m + 1} < \frac{(2m + 1)^2}{\pi^2} \sum_{k=1}^{m} \frac{1}{k^2} < m + \sum_{k=1}^{m} \cot^2 \frac{k\pi}{2m + 1}.$$

Use the formula of Exercise 1.49(c) to deduce the inequality

$$\frac{m(2m-1)\pi^2}{3(2m+1)^2} < \sum_{k=1}^{m} \frac{1}{k^2} < \frac{2m(m+1)\pi^2}{3(2m+1)^2}.$$

Now let $m \to \infty$ to obtain $\zeta(2) = \pi^2/6$.

8.47 Use an argument similar to that outlined in Exercise 8.46 to prove that $\zeta(4) = \pi^4/90$.

SUGGESTED REFERENCES FOR FURTHER STUDY

8.1 Hardy, G. H., *Divergent Series*. Oxford University Press, Oxford, 1949.

8.2 Hirschmann, I. I., *Infinite Series*. Holt, Rinehart and Winston, New York, 1962.

8.3 Knopp, K., *Theory and Application of Infinite Series*, 2nd ed. R. C. Young, translator. Hafner, New York, 1948.

CHAPTER 9

SEQUENCES
OF FUNCTIONS

9.1 POINTWISE CONVERGENCE OF SEQUENCES OF FUNCTIONS

This chapter deals with sequences $\{f_n\}$ whose terms are real- or complex-valued functions having a common domain on the real line $\mathbf{R}$ or in the complex plane $\mathbf{C}$. For each x in the domain we can form another sequence $\{f_n(x)\}$ whose terms are the corresponding function values. Let S denote the set of x for which this second sequence converges. The function f defined by the equation

$$f(x) = \lim_{n \to \infty} f_n(x), \qquad \text{if } x \in S,$$

is called the *limit function* of the sequence $\{f_n\}$, and we say that $\{f_n\}$ *converges pointwise* to f on the set S.

Our chief interest in this chapter is the following type of question: If each function of a sequence $\{f_n\}$ has a certain property, such as continuity, differentiability, or integrability, to what extent is this property transferred to the limit function? For example, if each function f_n is continuous at c, is the limit function f also continuous at c? We shall see that, in general, it is not. In fact, we shall find that pointwise convergence is usually not strong enough to transfer any of the properties mentioned above from the individual terms f_n to the limit function f. Therefore we are led to study stronger methods of convergence that *do* preserve these properties. The most important of these is the notion of *uniform* convergence.

Before we introduce uniform convergence, let us formulate one of our basic questions in another way. When we ask whether continuity of each f_n at c implies continuity of the limit function f at c, we are really asking whether the equation

$$\lim_{x \to c} f_n(x) = f_n(c),$$

implies the equation

$$\lim_{x \to c} f(x) = f(c). \tag{1}$$

But (1) can also be written as follows:

$$\lim_{x \to c} \lim_{n \to \infty} f_n(x) = \lim_{n \to \infty} \lim_{x \to c} f_n(x). \tag{2}$$

Therefore our question about continuity amounts to this: Can we interchange the limit symbols in (2)? We shall see that, in general, we cannot. First of all, the limit in (1) may not exist. Secondly, even if it does exist, it need not be equal to

218

$f(c)$. We encountered a similar situation in Chapter 8 in connection with iterated series when we found that $\sum_{m=1}^{\infty} \sum_{n=1}^{\infty} f(m, n)$ is not necessarily equal to $\sum_{n=1}^{\infty} \sum_{m=1}^{\infty} f(m, n)$.

The general question of whether we can reverse the order of two limit processes arises again and again in mathematical analysis. We shall find that uniform convergence is a far-reaching *sufficient* condition for the validity of interchanging certain limits, but it does not provide the complete answer to the question. We shall encounter examples in which the order of two limits can be interchanged although the sequence is not uniformly convergent.

9.2 EXAMPLES OF SEQUENCES OF REAL-VALUED FUNCTIONS

The following examples illustrate some of the possibilities that might arise when we form the limit function of a sequence of real-valued functions.

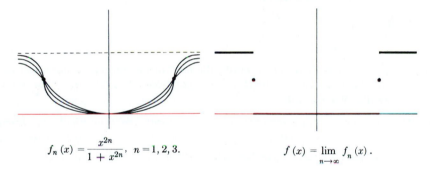

$$f_n(x) = \frac{x^{2n}}{1 + x^{2n}}, \quad n = 1, 2, 3. \qquad f(x) = \lim_{n \to \infty} f_n(x).$$

Figure 9.1

Example 1. *A sequence of continuous functions with a discontinuous limit function.* Let $f_n(x) = x^{2n}/(1 + x^{2n})$ if $x \in \mathbf{R}$, $n = 1, 2, \ldots$ The graphs of a few terms are shown in Fig. 9.1. In this case $\lim_{n \to \infty} f_n(x)$ exists for every real x, and the limit function f is given by

$$f(x) = \begin{cases} 0 & \text{if } |x| < 1, \\ \tfrac{1}{2} & \text{if } |x| = 1, \\ 1 & \text{if } |x| > 1. \end{cases}$$

Each f_n is continuous on $\mathbf{R}$, but f is discontinuous at $x = 1$ and $x = -1$.

Example 2. *A sequence of functions for which* $\lim_{n \to \infty} \int_0^1 f_n(x) \, dx \neq \int_0^1 \lim_{n \to \infty} f_n(x) \, dx$. Let $f_n(x) = n^2 x(1 - x)^n$ if $x \in \mathbf{R}$, $n = 1, 2, \ldots$ If $0 \le x \le 1$ the limit $f(x) = \lim_{n \to \infty} f_n(x)$ exists and equals 0. (See Fig. 9.2.) Hence $\int_0^1 f(x) \, dx = 0$. But

$$\int_0^1 f_n(x) \, dx = n^2 \int_0^1 x(1 - x)^n \, dx$$

$$= n^2 \int_0^1 (1 - t)t^n \, dt = \frac{n^2}{n + 1} - \frac{n^2}{n + 2} = \frac{n^2}{(n + 1)(n + 2)}$$

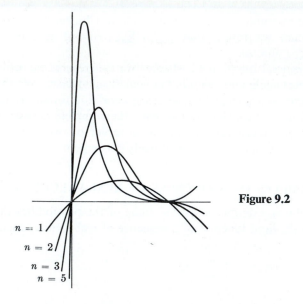

Figure 9.2

$n = 1$

$n = 2$

$n = 3$

$n = 5$

so $\lim_{n\to\infty} \int_0^1 f_n(x)\, dx = 1$. In other words, the limit of the integrals is not equal to the integral of the limit function. Therefore the operations of "limit" and "integration" cannot always be interchanged.

Example 3. *A sequence of differentiable functions $\{f_n\}$ with limit 0 for which $\{f_n'\}$ diverges.* Let $f_n(x) = (\sin nx)/\sqrt{n}$ if $x \in \mathbf{R}$, $n = 1, 2, \ldots$ Then $\lim_{n\to\infty} f_n(x) = 0$ for every x. But $f_n'(x) = \sqrt{n} \cos nx$, so $\lim_{n\to\infty} f_n'(x)$ does not exist for any x. (See Fig. 9.3.)

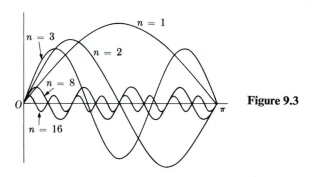

$n = 1$

$n = 3$

$n = 2$

$n = 8$

O

$n = 16$

π

Figure 9.3

9.3 DEFINITION OF UNIFORM CONVERGENCE

Let $\{f_n\}$ be a sequence of functions which converges pointwise on a set S to a limit function f. This means that for each point x in S and for each $\varepsilon > 0$, there exists an N (depending on both x and ε) such that

$$n > N \qquad \text{implies} \qquad |f_n(x) - f(x)| < \varepsilon.$$

If the same N works equally well for *every* point in S, the convergence is said to be *uniform* on S. That is, we have

Definition 9.1. *A sequence of functions $\{f_n\}$ is said to converge uniformly to f on a set S if, for every $\varepsilon > 0$, there exists an N (depending only on ε) such that $n > N$ implies*

$$|f_n(x) - f(x)| < \varepsilon, \qquad \textit{for every } x \textit{ in } S.$$

We denote this symbolically by writing

$$f_n \to f \textit{ uniformly on } S.$$

When each term of the sequence $\{f_n\}$ is real-valued, there is a useful geometric interpretation of uniform convergence. The inequality $|f_n(x) - f(x)| < \varepsilon$ is then equivalent to the *two* inequalities

$$f(x) - \varepsilon < f_n(x) < f(x) + \varepsilon. \qquad (3)$$

If (3) is to hold for all $n > N$ and for all x in S, this means that the entire graph of f_n (that is, the set $\{(x, y) : y = f_n(x), x \in S\}$) lies within a "band" of height 2ε situated symmetrically about the graph of f. (See Fig. 9.4.)

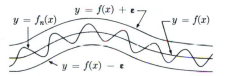

Figure 9.4

A sequence $\{f_n\}$ is said to be *uniformly bounded* on S if there exists a constant $M > 0$ such that $|f_n(x)| \leq M$ for all x in S and all n. The number M is called a *uniform bound* for $\{f_n\}$. If each individual function is bounded and if $f_n \to f$ uniformly on S, then it is easy to prove that $\{f_n\}$ is uniformly bounded on S. (See Exercise 9.1.) This observation often enables us to conclude that a sequence is *not* uniformly convergent. For instance, a glance at Fig. 9.2 tells us at once that the sequence of Example 2 cannot converge uniformly on any subset containing a neighborhood of the origin. However, the convergence in this example *is* uniform on every compact subinterval not containing the origin.

9.4 UNIFORM CONVERGENCE AND CONTINUITY

Theorem 9.2. *Assume that $f_n \to f$ uniformly on S. If each f_n is continuous at a point c of S, then the limit function f is also continuous at c.*

NOTE. If c is an accumulation point of S, the conclusion implies that

$$\lim_{x \to c} \lim_{n \to \infty} f_n(x) = \lim_{n \to \infty} \lim_{x \to c} f_n(x).$$

Proof. If c is an isolated point of S, then f is automatically continuous at c. Suppose, then, that c is an accumulation point of S. By hypothesis, for every $\varepsilon > 0$ there is an M such that $n \geq M$ implies

$$|f_n(x) - f(x)| < \frac{\varepsilon}{3} \qquad \text{for every } x \text{ in } S.$$

Since f_M is continuous at c, there is a neighborhood $B(c)$ such that $x \in B(c) \cap S$ implies

$$|f_M(x) - f_M(c)| < \frac{\varepsilon}{3}.$$

But

$$|f(x) - f(c)| \leq |f(x) - f_M(x)| + |f_M(x) - f_M(c)| + |f_M(c) - f(c)|.$$

If $x \in B(c) \cap S$, each term on the right is less than $\varepsilon/3$ and hence $|f(x) - f(c)| < \varepsilon$. This proves the theorem.

NOTE. Uniform convergence of $\{f_n\}$ is sufficient but not necessary to transmit continuity from the individual terms to the limit function. In Example 2 (Section 9.2), we have a nonuniformly convergent sequence of continuous functions with a continuous limit function.

9.5 THE CAUCHY CONDITION FOR UNIFORM CONVERGENCE

Theorem 9.3. *Let $\{f_n\}$ be a sequence of functions defined on a set S. There exists a function f such that $f_n \to f$ uniformly on S if, and only if, the following condition (called the Cauchy condition) is satisfied: For every $\varepsilon > 0$ there exists an N such that $m > N$ and $n > N$ implies*

$$|f_m(x) - f_n(x)| < \varepsilon, \qquad \text{for every } x \text{ in } S.$$

Proof. Assume that $f_n \to f$ uniformly on S. Then, given $\varepsilon > 0$, we can find N so that $n > N$ implies $|f_n(x) - f(x)| < \varepsilon/2$ for all x in S. Taking $m > N$, we also have $|f_m(x) - f(x)| < \varepsilon/2$, and hence $|f_m(x) - f_n(x)| < \varepsilon$ for every x in S.

Conversely, suppose the Cauchy condition is satisfied. Then, for each x in S, the sequence $\{f_n(x)\}$ converges. Let $f(x) = \lim_{n \to \infty} f_n(x)$ if $x \in S$. We must show that $f_n \to f$ uniformly on S. If $\varepsilon > 0$ is given, we can choose N so that $n > N$ implies $|f_n(x) - f_{n+k}(x)| < \varepsilon/2$ for every $k = 1, 2, \ldots$, and every x in S. Therefore, $\lim_{k \to \infty} |f_n(x) - f_{n+k}(x)| = |f_n(x) - f(x)| \leq \varepsilon/2$. Hence, $n > N$ implies $|f_n(x) - f(x)| < \varepsilon$ for every x in S. This proves that $f_n \to f$ uniformly on S.

NOTE. Pointwise and uniform convergence can be formulated in the more general setting of metric spaces. If f_n and f are functions from a nonempty set S to a metric space (T, d_T), we say that $f_n \to f$ uniformly on S, if, for every $\varepsilon > 0$, there is an N (depending only on ε) such that $n \geq N$ implies

$$d_T(f_n(x), f(x)) < \varepsilon \qquad \text{for all } x \text{ in } S.$$

Theorem 9.3 is valid in this more general setting and, if S is a metric space, Theorem 9.2 is also valid. The same proofs go through, with the appropriate replacement of the Euclidean metric by the metrics d_S and d_T. Since we are primarily interested in real- or complex-valued functions defined on subsets of $\mathbf{R}$ or of $\mathbf{C}$, we will not pursue this extension any further except to mention the following example.

Example. Consider the metric space $(B(S), d)$ of all bounded real-valued functions on a nonempty set S, with metric $d(f, g) = \|f - g\|$, where $\|f\| = \sup_{x \in S} |f(x)|$ is the sup norm. (See Exercise 4.66.) Then $f_n \to f$ in the metric space $(B(S), d)$ if and only if $f_n \to f$ uniformly on S. In other words, uniform convergence on S is the same as ordinary convergence in the metric space $(B(S), d)$.

9.6 UNIFORM CONVERGENCE OF INFINITE SERIES OF FUNCTIONS

Definition 9.4. *Given a sequence $\{f_n\}$ of functions defined on a set S. For each x in S, let*

$$s_n(x) = \sum_{k=1}^{n} f_k(x) \qquad (n = 1, 2, \ldots).\tag{4}$$

If there exists a function f such that $s_n \to f$ uniformly on S, we say the series $\sum f_n(x)$ converges uniformly on S and we write

$$\sum_{n=1}^{\infty} f_n(x) = f(x) \qquad (uniformly\ on\ S).$$

Theorem 9.5 *(Cauchy condition for uniform convergence of series). The infinite series $\sum f_n(x)$ converges uniformly on S if, and only if, for every $\varepsilon > 0$ there is an N such that $n > N$ implies*

$$\left| \sum_{k=n+1}^{n+p} f_k(x) \right| < \varepsilon, \qquad for\ each\ p = 1, 2, \ldots, and\ every\ x\ in\ S.$$

Proof. Define s_n by (4) and apply Theorem 9.3.

Theorem 9.6 (Weierstrass M-test). *Let $\{M_n\}$ be a sequence of nonnegative numbers such that*

$$0 \le |f_n(x)| \le M_n, \qquad for\ n = 1, 2, \ldots, and\ for\ every\ x\ in\ S.$$

Then $\sum f_n(x)$ converges uniformly on S if $\sum M_n$ converges.

Proof. Apply Theorems 8.11 and 9.5 in conjunction with the inequality

$$\left| \sum_{k=n+1}^{n+p} f_k(x) \right| \le \sum_{k=n+1}^{n+p} M_k.$$

Theorem 9.7. *Assume that $\sum f_n(x) = f(x)$ (uniformly on S). If each f_n is continuous at a point x_0 of S, then f is also continuous at x_0.*

Proof. Define s_n by (4). Continuity of each f_n at x_0 implies continuity of s_n at x_0, and the conclusion follows at once from Theorem 9.2.

NOTE. If x_0 is an accumulation point of S, this theorem permits us to interchange limits and infinite sums, as follows:

$$\lim_{x \to x_0} \sum_{n=1}^{\infty} f_n(x) = \sum_{n=1}^{\infty} \lim_{x \to x_0} f_n(x).$$

9.7 A SPACE-FILLING CURVE

We can apply Theorem 9.7 to construct a *space-filling curve*. This is a continuous curve in $\mathbf{R}^2$ that passes through every point of the unit square $[0, 1] \times [0, 1]$. Peano (1890) was the first to give an example of such a curve. The example to be presented here is due to I. J. Schoenberg (*Bulletin of the American Mathematical Society*, 1938) and can be described as follows:

Let ϕ be defined on the interval $[0, 2]$ by the following formulas:

$$\phi(t) = \begin{cases} 0, & \text{if } 0 \le t \le \tfrac{1}{3}, \text{ or if } \tfrac{5}{3} \le t \le 2, \\ 3t - 1, & \text{if } \tfrac{1}{3} \le t \le \tfrac{2}{3}, \\ 1, & \text{if } \tfrac{2}{3} \le t \le \tfrac{4}{3}, \\ -3t + 5, & \text{if } \tfrac{4}{3} \le t \le \tfrac{5}{3}. \end{cases}$$

Extend the definition of ϕ to all of $\mathbf{R}$ by the equation

$$\phi(t + 2) = \phi(t).$$

This makes ϕ periodic with period 2. (The graph of ϕ is shown in Fig. 9.5.)

$$\begin{array}{ccccccc} -2 & -1 & 0 & 1 & 2 & 3 & 4 \end{array}$$

Figure 9.5

Now define two functions f_1 and f_2 by the following equations:

$$f_1(t) = \sum_{n=1}^{\infty} \frac{\phi(3^{2n-2}t)}{2^n}, \qquad f_2(t) = \sum_{n=1}^{\infty} \frac{\phi(3^{2n-1}t)}{2^n}.$$

Both series converge absolutely for each real t and they converge *uniformly* on $\mathbf{R}$. In fact, since $|\phi(t)| \le 1$ for all t, the Weierstrass M-test is applicable with $M_n = 2^{-n}$. Since ϕ is continuous on $\mathbf{R}$, Theorem 9.7 tells us that f_1 and f_2 are also continuous on $\mathbf{R}$. Let $f = (f_1, f_2)$ and let Γ denote the image of the unit interval $[0, 1]$ under f. We will show that Γ "fills" the unit square, i.e., that $\Gamma = [0, 1] \times [0, 1]$.

First, it is clear that $0 \le f_1(t) \le 1$ and $0 \le f_2(t) \le 1$ for each t, since $\sum_{n=1}^{\infty} 2^{-n} = 1$. Hence, Γ is a subset of the unit square. Next, we must show that

$(a, b) \in \Gamma$ whenever $(a, b) \in [0, 1] \times [0, 1]$. For this purpose we write a and b in the binary system. That is, we write

$$a = \sum_{n=1}^{\infty} \frac{a_n}{2^n}, \qquad b = \sum_{n=1}^{\infty} \frac{b_n}{2^n},$$

where each a_n and each b_n is either 0 or 1. (See Exercise 1.22.) Now let

$$c = 2 \sum_{n=1}^{\infty} \frac{c_n}{3^n}, \qquad \text{where } c_{2n-1} = a_n \text{ and } c_{2n} = b_n, \; n = 1, 2, \ldots$$

Clearly, $0 \le c \le 1$ since $2 \sum_{n=1}^{\infty} 3^{-n} = 1$. We will show that $f_1(c) = a$ and that $f_2(c) = b$.

If we can prove that

$$\phi(3^k c) = c_{k+1}, \qquad \text{for each } k = 0, 1, 2, \ldots, \tag{5}$$

then we will have $\phi(3^{2n-2} c) = c_{2n-1} = a_n$ and $\phi(3^{2n-1} c) = c_{2n} = b_n$, and this will give us $f_1(c) = a, f_2(c) = b$. To prove (5), we write

$$3^k c = 2 \sum_{n=1}^{k} \frac{c_n}{3^{n-k}} + 2 \sum_{n=k+1}^{\infty} \frac{c_n}{3^{n-k}} = \text{(an even integer)} + d_k,$$

where $d_k = 2 \sum_{n=1}^{\infty} c_{n+k}/3^n$. Since ϕ has period 2, it follows that

$$\phi(3^k c) = \phi(d_k).$$

If $c_{k+1} = 0$, then we have $0 \le d_k \le 2 \sum_{n=2}^{\infty} 3^{-n} = \frac{1}{3}$, and hence $\phi(d_k) = 0$. Therefore, $\phi(3^k c) = c_{k+1}$ in this case. The only other case to consider is $c_{k+1} = 1$. But then we get $\frac{2}{3} \le d_k \le 1$ and hence $\phi(d_k) = 1$. Therefore, $\phi(3^k c) = c_{k+1}$ in all cases and this proves that $f_1(c) = a, f_2(c) = b$. Hence, Γ fills the unit square.

9.8 UNIFORM CONVERGENCE AND RIEMANN–STIELTJES INTEGRATION

Theorem 9.8. Let α be of bounded variation on $[a, b]$. Assume that each term of the sequence $\{f_n\}$ is a real-valued function such that $f_n \in R(\alpha)$ on $[a, b]$ for each $n = 1, 2, \ldots$ Assume that $f_n \to f$ uniformly on $[a, b]$ and define $g_n(x) = \int_a^x f_n(t) \, d\alpha(t)$ if $x \in [a, b], n = 1, 2, \ldots$ Then we have:

a) *$f \in R(\alpha)$ on $[a, b]$.*

b) *$g_n \to g$ uniformly on $[a, b]$, where $g(x) = \int_a^x f(t) \, d\alpha(t)$.*

NOTE. The conclusion implies that, for each x in $[a, b]$, we can write

$$\lim_{n \to \infty} \int_a^x f_n(t) \, d\alpha(t) = \int_a^x \lim_{n \to \infty} f_n(t) \, d\alpha(t).$$

This property is often described by saying that a uniformly convergent sequence can be integrated term by term.

Proof. We can assume that α is increasing with $\alpha(a) < \alpha(b)$. To prove (a), we will show that f satisfies Riemann's condition with respect to α on $[a, b]$. (See Theorem 7.19.)

Given $\varepsilon > 0$, choose N so that

$$|f(x) - f_N(x)| < \frac{\varepsilon}{3[\alpha(b) - \alpha(a)]}, \qquad \text{for all } x \text{ in } [a, b].$$

Then, for every partition P of $[a, b]$, we have

$$|U(P, f - f_N, \alpha)| \leq \frac{\varepsilon}{3} \qquad \text{and} \qquad |L(P, f - f_N, \alpha)| \leq \frac{\varepsilon}{3},$$

(using the notation of Definition 7.14). For this N, choose P_ε so that P finer than P_ε implies $U(P, f_N, \alpha) - L(P, f_N, \alpha) < \varepsilon/3$. Then for such P we have

$$U(P, f, \alpha) - L(P, f, \alpha) \leq U(P, f - f_N, \alpha) - L(P, f - f_N, \alpha)$$

$$+ U(P, f_N, \alpha) - L(P, f_N, \alpha)$$

$$< |U(P, f - f_N, \alpha)| + |L(P, f - f_N, \alpha)| + \frac{\varepsilon}{3} \leq \varepsilon.$$

This proves (a). To prove (b), let $\varepsilon > 0$ be given and choose N so that

$$|f_n(t) - f(t)| < \frac{\varepsilon}{2[\alpha(b) - \alpha(a)]},$$

for all $n > N$ and every t in $[a, b]$. If $x \in [a, b]$, we have

$$|g_n(x) - g(x)| \leq \int_a^x |f_n(t) - f(t)| \, d\alpha(t) \leq \frac{\alpha(x) - \alpha(a)}{\alpha(b) - \alpha(a)} \frac{\varepsilon}{2} \leq \frac{\varepsilon}{2} < \varepsilon.$$

This proves that $g_n \to g$ uniformly on $[a, b]$.

Theorem 9.9. *Let α be of bounded variation on $[a, b]$ and assume that $\sum f_n(x) = f(x)$ (uniformly on $[a, b]$), where each f_n is a real-valued function such that $f_n \in R(\alpha)$ on $[a, b]$. Then we have:*

a) $f \in R(\alpha)$ on $[a, b]$.

b) $\int_a^x \sum_{n=1}^\infty f_n(t) \, d\alpha(t) = \sum_{n=1}^\infty \int_a^x f_n(t) \, d\alpha(t)$ *(uniformly on $[a, b]$).*

Proof. Apply Theorem 9.8 to the sequence of partial sums.

NOTE. This theorem is described by saying that a uniformly convergent series can be integrated term by term.

9.9 NONUNIFORMLY CONVERGENT SEQUENCES THAT CAN BE INTEGRATED TERM BY TERM

Uniform convergence is a sufficient but not a necessary condition for term-by-term integration, as is seen by the following example.

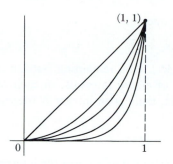

Figure 9.6

Example. Let $f_n(x) = x^n$ if $0 \le x \le 1$. (See Fig. 9.6.) The limit function f has the value 0 in $[0, 1)$ and $f(1) = 1$. Since this is a sequence of continuous functions with discontinuous limit, the convergence is not uniform on $[0, 1]$. Nevertheless, term-by-term integration on $[0, 1]$ leads to a correct result in this case. In fact, we have

$$\int_0^1 f_n(x)\,dx = \int_0^1 x^n\,dx = \frac{1}{n+1} \to 0 \text{ as } n \to \infty,$$

so $\lim_{n\to\infty} \int_0^1 f_n(x)\,dx = \int_0^1 f(x)\,dx = 0$.

The sequence in the foregoing example, although not uniformly convergent on $[0, 1]$, is uniformly convergent on every closed subinterval of $[0, 1]$ not containing 1. The next theorem is a general result which permits term-by-term integration in examples of this type. The added ingredient is that we assume that $\{f_n\}$ is uniformly bounded on $[a, b]$ and that the limit function f is integrable.

Definition 9.10. *A sequence of functions $\{f_n\}$ is said to be* **boundedly convergent** *on T if $\{f_n\}$ is pointwise convergent and uniformly bounded on T.*

Theorem 9.11. *Let $\{f_n\}$ be a boundedly convergent sequence on $[a, b]$. Assume that each $f_n \in R$ on $[a, b]$, and that the limit function $f \in R$ on $[a, b]$. Assume also that there is a partition P of $[a, b]$, say*

$$P = \{x_0, x_1, \ldots, x_m\},$$

such that, on every subinterval $[c, d]$ not containing any of the points x_k, the sequence $\{f_n\}$ converges uniformly to f. Then we have

$$\lim_{n\to\infty} \int_a^b f_n(t)\,dt = \int_a^b \lim_{n\to\infty} f_n(t)\,dt = \int_a^b f(t)\,dt. \tag{6}$$

Proof. Since f is bounded and $\{f_n\}$ is uniformly bounded, there is a positive number M such that $|f(x)| \le M$ and $|f_n(x)| \le M$ for all x in $[a, b]$ and all $n \ge 1$. Given $\varepsilon > 0$ such that $2\varepsilon < \|P\|$, let $h = \varepsilon/(2m)$, where m is the number of subintervals of P, and consider a new partition P' of $[a, b]$ given by

$$P' = \{x_0, x_0 + h, x_1 - h, x_1 + h, \ldots, x_{m-1} - h, x_{m-1} + h, x_m - h, x_m\}.$$

Since $|f - f_n|$ is integrable on $[a, b]$ and bounded by $2M$, the sum of the integrals of $|f - f_n|$ taken over the intervals

$$[x_0, x_0 + h], \quad [x_1 - h, x_1 + h], \quad \ldots, \quad [x_{m-1} - h, x_{m-1} + h], \quad [x_m - h, x_m],$$

is at most $2M(2mh) = 2M\varepsilon$. The remaining portion of $[a, b]$ (call it S) is the union of a finite number of closed intervals, in each of which $\{f_n\}$ is uniformly convergent to f. Therefore, there is an integer N (depending only on ε) such that for all x in S we have

$$|f(x) - f_n(x)| < \varepsilon \qquad \text{whenever } n \geq N.$$

Hence the sum of the integrals of $|f - f_n|$ over the intervals of S is at most $\varepsilon(b - a)$, so

$$\int_a^b |f(x) - f_n(x)| \, dx \leq (2M + b - a)\varepsilon \qquad \text{whenever } n \geq N.$$

This proves that $\int_a^b f_n(x) \, dx \to \int_a^b f(x) \, dx$ as $n \to \infty$.

There is a stronger theorem due to Arzelà which makes no reference whatever to uniform convergence.

Theorem 9.12 (Arzelà).   *Assume that $\{f_n\}$ is boundedly convergent on $[a,b]$ and suppose each f_n is Riemann-integrable on $[a, b]$. Assume also that the limit function f is Riemann-integrable on $[a, b]$. Then*

$$\lim_{n \to \infty} \int_a^b f_n(x) \, dx = \int_a^b \lim_{n \to \infty} f_n(x) \, dx = \int_a^b f(x) \, dx. \tag{7}$$

The proof of Arzelà's theorem is considerably more difficult than that of Theorem 9.11 and will not be given here. In the next chapter we shall prove a theorem on Lebesgue integrals which includes Arzelà's theorem as a special case. (See Theorem 10.29).

NOTE. It is easy to give an example of a boundedly convergent sequence $\{f_n\}$ of Riemann-integrable functions whose limit f is not Riemann-integrable. If $\{r_1, r_2, \ldots \}$ denotes the set of rational numbers in $[0, 1]$, define $f_n(x)$ to have the value 1 if $x = r_k$ for all $k = 1, 2, \ldots, n$, and put $f_n(x) = 0$ otherwise. Then the integral $\int_0^1 f_n(x) \, dx = 0$ for each n, but the pointwise limit function f is not Riemann-integrable on $[0, 1]$.

9.10 UNIFORM CONVERGENCE AND DIFFERENTIATION

By analogy with Theorems 9.2 and 9.8, one might expect the following result to hold: If $f_n \to f$ uniformly on $[a, b]$ and if f_n' exists for each n, then f' exists and $f_n' \to f'$ uniformly on $[a, b]$. However, Example 3 of Section 9.2 shows that this cannot be true. Although the sequence $\{f_n\}$ of Example 3 converges uniformly on **R**, the sequence $\{f_n'\}$ does not even converge pointwise on **R**. For example, $\{f_n'(0)\}$ diverges since $f_n'(0) = \sqrt{n}$. Therefore the analog of Theorems 9.2 and 9.8 for differentiation must take a different form.

Theorem 9.13. *Assume that each term of $\{f_n\}$ is a real-valued function having a finite derivative at each point of an open interval (a, b). Assume that for at least one point x_0 in (a, b) the sequence $\{f_n(x_0)\}$ converges. Assume further that there exists a function g such that $f'_n \to g$ uniformly on (a, b). Then:*

  a) *There exists a function f such that $f_n \to f$ uniformly on (a, b).*

  b) *For each x in (a, b) the derivative $f'(x)$ exists and equals $g(x)$.*

Proof. Assume that $c \in (a, b)$ and define a new sequence $\{g_n\}$ as follows:

$$g_n(x) = \begin{cases} \dfrac{f_n(x) - f_n(c)}{x - c} & \text{if } x \neq c, \\[2mm] f'_n(c) & \text{if } x = c. \end{cases} \tag{8}$$

The sequence $\{g_n\}$ so formed depends on the choice of c. Convergence of $\{g_n(c)\}$ follows from the hypothesis, since $g_n(c) = f'_n(c)$. We will prove next that $\{g_n\}$ converges uniformly on (a, b). If $x \neq c$, we have

$$g_n(x) - g_m(x) = \frac{h(x) - h(c)}{x - c}, \tag{9}$$

where $h(x) = f_n(x) - f_m(x)$. Now $h'(x)$ exists for each x in (a, b) and has the value $f'_n(x) - f'_m(x)$. Applying the Mean-Value Theorem in (9), we get

$$g_n(x) - g_m(x) = f'_n(x_1) - f'_m(x_1), \tag{10}$$

where x_1 lies between x and c. Since $\{f'_n\}$ converges uniformly on (a, b) (by hypothesis), we can use (10), together with the Cauchy condition (Theorem 9.3), to deduce that $\{g_n\}$ converges uniformly on (a, b).

Now we can show that $\{f_n\}$ converges uniformly on (a, b). Let us form the particular sequence $\{g_n\}$ corresponding to the special point $c = x_0$ for which $\{f_n(x_0)\}$ is assumed to converge. From (8) we can write

$$f_n(x) = f_n(x_0) + (x - x_0)g_n(x),$$

an equation which holds for every x in (a, b). Hence we have

$$f_n(x) - f_m(x) = f_n(x_0) - f_m(x_0) + (x - x_0)[g_n(x) - g_m(x_0)].$$

This equation, with the help of the Cauchy condition, establishes the uniform convergence of $\{f_n\}$ on (a, b). This proves (a).

To prove (b), return to the sequence $\{g_n\}$ defined by (8) for an *arbitrary* point c in (a, b) and let $G(x) = \lim_{n \to \infty} g_n(x)$. The hypothesis that f'_n exists means that $\lim_{x \to c} g_n(x) = g_n(c)$. In other words, each g_n is continuous at c. Since $g_n \to G$ uniformly on (a, b), the limit function G is also continuous at c. This means that

$$G(c) = \lim_{x \to c} G(x), \tag{11}$$

the existence of the limit being part of the conclusion. But, for $x \neq c$, we have

$$G(x) = \lim_{n \to \infty} g_n(x) = \lim_{n \to \infty} \frac{f_n(x) - f_n(c)}{x - c} = \frac{f(x) - f(c)}{x - c}.$$

Hence, (11) states that the derivative $f'(c)$ exists and equals $G(c)$. But

$$G(c) = \lim_{n \to \infty} g_n(c) = \lim_{n \to \infty} f_n'(c) = g(c);$$

hence $f'(c) = g(c)$. Since c is an arbitrary point of (a, b), this proves (b).

When we reformulate Theorem 9.13 in terms of series, we obtain

Theorem 9.14. *Assume that each f_n is a real-valued function defined on (a, b) such that the derivative $f_n'(x)$ exists for each x in (a, b). Assume that, for at least one point x_0 in (a, b), the series $\sum f_n(x_0)$ converges. Assume further that there exists a function g such that $\sum f_n'(x) = g(x)$ (uniformly on (a, b)). Then:*

a) *There exists a function f such that $\sum f_n(x) = f(x)$ (uniformly on (a, b)).*
b) *If $x \in (a, b)$, the derivative $f'(x)$ exists and equals $\sum f_n'(x)$.*

9.11 SUFFICIENT CONDITIONS FOR UNIFORM CONVERGENCE OF A SERIES

The importance of uniformly convergent series has been amply illustrated in some of the preceding theorems. Therefore it seems natural to seek some simple ways of testing a series for uniform convergence without resorting to the definition in each case. One such test, the *Weierstrass M-test*, was described in Theorem 9.6. There are other tests that may be useful when the M-test is not applicable. One of these is the analog of Theorem 8.28.

Theorem 9.15 (Dirichlet's test for uniform convergence). *Let $F_n(x)$ denote the nth partial sum of the series $\sum f_n(x)$, where each f_n is a complex-valued function defined on a set S. Assume that $\{F_n\}$ is uniformly bounded on S. Let $\{g_n\}$ be a sequence of real-valued functions such that $g_{n+1}(x) \leq g_n(x)$ for each x in S and for every $n = 1, 2, \ldots$, and assume that $g_n \to 0$ uniformly on S. Then the series $\sum f_n(x)g_n(x)$ converges uniformly on S.*

Proof. Let $s_n(x) = \sum_{k=1}^{n} f_k(x)g_k(x)$. By partial summation we have

$$s_n(x) = \sum_{k=1}^{n} F_k(x)\big(g_k(x) - g_{k+1}(x)\big) + g_{n+1}(x)F_n(x),$$

and hence if $n > m$, we can write

$$s_n(x) - s_m(x) = \sum_{k=m+1}^{n} F_k(x)\big(g_k(x) - g_{k+1}(x)\big) + g_{n+1}(x)F_n(x) - g_{m+1}(x)F_m(x).$$

Therefore, if M is a uniform bound for $\{F_n\}$, we have

$$|s_n(x) - s_m(x)| \leq M \sum_{k=m+1}^{n} (g_k(x) - g_{k+1}(x)) + Mg_{n+1}(x) + Mg_{m+1}(x)$$

$$= M(g_{m+1}(x) - g_{n+1}(x)) + Mg_{n+1}(x) + Mg_{m+1}(x)$$

$$= 2Mg_{m+1}(x).$$

Since $g_n \to 0$ uniformly on S, this inequality (together with the Cauchy condition) implies that $\sum f_n(x)g_n(x)$ converges uniformly on S.

The reader should have no difficulty in extending Theorem 8.29 (Abel's test) in a similar way so that it yields a test for uniform convergence. (Exercise 9.13.)

Example. Let $F_n(x) = \sum_{k=1}^{n} e^{ikx}$. In the last chapter (see Theorem 8.30), we derived the inequality $|F_n(x)| \leq 1/|\sin(x/2)|$, valid for every real $x \neq 2m\pi$ (m is an integer). Therefore, if $0 < \delta < \pi$, we have the estimate

$$|F_n(x)| \leq 1/\sin(\delta/2) \qquad \text{if } \delta \leq x \leq 2\pi - \delta.$$

Hence, $\{F_n\}$ is uniformly bounded on the interval $[\delta, 2\pi - \delta]$. If $\{g_n\}$ satisfies the conditions of Theorem 9.15, we can conclude that the series $\sum g_n(x)e^{inx}$ converges uniformly on $[\delta, 2\pi - \delta]$. In particular, if we take $g_n(x) = 1/n$, this establishes the uniform convergence of the series

$$\sum_{n=1}^{\infty} \frac{e^{inx}}{n}$$

on $[\delta, 2\pi - \delta]$ if $0 < \delta < \pi$. Note that the Weierstrass M-test cannot be used to establish uniform convergence in this case, since $|e^{inx}| = 1$.

9.12 UNIFORM CONVERGENCE AND DOUBLE SEQUENCES

As a different type of application of uniform convergence, we deduce the following theorem on double sequences which can be viewed as a converse to Theorem 8.39.

Theorem 9.16. *Let f be a double sequence and let $\mathbf{Z}^+$ denote the set of positive integers. For each $n = 1, 2, \ldots$, define a function g_n on $\mathbf{Z}^+$ as follows:*

$$g_n(m) = f(m, n), \qquad \text{if } m \in \mathbf{Z}^+.$$

Assume that $g_n \to g$ uniformly on $\mathbf{Z}^+$, where $g(m) = \lim_{n \to \infty} f(m, n)$. If the iterated limit $\lim_{m \to \infty} \left(\lim_{n \to \infty} f(m, n) \right)$ exists, then the double limit $\lim_{m,n \to \infty} f(m, n)$ also exists and has the same value.

Proof. Given $\varepsilon > 0$, choose N_1 so that $n > N_1$ implies

$$|f(m, n) - g(m)| < \frac{\varepsilon}{2}, \qquad \text{for every } m \text{ in } \mathbf{Z}^+.$$

Let $a = \lim_{m \to \infty} \left(\lim_{n \to \infty} f(m, n) \right) = \lim_{m \to \infty} g(m)$. For the same ε, choose N_2 so that $m > N_2$ implies $|g(m) - a| < \varepsilon/2$. Then, if N is the larger of N_1 and N_2, we have $|f(m, n) - a| < \varepsilon$ whenever both $m > N$ and $n > N$. In other words, $\lim_{m,n \to \infty} f(m, n) = a$.

9.13 MEAN CONVERGENCE

The functions in this section may be real- or complex-valued.

Definition 9.17 *Let $\{f_n\}$ be a sequence of Riemann-integrable functions defined on $[a, b]$. Assume that $f \in R$ on $[a, b]$. The sequence $\{f_n\}$ is said to converge in the mean to f on $[a, b]$, and we write*

$$\mathrm{l.i.m.}_{n \to \infty} f_n = f \qquad \text{on } [a, b],$$

if

$$\lim_{n \to \infty} \int_a^b |f_n(x) - f(x)|^2 \, dx = 0.$$

If the inequality $|f(x) - f_n(x)| < \varepsilon$ holds for every x in $[a, b]$, then we have $\int_a^b |f(x) - f_n(x)|^2 \, dx \le \varepsilon^2(b - a)$. Therefore, uniform convergence of $\{f_n\}$ to f on $[a, b]$ implies mean convergence, provided that each f_n is Riemann-integrable on $[a, b]$. A rather surprising fact is that convergence in the mean need not imply pointwise convergence at any point of the interval. This can be seen as follows: For each integer $n \ge 0$, subdivide the interval $[0, 1]$ into 2^n equal subintervals and let I_{2^n+k} denote that subinterval whose right endpoint is $(k + 1)/2^n$, where $k = 0, 1, 2, \ldots, 2^n - 1$. This yields a collection $\{I_1, I_2, \ldots\}$ of subintervals of $[0, 1]$, of which the first few are:

$$I_1 = [0, 1], \qquad I_2 = [0, \tfrac{1}{2}], \qquad I_3 = [\tfrac{1}{2}, 1],$$
$$I_4 = [0, \tfrac{1}{4}], \qquad I_5 = [\tfrac{1}{4}, \tfrac{1}{2}], \qquad I_6 = [\tfrac{1}{2}, \tfrac{3}{4}],$$

and so forth. Define f_n on $[0, 1]$ as follows:

$$f_n(x) = \begin{cases} 1 & \text{if } x \in I_n, \\ 0 & \text{if } x \in [0, 1] - I_n. \end{cases}$$

Then $\{f_n\}$ converges in the mean to 0, since $\int_0^1 |f_n(x)|^2 \, dx$ is the length of I_n, and this approaches 0 as $n \to \infty$. On the other hand, for each x in $[0, 1]$ we have

$$\limsup_{n \to \infty} f_n(x) = 1 \qquad \text{and} \qquad \liminf_{n \to \infty} f_n(x) = 0.$$

[Why?] Hence, $\{f_n(x)\}$ does not converge for any x in $[0, 1]$.

The next theorem illustrates the importance of mean convergence.

Theorem 9.18. *Assume that* l.i.m.$_{n \to \infty}$ $f_n = f$ *on* $[a, b]$. *If* $g \in R$ *on* $[a, b]$, *define*

$$h(x) = \int_a^x f(t)g(t) \, dt, \qquad h_n(x) = \int_a^x f_n(t)g(t) \, dt,$$

if $x \in [a, b]$. *Then* $h_n \to h$ *uniformly on* $[a, b]$.

Proof. The proof is based on the inequality

$$0 \le \left(\int_a^x |f(t) - f_n(t)| \, |g(t)| \, dt \right)^2$$

$$\le \left(\int_a^x |f(t) - f_n(t)|^2 \, dt \right) \left(\int_a^x |g(t)|^2 \, dt \right), \qquad (12)$$

which is a direct application of the Cauchy–Schwarz inequality for integrals. (See Exercise 7.16 for the statement of the Cauchy–Schwarz inequality and a sketch of its proof.) Given $\varepsilon > 0$, we can choose N so that $n > N$ implies

$$\int_a^b |f(t) - f_n(t)|^2 \, dt < \frac{\varepsilon^2}{A}, \qquad (13)$$

where $A = 1 + \int_a^b |g(t)|^2 \, dt$. Substituting (13) in (12), we find that $n > N$ implies $0 \le |h(x) - h_n(x)| < \varepsilon$ for every x in $[a, b]$.

This theorem is particularly useful in the theory of Fourier series. (See Theorem 11.16.) The following generalization is also of interest.

Theorem 9.19. *Assume that* l.i.m.$_{n \to \infty}$ $f_n = f$ *and* l.i.m.$_{n \to \infty}$ $g_n = g$ *on* $[a, b]$. *Define*

$$h(x) = \int_a^x f(t)g(t) \, dt, \qquad h_n(x) = \int_a^x f_n(t)g_n(t) \, dt,$$

if $x \in [a, b]$. *Then* $h_n \to h$ *uniformly on* $[a, b]$.

Proof. We have

$$h_n(x) - h(x) = \int_a^x (f - f_n)(g - g_n) \, dt$$

$$+ \left(\int_a^x f_n g \, dt - \int_a^x fg \, dt \right) + \left(\int_a^x fg_n \, dt - \int_a^x fg \, dt \right).$$

Applying the Cauchy–Schwarz inequality, we can write

$$0 \le \left(\int_a^x |f - f_n| \, |g - g_n| \, dt \right)^2 \le \left(\int_a^b |f - f_n|^2 \, dt \right) \left(\int_a^b |g - g_n|^2 \, dt \right).$$

The proof is now an easy consequence of Theorem 9.18.

9.14 POWER SERIES

An infinite series of the form

$$a_0 + \sum_{n=1}^{\infty} a_n(z - z_0)^n,$$

written more briefly as

$$\sum_{n=0}^{\infty} a_n(z - z_0)^n, \tag{14}$$

is called a power series in $z - z_0$. Here z, z_0, and a_n ($n = 0, 1, 2, \ldots$) are complex numbers. With every power series (14) there is associated a disk, called the *disk of convergence*, such that the series converges absolutely for every z interior to this disk and diverges for every z outside this disk. The center of the disk is at z_0 and its radius is called the *radius of convergence* of the power series. (The radius may be 0 or $+\infty$ in extreme cases.) The next theorem establishes the existence of the disk of convergence and provides us with a way of calculating its radius.

Theorem 9.20. *Given a power series* $\sum_{n=0}^{\infty} a_n(z - z_0)^n$, *let*

$$\lambda = \limsup_{n \to \infty} \sqrt[n]{|a_n|}, \qquad r = \frac{1}{\lambda},$$

(where $r = 0$ if $\lambda = +\infty$ and $r = +\infty$ if $\lambda = 0$). Then the series converges absolutely if $|z - z_0| < r$ and diverges if $|z - z_0| > r$. Furthermore, the series converges uniformly on every compact subset interior to the disk of convergence.

Proof. Applying the root test (Theorem 8.26), we have

$$\limsup_{n \to \infty} \sqrt[n]{|a_n(z - z_0)^n|} = \frac{|z - z_0|}{r},$$

and hence $\sum a_n(z - z_0)^n$ converges absolutely if $|z - z_0| < r$ and diverges if $|z - z_0| > r$.

To prove the second assertion, we simply observe that if T is a compact subset of the disk of convergence, there is a point p in T such that $z \in T$ implies

$$|z - z_0| \le |p - z_0| < r.$$

Hence, $|a_n(z - z_0)^n| \le |a_n(p - z_0)^n|$ for each z in T, and the Weierstrass M-test is applicable.

NOTE. If the limit $\lim_{n \to \infty} |a_n/a_{n+1}|$ exists (or if this limit is $+\infty$), its value is also equal to the radius of convergence of (14). (See Exercise 9.30.)

Example 1. The two series $\sum_{n=0}^{\infty} z^n$ and $\sum_{n=1}^{\infty} z^n/n^2$ have the same radius of convergence, namely, $r = 1$. On the boundary of the disk of convergence, the first converges nowhere, the second converges everywhere.

Example 2. The series $\sum_{n=1}^{\infty} z^n/n$ has radius of convergence $r = 1$, but it does not converge at $z = 1$. However, it does converge everywhere else on the boundary because of Dirichlet's test (Theorem 8.28).

These examples illustrate why Theorem 9.20 makes no assertion about the behavior of a power series on the *boundary* of the disk of convergence.

Theorem 9.21. *Assume that the power series $\sum_{n=0}^{\infty} a_n(z - z_0)^n$ converges for each z in $B(z_0; r)$. Then the function f defined by the equation*

$$f(z) = \sum_{n=0}^{\infty} a_n(z - z_0)^n, \qquad \text{if } z \in B(z_0; r), \tag{15}$$

is continuous on $B(z_0; r)$.

Proof. Since each point in $B(z_0; r)$ belongs to some compact subset of $B(z_0; r)$, the conclusion follows at once from Theorem 9.7.

NOTE. The series in (15) is said to *represent f* in $B(z_0; r)$. It is also called a *power series expansion* of f about z_0. Functions having power series expansions are continuous inside the disk of convergence. Much more than this is true, however. We will later prove that such functions have derivatives of every order inside the disk of convergence. The proof will make use of the following theorem:

Theorem 9.22. *Assume that $\sum a_n(z - z_0)^n$ converges if $z \in B(z_0; r)$. Suppose that the equation*

$$f(z) = \sum_{n=0}^{\infty} a_n(z - z_0)^n,$$

is known to be valid for each z in some open subset S of $B(z_0; r)$. Then, for each point z_1 in S, there exists a neighborhood $B(z_1; R) \subseteq S$ in which f has a power series expansion of the form

$$f(z) = \sum_{k=0}^{\infty} b_k(z - z_1)^k, \tag{16}$$

where

$$b_k = \sum_{n=k}^{\infty} \binom{n}{k} a_n(z_1 - z_0)^{n-k} \qquad (k = 0, 1, 2, \ldots). \tag{17}$$

Proof. If $z \in S$, we have

$$f(z) = \sum_{n=0}^{\infty} a_n(z - z_0)^n = \sum_{n=0}^{\infty} a_n(z - z_1 + z_1 - z_0)^n$$

$$= \sum_{n=0}^{\infty} a_n \sum_{k=0}^{n} \binom{n}{k} (z - z_1)^k(z_1 - z_0)^{n-k}$$

$$= \sum_{n=0}^{\infty} \sum_{k=0}^{\infty} c_n(k),$$

where

$$c_n(k) = \begin{cases} \dbinom{n}{k} a_n(z - z_1)^k(z_1 - z_0)^{n-k}, & \text{if } k \le n, \\ 0, & \text{if } k > n. \end{cases}$$

Now choose R so that $B(z_1; R) \subseteq S$ and assume that $z \in B(z_1; R)$. Then the iterated series $\sum_{n=0}^{\infty} \sum_{k=0}^{\infty} c_n(k)$ converges absolutely, since

$$\sum_{n=0}^{\infty} \sum_{k=0}^{\infty} |c_n(k)| = \sum_{n=0}^{\infty} |a_n|(|z - z_1| + |z_1 - z_0|)^n = \sum_{n=0}^{\infty} |a_n|(z_2 - z_0)^n, \quad (18)$$

where

$$z_2 = z_0 + |z - z_1| + |z_1 - z_0|.$$

But

$$|z_2 - z_0| < R + |z_1 - z_0| \le r,$$

and hence the series in (18) converges. Therefore, by Theorem 8.43, we can interchange the order of summation to obtain

$$f(z) = \sum_{k=0}^{\infty} \sum_{n=0}^{\infty} c_n(k) = \sum_{k=0}^{\infty} \sum_{n=k}^{\infty} \binom{n}{k} a_n(z - z_1)^k(z_1 - z_0)^{n-k}$$

$$= \sum_{k=0}^{\infty} b_k(z - z_1)^k,$$

where b_k is given by (17). This completes the proof.

NOTE. In the course of the proof we have shown that we may use any $R > 0$ that satisfies the condition

$$B(z_1; R) \subseteq S. \tag{19}$$

Theorem 9.23. *Assume that $\sum a_n(z - z_0)^n$ converges for each z in $B(z_0; r)$. Then the function f defined by the equation*

$$f(z) = \sum_{n=0}^{\infty} a_n(z - z_0)^n, \qquad \text{if } z \in B(z_0; r), \tag{20}$$

has a derivative $f'(z)$ for each z in $B(z_0; r)$, given by

$$f'(z) = \sum_{n=1}^{\infty} n a_n(z - z_0)^{n-1}. \tag{21}$$

NOTE. The series in (20) and (21) have the same radius of convergence.

Proof. Assume that $z_1 \in B(z_0; r)$ and expand f in a power series about z_1, as indicated in (16). Then, if $z \in B(z_1; R)$, $z \ne z_1$, we have

$$\frac{f(z) - f(z_1)}{z - z_1} = b_1 + \sum_{k=1}^{\infty} b_{k+1}(z - z_1)^k. \tag{22}$$

By continuity, the right member of (22) tends to b_1 as $z \to z_1$. Hence, $f'(z_1)$ exists and equals b_1. Using (17) to compute b_1, we find

$$b_1 = \sum_{n=1}^{\infty} n a_n (z_1 - z_0)^{n-1}.$$

Since z_1 is an arbitrary point of $B(z_0; r)$, this proves (21). The two series have the same radius of convergence because $\sqrt[n]{n} \to 1$ as $n \to \infty$.

NOTE. By repeated application of (21), we find that for each $k = 1, 2, \ldots,$ the derivative $f^{(k)}(z)$ exists in $B(z_0; r)$ and is given by the series

$$f^{(k)}(z) = \sum_{n=k}^{\infty} \frac{n!}{(n-k)!} a_n (z - z_0)^{n-k}. \tag{23}$$

If we put $z = z_0$ in (23), we obtain the important formula

$$f^{(k)}(z_0) = k! a_k \qquad (k = 1, 2, \ldots). \tag{24}$$

This equation tells us that if two power series $\sum a_n(z - z_0)^n$ and $\sum b_n(z - z_0)^n$ both represent the same function in a neighborhood $B(z_0; r)$, then $a_n = b_n$ for every n. That is, the power series expansion of a function f about a given point z_0 is uniquely determined (if it exists at all), and it is given by the formula

$$f(z) = \sum_{n=0}^{\infty} \frac{f^{(n)}(z_0)}{n!} (z - z_0)^n,$$

valid for each z in the disk of convergence.

9.15 MULTIPLICATION OF POWER SERIES

Theorem 9.24. *Given two power series expansions about the origin, say*

$$f(z) = \sum_{n=0}^{\infty} a_n z^n, \qquad \text{if } z \in B(0; r),$$

and

$$g(z) = \sum_{n=0}^{\infty} b_n z^n, \qquad \text{if } z \in B(0; R).$$

Then the product $f(z)g(z)$ is given by the power series

$$f(z)g(z) = \sum_{n=0}^{\infty} c_n z^n, \qquad \text{if } z \in B(0; r) \cap B(0; R),$$

where

$$c_n = \sum_{k=0}^{n} a_k b_{n-k} \qquad (n = 0, 1, 2, \ldots).$$

Proof. The Cauchy product of the two given series is

$$\sum_{n=0}^{\infty} \left(\sum_{k=0}^{n} a_k z^k b_{n-k} z^{n-k} \right) = \sum_{n=0}^{\infty} c_n z^n,$$

and the conclusion follows from Theorem 8.46 (Mertens' Theorem).

NOTE. If the two series are identical, we get

$$f(z)^2 = \sum_{n=0}^{\infty} c_n z^n,$$

where $c_n = \sum_{k=0}^{n} a_k a_{n-k} = \sum_{m_1 + m_2 = n} a_{m_1} a_{m_2}$. The symbol $\sum_{m_1 + m_2 = n}$ indicates that the summation is to be extended over all nonnegative integers m_1 and m_2 whose sum is n. Similarly, for any integer $p > 0$, we have

$$f(z)^p = \sum_{n=0}^{\infty} c_n(p) z^n,$$

where

$$c_n(p) = \sum_{m_1 + \cdots + m_p = n} a_{m_1} \cdots a_{m_p}.$$

9.16 THE SUBSTITUTION THEOREM

Theorem 9.25. *Given two power series expansions about the origin, say*

$$f(z) = \sum_{n=0}^{\infty} a_n z^n, \qquad \text{if } z \in B(0; r),$$

and

$$g(z) = \sum_{n=0}^{\infty} b_n z^n, \qquad \text{if } z \in B(0; R).$$

If, for a fixed z in $B(0; R)$, we have $\sum_{n=0}^{\infty} |b_n z^n| < r$, then for this z we can write

$$f[g(z)] = \sum_{k=0}^{\infty} c_k z^k,$$

where the coefficients c_k are obtained as follows: Define the numbers $b_k(n)$ by the equation

$$g(z)^n = \left(\sum_{k=0}^{\infty} b_k z^k \right)^n = \sum_{k=0}^{\infty} b_k(n) z^k.$$

Then $c_k = \sum_{n=0}^{\infty} a_n b_k(n)$ for $k = 0, 1, 2, \ldots$

NOTE. The series $\sum_{k=0}^{\infty} c_k z^k$ is the power series which arises *formally* by substituting the series for $g(z)$ in place of z in the expansion of f and then rearranging terms in increasing powers of z.

Proof. By hypothesis, we can choose z so that $\sum_{n=0}^{\infty} |b_n z^n| < r$. For this z we have $|g(z)| < r$ and hence we can write

$$f[g(z)] = \sum_{n=0}^{\infty} a_n g(z)^n = \sum_{n=0}^{\infty} \sum_{k=0}^{\infty} a_n b_k(n) z^k.$$

If we are allowed to interchange the order of summation, we obtain

$$f[g(z)] = \sum_{k=0}^{\infty} \left(\sum_{n=0}^{\infty} a_n b_k(n) \right) z^k = \sum_{k=0}^{\infty} c_k z^k,$$

which is the statement we set out to prove. To justify the interchange, we will establish the convergence of the series

$$\sum_{n=0}^{\infty} \sum_{k=0}^{\infty} |a_n b_k(n) z^k| = \sum_{n=0}^{\infty} |a_n| \sum_{k=0}^{\infty} |b_k(n) z^k|. \qquad (25)$$

Now each number $b_k(n)$ is a finite sum of the form

$$b_k(n) = \sum_{m_1 + \cdots + m_n = k} b_{m_1} \cdots b_{m_n},$$

and hence $|b_k(n)| \le \sum_{m_1 + \cdots + m_n = k} |b_{m_1}| \cdots |b_{m_n}|$. On the other hand, we have

$$\left(\sum_{k=0}^{\infty} |b_k| z^k \right)^n = \sum_{k=0}^{\infty} B_k(n) z^k,$$

where $B_k(n) = \sum_{m_1 + \cdots + m_n = k} |b_{m_1}| \cdots |b_{m_n}|$. Returning to (25), we have

$$\sum_{n=0}^{\infty} |a_n| \sum_{k=0}^{\infty} |b_k(n) z^k| \le \sum_{n=0}^{\infty} |a_n| \sum_{k=0}^{\infty} B_k(n) |z^k| = \sum_{n=0}^{\infty} |a_n| \left(\sum_{k=0}^{\infty} |b_k z^k| \right)^n,$$

and this establishes the convergence of (25).

9.17 RECIPROCAL OF A POWER SERIES

As an application of the substitution theorem, we will show that the reciprocal of a power series in z is again a power series in z, provided that the constant term is not 0.

Theorem 9.26. *Assume that we have*

$$p(z) = \sum_{n=0}^{\infty} p_n z^n, \qquad \text{if } z \in B(0; h),$$

where $p(0) \neq 0$. *Then there exists a neighborhood* $B(0; \delta)$ *in which the reciprocal of* p *has a power series expansion of the form*

$$\frac{1}{p(z)} = \sum_{n=0}^{\infty} q_n z^n.$$

Furthermore, $q_0 = 1/p_0$.

Proof. Without loss in generality we can assume that $p_0 = 1$. [Why?] Then $p(0) = 1$. Let $P(z) = 1 + \sum_{n=1}^{\infty} |p_n z^n|$ if $z \in B(0; h)$. By continuity, there exists a neighborhood $B(0; \delta)$ such that $|P(z) - 1| < 1$ if $z \in B(0; \delta)$. The conclusion follows by applying Theorem 9.25 with

$$f(z) = \frac{1}{1-z} = \sum_{n=0}^{\infty} z^n \quad \text{and} \quad g(z) = 1 - p(z) = \sum_{n=1}^{\infty} p_n z^n.$$

9.18 REAL POWER SERIES

If x, x_0, and a_n are real numbers, the series $\sum a_n(x - x_0)^n$ is called a *real power series*. Its disk of convergence intersects the real axis in an interval $(x_0 - r, x_0 + r)$ called the *interval of convergence*.

Each real power series defines a real-valued sum function whose value at each x in the interval of convergence is given by

$$f(x) = \sum_{n=0}^{\infty} a_n(x - x_0)^n.$$

The series is said to *represent f* in the interval of convergence, and it is called a *power-series expansion* of f about x_0.

Two problems concern us here:

1) Given the series, to find properties of the sum function f.

2) Given a function f, to find whether or not it can be represented by a power series.

It turns out that only rather special functions possess power-series expansions. Nevertheless, the class of such functions includes a large number of examples that arise in practice, so their study is of great importance.

Question (1) is answered by the theorems we have already proved for complex power series. A power series converges absolutely for each x in the open subinterval $(x_0 - r, x_0 + r)$ of convergence, and it converges uniformly on every compact subset of this interval. Since each term of the power series is continuous on **R**, the sum function f is continuous on every compact subset of the interval of convergence and hence f is continuous on $(x_0 - r, x_0 + r)$.

Because of uniform convergence, Theorem 9.9 tells us that we can integrate a power series term by term on every compact subinterval inside the interval of convergence. Thus, for every x in $(x_0 - r, x_0 + r)$ we have

$$\int_{x_0}^{x} f(t)\, dt = \sum_{n=0}^{\infty} a_n \int_{x_0}^{x} (t - x_0)^n\, dt = \sum_{n=0}^{\infty} \frac{a_n}{n+1} (x - x_0)^{n+1}.$$

The integrated series has the same radius of convergence.

The sum function has derivatives of every order in the interval of convergence and they can be obtained by differentiating the series term by term. Moreover,

$f^{(n)}(x_0) = n!a_n$ so the sum function is represented by the power series

$$f(x) = \sum_{n=0}^{\infty} \frac{f^{(n)}(x_0)}{n!} (x - x_0)^n. \tag{26}$$

We turn now to question (2). Suppose we are given a real-valued function f defined on some open interval $(x_0 - r, x_0 + r)$, and suppose f has derivatives of every order in this interval. Then we can certainly *form* the power series on the right of (26). Does this series converge for any x besides $x = x_0$? If so, is its sum equal to $f(x)$? In general, the answer to both questions is "No." (See Exercise 9.33 for a counter example.) A necessary and sufficient condition for answering both questions in the affirmative is given in the next section with the help of Taylor's formula (Theorem 5.19.)

9.19 THE TAYLOR'S SERIES GENERATED BY A FUNCTION

Definition 9.27. *Let f be a real-valued function defined on an interval I in* **R**. *If f has derivatives of every order at each point of I, we write $f \in C^{\infty}$ on I.*

If $f \in C^{\infty}$ on some neighborhood of a point c, the power series

$$\sum_{n=0}^{\infty} \frac{f^{(n)}(c)}{n!} (x - c)^n,$$

is called the *Taylor's series about c generated by f.* To indicate that f generates this series, we write

$$f(x) \sim \sum_{n=0}^{\infty} \frac{f^{(n)}(c)}{n!} (x - c)^n.$$

The question we are interested in is this: When can we replace the symbol $\sim$ by the symbol $=$? Taylor's formula states that if $f \in C^{\infty}$ on the closed interval $[a, b]$ and if $c \in [a, b]$, then, for every x in $[a, b]$ and for every n, we have

$$f(x) = \sum_{k=0}^{n-1} \frac{f^{(k)}(c)}{k!} (x - c)^k + \frac{f^{(n)}(x_1)}{n!} (x - c)^n, \tag{27}$$

where x_1 is some point between x and c. The point x_1 depends on x, c, and on n. Hence a necessary and sufficient condition for the Taylor's series to converge to $f(x)$ is that

$$\lim_{n \to \infty} \frac{f^{(n)}(x_1)}{n!} (x - c)^n = 0. \tag{28}$$

In practice it may be quite difficult to deal with this limit because of the unknown position of x_1. In some cases, however, a suitable upper bound can be obtained for $f^{(n)}(x_1)$ and the limit can be shown to be zero. Since $A^n/n! \to 0$ as $n \to \infty$ for all A, equation (28) will certainly hold if there is a positive constant M such that

$$|f^{(n)}(x)| \leq M^n,$$

for all x in $[a, b]$. In other words, the Taylor's series of a function f converges if the nth derivative $f^{(n)}$ grows no faster than the nth power of some positive number. This is stated more formally in the next theorem.

Theorem 9.28. *Assume that $f \in C^\infty$ on $[a, b]$ and let $c \in [a, b]$. Assume that there is a neighborhood $B(c)$ and a constant M (which might depend on c) such that $|f^{(n)}(x)| \leq M^n$ for every x in $B(c) \cap [a, b]$ and every $n = 1, 2, \ldots$ Then, for each x in $B(c) \cap [a, b]$, we have*

$$f(x) = \sum_{n=0}^{\infty} \frac{f^{(n)}(c)}{n!} (x - c)^n.$$

9.20 BERNSTEIN'S THEOREM

Another sufficient condition for convergence of the Taylor's series of f, formulated by S. Bernstein, will be proved in this section. To simplify the proof we first obtain another form of Taylor's formula in which the error term is expressed as an integral.

Theorem 9.29. *Assume f has a continuous derivative of order $n + 1$ in some open interval I containing c, and define $E_n(x)$ for x in I by the equation*

$$f(x) = \sum_{k=0}^{n} \frac{f^{(k)}(c)}{k!} (x - c)^k + E_n(x). \tag{29}$$

Then $E_n(x)$ is also given by the integral

$$E_n(x) = \frac{1}{n!} \int_c^x (x - t)^n f^{(n+1)}(t) \, dt. \tag{30}$$

Proof. The proof is by induction on n. For $n = 1$ we have

$$E_1(x) = f(x) - f(c) - f'(c)(x - c) = \int_c^x [f'(t) - f'(c)] \, dt = \int_c^x u(t) \, dv(t),$$

where $u(t) = f'(t) - f'(c)$ and $v(t) = t - x$. Integration by parts gives

$$\int_c^x u(t) \, dv(t) = u(x)v(x) - u(c)v(c) - \int_c^x v(t) \, du(t) = \int_c^x (x - t)f''(t) \, dt.$$

This proves (30) for $n = 1$. Now we assume (30) is true for n and prove it for $n + 1$. From (29) we have

$$E_{n+1}(x) = E_n(x) - \frac{f^{(n+1)}(c)}{(n + 1)!} (x - c)^{n+1}.$$

We write $E_n(x)$ as an integral and note that $(x - c)^{n+1} = (n + 1) \int_c^x (x - t)^n \, dt$ to obtain

$$E_{n+1}(x) = \frac{1}{n!} \int_c^x (x - t)^n f^{(n+1)}(t) \, dt - \frac{f^{(n+1)}(c)}{n!} \int_c^x (x - t)^n \, dt$$

$$= \frac{1}{n!} \int_c^x (x - t)^n \left[f^{(n+1)}(t) - f^{(n+1)}(c) \right] dt = \frac{1}{n!} \int_c^x u(t) \, dv(t),$$

where $u(t) = f^{(n+1)}(t) - f^{(n+1)}(c)$ and $v(t) = -(x - t)^{n+1}/(n + 1)$. Integration by parts gives us

$$E_{n+1}(x) = -\frac{1}{n!} \int_c^x v(t) \, du(t) = \frac{1}{(n + 1)!} \int_c^x (x - t)^{n+1} f^{(n+2)}(t) \, dt.$$

This proves (30).

NOTE. The change of variable $t = x + (c - x)u$ transforms the integral in (30) to the form

$$E_n(x) = \frac{(x - c)^{n+1}}{n!} \int_0^1 u^n f^{(n+1)}[x + (c - x)u] \, du. \tag{31}$$

Theorem 9.30 (Bernstein). *Assume f and all its derivatives are nonnegative on a compact interval $[b, b + r]$. Then, if $b \le x < b + r$, the Taylor's series*

$$\sum_{k=0}^{\infty} \frac{f^{(k)}(b)}{k!} (x - b)^k,$$

converges to $f(x)$.

Proof. By a translation we can assume $b = 0$. The result is trivial if $x = 0$ so we assume $0 < x < r$. We use Taylor's formula with remainder and write

$$f(x) = \sum_{k=0}^{n} \frac{f^{(k)}(0)}{k!} x^k + E_n(x). \tag{32}$$

We will prove that the error term satisfies the inequalities

$$0 \le E_n(x) \le \left(\frac{x}{r} \right)^{n+1} f(r). \tag{33}$$

This implies that $E_n(x) \to 0$ as $n \to \infty$ since $(x/r)^{n+1} \to 0$ if $0 < x < r$.
    To prove (33) we use (31) with $c = 0$ and find

$$E_n(x) = \frac{x^{n+1}}{n!} \int_0^1 u^n f^{(n+1)}(x - xu) \, du,$$

for each x in $[0, r]$. If $x \ne 0$, let

$$F_n(x) = \frac{E_n(x)}{x^{n+1}} = \frac{1}{n!} \int_0^1 u^n f^{(n+1)}(x - xu) \, du.$$

The function $f^{(n+1)}$ is monotonic increasing on $[0, r]$ since its derivative is non-negative. Therefore we have

$$f^{(n+1)}(x - xu) = f^{(n+1)}[x(1 - u)] \leq f^{(n+1)}[r(1 - u)],$$

if $0 \leq u \leq 1$, and this implies $F_n(x) \leq F_n(r)$ if $0 < x \leq r$. In other words, $E_n(x)/x^{n+1} \leq E_n(r)/r^{n+1}$, or

$$E_n(x) \leq \left(\frac{x}{r}\right)^{n+1} E_n(r). \tag{34}$$

Putting $x = r$ in (32), we see that $E_n(r) \leq f(r)$ since each term in the sum is nonnegative. Using this in (34), we obtain (33) which, in turn, completes the proof.

9.21 THE BINOMIAL SERIES

As an example illustrating the use of Bernstein's theorem, we will obtain the following expansion, known as the *binomial series*:

$$(1 + x)^a = \sum_{n=0}^{\infty} \binom{a}{n} x^n, \quad \text{if } -1 < x < 1, \tag{35}$$

where a is an arbitrary real number and $\binom{a}{n} = a(a - 1)\cdots(a - n + 1)/n!$. Bernstein's theorem is not directly applicable in this case. However we can argue as follows: Let $f(x) = (1 - x)^{-c}$, where $c > 0$ and $x < 1$. Then

$$f^{(n)}(x) = c(c + 1)\cdots(c + n - 1)(1 - x)^{-c-n},$$

and hence $f^{(n)}(x) \geq 0$ for each n, provided that $x < 1$. Applying Bernstein's theorem with $b = -1$ and $r = 2$ we find that $f(x)$ has a power series expansion about the point $b = -1$, convergent for $-1 \leq x < 1$. Therefore, by Theorem 9.22, $f(x)$ also has a power series expansion about 0, $f(x) = \sum_{k=0}^{\infty} f^{(k)}(0)x^k/k!$, convergent for $-1 < x < 1$. But $f^{(k)}(0) = \binom{-c}{k}(-1)^k k!$, so

$$\frac{1}{(1 - x)^c} = \sum_{k=0}^{\infty} \binom{-c}{k}(-1)^k x^k, \quad \text{if } -1 < x < 1. \tag{36}$$

Replacing c by $-a$ and x by $-x$ in (36) we find that (35) is valid for each $a < 0$. But now (35) can be extended to all real a by successive integration.

Of course, if a is a positive integer, say $a = m$, then $\binom{m}{n} = 0$ for $n > m$, and (35) reduces to a finite sum (the Binomial Theorem).

9.22 ABEL'S LIMIT THEOREM

If $-1 < x < 1$, integration of the geometric series

$$\frac{1}{1 - x} = \sum_{n=0}^{\infty} x^n$$

gives us the series expansion

$$\log(1 - x) = -\sum_{n=1}^{\infty} \frac{x^n}{n}, \tag{37}$$

also valid for $-1 < x < 1$. If we put $x = -1$ in the righthand side of (37), we obtain a convergent alternating series, namely, $\sum(-1)^{n+1}/n$. Can we also put $x = -1$ in the lefthand side of (37)? The next theorem answers this question in the affirmative.

Theorem 9.31 (Abel's limit theorem). *Assume that we have*

$$f(x) = \sum_{n=0}^{\infty} a_n x^n, \qquad if \; -r < x < r. \tag{38}$$

If the series also converges at $x = r$, then the limit $\lim_{x \to r-} f(x)$ exists and we have

$$\lim_{x \to r-} f(x) = \sum_{n=0}^{\infty} a_n r^n.$$

Proof. For simplicity, assume that $r = 1$ (this amounts to a change in scale). Then we are given that $f(x) = \sum a_n x^n$ for $-1 < x < 1$ and that $\sum a_n$ converges. Let us write $f(1) = \sum_{n=0}^{\infty} a_n$. We are to prove that $\lim_{x \to 1-} f(x) = f(1)$, or, in other words, that f is continuous from the left at $x = 1$.

If we multiply the series for $f(x)$ by the geometric series and use Theorem 9.24, we find

$$\frac{1}{1-x} f(x) = \sum_{n=0}^{\infty} c_n x^n, \qquad \text{where } c_n = \sum_{k=0}^{n} a_k.$$

Hence we have

$$f(x) - f(1) = (1 - x) \sum_{n=0}^{\infty} [c_n - f(1)] x^n, \qquad if \; -1 < x < 1. \tag{39}$$

By hypothesis, $\lim_{n \to \infty} c_n = f(1)$. Therefore, given $\varepsilon > 0$, we can find N such that $n \geq N$ implies $|c_n - f(1)| < \varepsilon/2$. If we split the sum (39) into two parts, we get

$$f(x) - f(1) = (1 - x) \sum_{n=0}^{N-1} [c_n - f(1)] x^n + (1 - x) \sum_{n=N}^{\infty} [c_n - f(1)] x^n. \tag{40}$$

Let M denote the largest of the N numbers $|c_n - f(1)|$, $n = 0, 1, 2, \ldots, N - 1$. If $0 < x < 1$, (40) gives us

$$|f(x) - f(1)| \leq (1 - x)NM + (1 - x) \frac{\varepsilon}{2} \sum_{n=N}^{\infty} x^n$$

$$= (1 - x)NM + (1 - x) \frac{\varepsilon}{2} \frac{x^N}{1 - x} < (1 - x)NM + \frac{\varepsilon}{2}.$$

Now let $\delta = \varepsilon/2NM$. Then $0 < 1 - x < \delta$ implies $|f(x) - f(1)| < \varepsilon$, which means $\lim_{x \to 1-} f(x) = f(1)$. This completes the proof.

Example. We may put $x = -1$ in (37) to obtain

$$\log 2 = \sum_{n=1}^{\infty} \frac{(-1)^{n+1}}{n}.$$

(See Exercise 8.18 for another derivation of this formula.)

As an application of Abel's theorem we can derive the following result on multiplication of series:

Theorem 9.32. *Let $\sum_{n=0}^{\infty} a_n$ and $\sum_{n=0}^{\infty} b_n$ be two convergent series and let $\sum_{n=0}^{\infty} c_n$ denote their Cauchy product. If $\sum_{n=0}^{\infty} c_n$ converges, we have*

$$\sum_{n=0}^{\infty} c_n = \left(\sum_{n=0}^{\infty} a_n \right) \left(\sum_{n=0}^{\infty} b_n \right).$$

NOTE. This result is similar to Theorem 8.46 except that we do not assume absolute convergence of either of the two given series. However, we *do* assume convergence of their Cauchy product.

Proof. The two power series $\sum a_n x^n$ and $\sum b_n x^n$ both converge for $x = 1$, and hence they converge in the neighborhood $B(0; 1)$. Keep $|x| < 1$ and write

$$\sum_{n=0}^{\infty} c_n x^n = \left(\sum_{n=0}^{\infty} a_n x^n \right) \left(\sum_{n=0}^{\infty} b_n x^n \right),$$

using Theorem 9.24. Now let $x \to 1-$ and apply Abel's theorem.

9.23 TAUBER'S THEOREM

The converse of Abel's limit theorem is false in general. That is, if f is given by (38), the limit $f(r-)$ may exist but yet the series $\sum a_n r^n$ may fail to converge. For example, take $a_n = (-1)^n$. Then $f(x) = 1/(1 + x)$ if $-1 < x < 1$ and $f(x) \to \frac{1}{2}$ as $x \to 1-$. However, $\sum(-1)^n$ diverges. A. Tauber (1897) discovered that by placing further restrictions on the coefficients a_n, one can obtain a converse to Abel's theorem. A large number of such results are now known and they are referred to as *Tauberian theorems*. The simplest of these, sometimes called *Tauber's first theorem*, is the following:

Theorem 9.33 (Tauber). *Let $f(x) = \sum_{n=0}^{\infty} a_n x^n$ for $-1 < x < 1$, and assume that $\lim_{n \to \infty} na_n = 0$. If $f(x) \to S$ as $x \to 1-$, then $\sum_{n=0}^{\infty} a_n$ converges and has sum S.*

Proof. Let $n\sigma_n = \sum_{k=0}^{n} k|a_k|$. Then $\sigma_n \to 0$ as $n \to \infty$. (See Note following Theorem 8.48.) Also, $\lim_{n \to \infty} f(x_n) = S$ if $x_n = 1 - 1/n$. Hence, given $\varepsilon > 0$, we can choose N so that $n \geq N$ implies

$$|f(x_n) - S| < \frac{\varepsilon}{3}, \qquad \sigma_n < \frac{\varepsilon}{3}, \qquad n|a_n| < \frac{\varepsilon}{3}.$$

Now let $s_n = \sum_{k=0}^{n} a_k$. Then, for $-1 < x < 1$, we can write

$$s_n - S = f(x) - S + \sum_{k=0}^{n} a_k(1 - x^k) - \sum_{k=n+1}^{\infty} a_k x^k.$$

Now keep x in $(0, 1)$. Then

$$(1 - x^k) = (1 - x)(1 + x + \cdots + x^{k-1}) \leq k(1 - x),$$

for each k. Therefore, if $n \geq N$ and $0 < x < 1$, we have

$$|s_n - S| \leq |f(x) - S| + (1 - x) \sum_{k=0}^{n} k|a_k| + \frac{\varepsilon}{3n(1 - x)}.$$

Taking $x = x_n = 1 - 1/n$, we find $|s_n - S| < \varepsilon/3 + \varepsilon/3 + \varepsilon/3 = \varepsilon$. This completes the proof.

NOTE. See Exercise 9.37 for another Tauberian theorem.

EXERCISES

Uniform convergence

9.1 Assume that $f_n \to f$ uniformly on S and that each f_n is bounded on S. Prove that $\{f_n\}$ is uniformly bounded on S.

9.2 Define two sequences $\{f_n\}$ and $\{g_n\}$ as follows:

$$f_n(x) = x \left(1 + \frac{1}{n}\right) \qquad \text{if } x \in \mathbf{R}, \quad n = 1, 2, \ldots,$$

$$g_n(x) = \begin{cases} \dfrac{1}{n} & \text{if } x = 0 \text{ or if } x \text{ is irrational,} \\[2mm] b + \dfrac{1}{n} & \text{if } x \text{ is rational, say } x = \dfrac{a}{b}, \quad b > 0. \end{cases}$$

Let $h_n(x) = f_n(x)g_n(x)$.

    a) Prove that both $\{f_n\}$ and $\{g_n\}$ converge uniformly on every bounded interval.

    b) Prove that $\{h_n\}$ does not converge uniformly on any bounded interval.

9.3 Assume that $f_n \to f$ uniformly on S, $g_n \to g$ uniformly on S.

    a) Prove that $f_n + g_n \to f + g$ uniformly on S.

    b) Let $h_n(x) = f_n(x)g_n(x)$, $h(x) = f(x)g(x)$, if $x \in S$. Exercise 9.2 shows that the assertion $h_n \to h$ uniformly on S is, in general, incorrect. Prove that it *is* correct if each f_n and each g_n is bounded on S.

9.4 Assume that $f_n \to f$ uniformly on S and suppose there is a constant $M > 0$ such that $|f_n(x)| \le M$ for all x in S and all n. Let g be continuous on the closure of the disk $B(0; M)$ and define $h_n(x) = g[f_n(x)]$, $h(x) = g[f(x)]$, if $x \in S$. Prove that $h_n \to h$ uniformly on S.

9.5 a) Let $f_n(x) = 1/(nx + 1)$ if $0 < x < 1$, $n = 1, 2, \ldots$ Prove that $\{f_n\}$ converges pointwise but not uniformly on $(0, 1)$.

  b) Let $g_n(x) = x/(nx + 1)$ if $0 < x < 1$, $n = 1, 2, \ldots$ Prove that $g_n \to 0$ uniformly on $(0, 1)$.

9.6 Let $f_n(x) = x^n$. The sequence $\{f_n\}$ converges pointwise but not uniformly on $[0, 1]$. Let g be continuous on $[0, 1]$ with $g(1) = 0$. Prove that the sequence $\{g(x)x^n\}$ converges uniformly on $[0, 1]$.

9.7 Assume that $f_n \to f$ uniformly on S, and that each f_n is continuous on S. If $x \in S$, let $\{x_n\}$ be a sequence of points in S such that $x_n \to x$. Prove that $f_n(x_n) \to f(x)$.

9.8 Let $\{f_n\}$ be a sequence of continuous functions defined on a compact set S and assume that $\{f_n\}$ converges pointwise on S to a limit function f. Prove that $f_n \to f$ uniformly on S if, and only if, the following two conditions hold:

  i) The limit function f is continuous on S.

  ii) For every $\varepsilon > 0$, there exists an $m > 0$ and a $\delta > 0$ such that $n > m$ and $|f_k(x) - f(x)| < \delta$ implies $|f_{k+n}(x) - f(x)| < \varepsilon$ for all x in S and all $k = 1, 2, \ldots$

Hint. To prove the sufficiency of (i) and (ii), show that for each x_0 in S there is a neighborhood $B(x_0)$ and an integer k (depending on x_0) such that

$$|f_k(x) - f(x)| < \delta \qquad \text{if } x \in B(x_0).$$

By compactness, a finite set of integers, say $A = \{k_1, \ldots, k_r\}$, has the property that, for each x in S, some k in A satisfies $|f_k(x) - f(x)| < \delta$. Uniform convergence is an easy consequence of this fact.

9.9 a) Use Exercise 9.8 to prove the following theorem of Dini: *If $\{f_n\}$ is a sequence of real-valued continuous functions converging pointwise to a continuous limit function f on a compact set S, and if $f_n(x) \ge f_{n+1}(x)$ for each x in S and every $n = 1, 2, \ldots,$ then $f_n \to f$ uniformly on S.*

  b) Use the sequence in Exercise 9.5(a) to show that compactness of S is essential in Dini's theorem.

9.10 Let $f_n(x) = n^c x(1 - x^2)^n$ for x real and $n \ge 1$. Prove that $\{f_n\}$ converges pointwise on $[0, 1]$ for every real c. Determine those c for which the convergence is uniform on $[0, 1]$ and those for which term-by-term integration on $[0, 1]$ leads to a correct result.

9.11 Prove that $\sum x^n(1 - x)$ converges pointwise but not uniformly on $[0, 1]$, whereas $\sum(-1)^n x^n(1 - x)$ converges uniformly on $[0, 1]$. This illustrates that uniform convergence of $\sum f_n(x)$ along with pointwise convergence of $\sum |f_n(x)|$ does not necessarily imply uniform convergence of $\sum |f_n(x)|$.

9.12 Assume that $g_{n+1}(x) \le g_n(x)$ for each x in T and each $n = 1, 2, \ldots,$ and suppose that $g_n \to 0$ uniformly on T. Prove that $\sum(-1)^{n+1} g_n(x)$ converges uniformly on T.

9.13 Prove Abel's test for uniform convergence: Let $\{g_n\}$ be a sequence of real-valued functions such that $g_{n+1}(x) \le g_n(x)$ for each x in T and for every $n = 1, 2, \ldots$ If $\{g_n\}$

is uniformly bounded on T and if $\sum f_n(x)$ converges uniformly on T, then $\sum f_n(x)g_n(x)$ also converges uniformly on T.

9.14 Let $f_n(x) = x/(1 + nx^2)$ if $x \in \mathbf{R}$, $n = 1, 2, \ldots$ Find the limit function f of the sequence $\{f_n\}$ and the limit function g of the sequence $\{f_n'\}$.

    a) Prove that $f'(x)$ exists for every x but that $f'(0) \neq g(0)$. For what values of x is $f'(x) = g(x)$?

    b) In what subintervals of $\mathbf{R}$ does $f_n \to f$ uniformly?

    c) In what subintervals of $\mathbf{R}$ does $f_n' \to g$ uniformly?

9.15 Let $f_n(x) = (1/n)e^{-n^2x^2}$ if $x \in \mathbf{R}$, $n = 1, 2, \ldots$ Prove that $f_n \to 0$ uniformly on $\mathbf{R}$, that $f_n' \to 0$ pointwise on $\mathbf{R}$, but that the convergence of $\{f_n'\}$ is not uniform on any interval containing the origin.

9.16 Let $\{f_n\}$ be a sequence of real-valued continuous functions defined on $[0, 1]$ and assume that $f_n \to f$ uniformly on $[0, 1]$. Prove or disprove

$$\lim_{n \to \infty} \int_0^{1-1/n} f_n(x)\, dx = \int_0^1 f(x)\, dx.$$

9.17 Mathematicians from Slobbovia decided that the Riemann integral was too complicated so they replaced it by the *Slobbovian integral*, defined as follows: If f is a function defined on the set $\mathbf{Q}$ of rational numbers in $[0, 1]$, the Slobbovian integral of f, denoted by $S(f)$, is defined to be the limit

$$S(f) = \lim_{n \to \infty} \frac{1}{n} \sum_{k=1}^n f\left(\frac{k}{n}\right),$$

whenever this limit exists. Let $\{f_n\}$ be a sequence of functions such that $S(f_n)$ exists for each n and such that $f_n \to f$ uniformly on $\mathbf{Q}$. Prove that $\{S(f_n)\}$ converges, that $S(f)$ exists, and that $S(f_n) \to S(f)$ as $n \to \infty$.

9.18 Let $f_n(x) = 1/(1 + n^2x^2)$ if $0 \le x \le 1$, $n = 1, 2, \ldots$ Prove that $\{f_n\}$ converges pointwise but not uniformly on $[0, 1]$. Is term-by-term integration permissible?

9.19 Prove that $\sum_{n=1}^\infty x/n^\alpha(1 + nx^2)$ converges uniformly on every finite interval in $\mathbf{R}$ if $\alpha > \frac{1}{2}$. Is the convergence uniform on $\mathbf{R}$?

9.20 Prove that the series $\sum_{n=1}^\infty ((-1)^n/\sqrt{n}) \sin(1 + (x/n))$ converges uniformly on every compact subset of $\mathbf{R}$.

9.21 Prove that the series $\sum_{n=0}^\infty (x^{2n+1}/(2n + 1) - x^{n+1}/(2n + 2))$ converges pointwise but not uniformly on $[0, 1]$.

9.22 Prove that $\sum_{n=1}^\infty a_n \sin nx$ and $\sum_{n=1}^\infty a_n \cos nx$ are uniformly convergent on $\mathbf{R}$ if $\sum_{n=1}^\infty |a_n|$ converges.

9.23 Let $\{a_n\}$ be a decreasing sequence of positive terms. Prove that the series $\sum a_n \sin nx$ converges uniformly on $\mathbf{R}$ if, and only if, $na_n \to 0$ as $n \to \infty$.

9.24 Given a convergent series $\sum_{n=1}^\infty a_n$. Prove that the Dirichlet series $\sum_{n=1}^\infty a_n n^{-s}$ converges uniformly on the half-infinite interval $0 \le s < +\infty$. Use this to prove that $\lim_{s \to 0+} \sum_{n=1}^\infty a_n n^{-s} = \sum_{n=1}^\infty a_n$.

9.25 Prove that the series $\zeta(s) = \sum_{n=1}^{\infty} n^{-s}$ converges uniformly on every half-infinite interval $1 + h \le s < +\infty$, where $h > 0$. Show that the equation

$$\zeta'(s) = - \sum_{n=1}^{\infty} \frac{\log n}{n^s}$$

is valid for each $s > 1$ and obtain a similar formula for the kth derivative $\zeta^{(k)}(s)$.

Mean convergence

9.26 Let $f_n(x) = n^{3/2}xe^{-n^2x^2}$. Prove that $\{f_n\}$ converges pointwise to 0 on $[-1, 1]$ but that $\text{l.i.m.}_{n\to\infty} f_n \ne 0$ on $[-1, 1]$.

9.27 Assume that $\{f_n\}$ converges pointwise to f on $[a, b]$ and that $\text{l.i.m.}_{n\to\infty} f_n = g$ on $[a, b]$. Prove that $f = g$ if both f and g are continuous on $[a, b]$.

9.28 Let $f_n(x) = \cos^n x$ if $0 \le x \le \pi$.

a) Prove that $\text{l.i.m.}_{n\to\infty} f_n = 0$ on $[0, \pi]$ but that $\{f_n(\pi)\}$ does not converge.

b) Prove that $\{f_n\}$ converges pointwise but not uniformly on $[0, \pi/2]$.

9.29 Let $f_n(x) = 0$ if $0 \le x \le 1/n$ or if $2/n \le x \le 1$, and let $f_n(x) = n$ if $1/n < x < 2/n$. Prove that $\{f_n\}$ converges pointwise to 0 on $[0, 1]$ but that $\text{l.i.m.}_{n\to\infty} f_n \ne 0$ on $[0, 1]$.

Power series

9.30 If r is the radius of convergence of $\sum a_n(z - z_0)^n$, where each $a_n \ne 0$, show that

$$\liminf_{n\to\infty} \left| \frac{a_n}{a_{n+1}} \right| \le r \le \limsup_{n\to\infty} \left| \frac{a_n}{a_{n+1}} \right|.$$

9.31 Given that the power series $\sum_{n=0}^{\infty} a_n z^n$ has radius of convergence 2. Find the radius of convergence of each of the following series:

a) $\sum_{n=0}^{\infty} a_n^k z^n$,

b) $\sum_{n=0}^{\infty} a_n z^{kn}$,

c) $\sum_{n=0}^{\infty} a_n z^{n^2}$.

In (a) and (b), k is a fixed positive integer.

9.32 Given a power series $\sum_{n=0}^{\infty} a_n x^n$ whose coefficients are related by an equation of the form

$$a_n + Aa_{n-1} + Ba_{n-2} = 0 \qquad (n = 2, 3, \dots).$$

Show that for any x for which the series converges, its sum is

$$\frac{a_0 + (a_1 + Aa_0)x}{1 + Ax + Bx^2}.$$

9.33 Let $f(x) = e^{-1/x^2}$ if $x \ne 0$, $f(0) = 0$.

a) Show that $f^{(n)}(0)$ exists for all $n \ge 1$.

b) Show that the Taylor's series about 0 generated by f converges everywhere on $\mathbf{R}$ but that it represents f only at the origin.

9.34 Show that the binomial series $(1 + x)^{\alpha} = \sum_{n=0}^{\infty} \binom{\alpha}{n} x^n$ exhibits the following be-havior at the points $x = \pm 1$.

    a) If $x = -1$, the series converges for $\alpha \geq 0$ and diverges for $\alpha < 0$.

    b) If $x = 1$, the series diverges for $\alpha \leq -1$, converges conditionally for α in the interval $-1 < \alpha < 0$, and converges absolutely for $\alpha \geq 0$.

9.35 Show that $\sum a_n x^n$ converges uniformly on $[0, 1]$ if $\sum a_n$ converges. Use this fact to give another proof of Abel's limit theorem.

9.36 If each $a_n \geq 0$ and if $\sum a_n$ diverges, show that $\sum a_n x^n \to +\infty$ as $x \to 1-$. (Assume $\sum a_n x^n$ converges for $|x| < 1$.)

9.37 If each $a_n \geq 0$ and if $\lim_{x \to 1-} \sum a_n x^n$ exists and equals A, prove that $\sum a_n$ converges and has sum A. (Compare with Theorem 9.33.)

9.38 For each real t, define $f_t(x) = xe^{xt}/(e^x - 1)$ if $x \in \mathbf{R}$, $x \neq 0$, $f_t(0) = 1$.

    a) Show that there is a disk $B(0; \delta)$ in which f_t is represented by a power series in x.

    b) Define $P_0(t), P_1(t), P_2(t), \ldots$, by the equation

$$f_t(x) = \sum_{n=0}^{\infty} P_n(t) \frac{x^n}{n!}, \quad \text{if } x \in B(0; \delta),$$

and use the identity

$$\sum_{n=0}^{\infty} P_n(t) \frac{x^n}{n!} = e^{tx} \sum_{n=0}^{\infty} P_n(0) \frac{x^n}{n!}$$

to prove that $P_n(t) = \sum_{k=0}^{n} \binom{n}{k} P_k(0) t^{n-k}$. This shows that each function P_n is a polynomial. These are the *Bernoulli polynomials*. The numbers $B_n = P_n(0)$ $(n = 0, 1, 2, \ldots)$ are called the *Bernoulli numbers*. Derive the following further properties:

    c) $B_0 = 1$, $B_1 = -\frac{1}{2}$, $\sum_{k=0}^{n-1} \binom{n}{k} B_k = 0$, if $n = 2, 3, \ldots$

    d) $P_n'(t) = nP_{n-1}(t)$, if $n = 1, 2, \ldots$

    e) $P_n(t + 1) - P_n(t) = nt^{n-1}$ if $n = 1, 2, \ldots$

    f) $P_n(1 - t) = (-1)^n P_n(t)$ g) $B_{2n+1} = 0$ if $n = 1, 2, \ldots$

    h) $1^n + 2^n + \cdots + (k - 1)^n = \dfrac{P_{n+1}(k) - P_{n+1}(0)}{n + 1}$ $(n = 2, 3, \ldots)$.

SUGGESTED REFERENCES FOR FURTHER STUDY

9.1 Hardy, G. H., *Divergent Series*. Oxford Univ. Press, Oxford, 1949.

9.2 Hirschmann, I. I., *Infinite Series*. Holt, Rinehart and Winston, New York, 1962.

9.3 Knopp, K., *Theory and Application of Infinite Series*, 2nd ed. R. C. Young, trans-lator. Hafner, New York, 1948.

CHAPTER 10

THE LEBESGUE INTEGRAL

10.1 INTRODUCTION

The Riemann integral $\int_a^b f(x)\,dx$, as developed in Chapter 7, is well motivated, simple to describe, and serves all the needs of elementary calculus. However, this integral does not meet all the requirements of advanced analysis. An extension, called the *Lebesgue integral*, is discussed in this chapter. It permits more general functions as integrands, it treats bounded and unbounded functions simultaneously, and it enables us to replace the interval $[a, b]$ by more general sets.

The Lebesgue integral also gives more satisfying convergence theorems. If a sequence of functions $\{f_n\}$ converges pointwise to a limit function f on $[a, b]$, it is desirable to conclude that

$$\lim_{n \to \infty} \int_a^b f_n(x)\,dx = \int_a^b f(x)\,dx$$

with a minimum of additional hypotheses. The definitive result of this type is Lebesgue's *dominated convergence theorem*, which permits term-by-term integration if each $\{f_n\}$ is Lebesgue-integrable and if the sequence is dominated by a Lebesgue-integrable function. (See Theorem 10.27.) Here Lebesgue integrals are essential. The theorem is false for Riemann integrals.

In Riemann's approach the interval of integration is subdivided into a finite number of subintervals. In Lebesgue's approach the interval is subdivided into more general types of sets called *measurable sets*. In a classic memoir, *Integrale, longueur, aire*, published in 1902, Lebesgue gave a definition of measure for point sets and applied this to develop his new integral.

Since Lebesgue's early work, both measure theory and integration theory have undergone many generalizations and modifications. The work of Young, Daniell, Riesz, Stone, and others has shown that the Lebesgue integral can be introduced by a method which does not depend on measure theory but which focuses directly on functions and their integrals. This chapter follows this approach, as outlined in Reference 10.10. The only concept required from measure theory is sets of measure zero, a simple idea introduced in Chapter 7. Later, we indicate briefly how measure theory can be developed with the help of the Lebesgue integral.

10.2 THE INTEGRAL OF A STEP FUNCTION

The approach used here is to define the integral first for step functions, then for a larger class (called upper functions) which contains limits of certain increasing sequences of step functions, and finally for an even larger class, the Lebesgue-integrable functions.

We recall that a function s, defined on a compact interval $[a, b]$, is called a step function if there is a partition $P = \{x_0, x_1, \ldots, x_n\}$ of $[a, b]$ such that s is constant on every open subinterval, say

$$s(x) = c_k \qquad \text{if } x \in (x_{k-1}, x_k).$$

A step function is Riemann-integrable on each subinterval $[x_{k-1}, x_k]$ and its integral over this subinterval is given by

$$\int_{x_{k-1}}^{x_k} s(x)\, dx = c_k(x_k - x_{k-1}),$$

regardless of the values of s at the endpoints. The Riemann integral of s over $[a, b]$ is therefore equal to the sum

$$\int_a^b s(x)\, dx = \sum_{k=1}^n c_k(x_k - x_{k-1}). \qquad (1)$$

NOTE. Lebesgue theory can be developed without prior knowledge of Riemann integration by using equation (1) as the *definition* of the integral of a step function. It should be noted that the sum in (1) is independent of the choice of P as long as s is constant on the open subintervals of P.

It is convenient to remove the restriction that the domain of a step function be compact.

Definition 10.1. *Let I denote a general interval (bounded, unbounded, open, closed, or half-open). A function s is called a step function on I if there is a compact subinterval $[a, b]$ of I such that s is a step function on $[a, b]$ and $s(x) = 0$ if $x \in I - [a, b]$. The integral of s over I, denoted by $\int_I s(x)\, dx$ or by $\int_I s$, is defined to be the integral of s over $[a, b]$, as given by (1).*

There are, of course, many compact intervals $[a, b]$ outside of which s vanishes, but the integral of s is independent of the choice of $[a, b]$.

The sum and product of two step functions is also a step function. The following properties of the integral for step functions are easily deduced from the foregoing definition:

$$\int_I (s + t) = \int_I s + \int_I t, \qquad \int_I cs = c \int_I s \qquad \text{for every constant } c,$$

$$\int_I s \le \int_I t \qquad \text{if } s(x) \le t(x) \quad \text{for all } x \text{ in } I.$$

Also, if I is expressed as the union of a finite set of subintervals, say $I = \bigcup_{r=1}^{p} [a_r, b_r]$, where no two subintervals have interior points in common, then

$$\int_I s(x)\, dx = \sum_{r=1}^{p} \int_{a_r}^{b_r} s(x)\, dx.$$

10.3 MONOTONIC SEQUENCES OF STEP FUNCTIONS

A sequence of real-valued functions $\{f_n\}$ defined on a set S is said to be *increasing* on S if

$$f_n(x) \le f_{n+1}(x) \qquad \text{for all } x \text{ in } S \text{ and all } n.$$

A *decreasing* sequence is one satisfying the reverse inequality.

NOTE. We remind the reader that a subset T of $\mathbf{R}$ is said to be of measure 0 if, for every $\varepsilon > 0$, T can be covered by a countable collection of intervals, the sum of whose lengths is less than ε. A property is said to hold *almost everywhere* on a set S (written: *a.e.* on S) if it holds everywhere on S except for a set of measure 0.

NOTATION. If $\{f_n\}$ is an *increasing sequence* of functions on S such that $f_n \to f$ almost everywhere on S, we indicate this by writing

$$f_n \nearrow f \qquad a.e. \text{ on } S.$$

Similarly, the notation $f_n \searrow f$ a.e. on S means that $\{f_n\}$ is a *decreasing sequence* on S which converges to f almost everywhere on S.

The next theorem is concerned with decreasing sequences of step functions on a general interval I.

Theorem 10.2. *Let $\{s_n\}$ be a decreasing sequence of nonnegative step functions such that $s_n \searrow 0$ a.e. on an interval I. Then*

$$\lim_{n \to \infty} \int_I s_n = 0.$$

Proof. The idea of the proof is to write

$$\int_I s_n = \int_A s_n + \int_B s_n,$$

where each of A and B is a finite union of intervals. The set A is chosen so that in its intervals the integrand is small if n is sufficiently large. In B the integrand need not be small but the sum of the lengths of its intervals will be small. To carry out this idea we proceed as follows.

There is a compact interval $[a, b]$ outside of which s_1 vanishes. Since

$$0 \le s_n(x) \le s_1(x) \qquad \text{for all } x \text{ in } I,$$

each s_n vanishes outside $[a, b]$. Now s_n is constant on each open subinterval of some partition of $[a, b]$. Let D_n denote the set of endpoints of these subintervals, and let $D = \bigcup_{n=1}^{\infty} D_n$. Since each D_n is a finite set, the union D is countable and therefore has measure 0. Let E denote the set of points in $[a, b]$ at which the sequence $\{s_n\}$ does *not* converge to 0. By hypothesis, E has measure 0 so the set

$$F = D \cup E$$

also has measure 0. Therefore, if $\varepsilon > 0$ is given we can cover F by a countable collection of open intervals $F_1, F_2, \ldots$, the sum of whose lengths is less than ε.

Now suppose $x \in [a, b] - F$. Then $x \notin E$, so $s_n(x) \to 0$ as $n \to \infty$. Therefore there is an integer $N = N(x)$ such that $s_N(x) < \varepsilon$. Also, $x \notin D$ so x is interior to some interval of constancy of s_N. Hence there is an open interval $B(x)$ such that $s_N(t) < \varepsilon$ for all t in $B(x)$. Since $\{s_n\}$ is decreasing, we also have

$$s_n(t) < \varepsilon \qquad \text{for all } n \geq N \text{ and all } t \text{ in } B(x). \tag{2}$$

The set of all intervals $B(x)$ obtained as x ranges through $[a, b] - F$, together with the intervals $F_1, F_2, \ldots$, form an open covering of $[a, b]$. Since $[a, b]$ is compact there is a finite subcover, say

$$[a, b] \subseteq \bigcup_{i=1}^{p} B(x_i) \cup \bigcup_{r=1}^{q} F_r.$$

Let N_0 denote the largest of the integers $N(x_1), \ldots, N(x_p)$. From (2) we see that

$$s_n(t) < \varepsilon \qquad \text{for all } n \geq N_0 \text{ and all } t \text{ in } \bigcup_{i=1}^{p} B(x_i). \tag{3}$$

Now define A and B as follows:

$$B = \bigcup_{r=1}^{q} F_r, \qquad A = [a, b] - B.$$

Then A is a finite union of disjoint intervals and we have

$$\int_I s_n = \int_a^b s_n = \int_A s_n + \int_B s_n.$$

First we estimate the integral over B. Let M be an upper bound for s_1 on $[a, b]$. Since $\{s_n\}$ is decreasing, we have $s_n(x) \leq s_1(x) \leq M$ for all x in $[a, b]$. The sum of the lengths of the intervals in B is less than ε, so we have

$$\int_B s_n \leq M\varepsilon.$$

Next we estimate the integral over A. Since $A \subseteq \bigcup_{i=1}^{p} B(x_i)$, the inequality in (3) shows that $s_n(x) < \varepsilon$ if $x \in A$ and $n \geq N_0$. The sum of the lengths of the intervals in A does not exceed $b - a$, so we have the estimate

$$\int_A s_n \leq (b - a)\varepsilon \qquad \text{if } n \geq N_0.$$

The two estimates together give us $\int_I s_n \leq (M + b - a)\varepsilon$ if $n \geq N_0$, and this shows that $\lim_{n \to \infty} \int_I s_n = 0$.

Theorem 10.3. *Let $\{t_n\}$ be a sequence of step functions on an interval I such that:*

a) *There is a function f such that $t_n \nearrow f$ a.e. on I,*
   and

b) *the sequence $\{\int_I t_n\}$ converges.*

Then for any step function t such that $t(x) \leq f(x)$ a.e. on I, we have

$$\int_I t \leq \lim_{n \to \infty} \int_I t_n. \tag{4}$$

Proof. Define a new sequence of nonnegative step functions $\{s_n\}$ on I as follows:

$$s_n(x) = \begin{cases} t(x) - t_n(x) & \text{if } t(x) \geq t_n(x), \\ 0 & \text{if } t(x) < t_n(x). \end{cases}$$

Note that $s_n(x) = \max \{t(x) - t_n(x), 0\}$. Now $\{s_n\}$ is decreasing on I since $\{t_n\}$ is increasing, and $s_n(x) \to \max \{t(x) - f(x), 0\}$ a.e. on I. But $t(x) \leq f(x)$ a.e. on I, and therefore $s_n \searrow 0$ a.e. on I. Hence, by Theorem 10.2, $\lim_{n \to \infty} \int_I s_n = 0$. But $s_n(x) \geq t(x) - t_n(x)$ for all x in I, so

$$\int_I s_n \geq \int_I t - \int_I t_n.$$

Now let $n \to \infty$ to obtain (4).

10.4 UPPER FUNCTIONS AND THEIR INTEGRALS

Let $S(I)$ denote the set of all step functions on an interval I. The integral has been defined for all functions in $S(I)$. Now we shall extend the definition to a larger class $U(I)$ which contains limits of certain increasing sequences of step functions. The functions in this class are called *upper functions* and they are defined as follows:

Definition 10.4. *A real-valued function f defined on an interval I is called an upper function on I, and we write $f \in U(I)$, if there exists an increasing sequence of step functions $\{s_n\}$ such that*

a) $s_n \nearrow f$ *a.e. on I,*
   and

b) $\lim_{n \to \infty} \int_I s_n$ *is finite.*

The sequence $\{s_n\}$ is said to generate f. The integral of f over I is defined by the equation

$$\int_I f = \lim_{n \to \infty} \int_I s_n. \tag{5}$$

NOTE. Since $\{\int_I s_n\}$ is an increasing sequence of real numbers, condition (b) is equivalent to saying that $\{\int_I s_n\}$ is bounded above.

The next theorem shows that the definition of the integral in (5) is unambiguous.

Theorem 10.5. *Assume $f \in U(I)$ and let $\{s_n\}$ and $\{t_m\}$ be two sequences generating f. Then*

$$\lim_{n \to \infty} \int_I s_n = \lim_{m \to \infty} \int_I t_m.$$

Proof. The sequence $\{t_m\}$ satisfies hypotheses (a) and (b) of Theorem 10.3. Also, for every n we have

$$s_n(x) \leq f(x) \qquad a.e. \text{ on } I,$$

so (4) gives us

$$\int_I s_n \leq \lim_{m \to \infty} \int_I t_m.$$

Since this holds for every n, we have

$$\lim_{n \to \infty} \int_I s_n \leq \lim_{m \to \infty} \int_I t_m.$$

The same argument, with the sequences $\{s_n\}$ and $\{t_m\}$ interchanged, gives the reverse inequality and completes the proof.

It is easy to see that every step function is an upper function and that its integral, as given by (5), is the same as that given by the earlier definition in Section 10.2. Further properties of the integral for upper functions are described in the next theorem.

Theorem 10.6. *Assume $f \in U(I)$ and $g \in U(I)$. Then:*
a) *$(f + g) \in U(I)$ and*

$$\int_I (f + g) = \int_I f + \int_I g.$$

b) *$cf \in U(I)$ for every constant $c \geq 0$, and*

$$\int_I cf = c \int_I f.$$

c) *$\int_I f \leq \int_I g$ if $f(x) \leq g(x)$ a.e. on I.*

NOTE. In part (b) the requirement $c \geq 0$ is essential. There are examples for which $f \in U(I)$ but $-f \notin U(I)$. (See Exercise 10.4.) However, if $f \in U(I)$ and if $s \in S(I)$, then $f - s \in U(I)$ since $f - s = f + (-s)$.

Proof. Parts (a) and (b) are easy consequences of the corresponding properties for step functions. To prove (c), let $\{s_m\}$ be a sequence which generates f, and let

$\{t_n\}$ be a sequence which generates g. Then $s_m \nearrow f$ and $t_n \nearrow g$ a.e. on I, and

$$\lim_{m \to \infty} \int_I s_m = \int_I f, \qquad \lim_{n \to \infty} \int_I t_n = \int_I g.$$

But for each m we have

$$s_m(x) \le f(x) \le g(x) = \lim_{n \to \infty} t_n(x) \quad a.e. \text{ on } I.$$

Hence, by Theorem 10.3,

$$\int_I s_m \le \lim_{n \to \infty} \int_I t_n = \int_I g.$$

Now, let $m \to \infty$ to obtain (c).

The next theorem describes an important consequence of part (c).

Theorem 10.7. *If $f \in U(I)$ and $g \in U(I)$, and if $f(x) = g(x)$ almost everywhere on I, then $\int_I f = \int_I g$.*

Proof. We have both inequalities $f(x) \le g(x)$ and $g(x) \le f(x)$ almost everywhere on I, so Theorem 10.6 (c) gives $\int_I f \le \int_I g$ and $\int_I g \le \int_I f$.

Definition 10.8. *Let f and g be real-valued functions defined on I. We define* max (f, g) *and* min (f, g) *to be the functions whose values at each x in I are equal to* max $\{f(x), g(x)\}$ *and* min $\{f(x), g(x)\}$, *respectively.*

The reader can easily verify the following properties of max and min:

a) max (f, g) + min $(f, g) = f + g$,

b) max $(f + h, g + h)$ = max $(f, g) + h$, and min $(f + h, g + h)$ = min $(f, g) + h$.

If $f_n \nearrow f$ a.e. on I, and if $g_n \nearrow g$ a.e. on I, then

c) max $(f_n, g_n) \nearrow$ max (f, g) a.e. on I, and min $(f_n, g_n) \nearrow$ min (f, g) a.e. on I.

Theorem 10.9. *If $f \in U(I)$ and $g \in U(I)$, then* max $(f, g) \in U(I)$ *and* min $(f, g) \in U(I)$.

Proof. Let $\{s_n\}$ and $\{t_n\}$ be sequences of step functions which generate f and g, respectively, and let $u_n =$ max (s_n, t_n), $v_n =$ min (s_n, t_n). Then u_n and v_n are step functions such that $u_n \nearrow$ max (f, g) and $v_n \nearrow$ min (f, g) a.e. on I.

To prove that min $(f, g) \in U(I)$, it suffices to show that the sequence $\{\int_I v_n\}$ is bounded above. But $v_n =$ min $(s_n, t_n) \le f$ a.e. on I, so $\int_I v_n \le \int_I f$. Therefore the sequence $\{\int_I v_n\}$ converges. But the sequence $\{\int_I u_n\}$ also converges since, by property (a), $u_n = s_n + t_n - v_n$ and hence

$$\int_I u_n = \int_I s_n + \int_I t_n - \int_I v_n \to \int_I f + \int_I g - \int_I \text{min} (f, g).$$

The next theorem describes an additive property of the integral with respect to the interval of integration.

Theorem 10.10. *Let I be an interval which is the union of two subintervals, say $I = I_1 \cup I_2$, where I_1 and I_2 have no interior points in common.*

a) *If $f \in U(I)$ and if $f \geq 0$ a.e. on I, then $f \in U(I_1), f \in U(I_2)$, and*

$$\int_I f = \int_{I_1} f + \int_{I_2} f. \tag{6}$$

b) *Assume $f_1 \in U(I_1), f_2 \in U(I_2)$, and let f be defined on I as follows:*

$$f(x) = \begin{cases} f_1(x) & \text{if } x \in I_1, \\ f_2(x) & \text{if } x \in I - I_1. \end{cases}$$

Then $f \in U(I)$ and

$$\int_I f = \int_{I_1} f_1 + \int_{I_2} f_2.$$

Proof. If $\{s_n\}$ is an increasing sequence of step functions which generates f on I, let $s_n^+(x) = \max \{s_n(x), 0\}$ for each x in I. Then $\{s_n^+\}$ is an increasing sequence of *nonnegative* step functions which generates f on I (since $f \geq 0$). Moreover, for every subinterval J of I we have $\int_J s_n^+ \leq \int_I s_n^+ \leq \int_I f$ so $\{s_n^+\}$ generates f on J. Also

$$\int_I s_n^+ = \int_{I_1} s_n^+ + \int_{I_2} s_n^+,$$

so we let $n \to \infty$ to obtain (a). The proof of (b) is left as an exercise.

NOTE. There is a corresponding theorem (which can be proved by induction) for an interval which is expressed as the union of a finite number of subintervals, no two of which have interior points in common.

10.5 RIEMANN-INTEGRABLE FUNCTIONS AS EXAMPLES OF UPPER FUNCTIONS

The next theorem shows that the class of upper functions includes all the Riemann-integrable functions.

Theorem 10.11. *Let f be defined and bounded on a compact interval $[a, b]$, and assume that f is continuous almost everywhere on $[a, b]$. Then $f \in U([a, b])$ and the integral of f, as a function in $U([a, b])$, is equal to the Riemann integral $\int_a^b f(x) \, dx$.*

Proof. Let $P_n = \{x_0, x_1, \ldots, x_{2^n}\}$ be a partition of $[a, b]$ into 2^n equal subintervals of length $(b - a)/2^n$. The subintervals of P_{n+1} are obtained by bisecting those of P_n. Let

$$m_k = \inf \{f(x) : x \in [x_{k-1}, x_k]\} \qquad \text{for } 1 \leq k \leq 2^n,$$

and define a step function s_n on $[a, b]$ as follows:

$$s_n(x) = m_k \quad \text{if } x_{k-1} < x \le x_k, \qquad s_n(a) = m_1.$$

Then $s_n(x) \le f(x)$ for all x in $[a, b]$. Also, $\{s_n\}$ is increasing because the inf of f in a subinterval of $[x_{k-1}, x_k]$ cannot be less than that in $[x_{k-1}, x_k]$.

Next, we prove that $s_n(x) \to f(x)$ at each interior point of continuity of f. Since the set of discontinuities of f on $[a, b]$ has measure 0, this will show that $s_n \to f$ almost everywhere on $[a, b]$. If f is continuous at x, then for every $\varepsilon > 0$ there is a δ (depending on x and on ε) such that $f(x) - \varepsilon < f(y) < f(x) + \varepsilon$ whenever $x - \delta < y < x + \delta$.   Let $m(\delta) = \inf \{f(y) : y \in (x - \delta, x + \delta)\}$.   Then $f(x) - \varepsilon \le m(\delta)$, so $f(x) \le m(\delta) + \varepsilon$.   Some partition P_N has a subinterval $[x_{k-1}, x_k]$ containing x and lying within the interval $(x - \delta, x + \delta)$. Therefore

$$s_N(x) = m_k \le f(x) \le m(\delta) + \varepsilon \le m_k + \varepsilon = s_N(x) + \varepsilon.$$

But $s_n(x) \le f(x)$ for all n and $s_N(x) \le s_n(x)$ for all $n \ge N$. Hence

$$s_n(x) \le f(x) \le s_n(x) + \varepsilon \qquad \text{if } n \ge N,$$

which shows that $s_n(x) \to f(x)$ as $n \to \infty$.

The sequence of integrals $\{\int_a^b s_n\}$ converges because it is an increasing sequence, bounded above by $M(b - a)$, where $M = \sup \{f(x) : x \in [a, b]\}$. Moreover,

$$\int_a^b s_n = \sum_{k=1}^{2^n} m_k(x_k - x_{k-1}) = L(P_n, f),$$

where $L(P_n, f)$ is a lower Riemann sum. Since the limit of an increasing sequence is equal to its supremum, the sequence $\{\int_a^b s_n\}$ converges to the Riemann integral of f over $[a, b]$. (The Riemann integral $\int_a^b f(x)\, dx$ exists because of Lebesgue's criterion, Theorem 7.48.)

NOTE. As already mentioned, there exist functions f in $U(I)$ such that $-f \notin U(I)$. Therefore the class $U(I)$ is actually larger than the class of Riemann-integrable functions on I, since $-f \in R$ on I if $f \in R$ on I.

10.6 THE CLASS OF LEBESGUE-INTEGRABLE FUNCTIONS ON A GENERAL INTERVAL

If u and v are upper functions, the difference $u - v$ is not necessarily an upper function. We eliminate this undesirable property by enlarging the class of integrable functions.

Definition 10.12. *We denote by $L(I)$ the set of all functions f of the form $f = u - v$, where $u \in U(I)$ and $v \in U(I)$. Each function f in $L(I)$ is said to be Lebesgue-integrable on I, and its integral is defined by the equation*

$$\int_I f = \int_I u - \int_I v. \tag{7}$$

If $f \in L(I)$ it is possible to write f as a difference of two upper functions $u - v$ in more than one way. The next theorem shows that the integral of f is independent of the choice of u and v.

Theorem 10.13. *Let* u, v, u_1, *and* v_1 *be functions in* $U(I)$ *such that* $u - v = u_1 - v_1$. *Then*

$$\int_I u - \int_I v = \int_I u_1 - \int_I v_1. \tag{8}$$

Proof. The functions $u + v_1$ and $u_1 + v$ are in $U(I)$ and $u + v_1 = u_1 + v$. Hence, by Theorem 10.6(a), we have $\int_I u + \int_I v_1 = \int_I u_1 + \int_I v$, which proves (8).

NOTE. If the interval I has endpoints a and b in the extended real number system $\mathbf{R}^*$, where $a \le b$, we also write

$$\int_a^b f \quad \text{or} \quad \int_a^b f(x)\,dx$$

for the Lebesgue integral $\int_I f$. We also define $\int_b^a f = - \int_a^b f$.

If $[a, b]$ is a compact interval, every function which is Riemann-integrable on $[a, b]$ is in $U([a, b])$ and therefore also in $L([a, b])$.

10.7 BASIC PROPERTIES OF THE LEBESGUE INTEGRAL

Theorem 10.14. *Assume* $f \in L(I)$ *and* $g \in L(I)$. *Then we have:*

a) $(af + bg) \in L(I)$ *for every real* a *and* b, *and*

$$\int_I (af + bg) = a \int_I f + b \int_I g.$$

b) $\int_I f \ge 0$ *if* $f(x) \ge 0$ *a.e. on* I.

c) $\int_I f \ge \int_I g$ *if* $f(x) \ge g(x)$ *a.e. on* I.

d) $\int_I f = \int_I g$ *if* $f(x) = g(x)$ *a.e. on* I.

Proof. Part (a) follows easily from Theorem 10.6. To prove (b) we write $f = u - v$, where $u \in U(I)$ and $v \in U(I)$. Then $u(x) \ge v(x)$ almost everywhere on I so, by Theorem 10.6(c), we have $\int_I u \ge \int_I v$ and hence

$$\int_I f = \int_I u - \int_I v \ge 0.$$

Part (c) follows by applying (b) to $f - g$, and part (d) follows by applying (c) twice.

Definition 10.15. *If* f *is a real-valued function, its positive part, denoted by* f^+, *and its negative part, denoted by* f^-, *are defined by the equations*

$$f^+ = \max(f, 0), \qquad f^- = \max(-f, 0).$$

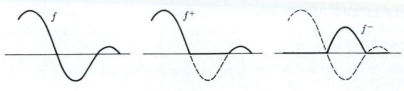

Figure 10.1

Note that f^+ and f^- are nonnegative functions and that

$$f = f^+ - f^-, \qquad |f| = f^+ + f^-.$$

Examples are shown in Fig. 10.1.

Theorem 10.16. *If f and g are in $L(I)$, then so are the functions f^+, f^-, $|f|$, max (f, g) and min (f, g). Moreover, we have*

$$\left| \int_I f \right| \le \int_I |f|. \tag{9}$$

Proof. Write $f = u - v$, where $u \in U(I)$ and $v \in U(I)$. Then

$$f^+ = \max (u - v, 0) = \max (u, v) - v.$$

But max $(u, v) \in U(I)$, by Theorem 10.9, and $v \in U(I)$, so $f^+ \in L(I)$. Since $f^- = f^+ - f$, we see that $f^- \in L(I)$. Finally, $|f| = f^+ + f^-$, so $|f| \in L(I)$.

Since $-|f(x)| \le f(x) \le |f(x)|$ for all x in I we have

$$- \int_I |f| \le \int_I f \le \int_I |f|,$$

which proves (9). To complete the proof we use the relations

$$\max (f, g) = \tfrac{1}{2}(f + g + |f - g|), \qquad \min (f, g) = \tfrac{1}{2}(f + g - |f - g|).$$

The next theorem describes the behavior of a Lebesgue integral when the interval of integration is translated, expanded or contracted, or reflected through the origin. We use the following notation, where c denotes any real number:

$$I + c = \{x + c : x \in I\}, \qquad cI = \{cx : x \in I\}.$$

Theorem 10.17. *Assume $f \in L(I)$. Then we have:*

a) *Invariance under translation. If $g(x) = f(x - c)$ for x in $I + c$, then $g \in L(I + c)$, and*

$$\int_{I+c} g = \int_I f.$$

b) *Behavior under expansion or contraction. If $g(x) = f(x/c)$ for x in cI, where $c > 0$, then $g \in L(cI)$ and*

$$\int_{cI} g = c \int_I f.$$

c) *Invariance under reflection. If $g(x) = f(-x)$ for x in $-I$, then $g \in L(-I)$ and*

$$\int_{-I} g = \int_{I} f.$$

NOTE. If I has endpoints $a < b$, where a and b are in the extended real number system $\mathbf{R}^*$, the formula in (a) can also be written as follows:

$$\int_{a+c}^{b+c} f(x - c)\, dx = \int_{a}^{b} f(x)\, dx.$$

Properties (b) and (c) can be combined into a single formula which includes both positive and negative values of c:

$$\int_{ca}^{cb} f(x/c)\, dx = |c| \int_{a}^{b} f(x)\, dx \qquad \text{if } c \neq 0.$$

Proof. In proving a theorem of this type, the procedure is always the same. First, we verify the theorem for step functions, then for upper functions, and finally for Lebesgue-integrable functions. At each step the argument is straightforward, so we omit the details.

Theorem 10.18. *Let I be an interval which is the union of two subintervals, say $I = I_1 \cup I_2$, where I_1 and I_2 have no interior points in common.*

a) *If $f \in L(I)$, then $f \in L(I_1)$, $f \in L(I_2)$, and*

$$\int_{I} f = \int_{I_1} f + \int_{I_2} f.$$

b) *Assume $f_1 \in L(I_1)$, $f_2 \in L(I_2)$, and let f be defined on I as follows:*

$$f(x) = \begin{cases} f_1(x) & \text{if } x \in I_1, \\ f_2(x) & \text{if } x \in I - I_1. \end{cases}$$

Then $f \in L(I)$ and $\int_I f = \int_{I_1} f_1 + \int_{I_2} f_2$.

Proof. Write $f = u - v$ where $u \in U(I)$ and $v \in U(I)$. Then $u = u^+ - u^-$ and $v = v^+ - v^-$, so $f = u^+ + v^- - (u^- + v^+)$. Now apply Theorem 10.10 to each of the nonnegative functions $u^+ + v^-$ and $u^- + v^+$ to deduce part (a). The proof of part (b) is left to the reader.

NOTE. There is an extension of Theorem 10.18 for an interval which can be expressed as the union of a finite number of subintervals, no two of which have interior points in common. The reader can formulate this for himself.

We conclude this section with two approximation properties that will be needed later. The first tells us that every Lebesgue-integrable function f is equal to an upper function u minus a nonnegative upper function v with a small integral. The second tells us that f is equal to a step function s plus an integrable function

g with a small integral. More precisely, we have:

Theorem 10.19. *Assume $f \in L(I)$ and let $\varepsilon > 0$ be given. Then:*

a) *There exist functions u and v in $U(I)$ such that $f = u - v$, where v is non-negative a.e. on I and $\int_I v < \varepsilon$.*

b) *There exists a step function s and a function g in $L(I)$ such that $f = s + g$, where $\int_I |g| < \varepsilon$.*

Proof. Since $f \in L(I)$, we can write $f = u_1 - v_1$ where u_1 and v_1 are in $U(I)$. Let $\{t_n\}$ be a sequence which generates v_1. Since $\int_I t_n \to \int_I v_1$, we can choose N so that $0 \le \int_I (v_1 - t_N) < \varepsilon$. Now let $v = v_1 - t_N$ and $u = u_1 - t_N$. Then both u and v are in $U(I)$ and $u - v = u_1 - v_1 = f$. Also, v is nonnegative *a.e.* on I and $\int_I v < \varepsilon$. This proves (a).

To prove (b) we use (a) to choose u and v in $U(I)$ so that $v \ge 0$ *a.e.* on I,

$$f = u - v \qquad \text{and} \qquad 0 \le \int_I v < \frac{\varepsilon}{2}.$$

Now choose a step function s such that $0 \le \int_I (u - s) < \varepsilon/2$. Then

$$f = u - v = s + (u - s) - v = s + g,$$

where $g = (u - s) - v$. Hence $g \in L(I)$ and

$$\int_I |g| \le \int_I |u - s| + \int_I |v| < \frac{\varepsilon}{2} + \frac{\varepsilon}{2} = \varepsilon.$$

10.8 LEBESGUE INTEGRATION AND SETS OF MEASURE ZERO

The theorems in this section show that the behavior of a Lebesgue-integrable function on a set of measure zero does not affect its integral.

Theorem 10.20. *Let f be defined on I. If $f = 0$ almost everywhere on I, then $f \in L(I)$ and $\int_I f = 0$.*

Proof. Let $s_n(x) = 0$ for all x in I. Then $\{s_n\}$ is an increasing sequence of step functions which converges to 0 everywhere on I. Hence $\{s_n\}$ converges to f almost everywhere on I. Since $\int_I s_n = 0$ the sequence $\{\int_I s_n\}$ converges. Therefore f is an upper function, so $f \in L(I)$ and $\int_I f = \lim_{n \to \infty} \int_I s_n = 0$.

Theorem 10.21. *Let f and g be defined on I. If $f \in L(I)$ and if $f = g$ almost everywhere on I, then $g \in L(I)$ and $\int_I f = \int_I g$.*

Proof. Apply Theorem 10.20 to $f - g$. Then $f - g \in L(I)$ and $\int_I (f - g) = 0$. Hence $g = f - (f - g) \in L(I)$ and $\int_I g = \int_I f - \int_I (f - g) = \int_I f$.

Example. Define f on the interval $[0, 1]$ as follows:

$$f(x) = \begin{cases} 1 & \text{if } x \text{ is rational} \\ 0 & \text{if } x \text{ is irrational.} \end{cases}$$

Then $f = 0$ almost everywhere on $[0, 1]$ so f is Lebesgue-integrable on $[0, 1]$ and its Lebesgue integral is 0. As noted in Chapter 7, this function is not Riemann-integrable on $[0, 1]$.

NOTE. Theorem 10.21 suggests a definition of the integral for functions that are defined almost everywhere on I. If g is such a function and if $g(x) = f(x)$ almost everywhere on I, where $f \in L(I)$, we say that $g \in L(I)$ and that

$$\int_I g = \int_I f.$$

10.9 THE LEVI MONOTONE CONVERGENCE THEOREMS

We turn next to convergence theorems concerning term-by-term integration of monotonic sequences of functions. We begin with three versions of a famous theorem of Beppo Levi. The first concerns sequences of step functions, the second sequences of upper functions, and the third sequences of Lebesgue-integrable functions. Although the theorems are stated for increasing sequences, there are corresponding results for decreasing sequences.

Theorem 10.22 (Levi theorem for step functions). *Let $\{s_n\}$ be a sequence of step functions such that*

a) *$\{s_n\}$ increases on an interval I, and*

b) *$\lim_{n \to \infty} \int_I s_n$ exists.*

Then $\{s_n\}$ converges almost everywhere on I to a limit function f in $U(I)$, and

$$\int_I f = \lim_{n \to \infty} \int_I s_n.$$

Proof. We can assume, without loss of generality, that the step functions s_n are nonnegative. (If not, consider instead the sequence $\{s_n - s_1\}$. If the theorem is true for $\{s_n - s_1\}$, then it is also true for $\{s_n\}$.) Let D be the set of x in I for which $\{s_n(x)\}$ diverges, and let $\varepsilon > 0$ be given. We will prove that D has measure 0 by showing that D can be covered by a countable collection of intervals, the sum of whose lengths is $< \varepsilon$.

Since the sequence $\{\int_I s_n\}$ converges it is bounded by some positive constant M. Let

$$t_n(x) = \left[\frac{\varepsilon}{2M} s_n(x) \right] \qquad \text{if } x \in I,$$

where $[y]$ denotes the greatest integer $\leq y$. Then $\{t_n\}$ is an increasing sequence of step functions and each function value $t_n(x)$ is a nonnegative integer.

If $\{s_n(x)\}$ converges, then $\{s_n(x)\}$ is bounded so $\{t_n(x)\}$ is bounded and hence $t_{n+1}(x) = t_n(x)$ for all sufficiently large n, since each $t_n(x)$ is an integer.

If $\{s_n(x)\}$ diverges, then $\{t_n(x)\}$ also diverges and $t_{n+1}(x) - t_n(x) \geq 1$ for infinitely many values of n. Let

$$D_n = \{x : x \in I \quad \text{and} \quad t_{n+1}(x) - t_n(x) \geq 1\}.$$

Then D_n is the union of a finite number of intervals, the sum of whose lengths we denote by $|D_n|$. Now

$$D \subseteq \bigcup_{n=1}^{\infty} D_n,$$

so if we prove that $\sum_{n=1}^{\infty} |D_n| < \varepsilon$, this will show that D has measure 0.

To do this we integrate the nonnegative step function $t_{n+1} - t_n$ over I and obtain the inequalities

$$\int_I (t_{n+1} - t_n) \geq \int_{D_n} (t_{n+1} - t_n) \geq \int_{D_n} 1 = |D_n|.$$

Hence for every $m \geq 1$ we have

$$\sum_{n=1}^{m} |D_n| \leq \sum_{n=1}^{m} \int_I (t_{n+1} - t_n) = \int_I t_{m+1} - \int_I t_1 \leq \int_I t_{m+1} \leq \frac{\varepsilon}{2M} \int_I s_{m+1} \leq \frac{\varepsilon}{2}.$$

Therefore $\sum_{n=1}^{\infty} |D_n| \leq \varepsilon/2 < \varepsilon$, so D has measure 0.

This proves that $\{s_n\}$ converges almost everywhere on I. Let

$$f(x) = \begin{cases} \lim_{n \to \infty} s_n(x) & \text{if } x \in I - D, \\ 0 & \text{if } x \in D. \end{cases}$$

Then f is defined everywhere on I and $s_n \to f$ almost everywhere on I. Therefore, $f \in U(I)$ and $\int_I f = \lim_{n \to \infty} \int_I s_n$.

Theorem 10.23 (Levi theorem for upper functions). *Let $\{f_n\}$ be a sequence of upper functions such that*

a) *$\{f_n\}$ increases almost everywhere on an interval I,*

   and

b) *$\lim_{n \to \infty} \int_I f_n$ exists.*

Then $\{f_n\}$ converges almost everywhere on I to a limit function f in $U(I)$, and

$$\int_I f = \lim_{n \to \infty} \int_I f_n.$$

Proof. For each k there is an increasing sequence of step functions $\{s_{n,k}\}$ which generates f_k. Define a new step function t_n on I by the equation

$$t_n(x) = \max \{s_{n,1}(x), s_{n,2}(x), \ldots, s_{n,n}(x)\}.$$

Then $\{t_n\}$ is increasing on I because

$$t_{n+1}(x) = \max \{s_{n+1,1}(x), \ldots, s_{n+1,n+1}(x)\} \geq \max \{s_{n,1}(x), \ldots, s_{n,n+1}(x)\}$$
$$\geq \max \{s_{n,1}(x), \ldots, s_{n,n}(x)\} = t_n(x).$$

But $s_{n,k}(x) \le f_k(x)$ and $\{f_k\}$ increases almost everywhere on I, so we have

$$t_n(x) \le \max \{f_1(x), \ldots, f_n(x)\} = f_n(x) \tag{10}$$

almost everywhere on I. Therefore, by Theorem 10.6(c) we obtain

$$\int_I t_n \le \int_I f_n. \tag{11}$$

But, by (b), $\{\int_I f_n\}$ is bounded above so the increasing sequence $\{\int_I t_n\}$ is also bounded above and hence converges. By the Levi theorem for step functions, $\{t_n\}$ converges almost everywhere on I to a limit function f in $U(I)$, and $\int_I f = \lim_{n \to \infty} \int_I t_n$. We prove next that $f_n \to f$ almost everywhere on I.

The definition of $t_n(x)$ implies $s_{n,k}(x) \le t_n(x)$ for all $k \le n$ and all x in I. Letting $n \to \infty$ we find

$$f_k(x) \le f(x) \quad \text{almost everywhere on } I. \tag{12}$$

Therefore the increasing sequence $\{f_k(x)\}$ is bounded above by $f(x)$ almost everywhere on I, so it converges almost everywhere on I to a limit function g satisfying $g(x) \le f(x)$ almost everywhere on I. But (10) states that $t_n(x) \le f_n(x)$ almost everywhere on I so, letting $n \to \infty$, we find $f(x) \le g(x)$ almost everywhere on I. In other words, we have

$$\lim_{n \to \infty} f_n(x) = f(x) \quad \text{almost everywhere on } I.$$

Finally, we show that $\int_I f = \lim_{n \to \infty} \int_I f_n$. Letting $n \to \infty$ in (11) we obtain

$$\int_I f \le \lim_{n \to \infty} \int_I f_n. \tag{13}$$

Now integrate (12), using Theorem 10.6(c) again, to get $\int_I f_k \le \int_I f$. Letting $k \to \infty$ we obtain $\lim_{k \to \infty} \int_I f_k \le \int_I f$ which, together with (13), completes the proof.

NOTE. The class $U(I)$ of upper functions was constructed from the class $S(I)$ of step functions by a certain process which we can call P. Beppo Levi's theorem shows that when process P is applied to $U(I)$ it again gives functions in $U(I)$. The next theorem shows that when P is applied to $L(I)$ it again gives functions in $L(I)$.

Theorem 10.24 (Levi theorem for sequences of Lebesgue-integrable functions). *Let $\{f_n\}$ be a sequence of functions in $L(I)$ such that*

a) *$\{f_n\}$ increases almost everywhere on I,*

   and

b) *$\lim_{n \to \infty} \int_I f_n$ exists.*

Then $\{f_n\}$ converges almost everywhere on I to a limit function f in $L(I)$, and

$$\int_I f = \lim_{n\to\infty} \int_I f_n.$$

We shall deduce this theorem from an equivalent result stated for *series* of functions.

Theorem 10.25 (Levi theorem for series of Lebesgue-integrable functions). *Let $\{g_n\}$ be a sequence of functions in $L(I)$ such that*

a) *each g_n is nonnegative almost everywhere on I,*

 and

b) *the series $\sum_{n=1}^{\infty} \int_I g_n$ converges.*

Then the series $\sum_{n=1}^{\infty} g_n$ converges almost everywhere on I to a sum function g in $L(I)$, and we have

$$\int_I g = \int_I \sum_{n=1}^{\infty} g_n = \sum_{n=1}^{\infty} \int_I g_n. \tag{14}$$

Proof. Since $g_n \in L(I)$, Theorem 10.19 tells us that for every $\varepsilon > 0$ we can write

$$g_n = u_n - v_n,$$

where $u_n \in U(I)$, $v_n \in U(I)$, $v_n \geq 0$ a.e. on I, and $\int_I v_n < \varepsilon$. Choose u_n and v_n corresponding to $\varepsilon = (\frac{1}{2})^n$. Then

$$u_n = g_n + v_n, \qquad \text{where} \int_I v_n < (\tfrac{1}{2})^n.$$

The inequality on $\int_I v_n$ assures us that the series $\sum_{n=1}^{\infty} \int_I v_n$ converges. Now $u_n \geq 0$ almost everywhere on I, so the partial sums

$$U_n(x) = \sum_{k=1}^{n} u_k(x)$$

form a sequence of upper functions $\{U_n\}$ which increases almost everywhere on I. Since

$$\int_I U_n = \int_I \sum_{k=1}^{n} u_k = \sum_{k=1}^{n} \int_I u_k = \sum_{k=1}^{n} \int_I g_k + \sum_{k=1}^{n} \int_I v_k,$$

the sequence of integrals $\{\int_I U_n\}$ converges because both series $\sum_{k=1}^{\infty} \int_I g_k$ and $\sum_{k=1}^{\infty} \int_I v_k$ converge. Therefore, by the Levi theorem for upper functions, the sequence $\{U_n\}$ converges almost everywhere on I to a limit function U in $U(I)$, and $\int_I U = \lim_{n\to\infty} \int_I U_n$. But

$$\int_I U_n = \sum_{k=1}^{n} \int_I u_k,$$

so

$$\int_I U = \sum_{k=1}^{\infty} \int_I u_k.$$

Similarly, the sequence of partial sums $\{V_n\}$ given by

$$V_n(x) = \sum_{k=1}^{n} v_k(x)$$

converges almost everywhere on I to a limit function V in $U(I)$ and

$$\int_I V = \sum_{k=1}^{\infty} \int_I v_k.$$

Therefore $U - V \in L(I)$ and the sequence $\{\sum_{k=1}^{n} g_k\} = \{U_n - V_n\}$ converges almost everywhere on I to $U - V$. Let $g = U - V$. Then $g \in L(I)$ and

$$\int_I g = \int_I U - \int_I V = \sum_{k=1}^{\infty} \int_I (u_k - v_k) = \sum_{k=1}^{\infty} \int_I g_k.$$

This completes the proof of Theorem 10.25.

Proof of Theorem 10.24. Assume $\{f_n\}$ satisfies the hypotheses of Theorem 10.24. Let $g_1 = f_1$ and let $g_n = f_n - f_{n-1}$ for $n \geq 2$, so that

$$f_n = \sum_{k=1}^{n} g_k.$$

Applying Theorem 10.25 to $\{g_n\}$, we find that $\sum_{n=1}^{\infty} g_n$ converges almost everywhere on I to a sum function g in $L(I)$, and Equation (14) holds. Therefore $f_n \to g$ almost everywhere on I and $\int_I g = \lim_{n \to \infty} \int_I f_n$.

In the following version of the Levi theorem for series, the terms of the series are not assumed to be nonnegative.

Theorem 10.26. *Let $\{g_n\}$ be a sequence of functions in $L(I)$ such that the series*

$$\sum_{n=1}^{\infty} \int_I |g_n|$$

is convergent. Then the series $\sum_{n=1}^{\infty} g_n$ converges almost everywhere on I to a sum function g in $L(I)$ and we have

$$\int_I \sum_{n=1}^{\infty} g_n = \sum_{n=1}^{\infty} \int_I g_n.$$

Proof. Write $g_n = g_n^+ - g_n^-$ and apply Theorem 10.25 to the sequences $\{g_n^+\}$ and $\{g_n^-\}$ separately.

The following examples illustrate the use of the Levi theorem for sequences.

Example 1. Let $f(x) = x^s$ for $x > 0$, $f(0) = 0$. Prove that the Lebesgue integral $\int_0^1 f(x)\, dx$ exists and has the value $1/(s + 1)$ if $s > -1$.

Solution. If $s \geq 0$, then f is bounded and Riemann-integrable on $[0, 1]$ and its Riemann integral is equal to $1/(s + 1)$.

If $s < 0$, then f is not bounded and hence not Riemann-integrable on $[0, 1]$. Define a sequence of functions $\{f_n\}$ as follows:

$$f_n(x) = \begin{cases} x^s & \text{if } x \geq 1/n, \\ 0 & \text{if } 0 \leq x < 1/n. \end{cases}$$

Then $\{f_n\}$ is increasing and $f_n \to f$ everywhere on $[0, 1]$. Each f_n is Riemann-integrable and hence Lebesgue-integrable on $[0, 1]$ and

$$\int_0^1 f_n(x)\, dx = \int_{1/n}^1 x^s\, dx = \frac{1}{s + 1}\left(1 - \frac{1}{n^{s+1}}\right).$$

If $s + 1 > 0$, the sequence $\{\int_0^1 f_n\}$ converges to $1/(s + 1)$. Therefore, the Levi theorem for sequences shows that $\int_0^1 f$ exists and equals $1/(s + 1)$.

Example 2. The same type of argument shows that the Lebesgue integral $\int_0^1 e^{-x} x^{y-1}\, dx$ exists for every real $y > 0$. This integral will be used later in discussing the Gamma function.

10.10 THE LEBESGUE DOMINATED CONVERGENCE THEOREM

Levi's theorems have many important consequences. The first is Lebesgue's dominated convergence theorem, the cornerstone of Lebesgue's theory of integration.

Theorem 10.27 (Lebesgue dominated convergence theorem). *Let $\{f_n\}$ be a sequence of Lebesgue-integrable functions on an interval I. Assume that*

a) *$\{f_n\}$ converges almost everywhere on I to a limit function f,*

   and

b) *there is a nonnegative function g in $L(I)$ such that, for all $n \geq 1$,*

$$|f_n(x)| \leq g(x) \quad \text{a.e. on } I.$$

Then the limit function $f \in L(I)$, the sequence $\{\int_I f_n\}$ converges and

$$\int_I f = \lim_{n \to \infty} \int_I f_n. \tag{15}$$

NOTE. Property (b) is described by saying that the sequence $\{f_n\}$ is *dominated* by g almost everywhere on I.

Proof. The idea of the proof is to obtain upper and lower bounds of the form

$$g_n(x) \leq f_n(x) \leq G_n(x) \tag{16}$$

where $\{g_n\}$ increases and $\{G_n\}$ decreases almost everywhere on I to the limit function f. Then we use the Levi theorem to show that $f \in L(I)$ and that $\int_I f = \lim_{n\to\infty} \int_I g_n = \lim_{n\to\infty} \int_I G_n$, from which we obtain (15).

To construct $\{g_n\}$ and $\{G_n\}$, we make repeated use of the Levi theorem for sequences in $L(I)$. First we define a sequence $\{G_{n,1}\}$ as follows:

$$G_{n,1}(x) = \max \{f_1(x), f_2(x), \ldots, f_n(x)\}.$$

Each function $G_{n,1} \in L(I)$, by Theorem 10.16, and the sequence $\{G_{n,1}\}$ is increasing on I. Since $|G_{n,1}(x)| \leq g(x)$ almost everywhere on I, we have

$$\left| \int_I G_{n,1} \right| \leq \int_I |G_{n,1}| \leq \int_I g. \tag{17}$$

Therefore the increasing sequence of numbers $\{\int_I G_{n,1}\}$ is bounded above by $\int_I g$, so $\lim_{n\to\infty} \int_I G_{n,1}$ exists. By the Levi theorem, the sequence $\{G_{n,1}\}$ converges almost everywhere on I to a function G_1 in $L(I)$, and

$$\int_I G_1 = \lim_{n\to\infty} \int_I G_{n,1} \leq \int_I g.$$

Because of (17) we also have the inequality $-\int_I g \leq \int_I G_1$. Note that if x is a point in I for which $G_{n,1}(x) \to G_1(x)$, then we also have

$$G_1(x) = \sup \{f_1(x), f_2(x), \ldots\}.$$

In the same way, for each fixed $r \geq 1$ we let

$$G_{n,r}(x) = \max \{f_r(x), f_{r+1}(x), \ldots, f_n(x)\}$$

for $n \geq r$. Then the sequence $\{G_{n,r}\}$ increases and converges almost everywhere on I to a limit function G_r in $L(I)$ with

$$-\int_I g \leq \int_I G_r \leq \int_I g.$$

Also, at those points for which $G_{n,r}(x) \to G_r(x)$ we have

$$G_r(x) = \sup \{f_r(x), f_{r+1}(x), \ldots\},$$

so

$$f_r(x) \leq G_r(x) \quad a.e. \text{ on } I.$$

Now we examine properties of the sequence $\{G_n(x)\}$. Since $A \subseteq B$ implies $\sup A \leq \sup B$, the sequence $\{G_r(x)\}$ decreases almost everywhere and hence converges almost everywhere on I. We show next that $G_n(x) \to f(x)$ whenever

$$\lim_{n\to\infty} f_n(x) = f(x). \tag{18}$$

If (18) holds, then for every $\varepsilon > 0$ there is an integer N such that

$$f(x) - \varepsilon < f_n(x) < f(x) + \varepsilon \quad \text{for all } n \geq N.$$

Hence, if $m \geq N$ we have

$$f(x) - \varepsilon \leq \sup \{f_m(x), f_{m+1}(x), \ldots\} \leq f(x) + \varepsilon.$$

In other words,

$$m \geq N \quad \text{implies} \quad f(x) - \varepsilon \leq G_m(x) \leq f(x) + \varepsilon,$$

and this implies that

$$\lim_{m \to \infty} G_m(x) = f(x) \quad \text{almost everywhere on } I. \tag{19}$$

On the other hand, the decreasing sequence of numbers $\{\int_I G_n\}$ is bounded below by $-\int_I g$, so it converges. By (19) and the Levi theorem, we see that $f \in L(I)$ and

$$\lim_{n \to \infty} \int_I G_n = \int_I f.$$

By applying the same type of argument to the sequence

$$g_{n,r}(x) = \min \{f_r(x), f_{r+1}(x), \ldots, f_n(x)\},$$

for $n \geq r$, we find that $\{g_{n,r}\}$ decreases and converges almost everywhere to a limit function g_r in $L(I)$, where

$$g_r(x) = \inf \{f_r(x), f_{r+1}(x), \ldots\} \quad a.e. \text{ on } I.$$

Also, almost everywhere on I we have $g_r(x) \leq f_r(x)$, $\{g_r\}$ increases, $\lim_{n \to \infty} g_n(x) = f(x)$, and

$$\lim_{n \to \infty} \int_I g_n = \int_I f.$$

Since (16) holds almost everywhere on I we have $\int_I g_n \leq \int_I f_n \leq \int_I G_n$. Letting $n \to \infty$ we find that $\{\int_I f_n\}$ converges and that

$$\lim_{n \to \infty} \int_I f_n = \int_I f.$$

10.11 APPLICATIONS OF LEBESGUE'S DOMINATED CONVERGENCE THEOREM

The first application concerns term-by-term integration of series and is a companion result to Levi's theorem on series.

Theorem 10.28. *Let $\{g_n\}$ be a sequence of functions in $L(I)$ such that:*

a) *each g_n is nonnegative almost everywhere on I, and*

b) *the series $\sum_{n=1}^{\infty} g_n$ converges almost everywhere on I to a function g which is bounded above by a function in $L(I)$.*

Then $g \in L(I)$, the series $\sum_{n=1}^{\infty} \int_I g_n$ converges, and we have

$$\int_I \sum_{n=1}^{\infty} g_n = \sum_{n=1}^{\infty} \int_I g_n.$$

Proof. Let

$$f_n(x) = \sum_{k=1}^{n} g_k(x) \qquad \text{if } x \in I.$$

Then $f_n \to g$ almost everywhere on I, and $\{f_n\}$ is dominated almost everywhere on I by the function in $L(I)$ which bounds g from above. Therefore, by the Lebesgue dominated convergence theorem, $g \in L(I)$, the sequence $\{\int_I f_n\}$ converges, and $\int_I g = \lim_{n\to\infty} \int_I f_n$. This proves the theorem.

The next application, sometimes called the *Lebesgue bounded convergence theorem*, refers to a bounded interval.

Theorem 10.29. *Let I be a bounded interval. Assume $\{f_n\}$ is a sequence of functions in $L(I)$ which is boundedly convergent almost everywhere on I. That is, assume there is a limit function f and a positive constant M such that*

$$\lim_{n\to\infty} f_n(x) = f(x) \qquad and \qquad |f_n(x)| \le M, \quad almost \ everywhere \ on \ I.$$

Then $f \in L(I)$ and $\lim_{n\to\infty} \int_I f_n = \int_I f$.

Proof. Apply Theorem 10.27 with $g(x) = M$ for all x in I. Then $g \in L(I)$, since I is a bounded interval.

NOTE. A special case of Theorem 10.29 is Arzelà's theorem stated earlier (Theorem 9.12). If $\{f_n\}$ is a boundedly convergent sequence of Riemann-integrable functions on a compact interval $[a, b]$, then each $f_n \in L([a, b])$, the limit function $f \in L([a, b])$, and we have

$$\lim_{n\to\infty} \int_a^b f_n = \int_a^b f.$$

If the limit function f is Riemann-integrable (as assumed in Arzelà's theorem), then the Lebesgue integral $\int_a^b f$ is the same as the Riemann integral $\int_a^b f(x) \, dx$.

The next theorem is often used to show that functions are Lebesgue-integrable.

Theorem 10.30. *Let $\{f_n\}$ be a sequence of functions in $L(I)$ which converges almost everywhere on I to a limit function f. Assume that there is a nonnegative function g in $L(I)$ such that*

$$|f(x)| \le g(x) \quad a.e. \ on \ I.$$

Then $f \in L(I)$.

Proof. Define a new sequence of functions $\{g_n\}$ on I as follows:

$$g_n = \max \{\min (f_n, g), -g\}.$$

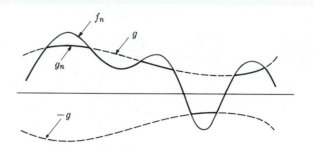

Figure 10.2

Geometrically, the function g_n is obtained from f_n by cutting off the graph of f_n from above by g and from below by $-g$, as shown by the example in Fig. 10.2. Then $|g_n(x)| \leq g(x)$ almost everywhere on I, and it is easy to verify that $g_n \to f$ almost everywhere on I. Therefore, by the Lebesgue dominated convergence theorem, $f \in L(I)$.

10.12 LEBESGUE INTEGRALS ON UNBOUNDED INTERVALS AS LIMITS OF INTEGRALS ON BOUNDED INTERVALS

Theorem 10.31. *Let f be defined on the half-infinite interval $I = [a, +\infty)$. Assume that f is Lebesgue-integrable on the compact interval $[a, b]$ for each $b \geq a$, and that there is a positive constant M such that*

$$\int_a^b |f| \leq M \qquad \text{for all } b \geq a. \tag{20}$$

Then $f \in L(I)$, the limit $\lim_{b \to +\infty} \int_a^b f$ exists, and

$$\int_a^{+\infty} f = \lim_{b \to +\infty} \int_a^b f. \tag{21}$$

Proof. Let $\{b_n\}$ be any increasing sequence of real numbers with $b_n \geq a$ such that $\lim_{n \to \infty} b_n = +\infty$. Define a sequence $\{f_n\}$ on I as follows:

$$f_n(x) = \begin{cases} f(x) & \text{if } a \leq x \leq b_n, \\ 0 & \text{otherwise.} \end{cases}$$

Each $f_n \in L(I)$ (by Theorem 10.18) and $f_n \to f$ on I. Hence, $|f_n| \to |f|$ on I. But $|f_n|$ is increasing and, by (20), the sequence $\{\int_I |f_n|\}$ is bounded above by M. Therefore $\lim_{n \to \infty} \int_I |f_n|$ exists. By the Levi theorem, the limit function $|f| \in L(I)$. Now each $|f_n| \leq |f|$ and $f_n \to f$ on I, so by the Lebesgue dominated convergence theorem, $f \in L(I)$ and $\lim_{n \to \infty} \int_I f_n = \int_I f$. Therefore

$$\lim_{n \to \infty} \int_a^{b_n} f = \int_a^{+\infty} f$$

for all sequences $\{b_n\}$ which increase to $+\infty$. This completes the proof.

There is, of course, a corresponding theorem for the interval $(-\infty, a]$ which concludes that

$$\int_{-\infty}^{a} f = \lim_{c \to -\infty} \int_{c}^{a} f,$$

provided that $\int_{c}^{a} |f| \le M$ for all $c \le a$. If $\int_{c}^{b} |f| \le M$ for all real c and b with $c \le b$, the two theorems together show that $f \in L(\mathbf{R})$ and that

$$\int_{-\infty}^{+\infty} f = \lim_{c \to -\infty} \int_{c}^{a} f + \lim_{b \to +\infty} \int_{a}^{b} f.$$

Example 1. Let $f(x) = 1/(1 + x^2)$ for all x in $\mathbf{R}$. We shall prove that $f \in L(\mathbf{R})$ and that $\int_{\mathbf{R}} f = \pi$. Now f is nonnegative, and if $c \le b$ we have

$$\int_{c}^{b} f = \int_{c}^{b} \frac{dx}{1 + x^2} = \arctan b - \arctan c \le \pi.$$

Therefore, $f \in L(\mathbf{R})$ and

$$\int_{-\infty}^{+\infty} f = \lim_{c \to -\infty} \int_{c}^{0} \frac{dx}{1 + x^2} + \lim_{b \to +\infty} \int_{0}^{b} \frac{dx}{1 + x^2} = \frac{\pi}{2} + \frac{\pi}{2} = \pi.$$

Example 2. In this example the limit on the right of (21) exists but $f \notin L(I)$. Let $I = [0, +\infty)$ and define f on I as follows:

$$f(x) = \frac{(-1)^n}{n} \quad \text{if } n - 1 \le x < n, \quad \text{for } n = 1, 2, \ldots$$

If $b > 0$, let $m = [b]$, the greatest integer $\le b$. Then

$$\int_{0}^{b} f = \int_{0}^{m} f + \int_{m}^{b} f = \sum_{n=1}^{m} \frac{(-1)^n}{n} + \frac{(b - m)(-1)^{m+1}}{m + 1}.$$

As $b \to +\infty$ the last term $\to 0$, and we find

$$\lim_{b \to +\infty} \int_{0}^{b} f = \sum_{n=1}^{\infty} \frac{(-1)^n}{n} = -\log 2.$$

Now we assume $f \in L(I)$ and obtain a contradiction. Let f_n be defined by

$$f_n(x) = \begin{cases} |f(x)| & \text{for } 0 \le x \le n, \\ 0 & \text{for } x > n. \end{cases}$$

Then $\{f_n\}$ increases and $f_n(x) \to |f(x)|$ everywhere on I. Since $f \in L(I)$ we also have $|f| \in L(I)$. But $|f_n(x)| \le |f(x)|$ everywhere on I so by the Lebesgue dominated convergence theorem the sequence $\{\int_{I} f_n\}$ converges. But this is a contradiction since

$$\int_{I} f_n = \int_{0}^{n} |f| = \sum_{k=1}^{n} \frac{1}{k} \to +\infty \quad \text{as } n \to \infty.$$

10.13 IMPROPER RIEMANN INTEGRALS

Definition 10.32. *If f is Riemann-integrable on* $[a, b]$ *for every* $b \geq a$, *and if the limit*

$$\lim_{b \to +\infty} \int_a^b f(x)\, dx \quad \text{exists,}$$

then f is said to be improper Riemann-integrable on $[a, +\infty)$ *and the improper Riemann integral of f, denoted by* $\int_a^{+\infty} f(x)\, dx$ *or* $\int_a^{\infty} f(x)\, dx$, *is defined by the equation*

$$\int_a^{+\infty} f(x)\, dx = \lim_{b \to +\infty} \int_a^b f(x)\, dx.$$

In Example 2 of the foregoing section the improper Riemann integral $\int_0^{+\infty} f(x)\, dx$ exists but f is not Lebesgue-integrable on $[0, +\infty)$. That example should be contrasted with the following theorem.

Theorem 10.33. *Assume f is Riemann-integrable on* $[a, b]$ *for every* $b \geq a$, *and assume there is a positive constant M such that*

$$\int_a^b |f(x)|\, dx \leq M \qquad \text{for every } b \geq a. \tag{22}$$

Then both f and $|f|$ *are improper Riemann-integrable on* $[a, +\infty)$. *Also, f is Lebesgue-integrable on* $[a, +\infty)$ *and the Lebesgue integral of f is equal to the improper Riemann integral of f.*

Proof. Let $F(b) = \int_a^b |f(x)|\, dx$. Then F is an increasing function which is bounded above by M, so $\lim_{b \to +\infty} F(b)$ exists. Therefore $|f|$ is improper Riemann-integrable on $[a, +\infty)$. Since

$$0 \leq |f(x)| - f(x) \leq 2|f(x)|,$$

the limit

$$\lim_{b \to +\infty} \int_a^b \{|f(x)| - f(x)\}\, dx$$

also exists; hence the limit $\lim_{b \to +\infty} \int_a^b f(x)\, dx$ exists. This proves that f is improper Riemann-integrable on $[a, +\infty)$. Now we use inequality (22), along with Theorem 10.31, to deduce that f is Lebesgue-integrable on $[a, +\infty)$ and that the Lebesgue integral of f is equal to the improper Riemann integral of f.

NOTE. There are corresponding results for improper Riemann integrals of the form

$$\int_{-\infty}^b f(x)\, dx = \lim_{a \to -\infty} \int_a^b f(x)\, dx,$$

$$\int_a^c f(x)\, dx = \lim_{b \to c-} \int_a^b f(x)\, dx,$$

and

$$\int_c^b f(x)\, dx = \lim_{a \to c+} \int_a^b f(x)\, dx,$$

which the reader can formulate for himself.

If both integrals $\int_{-\infty}^a f(x)\, dx$ and $\int_a^{+\infty} f(x)\, dx$ exist, we say that the integral $\int_{-\infty}^{+\infty} f(x)\, dx$ exists, and its value is defined to be their sum,

$$\int_{-\infty}^{+\infty} f(x)\, dx = \int_{-\infty}^a f(x)\, dx + \int_a^{+\infty} f(x)\, dx.$$

If the integral $\int_{-\infty}^{+\infty} f(x)\, dx$ exists, its value is also equal to the symmetric limit

$$\lim_{b \to +\infty} \int_{-b}^b f(x)\, dx.$$

However, it is important to realize that the symmetric limit might exist even when $\int_{-\infty}^{+\infty} f(x)\, dx$ does not exist (for example, take $f(x) = x$ for all x). In this case the symmetric limit is called the *Cauchy principal value* of $\int_{-\infty}^{+\infty} f(x)\, dx$. Thus $\int_{-\infty}^{+\infty} x\, dx$ has Cauchy principal value 0, but the integral does not exist.

Example 1. Let $f(x) = e^{-x} x^{y-1}$, where y is a fixed real number. Since $e^{-x/2} x^{y-1} \to 0$ as $x \to +\infty$, there is a constant M such that $e^{-x/2} x^{y-1} \le M$ for all $x \ge 1$. Then $e^{-x} x^{y-1} \le M e^{-x/2}$, so

$$\int_1^b |f(x)|\, dx \le M \int_0^b e^{-x/2}\, dx = 2M(1 - e^{-b/2}) < 2M.$$

Hence the integral $\int_1^{+\infty} e^{-x} x^{y-1}\, dx$ exists for every real y, both as an improper Riemann integral and as a Lebesgue integral.

Example 2. *The Gamma function integral.* Adding the integral of Example 1 to the integral $\int_0^1 e^{-x} x^{y-1}\, dx$ of Example 2 of Section 10.9, we find that the Lebesgue integral

$$\Gamma(y) = \int_0^{+\infty} e^{-x} x^{y-1}\, dx$$

exists for each real $y > 0$. The function Γ so defined is called the *Gamma function*. Example 4 below shows its relation to the Riemann zeta function.

NOTE. Many of the theorems in Chapter 7 concerning Riemann integrals can be converted into theorems on improper Riemann integrals. To illustrate the straight-forward manner in which some of these extensions can be made, consider the formula for integration by parts:

$$\int_a^b f(x) g'(x)\, dx = f(b) g(b) - f(a) g(a) - \int_a^b g(x) f'(x)\, dx.$$

Since b appears in three terms of this equation, there are three limits to consider as $b \to +\infty$. If two of these limits exist, the third also exists and we get the formula

$$\int_a^\infty f(x)g'(x) \, dx = \lim_{b \to +\infty} f(b)g(b) - f(a)g(a) - \int_a^\infty g(x)f'(x) \, dx.$$

Other theorems on Riemann integrals can be extended in much the same way to improper Riemann integrals. However, it is not necessary to develop the details of these extensions any further, since in any particular example, it suffices to apply the required theorem to a compact interval $[a, b]$ and then let $b \to +\infty$.

Example 3. *The functional equation* $\Gamma(y + 1) = y\Gamma(y)$. If $0 < a < b$, integration by parts gives

$$\int_a^b e^{-x}x^y \, dx = a^y e^{-a} - b^y e^{-b} + y \int_a^b e^{-x}x^{y-1} \, dx.$$

Letting $a \to 0+$ and $b \to +\infty$, we find $\Gamma(y + 1) = y\Gamma(y)$.

Example 4. *Integral representation for the Riemann zeta function.* The Riemann zeta function ζ is defined for $s > 1$ by the equation

$$\zeta(s) = \sum_{n=1}^\infty \frac{1}{n^s}.$$

This example shows how the Levi convergence theorem for series can be used to derive an integral representation,

$$\zeta(s)\Gamma(s) = \int_0^\infty \frac{x^{s-1}}{e^x - 1} \, dx.$$

The integral exists as a Lebesgue integral.

In the integral for $\Gamma(s)$ we make the change of variable $t = nx$, $n > 0$, to obtain

$$\Gamma(s) = \int_0^\infty e^{-t}t^{s-1} \, dt = n^s \int_0^\infty e^{-nx}x^{s-1} \, dx.$$

Hence, if $s > 0$, we have

$$n^{-s}\Gamma(s) = \int_0^\infty e^{-nx}x^{s-1} \, dx.$$

If $s > 1$, the series $\sum_{n=1}^\infty n^{-s}$ converges, so we have

$$\zeta(s)\Gamma(s) = \sum_{n=1}^\infty \int_0^\infty e^{-nx}x^{s-1} \, dx,$$

the series on the right being convergent. Since the integrand is nonnegative, Levi's convergence theorem (Theorem 10.25) tells us that the series $\sum_{n=1}^\infty e^{-nx} x^{s-1}$ converges almost everywhere to a sum function which is Lebesgue-integrable on $[0, +\infty)$ and that

$$\zeta(s)\Gamma(s) = \sum_{n=1}^\infty \int_0^\infty e^{-nx}x^{s-1}dx = \int_0^\infty \sum_{n=1}^\infty e^{-nx}x^{s-1} \, dx.$$

But if $x > 0$, we have $0 < e^{-x} < 1$ and hence,

$$\sum_{n=1}^{\infty} e^{-nx} = \frac{e^{-x}}{1 - e^{-x}} = \frac{1}{e^x - 1},$$

the series being a geometric series. Therefore we have

$$\sum_{n=1}^{\infty} e^{-nx} x^{s-1} = \frac{x^{s-1}}{e^x - 1}$$

almost everywhere on $[0, +\infty)$, in fact everywhere except at 0, so

$$\zeta(s)\Gamma(s) = \int_0^{\infty} \sum_{n=1}^{\infty} e^{-nx} x^{s-1} \, dx = \int_0^{\infty} \frac{x^{s-1}}{e^x - 1} \, dx.$$

10.14 MEASURABLE FUNCTIONS

Every function f which is Lebesgue-integrable on an interval I is the limit, almost everywhere on I, of a certain sequence of step functions. However, the converse is not true. For example, the constant function $f = 1$ is a limit of step functions on the real line $\mathbf{R}$, but this function is not in $L(\mathbf{R})$. Therefore, the class of functions which are limits of step functions is larger than the class of Lebesgue-integrable functions. The functions in this larger class are called *measurable functions*.

Definition 10.34. *A function f defined on I is called measurable on I, and we write $f \in M(I)$, if there exists a sequence of step functions $\{s_n\}$ on I such that*

$$\lim_{n \to \infty} s_n(x) = f(x) \quad \text{almost everywhere on } I.$$

NOTE. If f is measurable on I then f is measurable on every subinterval of I.

As already noted, every function in $L(I)$ is measurable on I, but the converse is not true. The next theorem provides a partial converse.

Theorem 10.35. *If $f \in M(I)$ and if $|f(x)| \le g(x)$ almost everywhere on I for some nonnegative g in $L(I)$, then $f \in L(I)$.*

Proof. There is a sequence of step functions $\{s_n\}$ such that $s_n(x) \to f(x)$ almost everywhere on I. Now apply Theorem 10.30 to deduce that $f \in L(I)$.

Corollary 1. *If $f \in M(I)$ and $|f| \in L(I)$, then $f \in L(I)$.*

Corollary 2. *If f is measurable and bounded on a bounded interval I, then $f \in L(I)$.*

Further properties of measurable functions are given in the next theorem.

Theorem 10.36. *Let φ be a real-valued function continuous on $\mathbf{R}^2$. If $f \in M(I)$ and $g \in M(I)$, define h on I by the equation*

$$h(x) = \varphi[f(x), g(x)].$$

Then $h \in M(I)$. *In particular,* $f + g$, $f \cdot g$, $|f|$, $\max (f, g)$, *and* $\min (f, g)$ *are in* $M(I)$. *Also,* $1/f \in M(I)$ *if* $f(x) \neq 0$ *almost everywhere on* I.

Proof. Let $\{s_n\}$ and $\{t_n\}$ denote sequences of step functions such that $s_n \to f$ and $t_n \to g$ almost everywhere on I. Then the function $u_n = \varphi(s_n, t_n)$ is a step function such that $u_n \to h$ almost everywhere on I. Hence $h \in M(I)$.

The next theorem shows that the class $M(I)$ cannot be enlarged by taking limits of functions in $M(I)$.

Theorem 10.37. *Let f be defined on I and assume that $\{f_n\}$ is a sequence of measurable functions on I such that $f_n(x) \to f(x)$ almost everywhere on I. Then f is measurable on I.*

Proof. Choose any positive function g in $L(I)$, for example, $g(x) = 1/(1 + x^2)$ for all x in I. Let

$$F_n(x) = g(x) \frac{f_n(x)}{1 + |f_n(x)|} \qquad \text{for } x \text{ in } I.$$

Then

$$F_n(x) \to \frac{g(x)f(x)}{1 + |f(x)|} \qquad \text{almost everywhere on } I.$$

Let $F(x) = g(x)f(x)/\{1 + |f(x)|\}$. Since each F_n is measurable on I and since $|F_n(x)| < g(x)$ for all x, Theorem 10.35 shows that each $F_n \in L(I)$. Also, $|F(x)| < g(x)$ for all x in I so, by Theorem 10.30, $F \in L(I)$ and hence $F \in M(I)$. Now we have

$$f(x)\{g(x) - |F(x)|\} = f(x)g(x)\left\{1 - \frac{|f(x)|}{1 + |f(x)|}\right\} = \frac{f(x)g(x)}{1 + |f(x)|} = F(x)$$

for all x in I, so

$$f(x) = \frac{F(x)}{g(x) - |F(x)|}.$$

Therefore $f \in M(I)$ since each of F, g, and $|F|$ is in $M(I)$ and $g(x) - |F(x)| > 0$ for all x in I.

NOTE. There exist nonmeasurable functions, but the foregoing theorems show that it is not easy to construct an example. The usual operations of analysis, applied to measurable functions, produce measurable functions. Therefore, every function which occurs in practice is likely to be measurable. (See Exercise 10.37 for an example of a nonmeasurable function.)

10.15 CONTINUITY OF FUNCTIONS DEFINED BY LEBESGUE INTEGRALS

Let f be a real-valued function of two variables defined on a subset of $\mathbf{R}^2$ of the form $X \times Y$, where each of X and Y is a general subinterval of $\mathbf{R}$. Many functions in analysis appear as integrals of the form

$$F(y) = \int_X f(x, y) \, dx.$$

We shall discuss three theorems which transmit continuity, differentiability, and integrability from the integrand f to the function F. The first theorem concerns continuity.

Theorem 10.38. *Let X and Y be two subintervals of $\mathbf{R}$, and let f be a function defined on $X \times Y$ and satisfying the following conditions:*

a) *For each fixed y in Y, the function f_y defined on X by the equation*

$$f_y(x) = f(x, y)$$

    is measurable on X.

b) *There exists a nonnegative function g in $L(X)$ such that, for each y in Y,*

$$|f(x, y)| \leq g(x) \qquad \textit{a.e. on } X.$$

c) *For each fixed y in Y,*

$$\lim_{t \to y} f(x, t) = f(x, y) \qquad \textit{a.e. on } X.$$

Then the Lebesgue integral $\int_X f(x, y) \, dx$ exists for each y in Y, and the function F defined by the equation

$$F(y) = \int_X f(x, y) \, dx$$

is continuous on Y. That is, if $y \in Y$ we have

$$\lim_{t \to y} \int_X f(x, t) \, dx = \int_X \lim_{t \to y} f(x, t) \, dx.$$

Proof. Since f_y is measurable on X and dominated almost everywhere on X by a nonnegative function g in $L(X)$, Theorem 10.35 shows that $f_y \in L(X)$. In other words, the Lebesgue integral $\int_X f(x, y) \, dx$ exists for each y in Y.

Now choose a fixed y in Y and let $\{y_n\}$ be any sequence of points in Y such that $\lim y_n = y$. We will prove that $\lim F(y_n) = F(y)$. Let $G_n(x) = f(x, y_n)$. Each $G_n \in L(X)$ and (c) shows that $G_n(x) \to f(x, y)$ almost everywhere on X. Note that $F(y_n) = \int_X G_n(x) \, dx$. Since (b) holds, the Lebesgue dominated convergence theorem shows that the sequence $\{F(y_n)\}$ converges and that

$$\lim_{n \to \infty} F(y_n) = \int_X f(x, y) \, dx = F(y).$$

Example 1. *Continuity of the Gamma function* $\Gamma(y) = \int_0^{+\infty} e^{-x}x^{y-1} \, dx$ *for* $y > 0$. We apply Theorem 10.38 with $X = [0, +\infty)$, $Y = (0, +\infty)$. For each $y > 0$ the integrand, as a function of x, is continuous (hence measurable) almost everywhere on X, so (a) holds. For each fixed $x > 0$, the integrand, as a function of y, is continuous on Y, so (c) holds. Finally, we verify (b), not on Y but on every compact subinterval $[a, b]$, where $0 < a < b$. For each y in $[a, b]$ the integrand is dominated by the function

$$g(x) = \begin{cases} x^{a-1} & \text{if } 0 < x \leq 1, \\ Me^{-x/2} & \text{if } x \geq 1, \end{cases}$$

where M is some positive constant. This g is Lebesgue-integrable on X, by Theorem 10.18, so Theorem 10.38 tells us that Γ is continuous on $[a, b]$. But since this is true for every subinterval $[a, b]$, it follows that Γ is continuous on $Y = (0, +\infty)$.

Example 2. *Continuity of*

$$F(y) = \int_0^{+\infty} e^{-xy} \frac{\sin x}{x} \, dx$$

for $y > 0$. In this example it is understood that the quotient $(\sin x)/x$ is to be replaced by 1 when $x = 0$. Let $X = [0, +\infty)$, $Y = (0, +\infty)$. Conditions (a) and (c) of Theorem 10.38 are satisfied. As in Example 1, we verify (b) on each subinterval $Y_a = [a, +\infty)$, $a > 0$. Since $|(\sin x)/x| \leq 1$, the integrand is dominated on Y_a by the function

$$g(x) = e^{-ax} \qquad \text{for } x \geq 0.$$

Since g is Lebesgue-integrable on X, F is continuous on Y_a for every $a > 0$; hence F is continuous on $Y = (0, +\infty)$.

To illustrate another use of the Lebesgue dominated convergence theorem we shall prove that $F(y) \to 0$ as $y \to +\infty$.

Let $\{y_n\}$ be any increasing sequence of real numbers such that $y_n \geq 1$ and $y_n \to +\infty$ as $n \to \infty$. We will prove that $F(y_n) \to 0$ as $n \to \infty$. Let

$$f_n(x) = e^{-xy_n} \frac{\sin x}{x} \qquad \text{for } x \geq 0.$$

Then $\lim_{n \to \infty} f_n(x) = 0$ almost everywhere on $[0, +\infty)$, in fact, for all x except 0. Now

$$y_n \geq 1 \quad \text{implies} \quad |f_n(x)| \leq e^{-x} \qquad \text{for all } x \geq 0.$$

Also, each f_n is Riemann-integrable on $[0, b]$ for every $b > 0$ and

$$\int_0^b |f_n| \leq \int_0^b e^{-x} \, dx < 1.$$

Therefore, by Theorem 10.33, f_n is Lebesgue-integrable on $[0, +\infty)$. Since the sequence $\{f_n\}$ is dominated by the function $g(x) = e^{-x}$ which is Lebesgue-integrable on $[0, +\infty)$, the Lebesgue dominated convergence theorem shows that the sequence $\{\int_0^{+\infty} f_n\}$ converges and that

$$\lim_{n \to \infty} \int_0^{+\infty} f_n = \int_0^{+\infty} \lim_{n \to \infty} f_n = 0.$$

But $\int_0^{+\infty} f_n = F(y_n)$, so $F(y_n) \to 0$ as $n \to \infty$. Hence, $F(y) \to 0$ as $y \to +\infty$.

NOTE. In much of the material that follows, we shall have occasion to deal with integrals involving the quotient $(\sin x)/x$. It will be understood that this quotient is to be replaced by 1 when $x = 0$. Similarly, a quotient of the form $(\sin xy)/x$ is to be replaced by y, its limit as $x \to 0$. More generally, if we are dealing with an integrand which has removable discontinuities at certain isolated points within the interval of integration, we will agree that these discontinuities are to be "removed" by redefining the integrand suitably at these exceptional points. At points where the integrand is not defined, we assign the value 0 to the integrand.

10.16 DIFFERENTIATION UNDER THE INTEGRAL SIGN

Theorem 10.39. *Let X and Y be two subintervals of* **R**, *and let f be a function defined on X $\times$ Y and satisfying the following conditions:*

a) *For each fixed y in Y, the function f_y defined on X by the equation $f_y(x) = f(x, y)$ is measurable on X, and $f_a \in L(X)$ for some a in Y.*

b) *The partial derivative $D_2 f(x, y)$ exists for each interior point (x, y) of $X \times Y$.*

c) *There is a nonnegative function G in L(X) such that*

$$|D_2 f(x, y)| \leq G(x) \qquad \text{for all interior points of } X \times Y.$$

Then the Lebesgue integral $\int_X f(x, y)\, dx$ exists for every y in Y, and the function F defined by

$$F(y) = \int_X f(x, y)\, dx$$

is differentiable at each interior point of Y. Moreover, its derivative is given by the formula

$$F'(y) = \int_X D_2 f(x, y)\, dx.$$

NOTE. The derivative $F'(y)$ is said to be obtained by *differentiation under the integral sign.*

Proof. First we establish the inequality

$$|f_y(x)| \leq |f_a(x)| + |y - a|\, G(x), \tag{23}$$

for all interior points (x, y) of $X \times Y$. The Mean-Value Theorem gives us

$$f(x, y) - f(x, a) = (y - a) D_2 f(x, c),$$

where c lies between a and y. Since $|D_2 f(x, c)| \leq G(x)$, this implies

$$|f(x, y)| \leq |f(x, a)| + |y - a| G(x),$$

which proves (23). Since f_y is measurable on X and dominated almost everywhere on X by a nonnegative function in $L(X)$, Theorem 10.35 shows that $f_y \in L(X)$. In other words, the integral $\int_X f(x, y) \, dx$ exists for each y in Y.

Now choose any sequence $\{y_n\}$ of points in Y such that each $y_n \neq y$ but $\lim y_n = y$. Define a sequence of functions $\{q_n\}$ on X by the equation

$$q_n(x) = \frac{f(x, y_n) - f(x, y)}{y_n - y}.$$

Then $q_n \in L(X)$ and $q_n(x) \to D_2 f(x, y)$ at each interior point of X. By the Mean-Value Theorem we have $q_n(x) = D_2 f(x, c_n)$, where c_n lies between y_n and y. Hence, by (c) we have $|q_n(x)| \leq G(x)$ almost everywhere on X. Lebesgue's dominated convergence theorem shows that the sequence $\{\int_X q_n\}$ converges, the integral $\int_X D_2 f(x, y) \, dx$ exists, and

$$\lim_{n \to \infty} \int_X q_n = \int_X \lim_{n \to \infty} q_n = \int_X D_2 f(x, y) \, dx.$$

But

$$\int_X q_n = \frac{1}{y_n - y} \int_X \{f(x, y_n) - f(x, y)\} \, dx = \frac{F(y_n) - F(y)}{y_n - y}.$$

Since this last quotient tends to a limit for all sequences $\{y_n\}$, it follows that $F'(y)$ exists and that

$$F'(y) = \lim_{n \to \infty} \int_X q_n = \int_X D_2 f(x, y) \, dx.$$

Example 1. *Derivative of the Gamma function.* The derivative $\Gamma'(y)$ exists for each $y > 0$ and is given by the integral

$$\Gamma'(y) = \int_0^{+\infty} e^{-x} x^{y-1} \log x \, dx,$$

obtained by differentiating the integral for $\Gamma(y)$ under the integral sign. This is a consequence of Theorem 10.39 because for each y in $[a, b]$, $0 < a < b$, the partial derivative $D_2(e^{-x} x^{y-1})$ is dominated *a.e.* by a function g which is integrable on $[0, +\infty)$. In fact,

$$D_2(e^{-x} x^{y-1}) = \frac{\partial}{\partial y} (e^{-x} x^{y-1}) = e^{-x} x^{y-1} \log x \qquad \text{if } x > 0,$$

so if $y \geq a$ the partial derivative is dominated (except at 0) by the function

$$g(x) = \begin{cases} x^{a-1}|\log x| & \text{if } 0 < x \leq 1, \\ Me^{-x/2} & \text{if } x > 1, \\ 0 & \text{if } x = 0, \end{cases}$$

where M is some positive constant. The reader can easily verify that g is Lebesgue-integrable on $[0, +\infty)$.

Example 2. *Evaluation of the integral*

$$F(y) = \int_0^{+\infty} e^{-xy} \frac{\sin x}{x} \, dx.$$

Applying Theorem 10.39, we find

$$F'(y) = -\int_0^{+\infty} e^{-xy} \sin x \, dx \qquad \text{if } y > 0.$$

(As in Example 1, we prove the result on every interval $Y_a = [a, +\infty)$, $a > 0$.) In this example, the Riemann integral $\int_0^b e^{-xy} \sin x \, dx$ can be calculated by the methods of elementary calculus (using integration by parts twice). This gives us

$$\int_0^b e^{-xy} \sin x \, dx = \frac{e^{-by}(-y \sin b - \cos b)}{1 + y^2} + \frac{1}{1 + y^2} \qquad (24)$$

for all real y. Letting $b \to +\infty$ we find

$$\int_0^{+\infty} e^{-xy} \sin x \, dx = \frac{1}{1 + y^2} \qquad \text{if } y > 0.$$

Therefore $F'(y) = -1/(1 + y^2)$ if $y > 0$. Integration of this equation gives us

$$F(y) - F(b) = -\int_b^y \frac{dt}{1 + t^2} = \arctan b - \arctan y, \qquad \text{for } y > 0, b > 0.$$

Now let $b \to +\infty$. Then $\arctan b \to \pi/2$ and $F(b) \to 0$ (see Example 2, Section 10.15), so $F(y) = \pi/2 - \arctan y$. In other words, we have

$$\int_0^{+\infty} e^{-xy} \frac{\sin x}{x} \, dx = \frac{\pi}{2} - \arctan y \qquad \text{if } y > 0. \qquad (25)$$

This equation is also valid if $y = 0$. That is, we have the formula

$$\int_0^{+\infty} \frac{\sin x}{x} \, dx = \frac{\pi}{2}. \qquad (26)$$

However, we cannot deduce this by putting $y = 0$ in (25) because we have not shown that F is continuous at 0. In fact, the integral in (26) exists as an improper Riemann integral. It does not exist as a Lebesgue integral. (See Exercise 10.9.)

Example 3. *Proof of the formula*

$$\int_0^{+\infty} \frac{\sin x}{x} \, dx = \lim_{b \to +\infty} \int_0^b \frac{\sin x}{x} \, dx = \frac{\pi}{2}.$$

Let $\{g_n\}$ be the sequence of functions defined for all real y by the equation

$$g_n(y) = \int_0^n e^{-xy} \frac{\sin x}{x} \, dx. \tag{27}$$

First we note that $g_n(n) \to 0$ as $n \to \infty$ since

$$|g_n(n)| \le \int_0^n e^{-xn} \, dx = \frac{1}{n} \int_0^{n^2} e^{-t} \, dt < \frac{1}{n}.$$

Now we differentiate (27) and use (24) to obtain

$$g_n'(y) = -\int_0^n e^{-xy} \sin x \, dx = -\frac{e^{-ny}(-y \sin n - \cos n) + 1}{1 + y^2},$$

an equation valid for all real y. This shows that $g_n'(y) \to -1/(1 + y^2)$ for all y and that

$$|g_n'(y)| \le \frac{e^{-y}(y + 1) + 1}{1 + y^2} \qquad \text{for all } y \ge 0.$$

Therefore the function f_n defined by

$$f_n(y) = \begin{cases} g_n'(y) & \text{if } 0 \le y \le n, \\ 0 & \text{if } y > n, \end{cases}$$

is Lebesgue-integrable on $[0, +\infty)$ and is dominated by the nonnegative function

$$g(y) = \frac{e^{-y}(y + 1) + 1}{1 + y^2}.$$

Also, g is Lebesgue-integrable on $[0, +\infty)$. Since $f_n(y) \to -1/(1 + y^2)$ on $[0, +\infty)$, the Lebesgue dominated convergence theorem implies

$$\lim_{n \to \infty} \int_0^{+\infty} f_n = -\int_0^{+\infty} \frac{dy}{1 + y^2} = -\frac{\pi}{2}.$$

But we have

$$\int_0^{+\infty} f_n = \int_0^n g_n'(y) \, dy = g_n(n) - g_n(0).$$

Letting $n \to \infty$, we find $g_n(0) \to \pi/2$.

Now if $b > 0$ and if $n = [b]$, we have

$$\int_0^b \frac{\sin x}{x} \, dx = \int_0^n \frac{\sin x}{x} \, dx + \int_n^b \frac{\sin x}{x} \, dx = g_n(0) + \int_n^b \frac{\sin x}{x} \, dx.$$

Since

$$0 \le \left| \int_n^b \frac{\sin x}{x} \, dx \right| \le \int_n^b \frac{1}{n} \, dx = \frac{b - n}{n} \le \frac{1}{n} \to 0 \qquad \text{as } b \to +\infty,$$

we have

$$\lim_{b \to +\infty} \int_0^b \frac{\sin x}{x} \, dx = \lim_{n \to \infty} g_n(0) = \frac{\pi}{2}.$$

This formula will be needed in Chapter 11 in the study of Fourier series.

10.17 INTERCHANGING THE ORDER OF INTEGRATION

Theorem 10.40. *Let X and Y be two subintervals of $\mathbf{R}$, and let k be a function which is defined, continuous, and bounded on $X \times Y$, say*

$$|k(x, y)| \le M \qquad \text{for all } (x, y) \text{ in } X \times Y.$$

Assume $f \in L(X)$ and $g \in L(Y)$. Then we have:

a) *For each y in Y, the Lebesgue integral $\int_X f(x)k(x, y) \, dx$ exists, and the function F defined on Y by the equation*

$$F(y) = \int_X f(x)k(x, y) \, dx$$

   is continuous on Y.

b) *For each x in X, the Lebesgue integral $\int_Y g(y)k(x, y) \, dy$ exists, and the function G defined on X by the equation*

$$G(x) = \int_Y g(y)k(x, y) \, dy$$

   is continuous on X.

c) *The two Lebesgue integrals $\int_Y g(y)F(y) \, dy$ and $\int_X f(x)G(x) \, dx$ exist and are equal. That is,*

$$\int_X f(x) \left[\int_Y g(y)k(x, y) \, dy \right] dx = \int_Y g(y) \left[\int_X f(x)k(x, y) \, dx \right] dy. \quad (28)$$

Proof. For each fixed y in Y, let $f_y(x) = f(x)k(x, y)$. Then f_y is measurable on X and satisfies the inequality

$$|f_y(x)| = |f(x)k(x, y)| \le M|f(x)| \qquad \text{for all } x \text{ in } X.$$

Also, since k is continuous on $X \times Y$ we have

$$\lim_{t \to y} f(x)k(x, t) = f(x)k(x, y) \qquad \text{for all } x \text{ in } X.$$

Therefore, part (a) follows from Theorem 10.38. A similar argument proves (b).

Now the product $f \cdot G$ is measurable on X and satisfies the inequality

$$|f(x)G(x)| \leq |f(x)| \int_Y |g(y)| \, |k(x, y)| \, dy \leq M' \, |f(x)|,$$

where $M' = M \int_Y |g(y)| \, dy$. By Theorem 10.35 we see that $f \cdot G \in L(X)$. A similar argument shows that $g \cdot F \in L(Y)$.

Next we prove (28). First we note that (28) is true if each of f and g is a step function. In this case, each of f and g vanishes outside a compact interval, so each is Riemann-integrable on that interval and (28) is an immediate consequence of Theorem 7.42.

Now we use Theorem 10.19(b) to approximate each of f and g by step functions. If $\varepsilon > 0$ is given, there are step functions s and t such that

$$\int_X |f - s| < \varepsilon \quad \text{and} \quad \int_Y |g - t| < \varepsilon.$$

Therefore we have

$$\int_X f \cdot G = \int_X s \cdot G + A_1, \tag{29}$$

where

$$|A_1| = \left| \int_X (f - s) \cdot G \right| \leq \int_X |f - s| \int_Y |g(y)| \, |k(x, y)| \, dy < \varepsilon M \int_Y |g|.$$

Also, we have

$$G(x) = \int_Y g(y)k(x, y) \, dy = \int_Y t(y)k(x, y) \, dy + A_2,$$

where

$$|A_2| = \left| \int_Y (g - t)k(x, y) \, dy \right| \leq M \int_Y |g - t| < \varepsilon M.$$

Therefore

$$\int_X s \cdot G = \int_X s(x) \left[\int_Y t(y)k(x, y) \, dy \right] dx + A_3,$$

where

$$|A_3| = \left| A_2 \int_X s(x) \, dx \right| \leq \varepsilon M \int_X |s|$$

$$\leq \varepsilon M \int_X \{|s - f| + |f|\} < \varepsilon^2 M + \varepsilon M \int_X |f|,$$

so (29) becomes

$$\int_X f \cdot G = \int_X s(x) \left[\int_Y t(y) k(x, y) \, dy \right] dx + A_1 + A_3. \tag{30}$$

Similarly, we find

$$\int_Y g \cdot F = \int_Y t(y) \left[\int_X s(x) k(x, y) \, dx \right] dy + B_1 + B_3, \tag{31}$$

where

$$|B_1| < \varepsilon M \int_X |f| \quad \text{and} \quad |B_3| \le \varepsilon M \int_Y |t| < \varepsilon^2 M + \varepsilon M \int_Y |g|.$$

But the iterated integrals on the right of (30) and (31) are equal, so we have

$$\left| \int_X f \cdot G - \int_Y g \cdot F \right| \le |A_1| + |A_3| + |B_1| + |B_3|$$

$$< 2\varepsilon^2 M + 2\varepsilon M \left\{ \int_X |f| + \int_Y |g| \right\}.$$

Since this holds for every $\varepsilon > 0$ we have $\int_X f \cdot G = \int_Y g \cdot F$, as required.

NOTE. A more general version of Theorem 10.40 will be proved in Chapter 15 using double integrals. (See Theorem 15.6.)

10.18 MEASURABLE SETS ON THE REAL LINE

Definition 10.41. *Given any nonempty subset S of* **R**. *The function* χ_S *defined by*

$$\chi_S(x) = \begin{cases} 1 & \text{if } x \in S, \\ 0 & \text{if } x \in \mathbf{R} - S, \end{cases}$$

is called the characteristic function of S. If S is empty we define $\chi_S(x) = 0$ *for all x.*

Theorem 10.42. *Let* $\mathbf{R} = (-\infty, +\infty)$. *Then we have:*

a) *If S has measure* 0, *then* $\chi_S \in L(\mathbf{R})$ *and* $\int_{\mathbf{R}} \chi_S = 0$.
b) *If* $\chi_S \in L(\mathbf{R})$ *and if* $\int_{\mathbf{R}} \chi_S = 0$, *then S has measure* 0.

Proof. Part (a) follows by taking $f = \chi_S$ in Theorem 10.20. To prove (b), let $f_n = \chi_S$ for all n. Then $|f_n| = \chi_S$ so

$$\sum_{n=1}^{\infty} \int_{\mathbf{R}} |f_n| = \sum_{n=1}^{\infty} \int_{\mathbf{R}} \chi_S = 0.$$

By the Levi theorem for absolutely convergent series, it follows that the series $\sum_{n=1}^{\infty} f_n(x)$ converges everywhere on $\mathbf{R}$ except for a set T of measure 0. If $x \in S$, the series cannot converge since each term is 1. If $x \notin S$, the series converges because each term is 0. Hence $T = S$, so S has measure 0.

Definition 10.43. *A subset S of $\mathbf{R}$ is called* measurable *if its characteristic function χ_S is measurable. If, in addition, χ_S is Lebesgue-integrable on $\mathbf{R}$, then the* measure *$\mu(S)$ of the set S is defined by the equation*

$$\mu(S) = \int_{\mathbf{R}} \chi_S.$$

If χ_S is measurable but not Lebesgue-integrable on $\mathbf{R}$, we define $\mu(S) = +\infty$. The function μ so defined is called Lebesgue measure.

Examples

1. Theorem 10.42 shows that a set S of measure zero is measurable and that $\mu(S) = 0$.
2. Every interval I (bounded or unbounded) is measurable. If I is a bounded interval with endpoints $a \le b$, then $\mu(I) = b - a$. If I is an unbounded interval, then $\mu(I) = +\infty$.
3. If A and B are measurable and $A \subseteq B$, then $\mu(A) \le \mu(B)$.

Theorem 10.44. a) *If S and T are measurable, so is $S - T$.*

b) *If $S_1, S_2, \ldots$, are measurable, so are $\bigcup_{i=1}^{\infty} S_i$ and $\bigcap_{i=1}^{\infty} S_i$.*

Proof. To prove (a) we note that the characteristic function of $S - T$ is $\chi_S - \chi_S\chi_T$. To prove (b), let

$$U_n = \bigcup_{i=1}^{n} S_i, \qquad V_n = \bigcap_{i=1}^{n} S_i, \qquad U = \bigcup_{i=1}^{\infty} S_i, \qquad V = \bigcap_{i=1}^{\infty} S_i.$$

Then we have

$$\chi_{U_n} = \max(\chi_{S_1}, \ldots, \chi_{S_n}) \qquad \text{and} \qquad \chi_{V_n} = \min(\chi_{S_1}, \ldots, \chi_{S_n}),$$

so each of U_n and V_n is measurable. Also, $\chi_U = \lim_{n \to \infty} \chi_{U_n}$ and $\chi_V = \lim_{n \to \infty} \chi_{V_n}$, so U and V are measurable.

Theorem 10.45. *If A and B are disjoint measurable sets, then*

$$\mu(A \cup B) = \mu(A) + \mu(B). \tag{32}$$

Proof. Let $S = A \cup B$. Since A and B are disjoint we have

$$\chi_S = \chi_A + \chi_B.$$

Suppose that χ_S is integrable. Since both χ_A and χ_B are measurable and satisfy

$$0 \le \chi_A(x) \le \chi_S(x), \qquad 0 \le \chi_B(x) \le \chi_S(x) \quad \text{for all } x,$$

Theorem 10.35 shows that both χ_A and χ_B are integrable. Therefore

$$\mu(S) = \int_R \chi_S = \int_R \chi_A + \int_R \chi_B = \mu(A) + \mu(B).$$

In this case (32) holds with both members finite.

If χ_S is not integrable then at least one of χ_A or χ_B is not integrable, in which case (32) holds with both members infinite.

The following extension of Theorem 10.45 can be proved by induction.

Theorem 10.46. *If $\{A_1, \ldots, A_n\}$ is a finite disjoint collection of measurable sets, then*

$$\mu\left(\bigcup_{i=1}^{n} A_i\right) = \sum_{i=1}^{n} \mu(A_i).$$

NOTE. This property is described by saying that Lebesgue measure is *finitely additive.* In the next theorem we prove that Lebesgue measure is *countably additive.*

Theorem 10.47. *If $\{A_1, A_2, \ldots\}$ is a countable disjoint collection of measurable sets, then*

$$\mu\left(\bigcup_{i=1}^{\infty} A_i\right) = \sum_{i=1}^{\infty} \mu(A_i). \tag{33}$$

Proof. Let $T_n = \bigcup_{i=1}^{n} A_i$, $\chi_n = \chi_{T_n}$, $T = \bigcup_{i=1}^{\infty} A_i$. Since μ is finitely additive, we have

$$\mu(T_n) = \sum_{i=1}^{n} \mu(A_i) \qquad \text{for each } n.$$

We are to prove that $\mu(T_n) \to \mu(T)$ as $n \to \infty$. Note that $\mu(T_n) \leq \mu(T_{n+1})$ so $\{\mu(T_n)\}$ is an increasing sequence.

We consider two cases. If $\mu(T)$ is finite, then χ_T and each χ_n is integrable. Also, the sequence $\{\mu(T_n)\}$ is bounded above by $\mu(T)$ so it converges. By the Lebesgue dominated convergence theorem, $\mu(T_n) \to \mu(T)$.

If $\mu(T) = +\infty$, then χ_T is not integrable. Theorem 10.24 implies that either some χ_n is not integrable or else every χ_n is integrable but $\mu(T_n) \to +\infty$. In either case (33) holds with both members infinite.

For a further study of measure theory and its relation to integration, the reader can consult the references at the end of this chapter.

10.19 THE LEBESGUE INTEGRAL OVER ARBITRARY SUBSETS OF R

Definition 10.48. *Let f be defined on a measurable subset S of R. Define a new function $\tilde{f}$ on R as follows:*

$$\tilde{f}(x) = \begin{cases} f(x) & \text{if } x \in S, \\ 0 & \text{if } x \in R - S. \end{cases}$$

If $\tilde{f}$ is Lebesgue-integrable on $\mathbf{R}$, we say that f is Lebesgue-integrable on S and we write $f \in L(S)$. The integral of f over S is defined by the equation

$$\int_S f = \int_{\mathbf{R}} \tilde{f}.$$

This definition immediately gives the following properties:

If $f \in L(S)$, then $f \in L(T)$ for every subset of T of S.
If S has finite measure, then $\mu(S) = \int_S 1$.

The following theorem describes a countably additive property of the Lebesgue integral. Its proof is left as an exercise for the reader.

Theorem 10.49. Let $\{A_1, A_2, \dots\}$ be a countable disjoint collection of sets in $\mathbf{R}$, and let $S = \bigcup_{i=1}^{\infty} A_i$. Let f be defined on S.

a) If $f \in L(S)$, then $f \in L(A_i)$ for each i and

$$\int_S f = \sum_{i=1}^{\infty} \int_{A_i} f.$$

b) If $f \in L(A_i)$ for each i and if the series in (a) converges, then $f \in L(S)$ and the equation in (a) holds.

10.20 LEBESGUE INTEGRALS OF COMPLEX-VALUED FUNCTIONS

If f is a complex-valued function defined on an interval I, then $f = u + iv$, where u and v are real. We say f is Lebesgue-integrable on I if both u and v are Lebesgue-integrable on I, and we define

$$\int_I f = \int_I u + i \int_I v.$$

Similarly, f is called measurable on I if both u and v are in $M(I)$.

It is easy to verify that sums and products of complex-valued measurable functions are also measurable. Moreover, since

$$|f| = (u^2 + v^2)^{1/2},$$

Theorem 10.36 shows that $|f|$ is measurable if f is.

Many of the theorems concerning Lebesgue integrals of real-valued functions can be extended to complex-valued functions. However, we do not discuss these extensions since, in any particular case, it usually suffices to write $f = u + iv$ and apply the theorems to u and v. The only result that needs to be formulated explicitly is the following.

Theorem 10.50. *If a complex-valued function f is Lebesgue-integrable on I, then* $|f| \in L(I)$ *and we have*

$$\left| \int_I f \right| \leq \int_I |f|.$$

Proof. Write $f = u + iv$. Since f is measurable and $|f| \leq |u| + |v|$, Theorem 10.35 shows that $|f| \in L(I)$.

Let $a = \int_I f$. Then $a = re^{i\theta}$, where $r = |a|$. We wish to prove that $r \leq \int_I |f|$. Let

$$b = \begin{cases} e^{-i\theta} & \text{if } r > 0, \\ 1 & \text{if } r = 0. \end{cases}$$

Then $|b| = 1$ and $r = ba = b \int_I f = \int_I bf$. Now write $bf = U + iV$, where U and V are real. Then $\int_I bf = \int_I U$, since $\int_I bf$ is real. Hence

$$r = \int_I bf = \int_I U \leq \int_I |U| \leq \int_I |bf| = \int_I |f|.$$

10.21 INNER PRODUCTS AND NORMS

This section introduces *inner products* and *norms*, concepts which play an important role in the theory of Fourier series, to be discussed in Chapter 11.

Definition 10.51. *Let f and g be two real-valued functions in L(I) whose product $f \cdot g$ is in L(I). Then the integral*

$$\int_I f(x)g(x)\,dx \tag{34}$$

is called the inner product of f and g, and is denoted by (f, g). If $f^2 \in L(I)$, the nonnegative number $(f, f)^{1/2}$, denoted by $\|f\|$, is called the L^2-norm of f.

NOTE. The integral in (34) resembles the sum $\sum_{k=1}^{n} x_k y_k$ which defines the dot product of two vectors $\mathbf{x} = (x_1, \ldots, x_n)$ and $\mathbf{y} = (y_1, \ldots, y_n)$. The function values $f(x)$ and $g(x)$ in (34) play the role of the components x_k and y_k, and integration takes the place of summation. The L^2-norm of f is analogous to the length of a vector.

The first theorem gives a sufficient condition for a function in $L(I)$ to have an L^2-norm.

Theorem 10.52. *If $f \in L(I)$ and if f is bounded almost everywhere on I, then $f^2 \in L(I)$.*

Proof. Since $f \in L(I)$, f is measurable and hence f^2 is measurable on I and satisfies the inequality $|f(x)|^2 \leq M|f(x)|$ almost everywhere on I, where M is an upper bound for $|f|$. By Theorem 10.35, $f^2 \in L(I)$.

10.22 THE SET $L^2(I)$ OF SQUARE-INTEGRABLE FUNCTIONS

Definition 10.53. *We denote by $L^2(I)$ the set of all real-valued measurable functions f on I such that $f^2 \in L(I)$. The functions in $L^2(I)$ are said to be square-integrable.*

NOTE. The set $L^2(I)$ is neither larger than nor smaller than $L(I)$. For example, the function given by

$$f(x) = x^{-1/2} \qquad \text{for } 0 < x \le 1, \quad f(0) = 0,$$

is in $L([0, 1])$ but not in $L^2([0, 1])$. Similarly, the function $g(x) = 1/x$ for $x \ge 1$ is in $L^2([1, +\infty))$ but not in $L([1, +\infty))$.

Theorem 10.54. *If $f \in L^2(I)$ and $g \in L^2(I)$, then $f \cdot g \in L(I)$ and $(af + bg) \in L^2(I)$ for every real a and b.*

Proof. Both f and g are measurable so $f \cdot g \in M(I)$. Since

$$|f(x)g(x)| \le \frac{f^2(x) + g^2(x)}{2},$$

Theorem 10.35 shows that $f \cdot g \in L(I)$. Also, $(af + bg) \in M(I)$ and

$$(af + bg)^2 = a^2 f^2 + 2abf \cdot g + b^2 g^2,$$

so $(af + bg) \in L^2(I)$.

Thus, the inner product (f, g) is defined for every pair of functions f and g in $L^2(I)$. The basic properties of inner products and norms are described in the next theorem.

Theorem 10.55. *If f, g, and h are in $L^2(I)$ and if c is real we have:*

a) $(f, g) = (g, f)$ *(commutativity).*

b) $(f + g, h) = (f, h) + (g, h)$ *(linearity).*

c) $(cf, g) = c(f, g)$ *(associativity).*

d) $\|cf\| = |c|\,\|f\|$ *(homogeneity).*

e) $|(f, g)| \le \|f\|\,\|g\|$ *(Cauchy–Schwarz inequality).*

f) $\|f + g\| \le \|f\| + \|g\|$ *(triangle inequality).*

Proof. Parts (a) through (d) are immediate consequences of the definition. Part (e) follows at once from the inequality

$$\int_I \left[\int_I |f(x)g(y) - g(x)f(y)|^2 \, dy \right] dx \ge 0.$$

To prove (f) we use (e) along with the relation

$$\|f + g\|^2 = (f + g, f + g) = (f, f) + 2(f, g) + (g, g) = \|f\|^2 + \|g\|^2 + 2(f, g).$$

NOTE. The notion of inner product can be extended to complex-valued functions f such that $|f| \in L^2(I)$. In this case, (f, g) is defined by the equation

$$(f, g) = \int_I f(x)\overline{g(x)} \, dx,$$

where the bar denotes the complex conjugate. The conjugate is introduced so that the inner product of f with itself will be a nonnegative quantity, namely, $(f, f) = \int_I |f|^2$. The L^2-norm of f is, as before, $\|f\| = (f, f)^{1/2}$.

Theorem 10.55 is also valid for complex functions, except that part (a) must be modified by writing

$$(f, g) = \overline{(g, f)}. \tag{35}$$

This implies the following companion result to part (b):

$$(f, g + h) = \overline{(g + h, f)} = \overline{(g, f)} + \overline{(h, f)} = (f, g) + (f, h).$$

In parts (c) and (d) the constant c can be complex. From (c) and (35) we obtain

$$(f, cg) = \overline{c}(f, g).$$

The Cauchy-Schwarz inequality and the triangle inequality are also valid for complex functions.

10.23 THE SET $L^2(I)$ AS A SEMIMETRIC SPACE

We recall (Definition 3.32) that a metric space is a set T together with a nonnegative function d on $T \times T$ satisfying the following properties for all points x, y, z in T:

1. $d(x, x) = 0$.   2. $d(x, y) > 0$ if $x \neq y$.
3. $d(x, y) = d(y, x)$.   4. $d(x, y) \leq d(x, z) + d(z, y)$.

We try to convert $L^2(I)$ into a metric space by defining the distance $d(f, g)$ between any two complex-valued functions in $L^2(I)$ by the equation

$$d(f, g) = \|f - g\| = \left(\int_I |f - g|^2 \right)^{1/2}.$$

This function satisfies properties 1, 3, and 4, but not 2. If f and g are functions in $L^2(I)$ which differ on a nonempty set of measure zero, then $f \neq g$ but $f - g = 0$ almost everywhere on I, so $d(f, g) = 0$.

A function d which satisfies 1, 3, and 4, but not 2, is called a *semimetric*. The set $L^2(I)$, together with the semimetric d, is called a *semimetric space*.

10.24 A CONVERGENCE THEOREM FOR SERIES OF FUNCTIONS IN $L^2(I)$

The following convergence theorem is analogous to the Levi theorem for series (Theorem 10.26).

Theorem 10.56. *Let $\{g_n\}$ be a sequence of functions in $L^2(I)$ such that the series*

$$\sum_{n=1}^{\infty} \|g_n\|$$

converges. Then the series of functions $\sum_{n=1}^{\infty} g_n$ converges almost everywhere on I to a function g in $L^2(I)$, and we have

$$\|g\| = \lim_{n \to \infty} \left\| \sum_{k=1}^{n} g_k \right\| \leq \sum_{k=1}^{\infty} \|g_k\|. \tag{36}$$

Proof. Let $M = \sum_{n=1}^{\infty} \|g_n\|$. The triangle inequality, extended to finite sums, gives us

$$\left\| \sum_{k=1}^{n} |g_k| \right\| \leq \sum_{k=1}^{n} \|g_k\| \leq M.$$

This implies

$$\int_I \left(\sum_{k=1}^{n} |g_k(x)| \right)^2 dx = \left\| \sum_{k=1}^{n} |g_k| \right\|^2 \leq M^2. \tag{37}$$

If $x \in I$, let

$$f_n(x) = \left(\sum_{k=1}^{n} |g_k(x)| \right)^2.$$

The sequence $\{f_n\}$ is increasing, each $f_n \in L(I)$ (since each $g_k \in L^2(I)$), and (37) shows that $\int_I f_n \leq M^2$. Therefore the sequence $\{\int_I f_n\}$ converges. By the Levi theorem for sequences (Theorem 10.24), there is a function f in $L(I)$ such that $f_n \to f$ almost everywhere on I, and

$$\int_I f = \lim_{n \to \infty} \int_I f_n \leq M^2.$$

Therefore the series $\sum_{k=1}^{\infty} g_k(x)$ converges absolutely almost everywhere on I. Let

$$g(x) = \lim_{n \to \infty} \sum_{k=1}^{n} g_k(x)$$

at those points where the limit exists, and let

$$G_n(x) = \left| \sum_{k=1}^{n} g_k(x) \right|^2.$$

Then each $G_n \in L(I)$ and $G_n(x) \to |g(x)|^2$ almost everywhere on I. Also,

$$G_n(x) \leq f_n(x) \leq f(x) \qquad a.e. \text{ on } I.$$

Therefore, by the Lebesgue dominated convergence theorem, $|g|^2 \in L(I)$ and

$$\int_I |g|^2 = \lim_{n \to \infty} \int_I G_n. \tag{38}$$

Since g is measurable, this shows that $g \in L^2(I)$. Also, we have

$$\int_I G_n = \int_I \left| \sum_{k=1}^n g_k \right|^2 = \left\| \sum_{k=1}^n g_k \right\|^2, \quad \text{and} \quad \int_I G_n \leq \int_I f_n \leq M^2,$$

so (38) implies

$$\| g \|^2 = \lim_{n \to \infty} \left\| \sum_{k=1}^n g_k \right\|^2 \leq M^2,$$

and this, in turn, implies (36).

10.25 THE RIESZ–FISCHER THEOREM

The convergence theorem which we have just proved can be used to prove that every Cauchy sequence in the semimetric space $L^2(I)$ converges to a function in $L^2(I)$. In other words, the semimetric space $L^2(I)$ is complete. This result, called the *Riesz–Fischer theorem*, plays an important role in the theory of Fourier series.

Theorem 10.57. *Let $\{f_n\}$ be a Cauchy sequence of complex-valued functions in $L^2(I)$. That is, assume that for every $\varepsilon > 0$ there is an integer N such that*

$$\| f_m - f_n \| < \varepsilon \quad \text{whenever } m \geq n \geq N. \tag{39}$$

Then there exists a function f in $L^2(I)$ such that

$$\lim_{n \to \infty} \| f_n - f \| = 0. \tag{40}$$

Proof. By applying (39) repeatedly we can find an increasing sequence of integers $n(1) < n(2) < \cdots$ such that

$$\| f_m - f_{n(k)} \| < \frac{1}{2^k} \quad \text{whenever } m \geq n(k).$$

Let $g_1 = f_{n(1)}$, and let $g_k = f_{n(k)} - f_{n(k-1)}$ for $k \geq 2$. Then the series $\sum_{k=1}^\infty \| g_k \|$ converges, since it is dominated by

$$\| f_{n(1)} \| + \sum_{k=2}^\infty \| f_{n(k)} - f_{n(k-1)} \| < \| f_{n(1)} \| + \sum_{k=1}^\infty \frac{1}{2^k} = \| f_{n(1)} \| + 1.$$

Each g_n is in $L^2(I)$. Hence, by Theorem 10.56, the series $\sum_{n=1}^\infty g_n$ converges almost everywhere on I to a function f in $L^2(I)$. To complete the proof we will show that $\| f_m - f \| \to 0$ as $m \to \infty$.

For this purpose we use the triangle inequality to write

$$\| f_m - f \| \leq \| f_m - f_{n(k)} \| + \| f_{n(k)} - f \|. \tag{41}$$

If $m \geq n(k)$, the first term on the right is $< 1/2^k$. To estimate the second term we note that

$$f - f_{n(k)} = \sum_{r=k+1}^\infty \{ f_{n(r)} - f_{n(r-1)} \},$$

and that the series $\sum_{r=k+1}^{\infty} \|f_{n(r)} - f_{n(r-1)}\|$ converges. Therefore, we can use inequality (36) of Theorem 10.56 to write

$$\|f - f_{n(k)}\| \le \sum_{r=k+1}^{\infty} \|f_{n(r)} - f_{n(r-1)}\| < \sum_{r=k+1}^{\infty} \frac{1}{2^{r-1}} = \frac{1}{2^{k-1}}.$$

Hence, (41) becomes

$$\|f_m - f\| \le \frac{1}{2^{k-1}} + \frac{1}{2^k} = \frac{3}{2^k} \qquad \text{if } m \ge n(k).$$

Since $n(k) \to \infty$ as $k \to \infty$, this shows that $\|f_m - f\| \to 0$ as $m \to \infty$.

NOTE. In the course of the proof we have shown that every Cauchy sequence of functions in $L^2(I)$ has a subsequence which converges pointwise almost everywhere on I to a limit function f in $L^2(I)$. However, it does not follow that the sequence $\{f_n\}$ itself converges pointwise almost everywhere to f on I. (A counterexample is described in Section 9.13.) Although $\{f_n\}$ converges to f in the semimetric space $L^2(I)$, this convergence is not the same as pointwise convergence.

EXERCISES

Upper functions

10.1 Prove that max $(f, g) + $ min $(f, g) = f + g$, and that

$$\max (f + h, g + h) = \max (f, g) + h, \qquad \min (f + h, g + h) = \min (f, g) + h.$$

10.2 Let $\{f_n\}$ and $\{g_n\}$ be increasing sequences of functions on an interval I. Let $u_n = $ max (f_n, g_n) and $v_n = $ min (f_n, g_n).

    a) Prove that $\{u_n\}$ and $\{v_n\}$ are increasing on I.

    b) If $f_n \nearrow f$ *a.e.* on I and if $g_n \nearrow g$ *a.e.* on I, prove that $u_n \nearrow$ max (f, g) and $v_n \nearrow$ min (f, g) *a.e.* on I.

10.3 Let $\{s_n\}$ be an increasing sequence of step functions which converges pointwise on an interval I to a limit function f. If I is unbounded and if $f(x) \ge 1$ almost everywhere on I, prove that the sequence $\{\int_I s_n\}$ diverges.

10.4 This exercise gives an example of an upper function f on the interval $I = [0, 1]$ such that $-f \notin U(I)$. Let $\{r_1, r_2, \dots\}$ denote the set of rational numbers in $[0, 1]$ and let $I_n = [r_n - 4^{-n}, r_n + 4^{-n}] \cap I$. Let $f(x) = 1$ if $x \in I_n$ for some n, and let $f(x) = 0$ otherwise.

    a) Let $f_n(x) = 1$ if $x \in I_n$, $f_n(x) = 0$ if $x \notin I_n$, and let $s_n = $ max $(f_1, \dots, f_n)$. Show that $\{s_n\}$ is an increasing sequence of step functions which generates f. This shows that $f \in U(I)$.

    b) Prove that $\int_I f \le 2/3$.

    c) If a step function s satisfies $s(x) \le -f(x)$ on I, show that $s(x) \le -1$ almost everywhere on I and hence $\int_I s \le -1$.

    d) Assume that $-f \in U(I)$ and use (b) and (c) to obtain a contradiction.

NOTE. In the following exercises, the integrand is to be assigned the value 0 at points where it is undefined.

Convergence theorems

10.5 If $f_n(x) = e^{-nx} - 2e^{-2nx}$, show that

$$\sum_{n=1}^{\infty} \int_0^{\infty} f_n(x)\, dx \neq \int_0^{\infty} \sum_{n=1}^{\infty} f_n(x)\, dx.$$

10.6 Justify the following equations:

a) $\displaystyle \int_0^1 \log \frac{1}{1-x}\, dx = \int_0^1 \sum_{n=1}^{\infty} \frac{x^n}{n}\, dx = \sum_{n=1}^{\infty} \frac{1}{n} \int_0^1 x^n dx = 1.$

b) $\displaystyle \int_0^1 \frac{x^{p-1}}{1-x} \log\left(\frac{1}{x}\right) dx = \sum_{n=0}^{\infty} \frac{1}{(n+p)^2} \quad (p > 0).$

10.7 Prove Tannery's convergence theorem for Riemann integrals: *Given a sequence of functions $\{f_n\}$ and an increasing sequence $\{p_n\}$ of real numbers such that $p_n \to +\infty$ as $n \to \infty$. Assume that*

a) $f_n \to f$ *uniformly on $[a, b]$ for every $b \geq a$.*

b) f_n *is Riemann-integrable on $[a, b]$ for every $b \geq a$.*

c) $|f_n(x)| \leq g(x)$ *almost everywhere on $[a, +\infty)$, where g is nonnegative and improper Riemann-integrable on $[a, +\infty)$.*

Then both f and $|f|$ are improper Riemann-integrable on $[a, +\infty)$, the sequence $\{\int_a^{p_n} f_n\}$ converges, and

$$\int_a^{+\infty} f(x)\, dx = \lim_{n \to \infty} \int_a^{p_n} f_n(x)\, dx.$$

d) Use Tannery's theorem to prove that

$$\lim_{n \to \infty} \int_0^n \left(1 - \frac{x}{n}\right)^n x^p\, dx = \int_0^{\infty} e^{-x} x^p\, dx, \quad \text{if } p > -1.$$

10.8 Prove Fatou's lemma: *Given a sequence $\{f_n\}$ of nonnegative functions in $L(I)$ such that (a) $\{f_n\}$ converges almost everywhere on I to a limit function f, and (b) $\int_I f_n \leq A$ for some $A > 0$ and all $n \geq 1$. Then the limit function $f \in L(I)$ and $\int_I f \leq A$.*

NOTE. It is not asserted that $\{\int_I f_n\}$ converges. (Compare with Theorem 10.24.)

Hint. Let $g_n(x) = \inf \{f_n(x), f_{n+1}(x), \ldots\}$. Then $g_n \nearrow f$ a.e. on I and $\int_I g_n \leq \int_I f_n \leq A$ so $\lim_{n \to \infty} \int_I g_n$ exists and is $\leq A$. Now apply Theorem 10.24.

Improper Riemann Integrals

10.9 a) If $p > 1$, prove that the integral $\int_1^{+\infty} x^{-p} \sin x\, dx$ exists both as an improper Riemann integral and as a Lebesgue integral. *Hint.* Integration by parts.

b) If $0 < p \le 1$, prove that the integral in (a) exists as an improper Riemann integral but not as a Lebesgue integral. *Hint.* Let

$$g(x) = \begin{cases} \dfrac{\sqrt{2}}{2x} & \text{if } n\pi + \dfrac{\pi}{4} \le x \le n\pi + \dfrac{3\pi}{4} \quad \text{for } n = 1, 2, \dots, \\ 0 & \text{otherwise,} \end{cases}$$

and show that

$$\int_1^{n\pi} x^{-p} |\sin x| \, dx \ge \int_\pi^{n\pi} g(x) \, dx \ge \frac{\sqrt{2}}{4} \sum_{k=2}^{n} \frac{1}{k}.$$

10.10 a) Use the trigonometric identity $\sin 2x = 2 \sin x \cos x$, along with the formula $\int_0^\infty \sin x/x \, dx = \pi/2$, to show that

$$\int_0^\infty \frac{\sin x \cos x}{x} \, dx = \frac{\pi}{4}.$$

b) Use integration by parts in (a) to derive the formula

$$\int_0^\infty \frac{\sin^2 x}{x^2} \, dx = \frac{\pi}{2}.$$

c) Use the identity $\sin^2 x + \cos^2 x = 1$, along with (b), to obtain

$$\int_0^\infty \frac{\sin^4 x}{x^2} \, dx = \frac{\pi}{4}.$$

d) Use the result of (c) to obtain

$$\int_0^\infty \frac{\sin^4 x}{x^4} \, dx = \frac{\pi}{3}.$$

10.11 If $a > 1$, prove that the integral $\int_a^{+\infty} x^p (\log x)^q \, dx$ exists, both as an improper Riemann integral and as a Lebesgue integral for all q if $p < -1$, or for $q < -1$ if $p = -1$.

10.12 Prove that each of the following integrals exists, both as an improper Riemann integral and as a Lebesgue integral.

a) $\displaystyle\int_1^\infty \sin^2 \frac{1}{x} \, dx$,

b) $\displaystyle\int_0^\infty x^p e^{-x^q} \, dx \quad (p > 0, q > 0)$.

10.13 Determine whether or not each of the following integrals exists, either as an improper Riemann integral or as a Lebesgue integral.

a) $\displaystyle\int_0^\infty e^{-(t^2 + t^{-2})} \, dt$,

b) $\displaystyle\int_0^\infty \frac{\cos x}{\sqrt{x}} \, dx$,

c) $\displaystyle\int_1^\infty \frac{\log x}{x(x^2 - 1)^{1/2}} \, dx$,

d) $\displaystyle\int_0^\infty e^{-x} \sin \frac{1}{x} \, dx$,

e) $\displaystyle\int_0^1 \log x \sin \frac{1}{x} \, dx$,

f) $\displaystyle\int_0^\infty e^{-x} \log (\cos^2 x) \, dx$.

10.14 Determine those values of p and q for which the following Lebesgue integrals exist.

a) $\displaystyle\int_0^1 x^p(1 - x^2)^q \, dx,$

b) $\displaystyle\int_0^\infty x^x e^{-x^p} \, dx,$

c) $\displaystyle\int_0^\infty \frac{x^{p-1} - x^{q-1}}{1 - x} \, dx,$

d) $\displaystyle\int_0^\infty \frac{\sin (x^p)}{x^q} \, dx,$

e) $\displaystyle\int_0^\infty \frac{x^{p-1}}{1 + x^q} \, dx,$

f) $\displaystyle\int_\pi^\infty (\log x)^p (\sin x)^{-1/3} \, dx.$

10.15 Prove that the following improper Riemann integrals have the values indicated (m and n denote positive integers).

a) $\displaystyle\int_0^\infty \frac{\sin^{2n+1} x}{x} \, dx = \frac{\pi(2n)!}{2^{2n+1}(n!)^2},$

b) $\displaystyle\int_1^\infty \frac{\log x}{x^{n+1}} \, dx = n^{-2},$

c) $\displaystyle\int_0^\infty x^n(1 + x)^{-n-m-1} \, dx = \frac{n!(m - 1)!}{(m + n)!}.$

10.16 Given that f is Riemann-integrable on $[0, 1]$, that f is periodic with period 1, and that $\int_0^1 f(x) \, dx = 0$. Prove that the improper Riemann integral $\int_1^{+\infty} x^{-s} f(x) \, dx$ exists if $s > 0$. *Hint.* Let $g(x) = \int_1^x f(t) \, dt$ and write $\int_1^b x^{-s} f(x) \, dx = \int_1^b x^{-s} \, dg(x)$.

10.17 Assume that $f \in R$ on $[a, b]$ for every $b > a > 0$. Define g by the equation $xg(x) = \int_1^x f(t) \, dt$ if $x > 0$, assume that the limit $\lim_{x \to +\infty} g(x)$ exists, and denote this limit by B. If a and b are fixed positive numbers, prove that

a) $\displaystyle\int_a^b \frac{f(x)}{x} \, dx = g(b) - g(a) + \int_a^b \frac{g(x)}{x} \, dx.$

b) $\displaystyle\lim_{T \to +\infty} \int_{aT}^{bT} \frac{f(x)}{x} \, dx = B \log \frac{b}{a}.$

c) $\displaystyle\int_1^\infty \frac{f(ax) - f(bx)}{x} \, dx = B \log \frac{a}{b} + \int_a^b \frac{f(t)}{t} \, dt.$

d) Assume that the limit $\lim_{x \to 0+} x \int_x^1 f(t)t^{-2} \, dt$ exists, denote this limit by A, and prove that

$$\int_0^1 \frac{f(ax) - f(bx)}{x} \, dx = A \log \frac{b}{a} - \int_a^b \frac{f(t)}{t} \, dt.$$

e) Combine (c) and (d) to deduce

$$\int_0^\infty \frac{f(ax) - f(bx)}{x} \, dx = (B - A) \log \frac{a}{b}$$

and use this result to evaluate the following integrals:

$$\int_0^\infty \frac{\cos ax - \cos bx}{x} \, dx, \qquad \int_0^\infty \frac{e^{-ax} - e^{-bx}}{x} \, dx.$$

Lebesgue integrals

10.18 Prove that each of the following exists as a Lebesgue integral.

a) $\int_0^1 \dfrac{x \log x}{(1 + x)^2}\, dx,$

b) $\int_0^1 \dfrac{x^p - 1}{\log x}\, dx \quad (p > -1),$

c) $\int_0^1 \log x \log (1 + x)\, dx,$

d) $\int_0^1 \dfrac{\log (1 - x)}{(1 - x)^{1/2}}\, dx.$

10.19 Assume that f is continuous on $[0, 1]$, $f(0) = 0$, $f'(0)$ exists. Prove that the Lebesgue integral $\int_0^1 f(x)x^{-3/2}\, dx$ exists.

10.20 Prove that the integrals in (a) and (c) exist as Lebesgue integrals but that those in (b) and (d) do not.

a) $\int_0^\infty x^2 e^{-x^8 \sin^2 x}\, dx,$

b) $\int_0^\infty x^3 e^{-x^8 \sin^2 x}\, dx,$

c) $\int_1^\infty \dfrac{dx}{1 + x^4 \sin^2 x},$

d) $\int_1^\infty \dfrac{dx}{1 + x^2 \sin^2 x}.$

Hint. Obtain upper and lower bounds for the integrals over suitably chosen neighborhoods of the points $n\pi$ $(n = 1, 2, 3, \ldots)$.

Functions defined by integrals

10.21 Determine the set S of those real values of y for which each of the following integrals exists as a Lebesgue integral.

a) $\int_0^\infty \dfrac{\cos xy}{1 + x^2}\, dx,$

b) $\int_0^\infty (x^2 + y^2)^{-1}\, dx,$

c) $\int_0^\infty \dfrac{\sin^2 xy}{x^2}\, dx,$

d) $\int_0^\infty e^{-x^2} \cos 2xy\, dx.$

10.22 Let $F(y) = \int_0^\infty e^{-x^2} \cos 2xy\, dx$ if $y \in \mathbf{R}$. Show that F satisfies the differential equation $F'(y) + 2y F(y) = 0$ and deduce that $F(y) = \frac{1}{2}\sqrt{\pi}e^{-y^2}$. (Use the result $\int_0^\infty e^{-x^2}\, dx = \frac{1}{2}\sqrt{\pi}$, derived in Exercise 7.19.)

10.23 Let $F(y) = \int_0^\infty \sin xy/x(x^2 + 1)\, dx$ if $y > 0$. Show that F satisfies the differential equation $F''(y) - F(y) + \pi/2 = 0$ and deduce that $F(y) = \frac{1}{2}\pi(1 - e^{-y})$. Use this result to deduce the following equations, valid for $y > 0$ and $a > 0$:

$$\int_0^\infty \dfrac{\sin xy}{x(x^2 + a^2)}\, dx = \dfrac{\pi}{2a^2}(1 - e^{-ay}), \qquad \int_0^\infty \dfrac{\cos xy}{x^2 + a^2}\, dx = \dfrac{\pi e^{-ay}}{2a},$$

$$\int_0^\infty \dfrac{x \sin xy}{x^2 + a^2}\, dx = \dfrac{\pi}{2}e^{-ay}; \quad \text{you may use} \int_0^\infty \dfrac{\sin x}{x}\, dx = \dfrac{\pi}{2}.$$

10.24 Show that $\int_1^\infty [\int_1^\infty f(x, y)\, dx]\, dy \neq \int_1^\infty [\int_1^\infty f(x, y)\, dy]\, dx$ if

a) $f(x, y) = \dfrac{x - y}{(x + y)^3}.$

b) $f(x, y) = \dfrac{x^2 - y^2}{(x^2 + y^2)^2}.$

10.25 Show that the order of integration cannot be interchanged in the following integrals:

a) $\int_0^1 \left[\int_0^1 \frac{x-y}{(x+y)^3} \, dx \right] dy,$

b) $\int_0^1 \left[\int_1^\infty (e^{-xy} - 2e^{-2xy}) \, dy \right] dx.$

10.26 Let $f(x, y) = \int_0^\infty dt/[(1 + x^2 t^2)(1 + y^2 t^2)]$ if $(x, y) \neq (0, 0)$. Show (by methods of elementary calculus) that $f(x, y) = \frac{1}{2}\pi(x + y)^{-1}$. Evaluate the iterated integral $\int_0^1 [\int_0^1 f(x, y) \, dx] \, dy$ to derive the formula:

$$\int_0^\infty \frac{(\arctan x)^2}{x^2} \, dx = \pi \log 2.$$

10.27 Let $f(y) = \int_0^\infty \sin x \cos xy/x \, dx$ if $y \geq 0$. Show (by methods of elementary calculus) that $f(y) = \pi/2$ if $0 \leq y < 1$ and that $f(y) = 0$ if $y > 1$. Evaluate the integral $\int_0^a f(y) \, dy$ to derive the formula

$$\int_0^\infty \frac{\sin ax \sin x}{x^2} \, dx = \begin{cases} \dfrac{\pi a}{2} & \text{if } 0 \leq a \leq 1, \\[2mm] \dfrac{\pi}{2} & \text{if } a \geq 1. \end{cases}$$

10.28 a) If $s > 0$ and $a > 0$, show that the series

$$\sum_{n=1}^\infty \frac{1}{n} \int_a^\infty \frac{\sin 2n\pi x}{x^s} \, dx$$

converges and prove that

$$\lim_{a \to +\infty} \sum_{n=1}^\infty \frac{1}{n} \int_a^\infty \frac{\sin 2n\pi x}{x^s} \, dx = 0.$$

b) Let $f(x) = \sum_{n=1}^\infty \sin (2n\pi x)/n$. Show that

$$\int_0^\infty \frac{f(x)}{x^s} \, dx = (2\pi)^{s-1} \zeta(2 - s) \int_0^\infty \frac{\sin t}{t^s} \, dt, \qquad \text{if } 0 < s < 1,$$

where ζ denotes the Riemann zeta function.

10.29 a) Derive the following formula for the nth derivative of the Gamma function:

$$\Gamma^{(n)}(x) = \int_0^\infty e^{-t} t^{x-1} (\log t)^n \, dt \qquad (x > 0).$$

b) When $x = 1$, show that this can be written as follows:

$$\Gamma^{(n)}(1) = \int_0^1 (t^2 + (-1)^n e^{t - 1/t}) e^{-t} t^{-2} (\log t)^n \, dt.$$

c) Use (b) to show that $\Gamma^{(n)}(1)$ has the same sign as $(-1)^n$.

In Exercises 10.30 and 10.31, Γ denotes the Gamma function.

10.30 Use the result $\int_0^\infty e^{-x^2} \, dx = \frac{1}{2}\sqrt{\pi}$ to prove that $\Gamma(\frac{1}{2}) = \sqrt{\pi}$. Prove that $\Gamma(n + 1) = n!$ and that $\Gamma(n + \frac{1}{2}) = (2n)! \sqrt{\pi}/4^n n!$ if $n = 0, 1, 2, \ldots$

10.31 a) Show that for $x > 0$ we have the series representation

$$\Gamma(x) = \sum_{n=0}^{\infty} \frac{(-1)^n}{n!} \frac{1}{n+x} + \sum_{n=0}^{\infty} c_n x^n,$$

where $c_n = (1/n!) \int_1^{\infty} t^{-1} e^{-t} (\log t)^n \, dt$. *Hint:* Write $\int_0^{\infty} = \int_0^1 + \int_1^{\infty}$ and use an appropriate power series expansion in each integral.

b) Show that the power series $\sum_{n=0}^{\infty} c_n z^n$ converges for every complex z and that the series $\sum_{n=0}^{\infty} [(-1)^n/n!]/(n+z)$ converges for every complex $z \neq 0, -1, -2, \ldots$

10.32 Assume that f is of bounded variation on $[0, b]$ for every $b > 0$, and that $\lim_{x \to +\infty} f(x)$ exists. Denote this limit by $f(\infty)$ and prove that

$$\lim_{y \to 0+} y \int_0^{\infty} e^{-xy} f(x) \, dx = f(\infty).$$

Hint. Use integration by parts.

10.33 Assume that f is of bounded variation on $[0, 1]$. Prove that

$$\lim_{y \to 0+} y \int_0^1 x^{y-1} f(x) \, dx = f(0+).$$

Measurable functions

10.34 If f is Lebesgue-integrable on an open interval I and if $f'(x)$ exists almost everywhere on I, prove that f' is measurable on I.

10.35 a) Let $\{s_n\}$ be a sequence of step functions such that $s_n \to f$ everywhere on $\mathbf{R}$. Prove that, for every real a,

$$f^{-1}((a, +\infty)) = \bigcup_{n=1}^{\infty} \bigcap_{k=n}^{\infty} s_k^{-1}\left(\left(a + \frac{1}{n}, +\infty\right)\right).$$

b) If f is measurable on $\mathbf{R}$, prove that for every open subset A of $\mathbf{R}$ the set $f^{-1}(A)$ is measurable.

10.36 This exercise describes an example of a nonmeasurable set in $\mathbf{R}$. If x and y are real numbers in the interval $[0, 1]$, we say that x and y are equivalent, written $x \sim y$, whenever $x - y$ is rational. The relation $\sim$ is an equivalence relation, and the interval $[0, 1]$ can be expressed as a disjoint union of subsets (called equivalence classes) in each of which no two distinct points are equivalent. Choose a point from each equivalence class and let E be the set of points so chosen. We assume that E is measurable and obtain a contradiction. Let $A = \{r_1, r_2, \ldots\}$ denote the set of rational numbers in $[-1, 1]$ and let $E_n = \{r_n + x : x \in E\}$.

a) Prove that each E_n is measurable and that $\mu(E_n) = \mu(E)$.

b) Prove that $\{E_1, E_2, \ldots\}$ is a disjoint collection of sets whose union contains $[0, 1]$ and is contained in $[-1, 2]$.

c) Use parts (a) and (b) along with the countable additivity of Lebesgue measure to obtain a contradiction.

10.37 Refer to Exercise 10.36 and prove that the characteristic function χ_E is not measurable. Let $f = \chi_E - \chi_{I-E}$ where $I = [0, 1]$. Prove that $|f| \in L(I)$ but that $f \notin M(I)$. (Compare with Corollary 1 of Theorem 10.35.)

Square-integrable functions

In Exercises 10.38 through 10.42 all functions are assumed to be in $L^2(I)$. The L^2-norm $\|f\|$ is defined by the formula, $\|f\| = (\int_I |f|^2)^{1/2}$.

10.38 If $\lim_{n\to\infty} \|f_n - f\| = 0$, prove that $\lim_{n\to\infty} \|f_n\| = \|f\|$.

10.39 If $\lim_{n\to\infty} \|f_n - f\| = 0$ and if $\lim_{n\to\infty} f_n(x) = g(x)$ almost everywhere on I, prove that $f(x) = g(x)$ almost everywhere on I.

10.40 If $f_n \to f$ uniformly on a compact interval I, and if each f_n is continuous on I, prove that $\lim_{n\to\infty} \|f_n - f\| = 0$.

10.41 If $\lim_{n\to\infty} \|f_n - f\| = 0$, prove that $\lim_{n\to\infty} \int_I f_n \cdot g = \int_I f \cdot g$ for every g in $L^2(I)$.

10.42 If $\lim_{n\to\infty} \|f_n - f\| = 0$ and $\lim_{n\to\infty} \|g_n - g\| = 0$, prove that $\lim_{n\to\infty} \int_I f_n \cdot g_n = \int_I f \cdot g$.

SUGGESTED REFERENCES FOR FURTHER STUDY

10.1 Asplund, E., and Bungart, L., *A First Course in Integration*. Holt, Rinehart and Winston, New York, 1966.

10.2 Bartle, R., *The Elements of Integration*. Wiley, New York, 1966.

10.3 Burkill, J. C., *The Lebesgue Integral*. Cambridge University Press, 1951.

10.4 Halmos, P., *Measure Theory*. Van Nostrand, New York, 1950.

10.5 Hawkins, T., *Lebesgue's Theory of Integration: Its Origin and Development*. University of Wisconsin Press, Madison, 1970.

10.6 Hildebrandt, T. H., *Introduction to the Theory of Integration*. Academic Press, New York, 1963.

10.7 Kestelman, H., *Modern Theories of Integration*. Oxford University Press, 1937.

10.8 Korevaar, J., *Mathematical Methods*, Vol. 1. Academic Press, New York, 1968.

10.9 Munroe, M. E., *Measure and Integration*, 2nd ed. Addison-Wesley, Reading, 1971.

10.10 Riesz, F., and Sz.-Nagy, B., *Functional Analysis*. L. Boron, translator. Ungar, New York, 1955.

10.11 Rudin, W., *Principles of Mathematical Analysis*, 2nd ed. McGraw-Hill, New York, 1964.

10.12 Shilov, G. E., and Gurevich, B. L., *Integral, Measure and Derivative: A Unified Approach*. Prentice-Hall, Englewood Cliffs, 1966.

10.13 Taylor, A. E., *General Theory of Functions and Integration*. Blaisdell, New York, 1965.

10.14 Zaanen, A. C., *Integration*. North-Holland, Amsterdam, 1967.

CHAPTER 11

FOURIER SERIES
AND FOURIER INTEGRALS

11.1 INTRODUCTION

In 1807, Fourier astounded some of his contemporaries by asserting that an "arbitrary" function could be expressed as a linear combination of sines and cosines. These linear combinations, now called *Fourier series*, have become an indispensable tool in the analysis of certain periodic phenomena (such as vibrations, and planetary and wave motion) which are studied in physics and engineering. Many important mathematical questions have also arisen in the study of Fourier series, and it is a remarkable historical fact that much of the development of modern mathematical analysis has been profoundly influenced by the search for answers to these questions. For a brief but excellent account of the history of this subject and its impact on the development of mathematics see Reference 11.1.

11.2 ORTHOGONAL SYSTEMS OF FUNCTIONS

The basic problems in the theory of Fourier series are best described in the setting of a more general discipline known as the theory of orthogonal functions. Therefore we begin by introducing some terminology concerning orthogonal functions.

NOTE. As in the previous chapter, we shall consider functions defined on a general subinterval I of $\mathbf{R}$. The interval may be bounded, unbounded, open, closed, or half-open. We denote by $L^2(I)$ the set of all complex-valued functions f which are measurable on I and are such that $|f|^2 \in L(I)$. The inner product (f, g) of two such functions, defined by

$$(f, g) = \int_I f(x)\overline{g(x)}\, dx,$$

always exists. The nonnegative number $\|f\| = (f, f)^{1/2}$ is the L^2-norm of f.

Definition 11.1. *Let* $S = \{\varphi_0, \varphi_1, \varphi_2, \ldots\}$ *be a collection of functions in* $L^2(I)$. *If*

$$(\varphi_n, \varphi_m) = 0 \qquad whenever\ m \neq n,$$

the collection S is said to be an orthogonal system on I. If, in addition, each φ_n *has norm 1, then S is said to be orthonormal on I.*

NOTE. Every orthogonal system for which each $\|\varphi_n\| \neq 0$ can be converted into an orthonormal system by dividing each φ_n by its norm.

We shall be particularly interested in the special trigonometric system $S = \{\varphi_0, \varphi_1, \varphi_2, \ldots\}$, where

$$\varphi_0(x) = \frac{1}{\sqrt{2\pi}}, \qquad \varphi_{2n-1}(x) = \frac{\cos nx}{\sqrt{\pi}}, \qquad \varphi_{2n}(x) = \frac{\sin nx}{\sqrt{\pi}} \qquad (1)$$

for $n = 1, 2, \ldots$ It is a simple matter to verify that S is orthonormal on any interval of length 2π. (See Exercise 11.1.) The system in (1) consists of real-valued functions. An orthonormal system of complex-valued functions on every interval of length 2π is given by

$$\varphi_n(x) = \frac{e^{inx}}{\sqrt{2\pi}} = \frac{\cos nx + i \sin nx}{\sqrt{2\pi}}, \qquad n = 0, 1, 2, \ldots$$

11.3 THE THEOREM ON BEST APPROXIMATION

One of the basic problems in the theory of orthogonal functions is to approximate a given function f in $L^2(I)$ as closely as possible by linear combinations of elements of an orthonormal system.   More precisely, let $S = \{\varphi_0, \varphi_1, \varphi_2, \ldots\}$ be orthonormal on I and let

$$t_n(x) = \sum_{k=0}^{n} b_k \varphi_k(x),$$

where $b_0, b_1, \ldots, b_n$ are arbitrary complex numbers. We use the norm $\|f - t_n\|$ as a measure of the error made in approximating f by t_n. The first task is to choose the constants $b_0, \ldots, b_n$ so that this error will be as small as possible. The next theorem shows that there is a unique choice of the constants that minimizes this error.

To motivate the results in the theorem we consider the most favorable case. If f is already a linear combination of $\varphi_0, \varphi_1, \ldots, \varphi_n$, say

$$f = \sum_{k=0}^{n} c_k \varphi_k,$$

then the choice $t_n = f$ will make $\|f - t_n\| = 0$. We can determine the constants $c_0, \ldots, c_n$ as follows. Form the inner product (f, φ_m), where $0 \le m \le n$. Using the properties of inner products we have

$$(f, \varphi_m) = \left(\sum_{k=0}^{n} c_k \varphi_k, \varphi_m \right) = \sum_{k=0}^{n} c_k (\varphi_k, \varphi_m) = c_m,$$

since $(\varphi_k, \varphi_m) = 0$ if $k \ne m$ and $(\varphi_m, \varphi_m) = 1$.   In other words, in the most favorable case we have $c_m = (f, \varphi_m)$ for $m = 0, 1, \ldots, n$. The next theorem shows that this choice of constants is best for all functions in $L^2(I)$.

Theorem 11.2. *Let* $\{\varphi_0, \varphi_1, \varphi_2, \ldots\}$ *be orthonormal on* I, *and assume that* $f \in L^2(I)$. *Define two sequences of functions* $\{s_n\}$ *and* $\{t_n\}$ *on* I *as follows:*

$$s_n(x) = \sum_{k=0}^{n} c_k \varphi_k(x), \qquad t_n(x) = \sum_{k=0}^{n} b_k \varphi_k(x),$$

where

$$c_k = (f, \varphi_k) \qquad \text{for } k = 0, 1, 2, \ldots, \tag{2}$$

and $b_0, b_1, b_2, \ldots$, *are arbitrary complex numbers. Then for each* n *we have*

$$\|f - s_n\| \leq \|f - t_n\|. \tag{3}$$

Moreover, equality holds in (3) *if, and only if,* $b_k = c_k$ *for* $k = 0, 1, \ldots, n$.

Proof. We shall deduce (3) from the equation

$$\|f - t_n\|^2 = \|f\|^2 - \sum_{k=0}^{n} |c_k|^2 + \sum_{k=0}^{n} |b_k - c_k|^2. \tag{4}$$

It is clear that (4) implies (3) because the right member of (4) has its smallest value when $b_k = c_k$ for each k. To prove (4), write

$$\|f - t_n\|^2 = (f - t_n, f - t_n) = (f, f) - (f, t_n) - (t_n, f) + (t_n, t_n).$$

Using the properties of inner products we find

$$(t_n, t_n) = \left(\sum_{k=0}^{n} b_k \varphi_k, \sum_{m=0}^{n} b_m \varphi_m \right)$$

$$= \sum_{k=0}^{n} \sum_{m=0}^{n} b_k \bar{b}_m (\varphi_k, \varphi_m) = \sum_{k=0}^{n} |b_k|^2,$$

and

$$(f, t_n) = \left(f, \sum_{k=0}^{n} b_k \varphi_k \right) = \sum_{k=0}^{n} \bar{b}_k (f, \varphi_k) = \sum_{k=0}^{n} \bar{b}_k c_k.$$

Also, $(t_n, f) = \overline{(f, t_n)} = \sum_{k=0}^{n} b_k \bar{c}_k$, and hence

$$\|f - t_n\|^2 = \|f\|^2 - \sum_{k=0}^{n} \bar{b}_k c_k - \sum_{k=0}^{n} b_k \bar{c}_k + \sum_{k=0}^{n} |b_k|^2$$

$$= \|f\|^2 - \sum_{k=0}^{n} |c_k|^2 + \sum_{k=0}^{n} (b_k - c_k)(\bar{b}_k - \bar{c}_k)$$

$$= \|f\|^2 - \sum_{k=0}^{n} |c_k|^2 + \sum_{k=0}^{n} |b_k - c_k|^2.$$

11.4 THE FOURIER SERIES OF A FUNCTION RELATIVE TO AN ORTHONORMAL SYSTEM

Definition 11.3. *Let* $S = \{\varphi_0, \varphi_1, \varphi_2, \ldots\}$ *be orthonormal on I and assume that* $f \in L^2(I)$. *The notation*

$$f(x) \sim \sum_{n=0}^{\infty} c_n \varphi_n(x) \tag{5}$$

will mean that the numbers $c_0, c_1, c_2, \ldots$ *are given by the formulas:*

$$c_n = (f, \varphi_n) = \int_I f(x)\overline{\varphi_n(x)}\, dx \qquad (n = 0, 1, 2, \ldots). \tag{6}$$

The series in (5) *is called the Fourier series of f relative to S, and the numbers* $c_0, c_1, c_2, \ldots$ *are called the Fourier coefficients of f relative to S.*

NOTE. When $I = [0, 2\pi]$ and S is the system of trigonometric functions described in (1), the series is called simply *the Fourier series generated by f.* We then write (5) in the form

$$f(x) \sim \frac{a_0}{2} + \sum_{n=1}^{\infty} (a_n \cos nx + b_n \sin nx),$$

the coefficients being given by the following formulas:

$$a_n = \frac{1}{\pi} \int_0^{2\pi} f(t) \cos nt\, dt, \qquad b_n = \frac{1}{\pi} \int_0^{2\pi} f(t) \sin nt\, dt. \tag{7}$$

In this case the integrals for a_n and b_n exist if $f \in L([0, 2\pi])$.

11.5 PROPERTIES OF THE FOURIER COEFFICIENTS

Theorem 11.4. *Let* $\{\varphi_0, \varphi_1, \varphi_2, \ldots\}$ *be orthonormal on I, assume that* $f \in L^2(I)$, *and suppose that*

$$f(x) \sim \sum_{n=0}^{\infty} c_n \varphi_n(x).$$

Then

a) *The series* $\sum |c_n|^2$ *converges and satisfies the inequality*

$$\sum_{n=0}^{\infty} |c_n|^2 \le \|f\|^2 \qquad (Bessel's\ inequality). \tag{8}$$

b) *The equation*

$$\sum_{n=0}^{\infty} |c_n|^2 = \|f\|^2 \qquad (Parseval's\ formula)$$

holds if, and only if, we also have

$$\lim_{n \to \infty} \| f - s_n \| = 0,$$

where $\{s_n\}$ is the sequence of partial sums defined by

$$s_n(x) = \sum_{k=0}^{n} c_k \varphi_k(x).$$

Proof. We take $b_k = c_k$ in (4) and observe that the left member is nonnegative. Therefore

$$\sum_{k=0}^{n} |c_k|^2 \leq \| f \|^2.$$

This establishes (a). To prove (b), we again put $b_k = c_k$ in (4) to obtain

$$\| f - s_n \|^2 = \| f \|^2 - \sum_{k=0}^{n} |c_k|^2.$$

Part (b) follows at once from this equation.

As a further consequence of part (a) of Theorem 11.4 we observe that the Fourier coefficients c_n tend to 0 as $n \to \infty$ (since $\sum |c_n|^2$ converges). In particular, when $\varphi_n(x) = e^{inx}/\sqrt{2\pi}$ and $I = [0, 2\pi]$ we find

$$\lim_{n \to \infty} \int_0^{2\pi} f(x) e^{-inx} \, dx = 0,$$

from which we obtain the important formulas

$$\lim_{n \to \infty} \int_0^{2\pi} f(x) \cos nx \, dx = \lim_{n \to \infty} \int_0^{2\pi} f(x) \sin nx \, dx = 0. \tag{9}$$

These formulas are also special cases of the Riemann–Lebesgue lemma (Theorem 11.6).

NOTE. The Parseval formula

$$\| f \|^2 = |c_0|^2 + |c_1|^2 + |c_2|^2 + \cdots$$

is analogous to the formula

$$\| \mathbf{x} \|^2 = x_1^2 + x_2^2 + \cdots + x_n^2$$

for the length of a vector $\mathbf{x} = (x_1, \ldots, x_n)$ in $\mathbf{R}^n$. Each of these can be regarded as a generalization of the Pythagorean theorem for right triangles.

11.6 THE RIESZ–FISCHER THEOREM

The converse to part (a) of Theorem 11.4 is called the Riesz–Fischer theorem.

Theorem 11.5. *Assume* $\{\varphi_0, \varphi_1, \dots\}$ *is orthonormal on I. Let* $\{c_n\}$ *be any sequence of complex numbers such that* $\sum |c_k|^2$ *converges. Then there is a function f in* $L^2(I)$ *such that*

a) $(f, \varphi_k) = c_k$ *for each* $k \geq 0$,

and

b) $\|f\|^2 = \displaystyle\sum_{k=0}^{\infty} |c_k|^2.$

Proof. Let

$$s_n(x) = \sum_{k=0}^{n} c_k \, \varphi_k(x).$$

We will prove that there is a function f in $L^2(I)$ such that $(f, \varphi_k) = c_k$ and such that

$$\lim_{n \to \infty} \|s_n - f\| = 0.$$

Part (b) of Theorem 11.4 then implies part (b) of Theorem 11.5.

First we note that $\{s_n\}$ is a Cauchy sequence in the semimetric space $L^2(I)$ because, if $m > n$ we have

$$\|s_n - s_m\|^2 = \sum_{k=n+1}^{m} \sum_{r=n+1}^{m} c_k \bar{c}_r (\varphi_k, \varphi_r)$$

$$= \sum_{k=n+1}^{m} |c_k|^2,$$

and the last sum can be made less than ε if m and n are sufficiently large. By Theorem 10.57 there is a function f in $L^2(I)$ such that

$$\lim_{n \to \infty} \|s_n - f\| = 0.$$

To show that $(f, \varphi_k) = c_k$ we note that $(s_n, \varphi_k) = c_k$ if $n \geq k$, and use the Cauchy–Schwarz inequality to obtain

$$|c_k - (f, \varphi_k)| = |(s_n, \varphi_k) - (f, \varphi_k)| = |(s_n - f, \varphi_k)| \leq \|s_n - f\|.$$

Since $\|s_n - f\| \to 0$ as $n \to \infty$ this proves (a).

NOTE. The proof of this theorem depends on the fact that the semimetric space $L^2(I)$ is complete. There is no corresponding theorem for functions whose squares are Riemann-integrable.

11.7 THE CONVERGENCE AND REPRESENTATION PROBLEMS FOR TRIGONOMETRIC SERIES

Consider the trigonometric Fourier series generated by a function f which is Lebesgue-integrable on the interval $I = [0, 2\pi]$, say

$$f(x) \sim \frac{a_0}{2} + \sum_{n=1}^{\infty} (a_n \cos nx + b_n \sin nx).$$

Two questions arise. Does the series converge at some point x in I? If it does converge at x, is its sum $f(x)$? The first question is called the convergence problem; the second, the representation problem. In general, the answer to both questions is "No." In fact, there exist Lebesgue-integrable functions whose Fourier series diverge everywhere, and there exist continuous functions whose Fourier series diverge on an uncountable set.

Ever since Fourier's time, an enormous literature has been published on these problems. The object of much of the research has been to find sufficient conditions to be satisfied by f in order that its Fourier series may converge, either throughout the interval or at particular points. We shall prove later that the convergence or divergence of the series at a particular point depends only on the behavior of the function in arbitrarily small neighborhoods of the point. (See Theorem 11.11, Riemann's localization theorem.)

The efforts of Fourier and Dirichlet in the early nineteenth century, followed by the contributions of Riemann, Lipschitz, Heine, Cantor, Du Bois-Reymond, Dini, Jordan, and de la Vallée–Poussin in the latter part of the century, led to the discovery of sufficient conditions of a wide scope for establishing convergence of the series, either at particular points, or generally, throughout the interval.

After the discovery by Lebesgue, in 1902, of his general theory of measure and integration, the field of investigation was considerably widened and the names chiefly associated with the subject since then are those of Fejér, Hobson, W. H. Young, Hardy, and Littlewood. Fejér showed, in 1903, that divergent Fourier series may be utilized by considering, instead of the sequence of partial sums $\{s_n\}$, the sequence of arithmetic means $\{\sigma_n\}$, where

$$\sigma_n(x) = \frac{s_0(x) + s_1(x) + \cdots + s_{n-1}(x)}{n}.$$

He established the remarkable theorem that the sequence $\{\sigma_n(x)\}$ is convergent and its limit is $\frac{1}{2}[f(x+) + f(x-)]$ at every point in $[0, 2\pi]$ where $f(x+)$ and $f(x-)$ exist, the only restriction on f being that it be Lebesgue-integrable on $[0, 2\pi]$ (Theorem 11.15.). Fejér also proved that every Fourier series, whether it converges or not, can be integrated term-by-term (Theorem 11.16.) The most striking result on Fourier series proved in recent times is that of Lennart Carleson, a Swedish mathematician, who proved that the Fourier series of a function in $L^2(I)$ converges almost everywhere on I. (*Acta Mathematica*, **116** (1966), pp. 135–157.)

In this chapter we shall deduce some of the sufficient conditions for convergence of a Fourier series at a particular point. Then we shall prove Fejér's theorems. The discussion rests on two fundamental limit formulas which will be discussed first. These limit formulas, which are also used in the theory of Fourier integrals, deal with integrals depending on a real parameter α, and we are interested in the behavior of these integrals as $\alpha \to +\infty$. The first of these is a generalization of (9) and is known as the Riemann–Lebesgue lemma.

11.8 THE RIEMANN–LEBESGUE LEMMA

Theorem 11.6. *Assume that $f \in L(I)$. Then, for each real β, we have*

$$\lim_{\alpha \to +\infty} \int_I f(t) \sin (\alpha t + \beta)\, dt = 0. \tag{10}$$

Proof. If f is the characteristic function of a compact interval $[a, b]$ the result is obvious since we have

$$\left| \int_a^b \sin (\alpha t + \beta)\, dt \right| = \left| \frac{\cos (a\alpha + \beta) - \cos (b\alpha + \beta)}{\alpha} \right| \le \frac{2}{\alpha}, \qquad \text{if } \alpha > 0.$$

The result also holds if f is constant on the open interval (a, b) and zero outside $[a, b]$, regardless of how we define $f(a)$ and $f(b)$. Therefore (10) is valid if f is a step function. But now it is easy to prove (10) for every Lebesgue-integrable function f.

If $\varepsilon > 0$ is given, there exists a step function s such that $\int_I |f - s| < \varepsilon/2$ (by Theorem 10.19(b)). Since (10) holds for step functions, there is a positive M such that

$$\left| \int_I s(t) \sin (\alpha t + \beta)\, dt \right| < \frac{\varepsilon}{2} \qquad \text{if } \alpha \ge M.$$

Therefore, if $\alpha \ge M$ we have

$$\left| \int_I f(t) \sin (\alpha t + \beta)\, dt \right| \le \left| \int_I (f(t) - s(t)) \sin (\alpha t + \beta)\, dt \right|$$

$$+ \left| \int_I s(t) \sin (\alpha t + \beta)\, dt \right|$$

$$\le \int_I |f(t) - s(t)|\, dt + \frac{\varepsilon}{2} < \frac{\varepsilon}{2} + \frac{\varepsilon}{2} = \varepsilon.$$

This completes the proof of the Riemann–Lebesgue lemma.

Example. Taking $\beta = 0$ and $\beta = \pi/2$, we find, if $f \in L(I)$,

$$\lim_{\alpha \to +\infty} \int_I f(t) \sin \alpha t\, dt = \lim_{\alpha \to +\infty} \int_I f(t) \cos \alpha t\, dt = 0.$$

As an application of the Riemann–Lebesgue lemma we derive a result that will be needed in our discussion of Fourier integrals.

Theorem 11.7. *If $f \in L(-\infty, +\infty)$, we have*

$$\lim_{\alpha \to +\infty} \int_{-\infty}^{\infty} f(t) \frac{1 - \cos \alpha t}{t} \, dt = \int_{0}^{\infty} \frac{f(t) - f(-t)}{t} \, dt, \tag{11}$$

whenever the Lebesgue integral on the right exists.

Proof. For each fixed α, the integral on the left of (11) exists as a Lebesgue integral since the quotient $(1 - \cos \alpha t)/t$ is continuous and bounded on $(-\infty, +\infty)$. (At $t = 0$ the quotient is to be replaced by 0, its limit as $t \to 0$.) Hence we can write

$$\int_{-\infty}^{\infty} f(t) \frac{1 - \cos \alpha t}{t} \, dt = \int_{0}^{\infty} f(t) \frac{1 - \cos \alpha t}{t} \, dt + \int_{-\infty}^{0} f(t) \frac{1 - \cos \alpha t}{t} \, dt$$

$$= \int_{0}^{\infty} [f(t) - f(-t)] \frac{1 - \cos \alpha t}{t} \, dt$$

$$= \int_{0}^{\infty} \frac{f(t) - f(-t)}{t} \, dt - \int_{0}^{\infty} \frac{f(t) - f(-t)}{t} \cos \alpha t \, dt.$$

When $\alpha \to +\infty$, the last integral tends to 0, by the Riemann–Lebesgue lemma.

11.9 THE DIRICHLET INTEGRALS

Integrals of the form $\int_{0}^{\delta} g(t)(\sin \alpha t)/t \, dt$ (called Dirichlet integrals) play an important role in the theory of Fourier series and also in the theory of Fourier integrals. The function g in the integrand is assumed to have a finite right-hand limit $g(0+) = \lim_{t \to 0+} g(t)$ and we are interested in formulating further conditions on g which will guarantee the validity of the following equation:

$$\lim_{\alpha \to +\infty} \frac{2}{\pi} \int_{0}^{\delta} g(t) \frac{\sin \alpha t}{t} \, dt = g(0+). \tag{12}$$

To get an idea why we might expect a formula like (12) to hold, let us first consider the case when g is constant $(g(t) = g(0+))$ on $[0, \delta]$. Then (12) is a trivial consequence of the equation $\int_{0}^{\infty} (\sin t)/t \, dt = \pi/2$ (see Example 3, Section 10.16), since

$$\int_{0}^{\delta} \frac{\sin \alpha t}{t} \, dt = \int_{0}^{\alpha \delta} \frac{\sin t}{t} \, dt \to \frac{\pi}{2} \qquad \text{as } \alpha \to +\infty.$$

More generally, if $g \in L([0, \delta])$, and if $0 < \varepsilon < \delta$, we have

$$\lim_{\alpha \to +\infty} \frac{2}{\pi} \int_{\varepsilon}^{\delta} g(t) \frac{\sin \alpha t}{t} \, dt = 0,$$

by the Riemann–Lebesgue lemma. Hence the validity of (12) is governed entirely by the *local behavior* of g near 0. Since $g(t)$ is nearly $g(0+)$ when t is near 0, there is some hope of proving (12) without placing too many additional restrictions on g. It would seem that continuity of g at 0 should certainly be enough to insure the existence of the limit in (12). Dirichlet showed that continuity of g on $[0, \delta]$ is sufficient to prove (12), if, in addition, g has only a finite number of maxima or minima on $[0, \delta]$. Jordan later proved (12) under the less restrictive condition that g be of bounded variation on $[0, \delta]$. However, all attempts to prove (12) under the sole hypothesis that g is continuous on $[0, \delta]$ have resulted in failure. In fact, Du Bois–Reymond discovered an example of a continuous function g for which the limit in (12) fails to exist. Jordan's result, and a related theorem due to Dini, will be discussed here.

Theorem 11.8 (Jordan). *If g is of bounded variation on $[0, \delta]$, then*

$$\lim_{\alpha \to +\infty} \frac{2}{\pi} \int_0^\delta g(t) \frac{\sin \alpha t}{t} \, dt = g(0+). \tag{13}$$

Proof. It suffices to consider the case in which g is increasing on $[0, \delta]$. If $\alpha > 0$ and if $0 < h < \delta$, we have

$$\int_0^\delta g(t) \frac{\sin \alpha t}{t} \, dt = \int_0^h [g(t) - g(0+)] \frac{\sin \alpha t}{t} \, dt$$

$$+ g(0+) \int_0^h \frac{\sin \alpha t}{t} \, dt + \int_h^\delta g(t) \frac{\sin \alpha t}{t} \, dt$$

$$= I_1(\alpha, h) + I_2(\alpha, h) + I_3(\alpha, h), \tag{14}$$

let us say. We can apply the Riemann–Lebesgue lemma to $I_3(\alpha, h)$ (since the integral $\int_h^\delta g(t)/t \, dt$ exists) and we find $I_3(\alpha, h) \to 0$ as $\alpha \to +\infty$. Also,

$$I_2(\alpha, h) = g(0+) \int_0^h \frac{\sin \alpha t}{t} \, dt$$

$$= g(0+) \int_0^{h\alpha} \frac{\sin t}{t} \, dt \to \frac{\pi}{2} g(0+) \qquad \text{as } \alpha \to +\infty.$$

Next, choose $M > 0$ so that $|\int_a^b (\sin t)/t \, dt| < M$ for every $b \geq a \geq 0$. It follows that $|\int_a^b (\sin \alpha t)/t \, dt| < M$ for every $b \geq a \geq 0$ if $\alpha > 0$. Now let $\varepsilon > 0$ be given and choose h in $(0, \delta)$ so that $|g(h) - g(0+)| < \varepsilon/(3M)$. Since

$$g(t) - g(0+) \geq 0 \qquad \text{if } 0 \leq t \leq h,$$

we can apply Bonnet's theorem (Theorem 7.37) in $I_1(\alpha, h)$ to write

$$I_1(\alpha, h) = \int_0^h [g(t) - g(0+)] \frac{\sin \alpha t}{t} \, dt = [g(h) - g(0+)] \int_c^h \frac{\sin \alpha t}{t} \, dt,$$

where $c \in [0, h]$. The definition of h gives us

$$|I_1(\alpha, h)| = |g(h) - g(0+)| \left| \int_c^h \frac{\sin \alpha t}{t} \, dt \right| < \frac{\varepsilon}{3M} M = \frac{\varepsilon}{3}. \tag{15}$$

For the same h we can choose A so that $\alpha \geq A$ implies

$$|I_3(\alpha, h)| < \frac{\varepsilon}{3} \quad \text{and} \quad \left| I_2(\alpha, h) - \frac{\pi}{2} g(0+) \right| < \frac{\varepsilon}{3}. \tag{16}$$

Then, for $\alpha \geq A$, we can combine (14), (15), and (16) to get

$$\left| \int_0^\delta g(t) \frac{\sin \alpha t}{t} \, dt - \frac{\pi}{2} g(0+) \right| < \varepsilon.$$

This proves (13).

A different kind of condition for the validity of (13) was found by Dini and can be described as follows:

Theorem 11.9 (Dini). *Assume that* $g(0+)$ *exists and suppose that for some* $\delta > 0$ *the Lebesgue integral*

$$\int_0^\delta \frac{g(t) - g(0+)}{t} \, dt$$

exists. Then we have

$$\lim_{\alpha \to +\infty} \frac{2}{\pi} \int_0^\delta g(t) \frac{\sin \alpha t}{t} \, dt = g(0+).$$

Proof. Write

$$\int_0^\delta g(t) \frac{\sin \alpha t}{t} \, dt = \int_0^\delta \frac{g(t) - g(0+)}{t} \sin \alpha t \, dt + g(0+) \int_0^{\alpha \delta} \frac{\sin t}{t} \, dt.$$

When $\alpha \to +\infty$, the first term on the right tends to 0 (by the Riemann–Lebesgue lemma) and the second term tends to $\frac{1}{2}\pi g(0+)$.

NOTE. If $g \in L([a, \delta])$ for every positive $a < \delta$, it is easy to show that Dini's condition is satisfied whenever g satisfies a "right-handed" Lipschitz condition at 0; that is, whenever there exist two positive constants M and p such that

$$|g(t) - g(0+)| < Mt^p, \quad \text{for every } t \text{ in } (0, \delta].$$

(See Exercise 11.21.) In particular, the Lipschitz condition holds with $p = 1$ whenever g has a righthand derivative at 0. It is of interest to note that there exist functions which satisfy Dini's condition but which do not satisfy Jordan's condition. Similarly, there are functions which satisfy Jordan's condition but not Dini's. (See Reference 11.10.)

11.10 AN INTEGRAL REPRESENTATION FOR THE PARTIAL SUMS OF A FOURIER SERIES

A function f is said to be *periodic* with period $p \neq 0$ if f is defined on **R** and if $f(x + p) = f(x)$ for all x. The next theorem expresses the partial sums of a Fourier series in terms of the function

$$D_n(t) = \tfrac{1}{2} + \sum_{k=1}^{n} \cos kt = \begin{cases} \dfrac{\sin{(n + \frac{1}{2})t}}{2 \sin t/2} & \text{if } t \neq 2m\pi \quad (m \text{ an integer}), \\[2mm] n + \tfrac{1}{2} & \text{if } t = 2m\pi \quad (m \text{ an integer}). \end{cases} \qquad (17)$$

This formula was discussed in Section 8.16 in connection with the partial sums of the geometric series. The function D_n is called *Dirichlet's kernel*.

Theorem 11.10. *Assume that $f \in L([0, 2\pi])$ and suppose that f is periodic with period 2π. Let $\{s_n\}$ denote the sequence of partial sums of the Fourier series generated by f, say*

$$s_n(x) = \frac{a_0}{2} + \sum_{k=1}^{n} (a_k \cos kx + b_k \sin kx), \qquad (n = 1, 2, \dots). \qquad (18)$$

Then we have the integral representation

$$s_n(x) = \frac{2}{\pi} \int_0^\pi \frac{f(x + t) + f(x - t)}{2} D_n(t)\, dt. \qquad (19)$$

Proof. The Fourier coefficients of f are given by the integrals in (7). Substituting these integrals in (18) we find

$$s_n(x) = \frac{1}{\pi} \int_0^{2\pi} f(t) \left\{ \frac{1}{2} + \sum_{k=1}^{n} (\cos kt \cos kx + \sin kt \sin kx) \right\} dt$$

$$= \frac{1}{\pi} \int_0^{2\pi} f(t) \left\{ \frac{1}{2} + \sum_{k=1}^{n} \cos k(t - x) \right\} dt = \frac{1}{\pi} \int_0^{2\pi} f(t) D_n(t - x)\, dt.$$

Since both f and D_n are periodic with period 2π, we can replace the interval of integration by $[x - \pi, x + \pi]$ and then make a translation $u = t - x$ to get

$$s_n(x) = \frac{1}{\pi} \int_{x-\pi}^{x+\pi} f(t) D_n(t - x)\, dt$$

$$= \frac{1}{\pi} \int_{-\pi}^{\pi} f(x + u) D_n(u)\, du.$$

Using the equation $D_n(-u) = D_n(u)$, we obtain (19).

11.11 RIEMANN'S LOCALIZATION THEOREM

Formula (19) tells us that the Fourier series generated by f will converge at a point x if, and only if, the following limit exists:

$$\lim_{n \to \infty} \frac{2}{\pi} \int_0^\pi \frac{f(x+t) + f(x-t)}{2} \frac{\sin(n+\tfrac{1}{2})t}{2 \sin \tfrac{1}{2}t} \, dt, \tag{20}$$

in which case the value of this limit will be the sum of the series. This integral is essentially a Dirichlet integral of the type discussed in the previous section, except that $2 \sin \tfrac{1}{2}t$ appears in the denominator rather than t. However, the Riemann–Lebesgue lemma allows us to replace $2 \sin \tfrac{1}{2}t$ by t in (20) without affecting either the existence or the value of the limit. More precisely, the Riemann–Lebesgue lemma implies

$$\lim_{n \to \infty} \frac{2}{\pi} \int_0^\pi \left(\frac{1}{t} - \frac{1}{2 \sin \tfrac{1}{2}t} \right) \frac{f(x+t) + f(x-t)}{2} \sin(n+\tfrac{1}{2})t \, dt = 0,$$

because the function F defined by the equation

$$F(t) = \begin{cases} \dfrac{1}{t} - \dfrac{1}{2 \sin \tfrac{1}{2}t} & \text{if } 0 < t \le \pi, \\ 0 & \text{if } t = 0, \end{cases}$$

is continuous on $[0, \pi]$. Therefore the convergence problem for Fourier series amounts to finding conditions on f which will guarantee the existence of the following limit:

$$\lim_{n \to \infty} \frac{2}{\pi} \int_0^\pi \frac{f(x+t) + f(x-t)}{2} \frac{\sin(n+\tfrac{1}{2})t}{t} \, dt. \tag{21}$$

Using the Riemann–Lebesgue lemma once more, we need only consider the limit in (21) when the integral $\int_0^\pi$ is replaced by $\int_0^\delta$, where δ is any positive number $< \pi$, because the integral $\int_\delta^\pi$ tends to 0 as $n \to \infty$. Therefore we can sum up the results of the previous section in the following theorem:

Theorem 11.11. *Assume that $f \in L([0, 2\pi])$ and suppose f has period 2π. Then the Fourier series generated by f will converge for a given value of x if, and only if, for some positive $\delta < \pi$ the following limit exists:*

$$\lim_{n \to \infty} \frac{2}{\pi} \int_0^\delta \frac{f(x+t) + f(x-t)}{2} \frac{\sin(n+\tfrac{1}{2})t}{t} \, dt, \tag{22}$$

in which case the value of this limit is the sum of the Fourier series.

This theorem is known as *Riemann's localization theorem.* It tells us that the convergence or divergence of a Fourier series at a particular point is governed entirely by the behavior of f in an arbitrarily small neighborhood of the point. This is rather surprising in view of the fact that the coefficients of the Fourier series depend on the values which the function assumes throughout the entire interval $[0, 2\pi]$.

11.12 SUFFICIENT CONDITIONS FOR CONVERGENCE OF A FOURIER SERIES AT A PARTICULAR POINT

Assume that $f \in L([0, 2\pi])$ and suppose that f has period 2π. Consider a fixed x in $[0, 2\pi]$ and a positive $\delta < \pi$. Let

$$g(t) = \frac{f(x + t) + f(x - t)}{2} \qquad \text{if } t \in [0, \delta],$$

and let

$$s(x) = g(0+) = \lim_{t \to 0+} \frac{f(x + t) + f(x - t)}{2}$$

whenever this limit exists. Note that $s(x) = f(x)$ if f is continuous at x.

By combining Theorem 11.11 with Theorems 11.8 and 11.9, respectively, we obtain the following sufficient conditions for convergence of a Fourier series.

Theorem 11.12 (Jordan's test). *If f is of bounded variation on the compact interval $[x - \delta, x + \delta]$ for some $\delta < \pi$, then the limit $s(x)$ exists and the Fourier series generated by f converges to $s(x)$.*

Theorem 11.13 (Dini's test). *If the limit $s(x)$ exists and if the Lebesgue integral*

$$\int_0^\delta \frac{g(t) - s(x)}{t} \, dt$$

exists for some $\delta < \pi$, then the Fourier series generated by f converges to $s(x)$.

11.13 CESÀRO SUMMABILITY OF FOURIER SERIES

Continuity of a function f is not a very fruitful hypothesis when it comes to studying convergence of the Fourier series generated by f. In 1873, Du Bois–Reymond gave an example of a function, continuous throughout the interval $[0, 2\pi]$, whose Fourier series fails to converge on an uncountable subset of $[0, 2\pi]$. On the other hand, continuity does suffice to establish Cesàro summability of the series. This result (due to Fejér) and some of its consequences will be discussed next.

Our first task is to obtain an integral representation for the arithmetic means of the partial sums of a Fourier series.

Theorem 11.14. *Assume that $f \in L([0, 2\pi])$ and suppose that f is periodic with period 2π. Let s_n denote the nth partial sum of the Fourier series generated by f and let*

$$\sigma_n(x) = \frac{s_0(x) + s_1(x) + \cdots + s_{n-1}(x)}{n} \qquad (n = 1, 2, \ldots). \tag{23}$$

Then we have the integral representation

$$\sigma_n(x) = \frac{1}{n\pi} \int_0^\pi \frac{f(x+t) + f(x-t)}{2} \frac{\sin^2 \frac{1}{2}nt}{\sin^2 \frac{1}{2}t} \, dt. \tag{24}$$

Proof. If we use the integral representation for $s_n(x)$ given in (19) and form the sum defining $\sigma_n(x)$, we immediately obtain the required result because of formula (16), Section 8.16.

NOTE. If we apply Theorem 11.14 to the constant function whose value is 1 at each point we find $\sigma_n(x) = s_n(x) = 1$ for each n and hence (24) becomes

$$\frac{1}{n\pi} \int_0^\pi \frac{\sin^2 \frac{1}{2}nt}{\sin^2 \frac{1}{2}t} \, dt = 1. \tag{25}$$

Therefore, given any number s, we can combine (25) with (24) to write

$$\sigma_n(x) - s = \frac{1}{n\pi} \int_0^\pi \left\{ \frac{f(x+t) + f(x-t)}{2} - s \right\} \frac{\sin^2 \frac{1}{2}nt}{\sin^2 \frac{1}{2}t} \, dt. \tag{26}$$

If we can choose a value of s such that the integral on the right of (26) tends to 0 as $n \to \infty$, it will follow that $\sigma_n(x) \to s$ as $n \to \infty$. The next theorem shows that it suffices to take $s = [f(x+) + f(x-)]/2$.

Theorem 11.15 (Fejér). *Assume that $f \in L([0, 2\pi])$ and suppose that f is periodic with period 2π. Define a function s by the following equation:*

$$s(x) = \lim_{t \to 0+} \frac{f(x+t) + f(x-t)}{2}, \tag{27}$$

whenever the limit exists. Then, for each x for which $s(x)$ is defined, the Fourier series generated by f is Cesàro summable and has $(C, 1)$ sum $s(x)$. That is, we have

$$\lim_{n \to \infty} \sigma_n(x) = s(x),$$

where $\{\sigma_n\}$ is the sequence of arithmetic means defined by (23). If, in addition, f is continuous on $[0, 2\pi]$, then the sequence $\{\sigma_n\}$ converges uniformly to f on $[0, 2\pi]$.

Proof. Let $g_x(t) = [f(x+t) + f(x-t)]/2 - s(x)$, whenever $s(x)$ is defined. Then $g_x(t) \to 0$ as $t \to 0+$. Therefore, given $\varepsilon > 0$, there is a positive $\delta < \pi$ such that $|g_x(t)| < \varepsilon/2$ whenever $0 < t < \delta$. Note that δ depends on x as well as on ε. However, if f is continuous on $[0, 2\pi]$, then f is uniformly continuous on $[0, 2\pi]$, and there exists a δ which serves equally well for every x in $[0, 2\pi]$. Now we use (26) and divide the interval of integration into two subintervals $[0, \delta]$ and $[\delta, \pi]$. On $[0, \delta]$ we have

$$\left| \frac{1}{n\pi} \int_0^\delta g_x(t) \frac{\sin^2 \frac{1}{2}nt}{\sin^2 \frac{1}{2}t} \, dt \right| \leq \frac{\varepsilon}{2n\pi} \int_0^\pi \frac{\sin^2 \frac{1}{2}nt}{\sin^2 \frac{1}{2}t} \, dt = \frac{\varepsilon}{2},$$

because of (25). On $[\delta, \pi]$ we have

$$\left|\frac{1}{n\pi}\int_\delta^\pi g_x(t)\,\frac{\sin^2 \frac{1}{2}nt}{\sin^2 \frac{1}{2}t}\,dt\right| \le \frac{1}{n\pi \sin^2 \frac{1}{2}\delta}\int_\delta^\pi |g_x(t)|\,dt \le \frac{I(x)}{n\pi \sin^2 \frac{1}{2}\delta},$$

where $I(x) = \int_0^\pi |g_x(t)|\,dt$. Now choose N so that $I(x)/(N\pi \sin^2 \frac{1}{2}\delta) < \varepsilon/2$. Then $n \ge N$ implies

$$|\sigma_n(x) - s(x)| = \left|\frac{1}{n\pi}\int_0^\pi g_x(t)\,\frac{\sin^2 \frac{1}{2}nt}{\sin^2 \frac{1}{2}t}\,dt\right| < \varepsilon.$$

In other words, $\sigma_n(x) \to s(x)$ as $n \to \infty$.

If f is continuous on $[0, 2\pi]$, then, by periodicity, f is bounded on $\mathbf{R}$ and there is an M such that $|g_x(t)| \le M$ for all x and t, and we may replace $I(x)$ by πM in the above argument. The resulting N is then independent of x and hence $\sigma_n \to s = f$ uniformly on $[0, 2\pi]$.

11.14 CONSEQUENCES OF FEJÉR'S THEOREM

Theorem 11.16. *Let f be continuous on $[0, 2\pi]$ and periodic with period 2π. Let $\{s_n\}$ denote the sequence of partial sums of the Fourier series generated by f, say*

$$f(x) \sim \frac{a_0}{2} + \sum_{n=1}^\infty (a_n \cos nx + b_n \sin nx). \tag{28}$$

Then we have:

a) $\mathrm{l.i.m.}_{n\to\infty}\, s_n = f$ *on* $[0, 2\pi]$.

b) $\dfrac{1}{\pi}\displaystyle\int_0^{2\pi} |f(x)|^2\,dx = \dfrac{a_0^2}{2} + \sum_{n=1}^\infty (a_n^2 + b_n^2)$ (*Parseval's formula*).

c) *The Fourier series can be integrated term by term. That is, for all x we have*

$$\int_0^x f(t)\,dt = \frac{a_0 x}{2} + \sum_{n=1}^\infty \int_0^x (a_n \cos nt + b_n \sin nt)\,dt,$$

   the integrated series being uniformly convergent on every interval, even if the Fourier series in (28) *diverges.*

d) *If the Fourier series in* (28) *converges for some x, then it converges to $f(x)$.*

Proof. Applying formula (3) of Theorem 11.2, with $t_n(x) = \sigma_n(x) = (1/n)\sum_{k=0}^{n-1} s_k(x)$, we obtain the inequality

$$\int_0^{2\pi} |f(x) - s_n(x)|^2\,dx \le \int_0^{2\pi} |f(x) - \sigma_n(x)|^2\,dx. \tag{29}$$

But, since $\sigma_n \to f$ uniformly on $[0, 2\pi]$, it follows that $\mathrm{l.i.m.}_{n\to\infty}\, \sigma_n = f$ on $[0, 2\pi]$, and (29) implies (a). Part (b) follows from (a) because of Theorem 11.4. Part (c) also follows from (a), by Theorem 9.18. Finally, if $\{s_n(x)\}$ converges for some x, then $\{\sigma_n(x)\}$ must converge to the same limit. But since $\sigma_n(x) \to f(x)$ it follows that $s_n(x) \to f(x)$, which proves (d).

11.15 THE WEIERSTRASS APPROXIMATION THEOREM

Fejér's theorem can also be used to prove a famous theorem of Weierstrass which states that every continuous function on a compact interval can be uniformly approximated by a polynomial. More precisely, we have:

Theorem 11.17. *Let f be real-valued and continuous on a compact interval $[a, b]$. Then for every $\varepsilon > 0$ there is a polynomial p (which may depend on ε) such that*

$$|f(x) - p(x)| < \varepsilon \qquad \text{for every } x \text{ in } [a, b]. \tag{30}$$

Proof. If $t \in [0, \pi)$, let $g(t) = f[a + t(b - a)/\pi]$; if $t \in [\pi, 2\pi]$, let $g(t) = f[a + (2\pi - t)(b - a)/\pi]$ and define g outside $[0, 2\pi]$ so that g has period 2π. For the ε given in the theorem, we can apply Fejér's theorem to find a function σ defined by an equation of the form

$$\sigma(t) = A_0 + \sum_{k=1}^{N} (A_k \cos kt + B_k \sin kt)$$

such that $|g(t) - \sigma(t)| < \varepsilon/2$ for every t in $[0, 2\pi]$. (Note that N, and hence σ, depends on ε.) Since σ is a finite sum of trigonometric functions, it generates a power series expansion about the origin which converges uniformly on every finite interval. The partial sums of this power series expansion constitute a sequence of polynomials, say $\{p_n\}$, such that $p_n \to \sigma$ uniformly on $[0, 2\pi]$. Hence, for the same ε, there exists an m such that

$$|p_m(t) - \sigma(t)| < \frac{\varepsilon}{2}, \qquad \text{for every } t \text{ in } [0, 2\pi].$$

Therefore we have

$$|p_m(t) - g(t)| < \varepsilon, \qquad \text{for every } t \text{ in } [0, 2\pi]. \tag{31}$$

Now define the polynomial p by the formula $p(x) = p_m[\pi(x - a)/(b - a)]$. Then inequality (31) becomes (30) when we put $t = \pi(x - a)/(b - a)$.

11.16 OTHER FORMS OF FOURIER SERIES

Using the formulas

$$2 \cos nx = e^{inx} + e^{-inx} \qquad \text{and} \qquad 2i \sin nx = e^{inx} - e^{-inx},$$

the Fourier series generated by f can be expressed in terms of complex exponentials as follows:

$$f(x) \sim \frac{a_0}{2} + \sum_{n=1}^{\infty} (a_n \cos nx + b_n \sin nx) = \frac{a_0}{2} + \sum_{n=1}^{\infty} (\alpha_n e^{inx} + \beta_n e^{-inx}),$$

where $\alpha_n = (a_n - ib_n)/2$ and $\beta_n = (a_n + ib_n)/2$. If we put $\alpha_0 = a_0/2$ and $\alpha_{-n} = \beta_n$, we can write the exponential form more briefly as follows:

$$f(x) \sim \sum_{n=-\infty}^{\infty} \alpha_n e^{inx}.$$

The formulas (7) for the coefficients now become

$$\alpha_n = \frac{1}{2\pi} \int_0^{2\pi} f(t) e^{-int} \, dt \qquad (n = 0, \pm 1, \pm 2, \ldots).$$

If f has period 2π, the interval of integration can be replaced by any other interval of length 2π.

More generally, if $f \in L([0, p])$ and if f has period p, we write

$$f(x) \sim \frac{a_0}{2} + \sum_{n=1}^{\infty} \left(a_n \cos \frac{2\pi nx}{p} + b_n \sin \frac{2\pi nx}{p} \right)$$

to mean that the coefficients are given by the formulas

$$a_n = \frac{2}{p} \int_0^p f(t) \cos \frac{2\pi nt}{p} \, dt,$$

$$b_n = \frac{2}{p} \int_0^p f(t) \sin \frac{2\pi nt}{p} \, dt \qquad (n = 0, 1, 2, \ldots).$$

In exponential form we can write

$$f(x) \sim \sum_{n=-\infty}^{\infty} \alpha_n e^{2\pi inx/p},$$

where

$$\alpha_n = \frac{1}{p} \int_0^p f(t) e^{-2\pi int/p} \, dt, \qquad \text{if } n = 0, \pm 1, \pm 2, \ldots.$$

All the convergence theorems for Fourier series of period 2π can also be applied to the case of a general period p by making a suitable change of scale.

11.17 THE FOURIER INTEGRAL THEOREM

The hypothesis of periodicity, which appears in all the convergence theorems dealing with Fourier series, is not as serious a restriction as it may appear to be at first sight. If a function is initially defined on a finite interval, say $[a, b]$, we can always extend the definition of f outside $[a, b]$ by imposing some sort of periodicity condition. For example, if $f(a) = f(b)$, we can define f everywhere on $(-\infty, +\infty)$ by requiring the equation $f(x + p) = f(x)$ to hold for every x, where $p = b - a$. (The condition $f(a) = f(b)$ can always be brought about by changing the value of f at one of the endpoints if necessary. This does not affect the existence or the values of the integrals which are used to compute the Fourier coefficients of f.) However, if the given function is already defined everywhere on $(-\infty, +\infty)$ and is *not* periodic, then there is no hope of obtaining a Fourier series which represents the function everywhere on $(-\infty, +\infty)$. Nevertheless, in such a case the function can sometimes be represented by an *infinite integral* rather than by an infinite series. These integrals, which are in many ways analogous to Fourier series, are known as *Fourier integrals*, and the theorem which gives sufficient conditions for representing a function by such an integral is known as the *Fourier integral theorem*. The basic tools used in the theory are, as in the case of Fourier series, the Dirichlet integrals and the Riemann–Lebesgue lemma.

Theorem 11.18 (Fourier integral theorem). *Assume that $f \in L(-\infty, +\infty)$. Suppose there is a point x in $\mathbf{R}$ and an interval $[x - \delta, x + \delta]$ about x such that either*

a) *f is of bounded variation on $[x - \delta, x + \delta]$,*

or else

b) *both limits $f(x+)$ and $f(x-)$ exist and both Lebesgue integrals*

$$\int_0^\delta \frac{f(x + t) - f(x+)}{t} \, dt \quad and \quad \int_0^\delta \frac{f(x - t) - f(x-)}{t} \, dt$$

exist.

Then we have the formula

$$\frac{f(x+) + f(x-)}{2} = \frac{1}{\pi} \int_0^\infty \left[\int_{-\infty}^\infty f(u) \cos v(u - x) \, du \right] dv, \qquad (32)$$

the integral $\int_0^\infty$ being an improper Riemann integral.

Proof. The first step in the proof is to establish the following formula:

$$\lim_{\alpha \to +\infty} \frac{1}{\pi} \int_{-\infty}^\infty f(x + t) \frac{\sin \alpha t}{t} \, dt = \frac{f(x+) + f(x-)}{2}. \qquad (33)$$

For this purpose we write

$$\int_{-\infty}^\infty f(x + t) \frac{\sin \alpha t}{\pi t} \, dt = \int_{-\infty}^{-\delta} + \int_{-\delta}^0 + \int_0^\delta + \int_\delta^\infty.$$

When $\alpha \to +\infty$, the first and fourth integrals on the right tend to 0, because of the Riemann–Lebesgue lemma. In the third integral, we can apply either Theorem 11.8 or Theorem 11.9 (depending on whether (a) or (b) is satisfied) to get

$$\lim_{\alpha \to +\infty} \int_0^\delta f(x + t) \frac{\sin \alpha t}{\pi t} \, dt = \frac{f(x+)}{2}.$$

Similarly, we have

$$\int_{-\delta}^0 f(x + t) \frac{\sin \alpha t}{\pi t} \, dt = \int_0^\delta f(x - t) \frac{\sin \alpha t}{\pi t} \, dt \to \frac{f(x-)}{2} \qquad \text{as } \alpha \to +\infty.$$

Thus we have established (33). If we make a translation, we get

$$\int_{-\infty}^{\infty} f(x + t) \frac{\sin \alpha t}{t} \, dt = \int_{-\infty}^{\infty} f(u) \frac{\sin \alpha(u - x)}{u - x} \, du,$$

and if we use the elementary formula

$$\frac{\sin \alpha(u - x)}{u - x} = \int_{0}^{\alpha} \cos v(u - x) \, dv,$$

the limit relation in (33) becomes

$$\lim_{\alpha \to +\infty} \frac{1}{\pi} \int_{-\infty}^{\infty} f(u) \left[\int_{0}^{\alpha} \cos v(u - x) \, dv \right] du = \frac{f(x+) + f(x-)}{2}. \qquad (34)$$

But the formula we seek to prove is (34) with only the order of integration reversed. By Theorem 10.40 we have

$$\int_{0}^{\alpha} \left[\int_{-\infty}^{\infty} f(u) \cos v(u - x) \, du \right] dv = \int_{-\infty}^{\infty} \left[\int_{0}^{\alpha} f(u) \cos v(u - x) \, dv \right] du$$

for every $\alpha > 0$, since the cosine function is everywhere continuous and bounded. Since the limit in (34) exists, this proves that

$$\lim_{\alpha \to +\infty} \frac{1}{\pi} \int_{0}^{\alpha} \left[\int_{-\infty}^{\infty} f(u) \cos v(u - x) \, du \right] dv = \frac{f(x+) + f(x-)}{2}.$$

By Theorem 10.40, the integral $\int_{-\infty}^{\infty} f(u) \cos v(u - x) \, du$ is a continuous function of v on $[0, \alpha]$, so the integral $\int_{0}^{\infty}$ in (32) exists as an improper Riemann integral. It need not exist as a Lebesgue integral.

11.18 THE EXPONENTIAL FORM OF THE FOURIER INTEGRAL THEOREM

Theorem 11.19. *If f satisfies the hypotheses of the Fourier integral theorem, then we have*

$$\frac{f(x+) + f(x-)}{2} = \frac{1}{2\pi} \lim_{\alpha \to +\infty} \int_{-\alpha}^{\alpha} \left[\int_{-\infty}^{\infty} f(u) e^{iv(u-x)} \, du \right] dv. \qquad (35)$$

Proof. Let $F(v) = \int_{-\infty}^{\infty} f(u) \cos v(u - x) \, du$. Then F is continuous on $(-\infty, +\infty)$, $F(v) = F(-v)$ and hence $\int_{-\alpha}^{0} F(v) \, dv = \int_{0}^{\alpha} F(-v) \, dv = \int_{0}^{\alpha} F(v) \, dv$. Therefore (32) becomes

$$\frac{f(x+) + f(x-)}{2} = \lim_{\alpha \to +\infty} \frac{1}{\pi} \int_{0}^{\alpha} F(v) \, dv = \lim_{\alpha \to +\infty} \frac{1}{2\pi} \int_{-\alpha}^{\alpha} F(v) \, dv. \qquad (36)$$

Now define G on $(-\infty, +\infty)$ by the equation

$$G(v) = \int_{-\infty}^{\infty} f(u) \sin v(u - x) \, du.$$

Then G is everywhere continuous and $G(v) = -G(-v)$. Hence $\int_{-\alpha}^{\alpha} G(v) \, dv = 0$ for every α, so $\lim_{\alpha \to +\infty} \int_{-\alpha}^{\alpha} G(v) \, dv = 0$. Combining this with (36) we find

$$\frac{f(x+) + f(x-)}{2} = \lim_{\alpha \to +\infty} \frac{1}{2\pi} \int_{-\alpha}^{\alpha} \{F(v) + iG(v)\} \, dv.$$

This is formula (35).

11.19 INTEGRAL TRANSFORMS

Many functions in analysis can be expressed as Lebesgue integrals or improper Riemann integrals of the form

$$g(y) = \int_{-\infty}^{\infty} K(x, y)f(x) \, dx. \tag{37}$$

A function g defined by an equation of this sort (in which y may be either real or complex) is called an *integral transform* of f. The function K which appears in the integrand is referred to as the *kernel* of the transform.

Integral transforms are employed very extensively in both pure and applied mathematics. They are especially useful in solving certain boundary value problems and certain types of integral equations. Some of the more commonly used transforms are listed below:

Exponential Fourier transform: $\quad \int_{-\infty}^{\infty} e^{-ixy}f(x) \, dx.$

Fourier cosine transform: $\quad \int_{0}^{\infty} \cos xy \, f(x) \, dx.$

Fourier sine transform: $\quad \int_{0}^{\infty} \sin xy \, f(x) \, dx.$

Laplace transform: $\quad \int_{0}^{\infty} e^{-xy}f(x) \, dx.$

Mellin transform: $\quad \int_{0}^{\infty} x^{y-1}f(x) \, dx.$

Since $e^{-ixy} = \cos xy - i \sin xy$, the sine and cosine transforms are merely special cases of the exponential Fourier transform in which the function f vanishes on the negative real axis. The Laplace transform is also related to the exponential Fourier transform. If we consider a complex value of y, say $y = u + iv$, where u and v are real, we can write

$$\int_0^\infty e^{-xy} f(x)\, dx = \int_0^\infty e^{-ixv} e^{-xu} f(x)\, dx = \int_0^\infty e^{-ixv} \phi_u(x)\, dx,$$

where $\phi_u(x) = e^{-xu} f(x)$. Therefore the Laplace transform can also be regarded as a special case of the exponential Fourier transform.

NOTE. An equation such as (37) is sometimes written more briefly in the form $g = \mathscr{K}(f)$ or $g = \mathscr{K} f$, where $\mathscr{K}$ denotes the "operator" which converts f into g. Since integration is involved in this equation, the operator $\mathscr{K}$ is referred to as an *integral operator*. It is clear that $\mathscr{K}$ is also a linear operator. That is,

$$\mathscr{K}(a_1 f_1 + a_2 f_2) = a_1 \mathscr{K} f_1 + a_2 \mathscr{K} f_2,$$

if a_1 and a_2 are constants. The operator defined by the Fourier transform is often denoted by $\mathscr{F}$ and that defined by the Laplace transform is denoted by $\mathscr{L}$.

The exponential form of the Fourier integral theorem can be expressed in terms of Fourier transforms as follows. Let g denote the Fourier transform of f, so that

$$g(u) = \int_{-\infty}^\infty f(t) e^{-itu}\, dt. \tag{38}$$

Then, at points of continuity of f, formula (35) becomes

$$f(x) = \lim_{\alpha \to +\infty} \frac{1}{2\pi} \int_{-\alpha}^\alpha g(u) e^{ixu}\, du, \tag{39}$$

and this is called the *inversion formula* for Fourier transforms. It tells us that a continuous function f satisfying the conditions of the Fourier integral theorem is uniquely determined by its Fourier transform g.

NOTE. If $\mathscr{F}$ denotes the operator defined by (38), it is customary to denote by $\mathscr{F}^{-1}$ the operator defined by (39). Equations (38) and (39) can be expressed symbolically by writing $g = \mathscr{F} f$ and $f = \mathscr{F}^{-1} g$. The inversion formula tells us how to solve the equation $g = \mathscr{F} f$ for f in terms of g.

Before we pursue the study of Fourier transforms any further, we introduce a new notion, the *convolution* of two functions. This can be interpreted as a special kind of integral transform in which the kernel $K(x, y)$ depends only on the *difference* $x - y$.

11.20 CONVOLUTIONS

Definition 11.20. *Given two functions f and g, both Lebesgue integrable on $(-\infty, +\infty)$, let S denote the set of x for which the Lebesgue integral*

$$h(x) = \int_{-\infty}^\infty f(t) g(x - t)\, dt \tag{40}$$

exists. *This integral defines a function h on S called the convolution of f and g. We also write h = f * g to denote this function.*

NOTE. It is easy to see (by a translation) that $f * g = g * f$ whenever the integral exists.

An important special case occurs when both f and g vanish on the negative real axis. In this case, $g(x - t) = 0$ if $t > x$, and (40) becomes

$$h(x) = \int_0^x f(t)g(x - t)\, dt. \tag{41}$$

It is clear that, in this case, the convolution will be defined at each point of an interval $[a, b]$ if both f and g are Riemann-integrable on $[a, b]$. However, this need not be so if we assume only that f and g are Lebesgue integrable on $[a, b]$. For example, let

$$f(t) = \frac{1}{\sqrt{t}} \quad \text{and} \quad g(t) = \frac{1}{\sqrt{1 - t}}, \quad \text{if } 0 < t < 1,$$

and let $f(t) = g(t) = 0$ if $t \le 0$ or if $t \ge 1$. Then f has an infinite discontinuity at $t = 0$. Nevertheless, the Lebesgue integral $\int_{-\infty}^{\infty} f(t)\, dt = \int_0^1 t^{-1/2}\, dt$ exists. Similarly, the Lebesgue integral $\int_{-\infty}^{\infty} g(t)\, dt = \int_0^1 (1 - t)^{-1/2}\, dt$ exists, although g has an infinite discontinuity at $t = 1$. However, when we form the convolution integral in (40) corresponding to $x = 1$, we find

$$\int_{-\infty}^{\infty} f(t)g(1 - t)\, dt = \int_0^1 t^{-1}\, dt.$$

Observe that the two discontinuities of f and g have "coalesced" into *one* discontinuity of such nature that the convolution integral does not exist.

This example shows that there may be certain points on the real axis at which the integral in (40) fails to exist, even though both f and g are Lebesgue-integrable on $(-\infty, +\infty)$. Let us refer to such points as "singularities" of h. It is easy to show that such singularities cannot occur unless *both* f and g have infinite discontinuities. More precisely, we have the following theorem:

Theorem 11.21. *Let $\mathbf{R} = (-\infty, +\infty)$. Assume that $f \in L(\mathbf{R})$, $g \in L(\mathbf{R})$, and that either f or g is bounded on $\mathbf{R}$. Then the convolution integral*

$$h(x) = \int_{-\infty}^{\infty} f(t)g(x - t)\, dt \tag{42}$$

exists for every x in $\mathbf{R}$, and the function h so defined is bounded on $\mathbf{R}$. If, in addition, the bounded function f or g is continuous on $\mathbf{R}$, then h is also continuous on $\mathbf{R}$ and $h \in L(\mathbf{R})$.

Proof. Since $f * g = g * f$, it suffices to consider the case in which g is bounded. Suppose $|g| \le M$. Then

$$|f(t)g(x - t)| \le M|f(t)|. \tag{43}$$

The reader can verify that for each x, the product $f(t)g(x - t)$ is a measurable function of t on $\mathbf{R}$, so Theorem 10.35 shows that the integral for $h(x)$ exists. The inequality (43) also shows that $|h(x)| \le M \int |f|$, so h is bounded on $\mathbf{R}$.

Now if g is also continuous on $\mathbf{R}$, then Theorem 10.40 shows that h is continuous on $\mathbf{R}$. Now for every compact interval $[a, b]$ we have

$$\int_a^b |h(x)| \, dx \le \int_a^b \left[\int_{-\infty}^\infty |f(t)| \, |g(x - t)| \, dt \right] dx$$

$$= \int_{-\infty}^\infty |f(t)| \left[\int_a^b |g(x - t)| \, dx \right] dt$$

$$= \int_{-\infty}^\infty |f(t)| \left[\int_{a-t}^{b-t} |g(y)| \, dy \right] dt$$

$$\le \int_{-\infty}^\infty |f(t)| \, dt \int_{-\infty}^\infty |g(y)| \, dy,$$

so, by Theorem 10.31, $h \in L(\mathbf{R})$.

Theorem 11.22. *Let* $\mathbf{R} = (-\infty, +\infty)$. *Assume that* $f \in L^2(\mathbf{R})$ *and* $g \in L^2(\mathbf{R})$. *Then the convolution integral* (42) *exists for each x in $\mathbf{R}$ and the function h is bounded on* $\mathbf{R}$.

Proof. For fixed x, let $g_x(t) = g(x - t)$. Then g_x is measurable on $\mathbf{R}$ and $g_x \in L^2(\mathbf{R})$, so Theorem 10.54 implies that the product $f \cdot g_x \in L(\mathbf{R})$. In other words, the convolution integral $h(x)$ exists. Now $h(x)$ is an inner product, $h(x) = (f, g_x)$, hence the Cauchy–Schwarz inequality shows that

$$|h(x)| \le \|f\| \, \|g_x\| = \|f\| \, \|g\|,$$

so h is bounded on $\mathbf{R}$.

11.21 THE CONVOLUTION THEOREM FOR FOURIER TRANSFORMS

The next theorem shows that the Fourier transform of a convolution $f * g$ is the product of the Fourier transforms of f and of g. In operator notation,

$$\mathscr{F}(f * g) = \mathscr{F}(f) \cdot \mathscr{F}(g).$$

Theorem 11.23. *Let* $\mathbf{R} = (-\infty, +\infty)$. *Assume that* $f \in L(\mathbf{R})$, $g \in L(\mathbf{R})$, *and that at least one of f or g is continuous and bounded on $\mathbf{R}$. Let h denote the convolution,*

$h = f * g$. *Then for every real u we have*

$$\int_{-\infty}^{\infty} h(x) e^{-ixu}\, dx = \left(\int_{-\infty}^{\infty} f(t) e^{-itu}\, dt \right) \left(\int_{-\infty}^{\infty} g(y) e^{-iyu}\, dy \right). \qquad (44)$$

The integral on the left exists both as a Lebesgue integral and as an improper Riemann integral.

Proof. Assume that g is continuous and bounded on $\mathbf{R}$. Let $\{a_n\}$ and $\{b_n\}$ be two increasing sequences of positive real numbers such that $a_n \to +\infty$ and $b_n \to +\infty$. Define a sequence of functions $\{f_n\}$ on $\mathbf{R}$ as follows:

$$f_n(t) = \int_{-a_n}^{b_n} e^{-iux}\, g(x - t)\, dx.$$

Since

$$\int_a^b |e^{-iux}\, g(x - t)|\, dt \le \int_{-\infty}^{\infty} |g|$$

for all compact intervals $[a, b]$, Theorem 10.31 shows that

$$\lim_{n \to \infty} f_n(t) = \int_{-\infty}^{\infty} e^{-iux}\, g(x - t)\, dx \qquad \text{for every real } t. \qquad (45)$$

The translation $y = x - t$ gives us

$$\int_{-\infty}^{\infty} e^{-iux}\, g(x - t)\, dx = e^{-iut} \int_{-\infty}^{\infty} e^{-iuy}\, g(y)\, dy,$$

and (45) shows that

$$\lim_{n \to \infty} f(t) f_n(t) = f(t) e^{-iut} \left(\int_{-\infty}^{\infty} e^{-iuy}\, g(y)\, dy \right)$$

for all t. Now f_n is continuous on $\mathbf{R}$ (by Theorem 10.38), so the product $f \cdot f_n$ is measurable on $\mathbf{R}$. Since

$$|f(t) f_n(t)| \le |f(t)| \int_{-\infty}^{\infty} |g|,$$

the product $f \cdot f_n$ is Lebesgue-integrable on $\mathbf{R}$, and the Lebesgue dominated convergence theorem shows that

$$\lim_{n \to \infty} \int_{-\infty}^{\infty} f(t) f_n(t)\, dt = \left(\int_{-\infty}^{\infty} f(t) e^{-iut}\, dt \right) \left(\int_{-\infty}^{\infty} e^{-iuy} g(y)\, dy \right). \qquad (46)$$

But

$$\int_{-\infty}^{\infty} f(t) f_n(t)\, dt = \int_{-\infty}^{\infty} f(t) \left[\int_{-a_n}^{b_n} e^{-iux} g(x - t)\, dx \right] dt.$$

Since the function k defined by $k(x, t) = g(x - t)$ is continuous and bounded on $\mathbf{R}^2$ and since the integral $\int_a^b e^{-iux}\, dx$ exists for every compact interval $[a, b]$, Theorem 10.40 permits us to reverse the order of integration and we obtain

$$\int_{-\infty}^{\infty} f(t)f_n(t)\, dt = \int_{-a_n}^{b_n} e^{-iux}\left[\int_{-\infty}^{\infty} f(t)g(x - t)\, dt \right] dx$$

$$= \int_{-a_n}^{b_n} e^{-iux}h(x)\, dx.$$

Therefore, (46) shows that

$$\lim_{n \to \infty} \int_{-a_n}^{b_n} h(x)e^{-iux}\, dx = \left(\int_{-\infty}^{\infty} f(t)e^{-iut}\, dt \right)\left(\int_{-\infty}^{\infty} g(y)e^{-iuy}\, dy \right),$$

which proves (44). The integral on the left also exists as an improper Riemann integral because the integrand is continuous and bounded on $\mathbf{R}$ and $\int_a^b |h(x)e^{-iux}|\, dx \le \int_{-\infty}^{\infty} |h|$ for every compact interval $[a, b]$.

As an application of the convolution theorem we shall derive the following property of the Gamma function.

Example. If $p > 0$ and $q > 0$, we have the formula

$$\int_0^1 x^{p-1}(1 - x)^{q-1}\, dx = \frac{\Gamma(p)\Gamma(q)}{\Gamma(p + q)}. \tag{47}$$

The integral on the left is called the *Beta function* and is usually denoted by $B(p, q)$. To prove (47) we let

$$f_p(t) = \begin{cases} t^{p-1}e^{-t} & \text{if } t > 0, \\ 0 & \text{if } t \le 0. \end{cases}$$

Then $f_p \in L(\mathbf{R})$ and $\int_{-\infty}^{\infty} f_p(t)\, dt = \int_0^{\infty} t^{p-1}e^{-t}\, dt = \Gamma(p)$. Let h denote the convolution, $h = f_p * f_q$. Taking $u = 0$ in the convolution formula (44) we find, if $p > 1$ or $q > 1$,

$$\int_{-\infty}^{\infty} h(x)\, dx = \int_{-\infty}^{\infty} f_p(t)\, dt \int_{-\infty}^{\infty} f_q(y)\, dy = \Gamma(p)\Gamma(q). \tag{48}$$

Now we calculate the integral on the left in another way. Since both f_p and f_q vanish on the negative real axis, we have

$$h(x) = \int_0^x f_p(t)f_q(x - t)\, dt = \begin{cases} e^{-x}\displaystyle\int_0^x t^{p-1}(x - t)^{q-1}\, dt & \text{if } x > 0, \\ 0 & \text{if } x \le 0. \end{cases}$$

The change of variable $t = ux$ gives us, for $x > 0$,

$$h(x) = e^{-x}x^{p+q-1}\int_0^1 u^{p-1}(1 - u)^{q-1}\, du = e^{-x}x^{p+q-1}B(p, q).$$

Therefore $\int_{-\infty}^{\infty} h(x)\, dx = B(p, q)\int_0^{\infty} e^{-x}x^{p+q-1}\, dx = B(p, q)\Gamma(p + q)$ which, when used in (48), proves (47) if $p > 1$ or $q > 1$. To obtain the result for $p > 0$, $q > 0$ use the relation $pB(p, q) = (p + q)B(p + 1, q)$.

11.22 THE POISSON SUMMATION FORMULA

We conclude this chapter with a discussion of an important formula, called *Poisson's summation formula*, which has many applications. The formula can be expressed in different ways. For the applications we have in mind, the following form is convenient.

Theorem 11.24. *Let f be a nonnegative function such that the integral $\int_{-\infty}^{\infty} f(x)\,dx$ exists as an improper Riemann integral. Assume also that f increases on $(-\infty, 0]$ and decreases on $[0, +\infty)$. Then we have*

$$\sum_{m=-\infty}^{+\infty} \frac{f(m+) + f(m-)}{2} = \sum_{n=-\infty}^{+\infty} \int_{-\infty}^{\infty} f(t)e^{-2\pi i n t}\,dt, \qquad (49)$$

each series being absolutely convergent.

Proof. The proof makes use of the Fourier expansion of the function F defined by the series

$$F(x) = \sum_{m=-\infty}^{+\infty} f(m + x). \qquad (50)$$

First we show that this series converges absolutely for each real x and that the convergence is uniform on the interval $[0, 1]$.

Since f decreases on $[0, +\infty)$ we have, for $x \geq 0$,

$$\sum_{m=0}^{N} f(m + x) \leq f(0) + \sum_{m=1}^{N} f(m) \leq f(0) + \int_{0}^{\infty} f(t)\,dt.$$

Therefore, by the Weierstrass M-test (Theorem 9.6), the series $\sum_{m=0}^{\infty} f(m + x)$ converges uniformly on $[0, +\infty)$. A similar argument shows that the series $\sum_{m=-\infty}^{-1} f(m + x)$ converges uniformly on $(-\infty, 1]$. Therefore the series in (50) converges for all x and the convergence is uniform on the intersection

$$(-\infty, 1] \cap [0, +\infty) = [0, 1].$$

The sum function F is periodic with period 1. In fact, we have $F(x + 1) = \sum_{m=-\infty}^{+\infty} f(m + x + 1)$, and this series is merely a rearrangement of that in (50). Since all its terms are nonnegative, it converges to the same sum. Hence

$$F(x + 1) = F(x).$$

Next we show that F is of bounded variation on every compact interval. If $0 \leq x \leq \frac{1}{2}$, then $f(m + x)$ is a decreasing function of x if $m \geq 0$, and an increasing function of x if $m < 0$. Therefore we have

$$F(x) = \sum_{m=0}^{\infty} f(m + x) - \sum_{m=-\infty}^{-1} \{-f(m + x)\},$$

so F is the difference of two decreasing functions. Therefore F is of bounded variation on $[0, \frac{1}{2}]$. A similar argument shows that F is also of bounded variation on $[-\frac{1}{2}, 0]$. By periodicity, F is of bounded variation on every compact interval.

Now consider the Fourier series (in exponential form) generated by F, say

$$F(x) \sim \sum_{n=-\infty}^{+\infty} \alpha_n e^{2\pi inx}.$$

Since F is of bounded variation on $[0, 1]$ it is Riemann-integrable on $[0, 1]$, and the Fourier coefficients are given by the formula

$$\alpha_n = \int_0^1 F(x)e^{-2\pi inx}\, dx. \tag{51}$$

Also, since F is of bounded variation on every compact interval, Jordan's test shows that the Fourier series converges for every x and that

$$\frac{F(x+) + F(x-)}{2} = \sum_{n=-\infty}^{\infty} \alpha_n e^{2\pi inx}. \tag{52}$$

To obtain the Poisson summation formula we express the coefficients α_n in another form. We use (50) in (51) and integrate term by term (justified by uniform convergence) to obtain

$$\alpha_n = \sum_{m=-\infty}^{+\infty} \int_0^1 f(m + x)e^{-2\pi inx}\, dx.$$

The change of variable $t = m + x$ gives us

$$\alpha_n = \sum_{m=-\infty}^{+\infty} \int_m^{m+1} f(t)e^{-2\pi int}\, dt = \int_{-\infty}^{\infty} f(t)e^{-2\pi int}\, dt,$$

since $e^{2\pi imn} = 1$. Using this in (52) we obtain

$$\frac{F(x+) + F(x-)}{2} = \sum_{n=-\infty}^{\infty} \left\{ \int_{-\infty}^{\infty} f(t)e^{-2\pi int}\, dt \right\} e^{2\pi inx}. \tag{53}$$

When $x = 0$ this reduces to (49).

NOTE. In Theorem 11.24 there are no continuity requirements on f. However, if f is continuous at each integer, then each term $f(m + x)$ in the series (50) is continuous at $x = 0$ and hence, because of uniform convergence, the sum function F is also continuous at 0. In this case, (49) becomes

$$\sum_{m=-\infty}^{+\infty} f(m) = \sum_{n=-\infty}^{+\infty} \int_{-\infty}^{\infty} f(t)e^{-2\pi int}\, dt. \tag{54}$$

The monotonicity requirements on f can be relaxed. For example, since each member of (49) depends linearly on f, if the theorem is true for f_1 and for f_2 then it is also true for any linear combination $a_1 f_1 + a_2 f_2$. In particular, the formula holds for a complex-valued function $f = u + iv$ if it holds for u and v separately.

Example 1. *Transformation formula for the theta function.* The theta function θ is defined for all $x > 0$ by the equation

$$\theta(x) = \sum_{n=-\infty}^{+\infty} e^{-\pi n^2 x}.$$

We shall use Poisson's formula to derive the transformation equation

$$\theta(x) = \frac{1}{\sqrt{x}} \theta\left(\frac{1}{x}\right) \qquad \text{for } x > 0. \tag{55}$$

For fixed $\alpha > 0$, let $f(x) = e^{-\alpha x^2}$ for all real x. This function satisfies all the hypothesis of Theorem 11.24 and is continuous everywhere. Therefore, Poisson's formula implies

$$\sum_{m=-\infty}^{+\infty} e^{-\alpha m^2} = \sum_{n=-\infty}^{+\infty} \int_{-\infty}^{\infty} e^{-\alpha t^2} e^{2\pi i n t} \, dt. \tag{56}$$

The left member is $\theta(\alpha/\pi)$. The integral on the right is equal to

$$\int_{-\infty}^{\infty} e^{-\alpha t^2} e^{2\pi i n t} \, dt = 2 \int_{0}^{\infty} e^{-\alpha t^2} \cos 2\pi n t \, dt = \frac{2}{\sqrt{\alpha}} \int_{0}^{\infty} e^{-x^2} \cos \frac{2\pi n x}{\sqrt{\alpha}} \, dx = \frac{2}{\sqrt{\alpha}} F\left(\frac{\pi n}{\sqrt{\alpha}}\right)$$

where

$$F(y) = \int_{0}^{\infty} e^{-x^2} \cos 2xy \, dx.$$

But $F(y) = \frac{1}{2}\sqrt{\pi} e^{-y^2}$ (see Exercise 10.22), so

$$\int_{-\infty}^{\infty} e^{-\alpha t^2} e^{2\pi i n t} \, dt = \left(\frac{\pi}{\alpha}\right)^{1/2} e^{-\pi^2 n^2/\alpha}.$$

Using this in (56) and taking $\alpha = \pi x$ we obtain (55).

Example 2. *Partial-fraction decomposition of coth x.* The hyperbolic cotangent, coth x, is defined for $x \neq 0$ by the equation

$$\coth x = \frac{e^{2x} + 1}{e^{2x} - 1}.$$

We shall use Poisson's formula to derive the so-called partial-fraction decomposition

$$\coth x = \frac{1}{x} + 2x \sum_{n=1}^{\infty} \frac{1}{x^2 + \pi^2 n^2} \tag{57}$$

for $x > 0$. For fixed $\alpha > 0$, let

$$f(x) = \begin{cases} e^{-\alpha x} & \text{if } x \geq 0, \\ 0 & \text{if } x < 0. \end{cases}$$

Then f clearly satisfies the hypotheses of Theorem 11.24. Also, f is continuous everywhere except at 0, where $f(0+) = 1$ and $f(0-) = 0$. Therefore, the Poisson formula implies

$$\frac{1}{2} + \sum_{m=1}^{\infty} e^{-m\alpha} = \sum_{n=-\infty}^{+\infty} \int_{0}^{\infty} e^{-\alpha t - 2\pi i n t} \, dt. \tag{58}$$

The sum on the left is a geometric series with sum $1/(e^\alpha - 1)$, and the integral on the right is equal to $1/(\alpha + 2\pi i n)$. Therefore (58) becomes

$$\frac{1}{2} + \frac{1}{e^\alpha - 1} = \frac{1}{\alpha} + \sum_{n=1}^{\infty} \left(\frac{1}{\alpha + 2\pi i n} + \frac{1}{\alpha - 2\pi i n} \right),$$

and this gives (57) when α is replaced by $2x$.

EXERCISES

Orthogonal systems

11.1 Verify that the trigonometric system in (1) is orthonormal on $[0, 2\pi]$.

11.2 A finite collection of functions $\{\varphi_0, \varphi_1, \ldots, \varphi_m\}$ is said to be *linearly independent* on $[a, b]$ if the equation

$$\sum_{k=0}^{m} c_k \varphi_k(x) = 0 \qquad \text{for all } x \text{ in } [a, b]$$

implies $c_0 = c_1 = \cdots = c_m = 0$. An infinite collection is called linearly independent on $[a, b]$ if every finite subset is linearly independent on $[a, b]$. Prove that every orthonormal system on $[a, b]$ is linearly independent on $[a, b]$.

11.3 This exercise describes the *Gram-Schmidt process* for converting any linearly independent system to an orthogonal system. Let $\{f_0, f_1, \ldots\}$ be a linearly independent system on $[a, b]$ (as defined in Exercise 11.2). Define a new system $\{g_0, g_1, \ldots\}$ recursively as follows:

$$g_0 = f_0, \quad g_{r+1} = f_{r+1} - \sum_{k=1}^{r} a_k g_k,$$

where $a_k = (f_{r+1}, g_k)/(g_k, g_k)$ if $\|g_k\| \neq 0$, and $a_k = 0$ if $\|g_k\| = 0$. Prove that g_{n+1} is orthogonal to each of $g_0, g_1, \ldots, g_n$ for every $n \geq 0$.

11.4 Refer to Exercise 11.3. Let $(f, g) = \int_{-1}^{1} f(t)g(t)\, dt$. Apply the Gram-Schmidt process to the system of polynomials $\{1, t, t^2, \ldots\}$ on the interval $[-1, 1]$ and show that

$$g_1(t) = t, \qquad g_2(t) = t^2 - \tfrac{1}{3}, \qquad g_3(t) = t^3 - \tfrac{3}{5}t, \qquad g_4(t) = t^4 - \tfrac{6}{7}t^2 + \tfrac{3}{35}.$$

11.5 a) Assume $f \in R$ on $[0, 2\pi]$, where f is real and has period 2π. Prove that for every $\varepsilon > 0$ there is a continuous function g of period 2π such that $\|f - g\| < \varepsilon$.
   Hint. Choose a partition P_ε of $[0, 2\pi]$ for which f satisfies Riemann's condition $U(P, f) - L(P, f) < \varepsilon$ and construct a piecewise linear g which agrees with f at the points of P_ε.

   b) Use part (a) to show that Theorem 11.16(a), (b) and (c) holds if f is Riemann integrable on $[0, 2\pi]$.

11.6 In this exercise all functions are assumed to be continuous on a compact interval $[a, b]$. Let $\{\varphi_0, \varphi_1, \ldots\}$ be an orthonormal system on $[a, b]$.

   a) Prove that the following three statements are equivalent.

1) $(f, \varphi_n) = (g, \varphi_n)$ for all n implies $f = g$. (Two distinct continuous functions cannot have the same Fourier coefficients.)

2) $(f, \varphi_n) = 0$ for all n implies $f = 0$. (The only continuous function orthogonal to every φ_n is the zero function.)

3) If T is an orthonormal set on $[a, b]$ such that $\{\varphi_0, \varphi_1, \dots\} \subseteq T$, then $\{\varphi_0, \varphi_1, \dots\} = T$. (We cannot enlarge the orthonormal set.) This property is described by saying that $\{\varphi_0, \varphi_1, \dots\}$ is *maximal* or *complete*.

b) Let $\varphi_n(x) = e^{inx}/\sqrt{2\pi}$ for n an integer, and verify that the set $\{\varphi_n : n \in \mathbf{Z}\}$ is complete on every interval of length 2π.

11.7 If $x \in \mathbf{R}$ and $n = 1, 2, \dots$, let $f_n(x) = (x^2 - 1)^n$ and define

$$\phi_0(x) = 1, \quad \phi_n(x) = \frac{1}{2^n n!} f_n^{(n)}(x).$$

It is clear that ϕ_n is a polynomial. This is called the *Legendre polynomial* of order n. The first few are

$$\phi_1(x) = x, \qquad\qquad \phi_2(x) = \tfrac{3}{2}x^2 - \tfrac{1}{2},$$
$$\phi_3(x) = \tfrac{5}{2}x^3 - \tfrac{3}{2}x, \qquad \phi_4(x) = \tfrac{35}{8}x^4 - \tfrac{15}{4}x^2 + \tfrac{3}{8}.$$

Derive the following properties of Legendre polynomials:

a) $\phi_n'(x) = x\phi_{n-1}'(x) + n\phi_{n-1}(x)$.

b) $\phi_n(x) = x\phi_{n-1}(x) + \dfrac{x^2 - 1}{n} \phi_{n-1}'(x)$.

c) $(n + 1)\phi_{n+1}(x) = (2n + 1)x\phi_n(x) - n\phi_{n-1}(x)$.

d) ϕ_n satisfies the differential equation $[(1 - x^2)y']' + n(n + 1)y = 0$.

e) $[(1 - x^2)\Delta(x)]' + [m(m + 1) - n(n + 1)]\phi_m(x)\phi_n(x) = 0$,
   where $\Delta = \phi_n\phi_m' - \phi_m\phi_n'$.

f) The set $\{\phi_0, \phi_1, \phi_2, \dots\}$ is orthogonal on $[-1, 1]$.

g) $\displaystyle\int_{-1}^{1} \phi_n^2 \, dx = \frac{2n - 1}{2n + 1} \int_{-1}^{1} \phi_{n-1}^2 \, dx.$

h) $\displaystyle\int_{-1}^{1} \phi_n^2 \, dx = \frac{2}{2n + 1}.$

NOTE. The polynomials

$$g_n(t) = \frac{2^n (n!)^2}{(2n)!} \phi_n(t)$$

arise by applying the Gram-Schmidt process to the system $\{1, t, t^2, \dots\}$ on the interval $[-1, 1]$. (See Exercise 11.4.)

Trigonometric Fourier series

11.8 Assume that $f \in L([-\pi, \pi])$ and that f has period 2π. Show that the Fourier series generated by f assumes the following special forms under the conditions stated:

a) If $f(-x) = f(x)$ when $0 < x < \pi$, then

$$f(x) \sim \frac{a_0}{2} + \sum_{n=1}^{\infty} a_n \cos nx, \qquad \text{where } a_n = \frac{2}{\pi} \int_0^{\pi} f(t) \cos nt \, dt.$$

b) If $f(-x) = -f(x)$ when $0 < x < \pi$, then

$$f(x) \sim \sum_{n=1}^{\infty} b_n \sin nx, \qquad \text{where } b_n = \frac{2}{\pi} \int_0^{\pi} f(t) \sin nt \, dt.$$

In Exercises 11.9 through 11.15, show that each of the expansions is valid in the range indicated. *Suggestion.* Use Exercise 11.8 and Theorem 11.16(c) when possible.

11.9 a) $x = \pi - 2 \sum_{n=1}^{\infty} \frac{\sin nx}{n}$, $\qquad\qquad$ if $0 < x < 2\pi$.

b) $\dfrac{x^2}{2} = \pi x - \dfrac{\pi^2}{3} + 2 \sum_{n=1}^{\infty} \dfrac{\cos nx}{n^2}$, $\qquad$ if $0 \le x \le 2\pi$.

NOTE. When $x = 0$ this gives $\zeta(2) = \pi^2/6$.

11.10 a) $\dfrac{\pi}{4} = \sum_{n=1}^{\infty} \dfrac{\sin (2n - 1)x}{2n - 1}$, $\qquad\qquad$ if $0 < x < \pi$.

b) $x = \dfrac{\pi}{2} - \dfrac{4}{\pi} \sum_{n=1}^{\infty} \dfrac{\cos (2n - 1)x}{(2n - 1)^2}$, $\qquad$ if $0 \le x \le \pi$.

11.11 a) $x = 2 \sum_{n=1}^{\infty} \dfrac{(-1)^{n-1} \sin nx}{n}$, $\qquad\qquad$ if $-\pi < x < \pi$.

b) $x^2 = \dfrac{\pi^2}{3} + 4 \sum_{n=1}^{\infty} \dfrac{(-1)^n \cos nx}{n^2}$, $\qquad\qquad$ if $-\pi \le x \le \pi$.

11.12 $x^2 = \dfrac{4}{3} \pi^2 + 4 \sum_{n=1}^{\infty} \left(\dfrac{\cos nx}{n^2} - \dfrac{\pi \sin nx}{n} \right)$, $\qquad$ if $0 < x < 2\pi$.

11.13 a) $\cos x = \dfrac{8}{\pi} \sum_{n=1}^{\infty} \dfrac{n \sin 2nx}{4n^2 - 1}$, $\qquad\qquad$ if $0 < x < \pi$.

b) $\sin x = \dfrac{2}{\pi} - \dfrac{4}{\pi} \sum_{n=1}^{\infty} \dfrac{\cos 2nx}{4n^2 - 1}$, $\qquad\qquad$ if $0 < x < \pi$.

11.14 a) $x \cos x = -\tfrac{1}{2} \sin x + 2 \sum_{n=2}^{\infty} \dfrac{(-1)^n n \sin nx}{n^2 - 1}$, $\qquad$ if $-\pi < x < \pi$.

b) $x \sin x = 1 - \tfrac{1}{2} \cos x - 2 \sum_{n=2}^{\infty} \dfrac{(-1)^n \cos nx}{n^2 - 1}$, $\qquad$ if $-\pi \le x \le \pi$.

11.15 a) $\log \left| \sin \frac{x}{2} \right| = -\log 2 - \sum_{n=1}^{\infty} \frac{\cos nx}{n}$,   if $x \neq 2k\pi$ (k an integer).

b) $\log \left| \cos \frac{x}{2} \right| = -\log 2 - \sum_{n=1}^{\infty} \frac{(-1)^n \cos nx}{n}$,   if $x \neq (2k + 1)\pi$.

c) $\log \left| \tan \frac{x}{2} \right| = -2 \sum_{n=1}^{\infty} \frac{\cos (2n - 1)x}{2n - 1}$,   if $x \neq k\pi$.

11.16 a) Find a continuous function on $[-\pi, \pi]$ which generates the Fourier series $\sum_{n=1}^{\infty} (-1)^n n^{-3} \sin nx$. Then use Parseval's formula to prove that $\zeta(6) = \pi^6/945$.

b) Use an appropriate Fourier series in conjunction with Parseval's formula to show that $\zeta(4) = \pi^4/90$.

11.17 Assume that f has a continuous derivative on $[0, 2\pi]$, that $f(0) = f(2\pi)$, and that $\int_0^{2\pi} f(t)\, dt = 0$. Prove that $\|f'\| \geq \|f\|$, with equality if and only if $f(x) = a \cos x + b \sin x$. *Hint.* Use Parseval's formula.

11.18 A sequence $\{\bar{B}_n\}$ of periodic functions (of period 1) is defined on **R** as follows:

$$\bar{B}_{2n}(x) = (-1)^{n+1} \frac{2(2n)!}{(2\pi)^{2n}} \sum_{k=1}^{\infty} \frac{\cos 2\pi kx}{k^{2n}} \qquad (n = 1, 2, \ldots),$$

$$\bar{B}_{2n+1}(x) = (-1)^{n+1} \frac{2(2n + 1)!}{(2\pi)^{2n+1}} \sum_{k=1}^{\infty} \frac{\sin 2\pi kx}{k^{2n+1}} \qquad (n = 0, 1, 2, \ldots).$$

($\bar{B}_n$ is called the *Bernoulli function* of order n.) Show that:

a) $\bar{B}_1(x) = x - [x] - \frac{1}{2}$ if x is not an integer. ($[x]$ is the greatest integer $\leq x$.)

b) $\int_0^1 \bar{B}_n(x)\, dx = 0$ if $n \geq 1$ and $\bar{B}_n'(x) = n\bar{B}_{n-1}(x)$ if $n \geq 2$.

c) $\bar{B}_n(x) = P_n(x)$ if $0 < x < 1$, where P_n is the nth Bernoulli polynomial. (See Exercise 9.38 for the definition of P_n.)

d) $\bar{B}_n(x) = -\frac{n!}{(2\pi i)^n} \sum_{\substack{k=-\infty \\ k \neq 0}}^{\infty} \frac{e^{2\pi ikx}}{k^n}$ $(n = 1, 2, \ldots)$.

11.19 Let f be the function of period 2π whose values on $[-\pi, \pi]$ are

$f(x) = 1$ if $0 < x < \pi$, $f(x) = -1$ if $-\pi < x < 0$,

$f(x) = 0$ if $x = 0$ or $x = \pi$.

a) Show that

$$f(x) = \frac{4}{\pi} \sum_{n=1}^{\infty} \frac{\sin (2n - 1)x}{2n - 1}, \qquad \text{for every } x.$$

This is one example of a class of Fourier series which have a curious property known as *Gibbs' phenomenon*. This exercise is designed to illustrate this phenomenon. In that which follows, $s_n(x)$ denotes the nth partial sum of the series in part (a).

b) Show that

$$s_n(x) = \frac{2}{\pi} \int_0^x \frac{\sin 2nt}{\sin t} \, dt.$$

c) Show that, in $(0, \pi)$, s_n has local maxima at $x_1, x_3, \ldots, x_{2n-1}$ and local minima at $x_2, x_4, \ldots, x_{2n-2}$, where $x_m = \frac{1}{2}m\pi/n$ $(m = 1, 2, \ldots, 2n - 1)$.

d) Show that $s_n(\frac{1}{2}\pi/n)$ is the largest of the numbers

$$s_n(x_m) \qquad (m = 1, 2, \ldots, 2n - 1).$$

e) Interpret $s_n(\frac{1}{2}\pi/n)$ as a Riemann sum and prove that

$$\lim_{n \to \infty} s_n\left(\frac{\pi}{2n}\right) = \frac{2}{\pi} \int_0^\pi \frac{\sin t}{t} \, dt.$$

The value of the limit in (e) is about 1.179. Thus, although f has a jump equal to 2 at the origin, the graphs of the approximating curves s_n tend to approximate a vertical segment of length 2.358 in the vicinity of the origin. This is the Gibbs phenomenon.

11.20 If $f(x) \sim a_0/2 + \sum_{n=1}^\infty (a_n \cos nx + b_n \sin nx)$ and if f is of bounded variation on $[0, 2\pi]$, show that $a_n = O(1/n)$ and $b_n = O(1/n)$. *Hint.* Write $f = g - h$, where g and h are increasing on $[0, 2\pi]$. Then

$$a_n = \frac{1}{n\pi} \int_0^{2\pi} g(x) \, d(\sin nx) - \frac{1}{n\pi} \int_0^{2\pi} h(x) \, d(\sin nx).$$

Now apply Theorem 7.31.

11.21 Suppose $g \in L([a, \delta])$ for every a in $(0, \delta)$ and assume that g satisfies a "right-handed" Lipschitz condition at 0. (See the Note following Theorem 11.9.) Show that the Lebesgue integral $\int_0^\delta |g(t) - g(0+)|/t \, dt$ exists.

11.22 Use Exercise 11.21 to prove that differentiability of f at a point implies convergence of its Fourier series at the point.

11.23 Let g be continuous on $[0, 1]$ and assume that $\int_0^1 t^n g(t) \, dt = 0$ for $n = 0, 1, 2, \ldots$. Show that:

a) $\int_0^1 g(t)^2 \, dt = \int_0^1 g(t)(g(t) - P(t)) \, dt$ for every polynomial P.

b) $\int_0^1 g(t)^2 \, dt = 0$. *Hint.* Use Theorem 11.17.

c) $g(t) = 0$ for every t in $[0, 1]$.

11.24 Use the Weierstrass approximation theorem to prove each of the following statements.

a) If f is continuous on $[1, +\infty)$ and if $f(x) \to a$ as $x \to +\infty$, then f can be uniformly approximated on $[1, +\infty)$ by a function g of the form $g(x) = p(1/x)$, where p is a polynomial.

b) If f is continuous on $[0, +\infty)$ and if $f(x) \to a$ as $x \to +\infty$, then f can be uniformly approximated on $[0, +\infty)$ by a function g of the form $g(x) = p(e^{-x})$, where p is a polynomial.

11.25 Assume that $f(x) \sim a_0/2 + \sum_{n=1}^\infty (a_n \cos nx + b_n \sin nx)$ and let $\{\sigma_n\}$ be the sequence of arithmetic means of the partial sums of this series, as it was given in (23).

Show that:

a) $\sigma_n(x) = \dfrac{a_0}{2} + \displaystyle\sum_{k=1}^{n-1}\left(1 - \dfrac{k}{n}\right)(a_k \cos kx + b_k \sin kx).$

b) $\displaystyle\int_0^{2\pi} |f(x) - \sigma_n(x)|^2\, dx = \int_0^{2\pi} |f(x)|^2\, dx$

$$- \frac{\pi}{2} a_0^2 - \pi \sum_{k=1}^{n-1}(a_k^2 + b_k^2) + \frac{\pi}{n^2}\sum_{k=1}^{n-1} k^2(a_k^2 + b_k^2).$$

c) If f is continuous on $[0, 2\pi]$ and has period 2π, then

$$\lim_{n\to\infty} \frac{\pi}{n^2}\sum_{k=1}^{n} k^2(a_k^2 + b_k^2) = 0.$$

11.26 Consider the Fourier series (in exponential form) generated by a function f which is continuous on $[0, 2\pi]$ and periodic with period 2π, say

$$f(x) \sim \sum_{n=-\infty}^{+\infty} \alpha_n e^{inx}.$$

Assume also that the derivative $f' \in R$ on $[0, 2\pi]$.

a) Prove that the series $\sum_{n=-\infty}^{+\infty} n^2|\alpha_n|^2$ converges; then use the Cauchy–Schwarz inequality to deduce that $\sum_{n=-\infty}^{+\infty} |\alpha_n|$ converges.

b) From (a), deduce that the series $\sum_{n=-\infty}^{+\infty} \alpha_n e^{inx}$ converges uniformly to a continuous sum function g on $[0, 2\pi]$. Then prove that $f = g$.

Fourier integrals

11.27 If f satisfies the hypotheses of the Fourier integral theorem, show that:

a) If f is even, that is, if $f(-t) = f(t)$ for every t, then

$$\frac{f(x+) + f(x-)}{2} = \frac{2}{\pi}\lim_{\alpha\to+\infty}\int_0^\alpha \cos vx\left[\int_0^\infty f(u)\cos vu\, du\right] dv.$$

b) If f is odd, that is, if $f(-t) = -f(t)$ for every t, then

$$\frac{f(x+) + f(x-)}{2} = \frac{2}{\pi}\lim_{\alpha\to+\infty}\int_0^\alpha \sin vx\left[\int_0^\infty f(u)\sin vu\, du\right] dv.$$

Use the Fourier integral theorem to evaluate the improper integrals in Exercises 11.28 through 11.30. *Suggestion.* Use Exercise 11.27 when possible.

11.28 $\dfrac{2}{\pi}\displaystyle\int_0^\infty \frac{\sin v \cos vx}{v}\, dv = \begin{cases} 1 & \text{if } -1 < x < 1, \\ 0 & \text{if } |x| > 1, \\ \frac{1}{2} & \text{if } |x| = 1. \end{cases}$

11.29 $\displaystyle\int_0^\infty \frac{\cos ax}{b^2 + x^2}\, dx = \frac{\pi}{2b} e^{-|a|b}, \quad \text{if } b > 0.$

Hint. Apply Exercise 11.27 with $f(u) = e^{-b|u|}$.

11.30 $\displaystyle\int_0^\infty \frac{x \sin ax}{1 + x^2}\, dx = \frac{a}{|a|}\frac{\pi}{2} e^{-|a|}$,   if $a \neq 0$.

11.31 a) Prove that

$$\frac{\Gamma(p)\Gamma(p)}{\Gamma(2p)} = 2 \int_0^{1/2} x^{p-1}(1-x)^{p-1}\, dx.$$

b) Make a suitable change of variable in (a) and derive the duplication formula for the Gamma function:

$$\Gamma(2p)\Gamma(\tfrac{1}{2}) = 2^{2p-1}\Gamma(p)\Gamma(p + \tfrac{1}{2}).$$

NOTE. In Exercise 10.30 it is shown that $\Gamma(\tfrac{1}{2}) = \sqrt{\pi}$.

11.32 If $f(x) = e^{-x^2/2}$ and $g(x) = xf(x)$ for all x, prove that

$$f(y) = \sqrt{\frac{2}{\pi}}\int_0^\infty f(x)\cos xy\, dx \qquad \text{and} \qquad g(y) = \sqrt{\frac{2}{\pi}}\int_0^\infty g(x)\sin xy\, dx.$$

11.33 This exercise describes another form of Poisson's summation formula. Assume that f is nonnegative, decreasing, and continuous on $[0, +\infty)$ and that $\int_0^\infty f(x)\, dx$ exists as an improper Riemann integral. Let

$$g(y) = \sqrt{\frac{2}{\pi}}\int_0^\infty f(x)\cos xy\, dx.$$

If α and β are positive numbers such that $\alpha\beta = 2\pi$, prove that

$$\sqrt{\alpha}\left\{\tfrac{1}{2}f(0) + \sum_{m=1}^\infty f(m\alpha)\right\} = \sqrt{\beta}\left\{\tfrac{1}{2}g(0) + \sum_{n=1}^\infty g(n\beta)\right\}.$$

11.34 Prove that the transformation formula (55) for $\theta(x)$ can be put in the form

$$\sqrt{\alpha}\left\{\tfrac{1}{2} + \sum_{m=1}^\infty e^{-\alpha^2 m^2/2}\right\} = \sqrt{\beta}\left\{\tfrac{1}{2} + \sum_{n=1}^\infty e^{-\beta^2 n^2/2}\right\},$$

where $\alpha\beta = 2\pi$, $\alpha > 0$.

11.35 If $s > 1$, prove that

$$\pi^{-s/2}\,\Gamma\!\left(\frac{s}{2}\right) n^{-s} = \int_0^\infty e^{-\pi n^2 x}x^{s/2-1}\, dx$$

and derive the formula

$$\pi^{-s/2}\,\Gamma\!\left(\frac{s}{2}\right)\zeta(s) = \int_0^\infty \psi(x)x^{s/2-1}\, dx,$$

where $2\psi(x) = \theta(x) - 1$. Use this and the transformation formula for $\theta(x)$ to prove that

$$\pi^{-s/2} \, \Gamma\!\left(\frac{s}{2}\right) \zeta(s) = \frac{1}{s(s-1)} + \int_1^{\infty} (x^{s/2-1} + x^{(1-s)/2-1})\psi(x) \, dx.$$

Laplace transforms

Let c be a positive number such that the integral $\int_0^{\infty} e^{-ct}|f(t)| \, dt$ exists as an improper Riemann integral. Let $z = x + iy$, where $x > c$. It is easy to show that the integral

$$F(z) = \int_0^{\infty} e^{-zt} f(t) \, dt$$

exists both as an improper Riemann integral and as a Lebesgue integral. The function F so defined is called the *Laplace transform* of f, denoted by $\mathscr{L}(f)$. The following exercises describe some properties of Laplace transforms.

11.36 Verify the entries in the following table of Laplace transforms.

$f(t)$	$F(z) = \int_0^{\infty} e^{-zt} f(t) \, dt$	$z = x + iy$
$e^{\alpha t}$	$(z - \alpha)^{-1}$	$(x > \alpha)$
$\cos \alpha t$	$z/(z^2 + \alpha^2)$	$(x > 0)$
$\sin \alpha t$	$\alpha/(z^2 + \alpha^2)$	$(x > 0)$
$t^p e^{\alpha t}$	$\Gamma(p + 1)/(z - \alpha)^{p+1}$	$(x > \alpha, p > 0)$

11.37 Show that the convolution $h = f * g$ assumes the form

$$h(t) = \int_0^t f(x)g(t - x) \, dx$$

when both f and g vanish on the negative real axis. Use the convolution theorem for Fourier transforms to prove that $\mathscr{L}(f * g) = \mathscr{L}(f) \cdot \mathscr{L}(g)$.

11.38 Assume f is continuous on $(0, +\infty)$ and let $F(z) = \int_0^{\infty} e^{-zt} f(t) \, dt$ for $z = x + iy$, $x > c > 0$. If $s > c$ and $a > 0$ prove that:

a) $F(s + a) = a \int_0^{\infty} g(t)e^{-at} \, dt$, where $g(x) = \int_0^x e^{-st} f(t) \, dt$.

b) If $F(s + na) = 0$ for $n = 0, 1, 2, \ldots$, then $f(t) = 0$ for $t > 0$. *Hint.* Use Exercise 11.23.

c) If h is continuous on $(0, +\infty)$ and if f and h have the same Laplace transform, then $f(t) = h(t)$ for every $t > 0$.

11.39 Let $F(z) = \int_0^{\infty} e^{-zt} f(t) \, dt$ for $z = x + iy$, $x > c > 0$. Let t be a point at which f satisfies one of the "local" conditions (a) or (b) of the Fourier integral theorem (Theorem 11.18). Prove that for each $a > c$ we have

$$\frac{f(t+) + f(t-)}{2} = \frac{1}{2\pi} \lim_{T \to +\infty} \int_{-T}^{T} e^{(a+iv)t} F(a + iv) \, dv.$$

This is called the *inversion formula for Laplace transforms*. The limit on the right is usually evaluated with the help of residue calculus, as described in Section 16.26. *Hint.* Let $g(t) = e^{-at} f(t)$ for $t \geq 0$, $g(t) = 0$ for $t < 0$, and apply Theorem 11.19 to g.

SUGGESTED REFERENCES FOR FURTHER STUDY

11.1 Carslaw, H. S., *Introduction to the Theory of Fourier's Series and Integrals*, 3rd ed. Macmillan, London, 1930.

11.2 Edwards, R. E., *Fourier Series, A Modern Introduction*, Vol. 1. Holt, Rinehart and Winston, New York, 1967.

11.3 Hardy, G. H., and Rogosinski, W. W., *Fourier Series*. Cambridge University Press, 1950.

11.4 Hobson, E. W., *The Theory of Functions of a Real Variable and the Theory of Fourier's Series*, Vol. 1, 3rd ed. Cambridge University Press, 1927.

11.5 Indritz, J., *Methods in Analysis*. Macmillan, New York, 1963.

11.6 Jackson, D., *Fourier Series and Orthogonal Polynomials*. Carus Monograph No. 6. Open Court, New York, 1941.

11.7 Rogosinski, W. W., *Fourier Series*. H. Cohn and F. Steinhardt, translators. Chelsea, New York, 1950.

11.8 Titchmarsh, E. C., *Theory of Fourier Integrals*. Oxford University Press, 1937.

11.9 Wiener, N., *The Fourier Integral*. Cambridge University Press, 1933.

11.10 Zygmund, A., *Trigonometrical Series*, 2nd ed. Cambridge University Press, 1968.

CHAPTER 12

MULTIVARIABLE DIFFERENTIAL CALCULUS

12.1 INTRODUCTION

Partial derivatives of functions from $\mathbf{R}^n$ to $\mathbf{R}^1$ were discussed briefly in Chapter 5. We also introduced derivatives of vector-valued functions from $\mathbf{R}^1$ to $\mathbf{R}^n$. This chapter extends derivative theory to functions from $\mathbf{R}^n$ to $\mathbf{R}^m$.

As noted in Section 5.14, the partial derivative is a somewhat unsatisfactory generalization of the usual derivative because existence of all the partial derivatives $D_1 f, \ldots, D_n f$ at a particular point does not necessarily imply continuity of f at that point. The trouble with partial derivatives is that they treat a function of several variables as a function of one variable at a time. The partial derivative describes the rate of change of a function in the direction of each coordinate axis. There is a slight generalization, called the *directional derivative*, which studies the rate of change of a function in an arbitrary direction. It applies to both real- and vector-valued functions.

12.2 THE DIRECTIONAL DERIVATIVE

Let S be a subset of $\mathbf{R}^n$, and let $\mathbf{f} : S \to \mathbf{R}^m$ be a function defined on S with values in $\mathbf{R}^m$. We wish to study how $\mathbf{f}$ changes as we move from a point $\mathbf{c}$ in S along a line segment to a nearby point $\mathbf{c} + \mathbf{u}$, where $\mathbf{u} \neq \mathbf{0}$. Each point on the segment can be expressed as $\mathbf{c} + h\mathbf{u}$, where h is real. The vector $\mathbf{u}$ describes the direction of the line segment. We assume that $\mathbf{c}$ is an *interior point* of S. Then there is an n-ball $B(\mathbf{c}; r)$ lying in S, and, if h is small enough, the line segment joining $\mathbf{c}$ to $\mathbf{c} + h\mathbf{u}$ will lie in $B(\mathbf{c}; r)$ and hence in S.

Definition 12.1. *The directional derivative of* $\mathbf{f}$ *at* $\mathbf{c}$ *in the direction* $\mathbf{u}$, *denoted by the symbol* $\mathbf{f}'(\mathbf{c}; \mathbf{u})$, *is defined by the equation*

$$\mathbf{f}'(\mathbf{c}; \mathbf{u}) = \lim_{h \to 0} \frac{\mathbf{f}(\mathbf{c} + h\mathbf{u}) - \mathbf{f}(\mathbf{c})}{h}, \tag{1}$$

whenever the limit on the right exists.

NOTE. Some authors require that $\|\mathbf{u}\| = 1$, but this is not assumed here.

Examples

  1. The definition in (1) is meaningful if $\mathbf{u} = \mathbf{0}$. In this case $\mathbf{f}'(\mathbf{c}; \mathbf{0})$ exists and equals $\mathbf{0}$ for every $\mathbf{c}$ in S.

2. If $\mathbf{u} = \mathbf{u}_k$, the kth unit coordinate vector, then $\mathbf{f}'(\mathbf{c}; \mathbf{u}_k)$ is called a *partial derivative* and is denoted by $D_k\mathbf{f}(\mathbf{c})$. When $\mathbf{f}$ is real-valued this agrees with the definition given in Chapter 5.

3. If $\mathbf{f} = (f_1, \ldots, f_m)$, then $\mathbf{f}'(\mathbf{c}; \mathbf{u})$ exists if and only if $f'_k(\mathbf{c}; \mathbf{u})$ exists for each $k = 1, 2, \ldots, m$, in which case

$$\mathbf{f}'(\mathbf{c}; \mathbf{u}) = (f'_1(\mathbf{c}; \mathbf{u}), \ldots, f'_m(\mathbf{c}; \mathbf{u})).$$

In particular, when $\mathbf{u} = \mathbf{u}_k$ we find

$$D_k\mathbf{f}(\mathbf{c}) = (D_k f_1(\mathbf{c}), \ldots, D_k f_m(\mathbf{c})). \tag{2}$$

4. If $\mathbf{F}(t) = \mathbf{f}(\mathbf{c} + t\mathbf{u})$, then $\mathbf{F}'(0) = \mathbf{f}'(\mathbf{c}; \mathbf{u})$. More generally, $\mathbf{F}'(t) = \mathbf{f}'(\mathbf{c} + t\mathbf{u}; \mathbf{u})$ if either derivative exists.

5. If $f(\mathbf{x}) = \|\mathbf{x}\|^2$, then

$$F(t) = f(\mathbf{c} + t\mathbf{u}) = (\mathbf{c} + t\mathbf{u}) \cdot (\mathbf{c} + t\mathbf{u})$$

$$= \|\mathbf{c}\|^2 + 2t\mathbf{c} \cdot \mathbf{u} + t^2\|\mathbf{u}\|^2,$$

so $F'(t) = 2\mathbf{c} \cdot \mathbf{u} + 2t\|\mathbf{u}\|^2$; hence $F'(0) = f'(\mathbf{c}; \mathbf{u}) = 2\mathbf{c} \cdot \mathbf{u}$.

6. *Linear functions.* A function $\mathbf{f} : \mathbf{R}^n \to \mathbf{R}^m$ is called *linear* if $\mathbf{f}(a\mathbf{x} + b\mathbf{y}) = a\mathbf{f}(\mathbf{x}) + b\mathbf{f}(\mathbf{y})$ for every $\mathbf{x}$ and $\mathbf{y}$ in $\mathbf{R}^n$ and every pair of scalars a and b. If $\mathbf{f}$ is linear, the quotient on the right of (1) simplifies to $\mathbf{f}(\mathbf{u})$, so $\mathbf{f}'(\mathbf{c}; \mathbf{u}) = \mathbf{f}(\mathbf{u})$ for every $\mathbf{c}$ and every $\mathbf{u}$.

12.3 DIRECTIONAL DERIVATIVES AND CONTINUITY

If $\mathbf{f}'(\mathbf{c}; \mathbf{u})$ exists in every direction $\mathbf{u}$, then in particular all the partial derivatives $D_1\mathbf{f}(\mathbf{c}), \ldots, D_n\mathbf{f}(\mathbf{c})$ exist. However, the converse is not true. For example, consider the real-valued function $f : \mathbf{R}^2 \to \mathbf{R}^1$ given by

$$f(x, y) = \begin{cases} x + y & \text{if } x = 0 \text{ or } y = 0, \\ 1 & \text{otherwise.} \end{cases}$$

Then $D_1 f(0, 0) = D_2 f(0, 0) = 1$. Nevertheless, if we consider any other direction $\mathbf{u} = (a_1, a_2)$, where $a_1 \neq 0$ and $a_2 \neq 0$, then

$$\frac{f(\mathbf{0} + h\mathbf{u}) - f(\mathbf{0})}{h} = \frac{f(h\mathbf{u})}{h} = \frac{1}{h},$$

and this does not tend to a limit as $h \to 0$.

A rather surprising fact is that a function can have a finite directional derivative $\mathbf{f}'(\mathbf{c}; \mathbf{u})$ for *every* $\mathbf{u}$ but may fail to be continuous at $\mathbf{c}$. For example, let

$$f(x, y) = \begin{cases} xy^2/(x^2 + y^4) & \text{if } x \neq 0, \\ 0 & \text{if } x = 0. \end{cases}$$

Let $\mathbf{u} = (a_1, a_2)$ be any vector in $\mathbf{R}^2$. Then we have

$$\frac{f(\mathbf{0} + h\mathbf{u}) - f(\mathbf{0})}{h} = \frac{f(ha_1, ha_2)}{h} = \frac{a_1 a_2^2}{a_1^2 + h^2 a_2^4},$$

and hence

$$f'(\mathbf{0}; \mathbf{u}) = \begin{cases} a_2^2/a_1 & \text{if } a_1 \neq 0, \\ 0 & \text{if } a_1 = 0. \end{cases}$$

Thus, $f'(\mathbf{0}; \mathbf{u})$ exists for all $\mathbf{u}$. On the other hand, the function f takes the value $\frac{1}{2}$ at each point of the parabola $x = y^2$ (except at the origin), so f is not continuous at $(0, 0)$, since $f(0, 0) = 0$.

Thus we see that even the existence of all directional derivatives at a point fails to imply continuity at that point. For this reason, directional derivatives, like partial derivatives, are a somewhat unsatisfactory extension of the one-dimensional concept of derivative. We turn now to a more suitable generalization which implies continuity and, at the same time, extends the principal theorems of one-dimensional derivative theory to functions of several variables. This is called the *total derivative*.

12.4 THE TOTAL DERIVATIVE

In the one-dimensional case, a function f with a derivative at c can be approximated near c by a linear polynomial. In fact, if $f'(c)$ exists, let $E_c(h)$ denote the difference

$$E_c(h) = \frac{f(c + h) - f(c)}{h} - f'(c) \qquad \text{if } h \neq 0, \tag{3}$$

and let $E_c(0) = 0$. Then we have

$$f(c + h) = f(c) + f'(c)h + hE_c(h), \tag{4}$$

an equation which holds also for $h = 0$. This is called the *first-order Taylor formula* for approximating $f(c + h) - f(c)$ by $f'(c)h$. The error committed is $hE_c(h)$. From (3) we see that $E_c(h) \to 0$ as $h \to 0$. The error $hE_c(h)$ is said to be of *smaller order* than h as $h \to 0$.

We focus attention on two properties of formula (4). First, the quantity $f'(c)h$ is a *linear* function of h. That is, if we write $T_c(h) = f'(c)h$, then

$$T_c(ah_1 + bh_2) = aT_c(h_1) + bT_c(h_2).$$

Second, the error term $hE_c(h)$ is of smaller order than h as $h \to 0$. The total derivative of a function $\mathbf{f}$ from $\mathbf{R}^n$ to $\mathbf{R}^m$ will now be defined in such a way that it preserves these two properties.

Let $\mathbf{f}: S \to \mathbf{R}^m$ be a function defined on a set S in $\mathbf{R}^n$ with values in $\mathbf{R}^m$. Let $\mathbf{c}$ be an interior point of S, and let $B(\mathbf{c}; r)$ be an n-ball lying in S. Let $\mathbf{v}$ be a point in $\mathbf{R}^n$ with $\|\mathbf{v}\| < r$, so that $\mathbf{c} + \mathbf{v} \in B(\mathbf{c}; r)$.

Definition 12.2. *The function $\mathbf{f}$ is said to be differentiable at $\mathbf{c}$ if there exists a linear function $\mathbf{T_c}: \mathbf{R}^n \to \mathbf{R}^m$ such that*

$$\mathbf{f(c + v)} = \mathbf{f(c)} + \mathbf{T_c(v)} + \|\mathbf{v}\|\, \mathbf{E_c(v)}, \tag{5}$$

where $\mathbf{E_c(v)} \to \mathbf{0}$ as $\mathbf{v} \to \mathbf{0}$.

NOTE. Equation (5) is called a *first-order Taylor formula*. It is to hold for all $\mathbf{v}$ in $\mathbf{R}^n$ with $\|\mathbf{v}\| < r$. The linear function $\mathbf{T_c}$ is called the *total derivative* of $\mathbf{f}$ at $\mathbf{c}$. We also write (5) in the form

$$\mathbf{f}(\mathbf{c} + \mathbf{v}) = \mathbf{f}(\mathbf{c}) + \mathbf{T_c}(\mathbf{v}) + o(\|\mathbf{v}\|) \qquad \text{as } \mathbf{v} \to \mathbf{0}.$$

The next theorem shows that if the total derivative exists, it is unique. It also relates the total derivative to directional derivatives.

Theorem 12.3. *Assume $\mathbf{f}$ is differentiable at $\mathbf{c}$ with total derivative $\mathbf{T_c}$. Then the directional derivative $\mathbf{f}'(\mathbf{c}; \mathbf{u})$ exists for every $\mathbf{u}$ in $\mathbf{R}^n$ and we have*

$$\mathbf{T_c}(\mathbf{u}) = \mathbf{f}'(\mathbf{c}; \mathbf{u}). \qquad (6)$$

Proof. If $\mathbf{v} = \mathbf{0}$ then $\mathbf{f}'(\mathbf{c}; \mathbf{0}) = \mathbf{0}$ and $\mathbf{T_c}(\mathbf{0}) = \mathbf{0}$. Therefore we can assume that $\mathbf{v} \neq \mathbf{0}$. Take $\mathbf{v} = h\mathbf{u}$ in Taylor's formula (5), with $h \neq 0$, to get

$$\mathbf{f}(\mathbf{c} + h\mathbf{u}) - \mathbf{f}(\mathbf{c}) = \mathbf{T_c}(h\mathbf{u}) + \|h\mathbf{u}\| \, \mathbf{E_c}(\mathbf{v}) = h\mathbf{T_c}(\mathbf{u}) + |h| \, \|\mathbf{u}\| \, \mathbf{E_c}(\mathbf{v}).$$

Now divide by h and let $h \to 0$ to obtain (6).

Theorem 12.4. *If $\mathbf{f}$ is differentiable at $\mathbf{c}$, then $\mathbf{f}$ is continuous at $\mathbf{c}$.*

Proof. Let $\mathbf{v} \to \mathbf{0}$ in the Taylor formula (5). The error term $\|\mathbf{v}\| \, \mathbf{E_c}(\mathbf{v}) \to \mathbf{0}$; the linear term $\mathbf{T_c}(\mathbf{v})$ also tends to $\mathbf{0}$ because if $\mathbf{v} = v_1\mathbf{u}_1 + \cdots + v_n\mathbf{u}_n$, where $\mathbf{u}_1, \ldots, \mathbf{u}_n$ are the unit coordinate vectors, then by linearity we have

$$\mathbf{T_c}(\mathbf{u}) = v_1\mathbf{T_c}(\mathbf{u}_1) + \cdots + v_n\mathbf{T_c}(\mathbf{u}_n),$$

and each term on the right tends to $\mathbf{0}$ as $\mathbf{v} \to \mathbf{0}$.

NOTE. The total derivative $\mathbf{T_c}$ is also written as $\mathbf{f}'(\mathbf{c})$ to resemble the notation used in the one-dimensional theory. With this notation, the Taylor formula (5) takes the form

$$\mathbf{f}(\mathbf{c} + \mathbf{v}) = \mathbf{f}(\mathbf{c}) + \mathbf{f}'(\mathbf{c})(\mathbf{v}) + \|\mathbf{v}\| \, \mathbf{E_c}(\mathbf{v}), \qquad (7)$$

where $\mathbf{E_c}(\mathbf{v}) \to \mathbf{0}$ as $\mathbf{v} \to \mathbf{0}$. However, it should be realized that $\mathbf{f}'(\mathbf{c})$ is a *linear function*, not a number. It is defined everywhere on $\mathbf{R}^n$; the vector $\mathbf{f}'(\mathbf{c})(\mathbf{v})$ is the value of $\mathbf{f}'(\mathbf{c})$ at $\mathbf{v}$.

Example. If $\mathbf{f}$ is itself a linear function, then $\mathbf{f}(\mathbf{c} + \mathbf{v}) = \mathbf{f}(\mathbf{c}) + \mathbf{f}(\mathbf{v})$, so the derivative $\mathbf{f}'(\mathbf{c})$ exists for every $\mathbf{c}$ and equals $\mathbf{f}$. In other words, the total derivative of a linear function is the function itself.

12.5 THE TOTAL DERIVATIVE EXPRESSED IN TERMS OF PARTIAL DERIVATIVES

The next theorem shows that the vector $\mathbf{f}'(\mathbf{c})(\mathbf{v})$ is a linear combination of the partial derivatives of $\mathbf{f}$.

Theorem 12.5. *Let $\mathbf{f}: S \to \mathbf{R}^m$ be differentiable at an interior point $\mathbf{c}$ of S, where $S \subseteq \mathbf{R}^n$. If $\mathbf{v} = v_1\mathbf{u}_1 + \cdots + v_n\mathbf{u}_n$, where $\mathbf{u}_1, \ldots, \mathbf{u}_n$ are the unit coordinate* *vectors in* $\mathbf{R}^n$, *then*

$$\mathbf{f}'(\mathbf{c})(\mathbf{v}) = \sum_{k=1}^{n} v_k D_k \mathbf{f}(\mathbf{c}).$$

In particular, if f is real-valued $(m = 1)$ *we have*

$$f'(\mathbf{c})(\mathbf{v}) = \nabla f(\mathbf{c}) \cdot \mathbf{v}, \tag{8}$$

the dot product of $\mathbf{v}$ *with the vector* $\nabla f(\mathbf{c}) = (D_1 f(\mathbf{c}), \dots, D_n f(\mathbf{c}))$.

Proof. We use the linearity of $\mathbf{f}'(\mathbf{c})$ to write

$$\mathbf{f}'(\mathbf{c})(\mathbf{v}) = \sum_{k=1}^{n} \mathbf{f}'(\mathbf{c})(v_k \mathbf{u}_k) = \sum_{k=1}^{n} v_k \mathbf{f}'(\mathbf{c})(\mathbf{u}_k)$$

$$= \sum_{k=1}^{n} v_k \mathbf{f}'(\mathbf{c}; \mathbf{u}_k) = \sum_{k=1}^{n} v_k D_k \mathbf{f}(\mathbf{c}).$$

NOTE. The vector $\nabla f(\mathbf{c})$ in (8) is called the *gradient vector* of f at $\mathbf{c}$. It is defined at each point where the partials $D_1 f, \dots, D_n f$ exist. The Taylor formula for real-valued f now takes the form

$$f(\mathbf{c} + \mathbf{v}) = f(\mathbf{c}) + \nabla f(\mathbf{c}) \cdot \mathbf{v} + o(\|\mathbf{v}\|) \qquad \text{as } \mathbf{v} \to \mathbf{0}.$$

12.6 AN APPLICATION TO COMPLEX-VALUED FUNCTIONS

Let $f = u + iv$ be a complex-valued function of a complex variable. Theorem 5.22 showed that a necessary condition for f to have a derivative at a point c is that the four partials $D_1 u, D_2 u, D_1 v, D_2 v$ exist at c and satisfy the Cauchy–Riemann equations:

$$D_1 u(c) = D_2 v(c), \qquad D_1 v(c) = -D_2 u(c).$$

Also, an example showed that the equations by themselves are not sufficient for existence of $f'(c)$. The next theorem shows that the Cauchy–Riemann equations, along with differentiability of u and v, imply existence of $f'(c)$.

Theorem 12.6. *Let u and v be two real-valued functions defined on a subset S of the complex plane. Assume also that u and v are differentiable at an interior point c of S and that the partial derivatives satisfy the Cauchy–Riemann equations at c. Then the function $f = u + iv$ has a derivative at c. Moreover,*

$$f'(c) = D_1 u(c) + i D_1 v(c).$$

Proof. We have $f(z) - f(c) = u(z) - u(c) + i\{v(z) - v(c)\}$ for each z in S. Since each of u and v is differentiable at c, for z sufficiently near to c we have

$$u(z) - u(c) = \nabla u(c) \cdot (z - c) + o(\|z - c\|)$$

and

$$v(z) - v(c) = \nabla v(c) \cdot (z - c) + o(\|z - c\|).$$

Here we use vector notation and consider complex numbers as vectors in $\mathbf{R}^2$. We then have

$$f(z) - f(c) = \{\nabla u(c) + i\,\nabla v(c)\} \cdot (z - c) + o(\|z - c\|).$$

Writing $z = x + iy$ and $c = a + ib$, we find

$$\{\nabla u(c) + i\,\nabla v(c)\} \cdot (z - c)$$
$$= D_1 u(c)(x - a) + D_2 u(c)(y - b) + i\,\{D_1 v(c)(x - a) + D_2 v(c)(y - b)\}$$
$$= D_1 u(c)\{(x - a) + i(y - b)\} + iD_1 v(c)\{(x - a) + i(y - b)\},$$

because of the Cauchy–Riemann equations. Hence

$$f(z) - f(c) = \{D_1 u(c) + iD_1 v(c)\}\,(z - c) + o(\|z - c\|).$$

Dividing by $z - c$ and letting $z \to c$ we see that $f'(c)$ exists and is equal to

$$D_1 u(c) + iD_1 v(c).$$

12.7 THE MATRIX OF A LINEAR FUNCTION

In this section we digress briefly to record some elementary facts from linear algebra that are useful in certain calculations with derivatives.

Let $\mathbf{T} : \mathbf{R}^n \to \mathbf{R}^m$ be a linear function. (In our applications, $\mathbf{T}$ will be the total derivative of a function $\mathbf{f}$.) We will show that $\mathbf{T}$ determines an $m \times n$ matrix of scalars (see (9) below) which is obtained as follows:

Let $\mathbf{u}_1, \ldots, \mathbf{u}_n$ denote the unit coordinate vectors in $\mathbf{R}^n$. If $\mathbf{x} \in \mathbf{R}^n$ we have $\mathbf{x} = x_1\mathbf{u}_1 + \cdots + x_n\mathbf{u}_n$ so, by linearity,

$$\mathbf{T}(\mathbf{x}) = \sum_{k=1}^{n} x_k \mathbf{T}(\mathbf{u}_k).$$

Therefore $\mathbf{T}$ is completely determined by its action on the coordinate vectors $\mathbf{u}_1, \ldots, \mathbf{u}_n$.

Now let $\mathbf{e}_1, \ldots, \mathbf{e}_m$ denote the unit coordinate vectors in $\mathbf{R}^m$. Since $\mathbf{T}(\mathbf{u}_k) \in \mathbf{R}^m$, we can write $\mathbf{T}(\mathbf{u}_k)$ as a linear combination of $\mathbf{e}_1, \ldots, \mathbf{e}_m$, say

$$\mathbf{T}(\mathbf{u}_k) = \sum_{i=1}^{m} t_{ik}\mathbf{e}_i.$$

The scalars $t_{1k}, \ldots, t_{mk}$ are the coordinates of $\mathbf{T}(\mathbf{u}_k)$. We display these scalars vertically as follows:

$$\begin{bmatrix} t_{1k} \\ t_{2k} \\ \vdots \\ t_{mk} \end{bmatrix}.$$

This array is called a *column vector*. We form the column vector for each of $T(u_1), \ldots, T(u_n)$ and place them side by side to obtain the rectangular array

$$\begin{bmatrix} t_{11} & t_{12} & \cdots & t_{1n} \\ t_{21} & t_{22} & \cdots & t_{2n} \\ \vdots & \vdots & & \vdots \\ t_{m1} & t_{m2} & \cdots & t_{mn} \end{bmatrix}. \tag{9}$$

This is called the *matrix* of* T and is denoted by $m(T)$. It consists of m rows and n columns. The numbers going down the kth column are the components of $T(u_k)$. We also use the notation

$$m(T) = [t_{ik}]_{i,k=1}^{m,n} \quad \text{or} \quad m(T) = (t_{ik})$$

to denote the matrix in (9).

Now let $T : \mathbf{R}^n \to \mathbf{R}^m$ and $S : \mathbf{R}^m \to \mathbf{R}^p$ be two linear functions, with the domain of S containing the range of T. Then we can form the composition $S \circ T$ defined by

$$(S \circ T)(x) = S[T(x)] \quad \text{for all } x \text{ in } \mathbf{R}^n.$$

The composition $S \circ T$ is also linear and it maps $\mathbf{R}^n$ into $\mathbf{R}^p$.

Let us calculate the matrix $m(S \circ T)$. Denote the unit coordinate vectors in $\mathbf{R}^n$, $\mathbf{R}^m$, and $\mathbf{R}^p$, respectively, by

$$u_1, \ldots, u_n, \quad e_1, \ldots, e_m, \quad \text{and} \quad w_1, \ldots, w_p.$$

Suppose that S and T have matrices (s_{ij}) and (t_{ij}), respectively. This means that

$$S(e_k) = \sum_{i=1}^{p} s_{ik} w_i \quad \text{for } k = 1, 2, \ldots, m$$

and

$$T(u_j) = \sum_{k=1}^{m} t_{kj} e_k \quad \text{for } j = 1, 2, \ldots, n.$$

Then

$$(S \circ T)(u_j) = S[T(u_j)] = \sum_{k=1}^{m} t_{kj} S(e_k) = \sum_{k=1}^{m} t_{kj} \sum_{i=1}^{p} s_{ik} w_i$$

$$= \sum_{i=1}^{p} \left(\sum_{k=1}^{m} s_{ik} t_{kj} \right) w_i$$

so

$$m(S \circ T) = \left[\sum_{k=1}^{m} s_{ik} t_{kj} \right]_{i,j=1}^{p,n}.$$

In other words, $m(S \circ T)$ is a $p \times n$ matrix whose entry in the ith row and jth

---

* More precisely, the matrix of T relative to the given bases $u_1, \ldots, u_n$ of $\mathbf{R}^n$ and $e_1, \ldots, e_m$ of $\mathbf{R}^m$.

column is

$$\sum_{k=1}^{m} s_{ik} t_{kj},$$

the dot product of the ith row of $m(\mathbf{S})$ with the jth column of $m(\mathbf{T})$. This matrix is also called the *product* $m(\mathbf{S})m(\mathbf{T})$. Thus, $m(\mathbf{S} \circ \mathbf{T}) = m(\mathbf{S})m(\mathbf{T})$.

12.8 THE JACOBIAN MATRIX

Next we show how matrices arise in connection with total derivatives.

Let $\mathbf{f}$ be a function with values in $\mathbf{R}^m$ which is differentiable at a point $\mathbf{c}$ in $\mathbf{R}^n$, and let $\mathbf{T} = \mathbf{f}'(\mathbf{c})$ be the total derivative of $\mathbf{f}$ at $\mathbf{c}$. To find the matrix of $\mathbf{T}$ we consider its action on the unit coordinate vectors $\mathbf{u}_1, \ldots, \mathbf{u}_n$. By Theorem 12.3 we have

$$\mathbf{T}(\mathbf{u}_k) = \mathbf{f}'(\mathbf{c}; \mathbf{u}_k) = D_k \mathbf{f}(\mathbf{c}).$$

To express this as a linear combination of the unit coordinate vectors $\mathbf{e}_1, \ldots, \mathbf{e}_m$ of $\mathbf{R}^m$ we write $\mathbf{f} = (f_1, \ldots, f_m)$ so that $D_k \mathbf{f} = (D_k f_1, \ldots, D_k f_m)$, and hence

$$\mathbf{T}(\mathbf{u}_k) = D_k \mathbf{f}(\mathbf{c}) = \sum_{i=1}^{m} D_k f_i(\mathbf{c}) \mathbf{e}_i.$$

Therefore the matrix of $\mathbf{T}$ is $m(\mathbf{T}) = (D_k f_i(\mathbf{c}))$. This is called the *Jacobian matrix* of $\mathbf{f}$ at $\mathbf{c}$ and is denoted by $\mathbf{Df}(\mathbf{c})$. That is,

$$\mathbf{Df}(\mathbf{c}) = \begin{bmatrix} D_1 f_1(\mathbf{c}) & D_2 f_1(\mathbf{c}) & \cdots & D_n f_1(\mathbf{c}) \\ D_1 f_2(\mathbf{c}) & D_2 f_2(\mathbf{c}) & \cdots & D_n f_2(\mathbf{c}) \\ \vdots & \vdots & & \vdots \\ D_1 f_m(\mathbf{c}) & D_2 f_m(\mathbf{c}) & \cdots & D_n f_m(\mathbf{c}) \end{bmatrix}. \tag{10}$$

The entry in the ith row and kth column is $D_k f_i(\mathbf{c})$. Thus, to get the entries in the kth column, differentiate the components of $\mathbf{f}$ with respect to the kth coordinate vector. The Jacobian matrix $\mathbf{Df}(\mathbf{c})$ is defined at each point $\mathbf{c}$ in $\mathbf{R}^n$ where all the partial derivatives $D_k f_i(\mathbf{c})$ exist.

The kth row of the Jacobian matrix (10) is a vector in $\mathbf{R}^n$ called the *gradient vector* of f_k, denoted by $\nabla f_k(\mathbf{c})$. That is,

$$\nabla f_k(\mathbf{c}) = (D_1 f_k(\mathbf{c}), \ldots, D_n f_k(\mathbf{c})).$$

In the special case when f is real-valued ($m = 1$), the Jacobian matrix consists of only one row. In this case $\mathbf{Df}(\mathbf{c}) = \nabla f(\mathbf{c})$, and Equation (8) of Theorem 12.5 shows that the directional derivative $f'(\mathbf{c}; \mathbf{v})$ is the dot product of the gradient vector $\nabla f(\mathbf{c})$ with the direction $\mathbf{v}$.

For a vector-valued function $\mathbf{f} = (f_1, \ldots, f_m)$ we have

$$\mathbf{f}'(\mathbf{c})(\mathbf{v}) = \mathbf{f}'(\mathbf{c}; \mathbf{v}) = \sum_{k=1}^{m} f_k'(\mathbf{c}; \mathbf{v}) \mathbf{e}_k = \sum_{k=1}^{m} \{\nabla f_k(\mathbf{c}) \cdot \mathbf{v}\} \mathbf{e}_k, \tag{11}$$

so the vector $\mathbf{f}'(\mathbf{c})(\mathbf{v})$ has components

$$(\nabla f_1(\mathbf{c}) \cdot \mathbf{v}, \ldots, \nabla f_m(\mathbf{c}) \cdot \mathbf{v}).$$

Thus, the components of $\mathbf{f}'(\mathbf{c})(\mathbf{v})$ are obtained by taking the dot product of the successive rows of the Jacobian matrix with the vector $\mathbf{v}$. If we regard $\mathbf{f}'(\mathbf{c})(\mathbf{v})$ as an $m \times 1$ matrix, or column vector, then $\mathbf{f}'(\mathbf{c})(\mathbf{v})$ is equal to the matrix product $\mathbf{Df}(\mathbf{c})\mathbf{v}$, where $\mathbf{Df}(\mathbf{c})$ is the $m \times n$ Jacobian matrix and $\mathbf{v}$ is regarded as an $n \times 1$ matrix, or column vector.

NOTE. Equation (11), used in conjunction with the triangle inequality and the Cauchy–Schwarz inequality, gives us

$$\| \mathbf{f}'(\mathbf{c})(\mathbf{v}) \| = \left\| \sum_{k=1}^{m} \{\nabla f_k(\mathbf{c}) \cdot \mathbf{v}\} \mathbf{e}_k \right\| \le \sum_{k=1}^{m} |\nabla f_k(\mathbf{c}) \cdot \mathbf{v}| \le \|\mathbf{v}\| \sum_{k=1}^{m} \|\nabla f_k(\mathbf{c})\|.$$

Therefore we have

$$\|\mathbf{f}'(\mathbf{c})(\mathbf{v})\| \le M\|\mathbf{v}\|, \tag{12}$$

where $M = \sum_{k=1}^{m} \|\nabla f_k(\mathbf{c})\|$. This inequality will be used in the proof of the chain rule. It also shows that $\mathbf{f}'(\mathbf{c})(\mathbf{v}) \to \mathbf{0}$ as $\mathbf{v} \to \mathbf{0}$.

12.9 THE CHAIN RULE

Let $\mathbf{f}$ and $\mathbf{g}$ be functions such that the composition $\mathbf{h} = \mathbf{f} \circ \mathbf{g}$ is defined in a neighborhood of a point $\mathbf{a}$. The chain rule tells us how to compute the total derivative of $\mathbf{h}$ in terms of total derivatives of $\mathbf{f}$ and of $\mathbf{g}$.

Theorem 12.7. *Assume that $\mathbf{g}$ is differentiable at $\mathbf{a}$, with total derivative $\mathbf{g}'(\mathbf{a})$. Let $\mathbf{b} = \mathbf{g}(\mathbf{a})$ and assume that $\mathbf{f}$ is differentiable at $\mathbf{b}$, with total derivative $\mathbf{f}'(\mathbf{b})$. Then the composite function $\mathbf{h} = \mathbf{f} \circ \mathbf{g}$ is differentiable at $\mathbf{a}$, and the total derivative $\mathbf{h}'(\mathbf{a})$ is given by*

$$\mathbf{h}'(\mathbf{a}) = \mathbf{f}'(\mathbf{b}) \circ \mathbf{g}'(\mathbf{a}),$$

the composition of the linear functions $\mathbf{f}'(\mathbf{b})$ and $\mathbf{g}'(\mathbf{a})$.

Proof. We consider the difference $\mathbf{h}(\mathbf{a} + \mathbf{y}) - \mathbf{h}(\mathbf{a})$ for small $\|\mathbf{y}\|$, and show that we have a first-order Taylor formula. We have

$$\mathbf{h}(\mathbf{a} + \mathbf{y}) - \mathbf{h}(\mathbf{a}) = \mathbf{f}[\mathbf{g}(\mathbf{a} + \mathbf{y})] - \mathbf{f}[\mathbf{g}(\mathbf{a})] = \mathbf{f}(\mathbf{b} + \mathbf{v}) - \mathbf{f}(\mathbf{b}), \tag{13}$$

where $\mathbf{b} = \mathbf{g}(\mathbf{a})$ and $\mathbf{v} = \mathbf{g}(\mathbf{a} + \mathbf{y}) - \mathbf{b}$. The Taylor formula for $\mathbf{g}(\mathbf{a} + \mathbf{y})$ implies

$$\mathbf{v} = \mathbf{g}'(\mathbf{a})(\mathbf{y}) + \|\mathbf{y}\| \, \mathbf{E}_\mathbf{a}(\mathbf{y}), \qquad \text{where } \mathbf{E}_\mathbf{a}(\mathbf{y}) \to \mathbf{0} \text{ as } \mathbf{y} \to \mathbf{0}. \tag{14}$$

The Taylor formula for $\mathbf{f}(\mathbf{b} + \mathbf{v})$ implies

$$\mathbf{f}(\mathbf{b} + \mathbf{v}) - \mathbf{f}(\mathbf{b}) = \mathbf{f}'(\mathbf{b})(\mathbf{v}) + \|\mathbf{v}\| \, \mathbf{E}_\mathbf{b}(\mathbf{v}), \qquad \text{where } \mathbf{E}_\mathbf{b}(\mathbf{v}) \to \mathbf{0} \text{ as } \mathbf{v} \to \mathbf{0}. \tag{15}$$

Using (14) in (15) we find

$$\mathbf{f(b + v) - f(b) = f'(b)[g'(a)(y)] + f'(b)[\|y\| \ E_a(y)] + \|v\| \ E_b(v)}$$

$$\mathbf{= f'(b)[g'(a)(y)] + \|y\| \ E(y)}, \tag{16}$$

where $\mathbf{E(0) = 0}$ and

$$\mathbf{E(y) = f'(b)[E_a(y)] + \dfrac{\|v\|}{\|y\|} \ E_b(v)} \qquad \text{if } \mathbf{y \neq 0}. \tag{17}$$

To complete the proof we need to show that $\mathbf{E(y) \to 0}$ as $\mathbf{y \to 0}$.

The first term on the right of (17) tends to $\mathbf{0}$ as $\mathbf{y \to 0}$ because $\mathbf{E_a(y) \to 0}$. In the second term, the factor $\mathbf{E_b(v) \to 0}$ because $\mathbf{v \to 0}$ as $\mathbf{y \to 0}$. Now we show that the quotient $\|\mathbf{v}\|/\|\mathbf{y}\|$ remains bounded as $\mathbf{y \to 0}$. Using (14) and (12) to estimate the numerator we find

$$\mathbf{\|v\| \leq \|g'(a)(y)\| + \|y\| \ \|E_a(y)\| \leq \|y\|\{M + \|E_a(y)\|\}},$$

where $M = \sum_{k=1}^{m} \|\nabla g_k(\mathbf{a})\|$. Hence

$$\dfrac{\|\mathbf{v}\|}{\|\mathbf{y}\|} \leq M + \|\mathbf{E_a(y)}\|,$$

so $\|\mathbf{v}\|/\|\mathbf{y}\|$ remains bounded as $\mathbf{y \to 0}$. Using (13) and (16) we obtain the Taylor formula

$$\mathbf{h(a + y) - h(a) = f'(b)[g'(a)(y)] + \|y\| \ E(y)},$$

where $\mathbf{E(y) \to 0}$ as $\mathbf{y \to 0}$. This proves that $\mathbf{h}$ is differentiable at $\mathbf{a}$ and that its total derivative at $\mathbf{a}$ is the composition $\mathbf{f'(b) \circ g'(a)}$.

12.10 MATRIX FORM OF THE CHAIN RULE

The chain rule states that

$$\mathbf{h'(a) = f'(b) \circ g'(a)}, \tag{18}$$

where $\mathbf{h = f \circ g}$ and $\mathbf{b = g(a)}$. Since the matrix of a composition is the product of the corresponding matrices, (18) implies the following relation for Jacobian matrices:

$$\mathbf{Dh(a) = Df(b)Dg(a)}. \tag{19}$$

This is called the *matrix form of the chain rule*. It can also be written as a set of scalar equations by expressing each matrix in terms of its entries.

Specifically, suppose that $\mathbf{a} \in \mathbf{R}^p$, $\mathbf{b = g(a)} \in \mathbf{R}^n$, and $\mathbf{f(b)} \in \mathbf{R}^m$. Then $\mathbf{h(a)} \in \mathbf{R}^m$ and we can write

$$\mathbf{g} = (g_1, \ldots, g_n), \qquad \mathbf{f} = (f_1, \ldots, f_m), \qquad \mathbf{h} = (h_1, \ldots, h_m).$$

Then $\mathbf{Dh(a)}$ is an $m \times p$ matrix, $\mathbf{Df(b)}$ is an $m \times n$ matrix, and $\mathbf{Dg(a)}$ is an $n \times p$

matrix, given by

$$\mathbf{Dh(a)} = [D_j h_i(\mathbf{a})]_{i,j=1}^{m,p}, \qquad \mathbf{Df(b)} = [D_k f_i(\mathbf{b})]_{i,k=1}^{m,n}, \qquad \mathbf{Dg(a)} = [D_j g_k(\mathbf{a})]_{k,j=1}^{n,p}.$$

The matrix equation (19) is equivalent to the mp scalar equations

$$D_j h_i(\mathbf{a}) = \sum_{k=1}^{n} D_k f_i(\mathbf{b}) D_j g_k(\mathbf{a}), \qquad \text{for } i = 1, 2, \ldots, m \quad \text{and} \quad j = 1, 2, \ldots, p. \quad (20)$$

These equations express the partial derivatives of the components of $\mathbf{h}$ in terms of the partial derivatives of the components of $\mathbf{f}$ and $\mathbf{g}$.

The equations in (20) can be put in a form that is easier to remember. Write $\mathbf{y} = \mathbf{f(x)}$ and $\mathbf{x} = \mathbf{g(t)}$. Then $\mathbf{y} = \mathbf{f[g(t)]} = \mathbf{h(t)}$, and (20) becomes

$$\frac{\partial y_i}{\partial t_j} = \sum_{k=1}^{n} \frac{\partial y_i}{\partial x_k} \frac{\partial x_k}{\partial t_j}, \qquad (21)$$

where

$$\frac{\partial y_i}{\partial t_j} = D_j h_i, \qquad \frac{\partial y_i}{\partial x_k} = D_k f_i, \qquad \text{and} \qquad \frac{\partial x_k}{\partial t_j} = D_j g_k.$$

Examples. Suppose $m = 1$. Then both f and $h = f \circ \mathbf{g}$ are real-valued and there are p equations in (20), one for each of the partial derivatives of h:

$$D_j h(\mathbf{a}) = \sum_{k=1}^{n} D_k f(\mathbf{b}) D_j g_k(\mathbf{a}), \qquad j = 1, 2, \ldots, p.$$

The right member is the dot product of the two vectors $\nabla f(\mathbf{b})$ and $D_j \mathbf{g(a)}$. In this case Equation (21) takes the form

$$\frac{\partial y}{\partial t_j} = \sum_{k=1}^{n} \frac{\partial y}{\partial x_k} \frac{\partial x_k}{\partial t_j}, \qquad j = 1, 2, \ldots, p.$$

In particular, if $p = 1$ we get only one equation,

$$h'(\mathbf{a}) = \sum_{k=1}^{n} D_k f(\mathbf{b}) g_k'(\mathbf{a}) = \nabla f(\mathbf{b}) \cdot \mathbf{Dg(a)},$$

where the Jacobian matrix $\mathbf{Dg(a)}$ is a column vector.

The chain rule can be used to give a simple proof of the following theorem for differentiating an integral with respect to a parameter which appears both in the integrand and in the limits of integration.

Theorem 12.8. *Let f and $D_2 f$ be continuous on a rectangle $[a, b] \times [c, d]$. Let p and q be differentiable on $[c, d]$, where $p(y) \in [a, b]$ and $q(y) \in [a, b]$ for each y in $[c, d]$. Define F by the equation*

$$F(y) = \int_{p(y)}^{q(y)} f(x, y) \, dx, \qquad \text{if } y \in [c, d].$$

Then $F'(y)$ exists for each y in (c, d) and is given by

$$F'(y) = \int_{p(y)}^{q(y)} D_2 f(x, y) \, dx + f(q(y), y)q'(y) - f(p(y), y)p'(y).$$

Proof. Let $G(x_1, x_2, x_3) = \int_{x_1}^{x_2} f(t, x_3) \, dt$ whenever x_1 and x_2 are in $[a, b]$ and $x_3 \in [c, d]$. Then F is the composite function given by $F(y) = G(p(y), q(y), y)$. The chain rule implies

$$F'(y) = D_1 G(p(y), q(y), y)p'(y) + D_2 G(p(y), q(y), y)q'(y) + D_3 G(p(y), q(y), y).$$

By Theorem 7.32, we have $D_1 G(x_1, x_2, x_3) = -f(x_1, x_3)$ and $D_2 G(x_1, x_2, x_3) = f(x_2, x_3)$. By Theorem 7.40, we also have

$$D_3 G(x_1, x_2, x_3) = \int_{x_1}^{x_2} D_2 f(t, x_3) \, dt.$$

Using these results in the formula for $F'(y)$ we obtain the theorem.

12.11 THE MEAN-VALUE THEOREM FOR DIFFERENTIABLE FUNCTIONS

The Mean-Value Theorem for functions from $\mathbf{R}^1$ to $\mathbf{R}^1$ states that

$$f(y) - f(x) = f'(z)(y - x), \tag{22}$$

where z lies between x and y. This equation is false, in general, for vector-valued functions from $\mathbf{R}^n$ to $\mathbf{R}^m$, when $m > 1$. (See Exercise 12.19.) However, we will show that a correct equation is obtained by taking the dot product of each member of (22) with any vector in $\mathbf{R}^m$, provided z is suitably chosen. This gives a useful generalization of the Mean-Value Theorem for vector-valued functions.

In the statement of the theorem we use the notation $L(\mathbf{x}, \mathbf{y})$ to denote the line segment joining two points $\mathbf{x}$ and $\mathbf{y}$ in $\mathbf{R}^n$. That is,

$$L(\mathbf{x}, \mathbf{y}) = \{t\mathbf{x} + (1 - t)\mathbf{y} : 0 \le t \le 1\}.$$

Theorem 12.9 (*Mean-Value Theorem.*) *Let S be an open subset of $\mathbf{R}^n$ and assume that $\mathbf{f} : S \to \mathbf{R}^m$ is differentiable at each point of S. Let $\mathbf{x}$ and $\mathbf{y}$ be two points in S such that $L(\mathbf{x}, \mathbf{y}) \subseteq S$. Then for every vector $\mathbf{a}$ in $\mathbf{R}^m$ there is a point $\mathbf{z}$ in $L(\mathbf{x}, \mathbf{y})$ such that*

$$\mathbf{a} \cdot \{\mathbf{f}(\mathbf{y}) - \mathbf{f}(\mathbf{x})\} = \mathbf{a} \cdot \{\mathbf{f}'(\mathbf{z})(\mathbf{y} - \mathbf{x})\}. \tag{23}$$

Proof. Let $\mathbf{u} = \mathbf{y} - \mathbf{x}$. Since S is open and $L(\mathbf{x}, \mathbf{y}) \subseteq S$, there is a $\delta > 0$ such that $\mathbf{x} + t\mathbf{u} \in S$ for all real t in the interval $(-\delta, 1 + \delta)$. Let $\mathbf{a}$ be a fixed vector in $\mathbf{R}^m$ and let F be the real-valued function defined on $(-\delta, 1 + \delta)$ by the equation

$$F(t) = \mathbf{a} \cdot \mathbf{f}(\mathbf{x} + t\mathbf{u}).$$

Then F is differentiable on $(-\delta, 1 + \delta)$ and its derivative is given by

$$F'(t) = \mathbf{a} \cdot \mathbf{f}'(\mathbf{x} + t\mathbf{u}; \mathbf{u}) = \mathbf{a} \cdot \{\mathbf{f}'(\mathbf{x} + t\mathbf{u})(\mathbf{u})\}.$$

By the usual Mean-Value Theorem we have

$$F(1) - F(0) = F'(\theta), \qquad \text{where } 0 < \theta < 1.$$

Now

$$F'(\theta) = \mathbf{a} \cdot \{\mathbf{f}'(\mathbf{x} + \theta\mathbf{u})(\mathbf{u})\} = \mathbf{a} \cdot \{\mathbf{f}'(\mathbf{z})(\mathbf{y} - \mathbf{x})\},$$

where $\mathbf{z} = \mathbf{x} + \theta\mathbf{u} \in L(\mathbf{x}, \mathbf{y})$. But $F(1) - F(0) = \mathbf{a} \cdot \{\mathbf{f}(\mathbf{y}) - \mathbf{f}(\mathbf{x})\}$, so we obtain (23). Of course, the point $\mathbf{z}$ depends on F, and hence on $\mathbf{a}$.

NOTE. If S is convex, then $L(\mathbf{x}, \mathbf{y}) \subseteq S$ for all $\mathbf{x}$, $\mathbf{y}$ in S so (23) holds for all $\mathbf{x}$ and $\mathbf{y}$ in S.

Examples

1. If f is real-valued ($m = 1$) we can take $a = 1$ in (23) to obtain

$$f(\mathbf{y}) - f(\mathbf{x}) = f'(\mathbf{z})(\mathbf{y} - \mathbf{x}) = \nabla f(\mathbf{z}) \cdot (\mathbf{y} - \mathbf{x}). \tag{24}$$

2. If $\mathbf{f}$ is vector-valued and if $\mathbf{a}$ is a unit vector in $\mathbf{R}^m$, $\|\mathbf{a}\| = 1$, Eq. (23) and the Cauchy–Schwarz inequality give us

$$\|\mathbf{f}(\mathbf{y}) - \mathbf{f}(\mathbf{x})\| \le \|\mathbf{f}'(\mathbf{z})(\mathbf{y} - \mathbf{x})\|.$$

Using (12) we obtain the inequality

$$\|\mathbf{f}(\mathbf{y}) - \mathbf{f}(\mathbf{x})\| \le M\|\mathbf{y} - \mathbf{x}\|,$$

where $M = \sum_{k=1}^{m} \|\nabla f_k(\mathbf{z})\|$. Note that M depends on $\mathbf{z}$ and hence on $\mathbf{x}$ and $\mathbf{y}$.

3. If S is convex and if all the partial derivatives $D_j f_k$ are bounded on S, then there is a constant $A > 0$ such that

$$\|\mathbf{f}(\mathbf{y}) - \mathbf{f}(\mathbf{x})\| \le A\|\mathbf{y} - \mathbf{x}\|.$$

In other words, $\mathbf{f}$ satisfies a Lipschitz condition on S.

The Mean-Value Theorem gives a simple proof of the following result concerning functions with zero total derivative.

Theorem 12.10. *Let S be an open connected subset of $\mathbf{R}^n$, and let $\mathbf{f} : S \to \mathbf{R}^m$ be differentiable at each point of S. If $\mathbf{f}'(\mathbf{c}) = \mathbf{0}$ for each $\mathbf{c}$ in S, then $\mathbf{f}$ is constant on S.*

Proof. Since S is open and connected, it is polygonally connected . (See Section 4.18.) Therefore, every pair of points $\mathbf{x}$ and $\mathbf{y}$ in S can be joined by a polygonal arc lying in S. Denote the vertices of this arc by $\mathbf{p}_1, \ldots, \mathbf{p}_r$, where $\mathbf{p}_1 = \mathbf{x}$ and $\mathbf{p}_r = \mathbf{y}$. Since each segment $L(\mathbf{p}_{i+1}, \mathbf{p}_i) \subseteq S$, the Mean-Value Theorem shows that

$$\mathbf{a} \cdot \{\mathbf{f}(\mathbf{p}_{i+1}) - \mathbf{f}(\mathbf{p}_i)\} = 0,$$

for every vector $\mathbf{a}$. Adding these equations for $i = 1, 2, \ldots, r - 1$, we find

$$\mathbf{a} \cdot \{\mathbf{f}(\mathbf{y}) - \mathbf{f}(\mathbf{x})\} = 0,$$

for every $\mathbf{a}$. Taking $\mathbf{a} = \mathbf{f}(\mathbf{y}) - \mathbf{f}(\mathbf{x})$ we find $\mathbf{f}(\mathbf{x}) = \mathbf{f}(\mathbf{y})$, so $\mathbf{f}$ is constant on S.

12.12 A SUFFICIENT CONDITION FOR DIFFERENTIABILITY

Up to now we have been deriving consequences of the hypothesis that a function is differentiable. We have also seen that neither the existence of all partial derivatives nor the existence of all directional derivatives suffices to establish differentiability (since neither implies continuity). The next theorem shows that continuity of all but one of the partials does imply differentiability.

Theorem 12.11. *Assume that one of the partial derivatives $D_1\mathbf{f}, \ldots, D_n\mathbf{f}$ exists at $\mathbf{c}$ and that the remaining $n - 1$ partial derivatives exist in some n-ball $B(\mathbf{c})$ and are continuous at $\mathbf{c}$. Then $\mathbf{f}$ is differentiable at $\mathbf{c}$.*

Proof. First we note that a vector-valued function $\mathbf{f} = (f_1, \ldots, f_m)$ is differentiable at $\mathbf{c}$ if, and only if, each component f_k is differentiable at $\mathbf{c}$. (The proof of this is an easy exercise.) Therefore, it suffices to prove the theorem when $\mathbf{f}$ is real-valued.

For the proof we suppose that $D_1 f(\mathbf{c})$ exists and that the continuous partials are $D_2 f, \ldots, D_n f$.

The only candidate for $f'(\mathbf{c})$ is the gradient vector $\nabla f(\mathbf{c})$. We will prove that

$$f(\mathbf{c} + \mathbf{v}) - f(\mathbf{c}) = \nabla f(\mathbf{c}) \cdot \mathbf{v} + o(\|\mathbf{v}\|) \qquad \text{as } \mathbf{v} \to \mathbf{0},$$

and this will prove the theorem. The idea is to express the difference $f(\mathbf{c} + \mathbf{v}) - f(\mathbf{c})$ as a sum of n terms, where the kth term is an approximation to $D_k f(\mathbf{c})v_k$.

For this purpose we write $\mathbf{v} = \lambda\mathbf{y}$, where $\|\mathbf{y}\| = 1$ and $\lambda = \|\mathbf{v}\|$. We keep λ small enough so that $\mathbf{c} + \mathbf{v}$ lies in the ball $B(\mathbf{c})$ in which the partial derivatives $D_2 f, \ldots, D_n f$ exist. Expressing $\mathbf{y}$ in terms of its components we have

$$\mathbf{y} = y_1\mathbf{u}_1 + \cdots + y_n\mathbf{u}_n,$$

where $\mathbf{u}_k$ is the kth unit coordinate vector. Now we write the difference $f(\mathbf{c} + \mathbf{v}) - f(\mathbf{c})$ as a telescoping sum,

$$f(\mathbf{c} + \mathbf{v}) - f(\mathbf{c}) = f(\mathbf{c} + \lambda\mathbf{y}) - f(\mathbf{c}) = \sum_{k=1}^{n} \{f(\mathbf{c} + \lambda\mathbf{v}_k) - f(\mathbf{c} + \lambda\mathbf{v}_{k-1})\}, \quad (25)$$

where

$$\mathbf{v}_0 = \mathbf{0}, \quad \mathbf{v}_1 = y_1\mathbf{u}_1, \quad \mathbf{v}_2 = y_1\mathbf{u}_1 + y_2\mathbf{u}_2, \ldots, \mathbf{v}_n = y_1\mathbf{u}_1 + \cdots + y_n\mathbf{u}_n.$$

The first term in the sum is $f(\mathbf{c} + \lambda y_1\mathbf{u}_1) - f(\mathbf{c})$. Since the two points $\mathbf{c}$ and $\mathbf{c} + \lambda y_1\mathbf{u}_1$ differ only in their first component, and since $D_1 f(\mathbf{c})$ exists, we can write

$$f(\mathbf{c} + \lambda y_1\mathbf{u}_1) - f(\mathbf{c}) = \lambda y_1 D_1 f(\mathbf{c}) + \lambda y_1 E_1(\lambda),$$

where $E_1(\lambda) \to 0$ as $\lambda \to 0$.

For $k \geq 2$, the kth term in the sum is

$$f(\mathbf{c} + \lambda\mathbf{v}_{k-1} + \lambda y_k\mathbf{u}_k) - f(\mathbf{c} + \lambda\mathbf{v}_{k-1}) = f(\mathbf{b}_k + \lambda y_k\mathbf{u}_k) - f(\mathbf{b}_k),$$

where $\mathbf{b}_k = \mathbf{c} + \lambda\mathbf{v}_{k-1}$. The two points $\mathbf{b}_k$ and $\mathbf{b}_k + \lambda y_k\mathbf{u}_k$ differ only in their kth component, and we can apply the one-dimensional Mean-Value Theorem for derivatives to write

$$f(\mathbf{b}_k + \lambda y_k \mathbf{u}_k) - f(\mathbf{b}_k) = \lambda y_k D_k f(\mathbf{a}_k), \tag{26}$$

where $\mathbf{a}_k$ lies on the line segment joining $\mathbf{b}_k$ to $\mathbf{b}_k + \lambda y_k \mathbf{u}_k$. Note that $\mathbf{b}_k \to \mathbf{c}$ and hence $\mathbf{a}_k \to \mathbf{c}$ as $\lambda \to 0$. Since each $D_k f$ is continuous at $\mathbf{c}$ for $k \geq 2$ we can write

$$D_k f(\mathbf{a}_k) = D_k f(\mathbf{c}) + E_k(\lambda), \qquad \text{where } E_k(\lambda) \to 0 \text{ as } \lambda \to 0.$$

Using this in (26) we find that (25) becomes

$$f(\mathbf{c} + \mathbf{v}) - f(\mathbf{c}) = \lambda \sum_{k=1}^{n} D_k f(\mathbf{c}) y_k + \lambda \sum_{k=1}^{n} y_k E_k(\lambda)$$

$$= \nabla f(\mathbf{c}) \cdot \mathbf{v} + \|\mathbf{v}\| E(\lambda),$$

where

$$E(\lambda) = \sum_{k=1}^{n} y_k E_k(\lambda) \to 0 \text{ as } \|\mathbf{v}\| \to 0.$$

This completes the proof.

NOTE. Continuity of at least $n - 1$ of the partials $D_1 \mathbf{f}, \ldots, D_n \mathbf{f}$ at $\mathbf{c}$, although sufficient, is by no means necessary for differentiability of $\mathbf{f}$ at $\mathbf{c}$. (See Exercises 12.5 and 12.6.)

12.13 A SUFFICIENT CONDITION FOR EQUALITY OF MIXED PARTIAL DERIVATIVES

The partial derivatives $D_1 \mathbf{f}, \ldots, D_n \mathbf{f}$ of a function from $\mathbf{R}^n$ to $\mathbf{R}^m$ are themselves functions from $\mathbf{R}^n$ to $\mathbf{R}^m$ and they, in turn, can have partial derivatives. These are called *second-order* partial derivatives. We use the notation introduced in Chapter 5 for real-valued functions:

$$D_{r,k} \mathbf{f} = D_r(D_k \mathbf{f}) = \frac{\partial^2 \mathbf{f}}{\partial x_r \partial x_k}.$$

Higher-order partial derivatives are similarly defined.

The example

$$f(x, y) = \begin{cases} xy(x^2 - y^2)/(x^2 + y^2) & \text{if } (x, y) \neq (0, 0), \\ 0 & \text{if } (x, y) = (0, 0), \end{cases}$$

shows that $D_{1,2}f(x, y)$ is not necessarily the same as $D_{2,1}f(x, y)$. In fact, in this example we have

$$D_1 f(x, y) = \frac{y(x^4 + 4x^2y^2 - y^4)}{(x^2 + y^2)^2}, \qquad \text{if } (x, y) \neq (0, 0),$$

and $D_1 f(0, 0) = 0$. Hence, $D_1 f(0, y) = -y$ for all y and therefore

$$D_{2,1} f(0, y) = -1, \qquad D_{2,1} f(0, 0) = -1.$$

On the other hand, we have

$$D_2 f(x, y) = \frac{x(x^4 - 4x^2 y^2 - y^4)}{(x^2 + y^2)^2}, \qquad \text{if } (x, y) \neq (0, 0),$$

and $D_2 f(0, 0) = 0$, so that $D_2 f(x, 0) = x$ for all x. Therefore, $D_{1,2} f(x, 0) = 1$, $D_{1,2} f(0, 0) = 1$, and we see that $D_{2,1} f(0, 0) \neq D_{1,2} f(0, 0)$.

The next theorem gives us a criterion for determining when the two mixed partials $D_{1,2} \mathbf{f}$ and $D_{2,1} \mathbf{f}$ will be equal.

Theorem 12.12. *If both partial derivatives $D_r \mathbf{f}$ and $D_k \mathbf{f}$ exist in an n-ball $B(\mathbf{c}; \delta)$ and if both are differentiable at $\mathbf{c}$, then*

$$D_{r,k} \mathbf{f}(\mathbf{c}) = D_{k,r} \mathbf{f}(\mathbf{c}). \tag{27}$$

Proof. If $\mathbf{f} = (f_1, \ldots, f_m)$, then $D_k \mathbf{f} = (D_k f_1, \ldots, D_k f_m)$. Therefore it suffices to prove the theorem for real-valued f. Also, since only two components are involved in (27), it suffices to consider the case $n = 2$. For simplicity, we assume that $\mathbf{c} = (0, 0)$. We shall prove that

$$D_{1,2} f(0, 0) = D_{2,1} f(0, 0).$$

Choose $h \neq 0$ so that the square with vertices $(0, 0)$, $(h, 0)$, (h, h), and $(0, h)$ lies in the 2-ball $B(\mathbf{0}; \delta)$. Consider the quantity

$$\Delta(h) = f(h, h) - f(h, 0) - f(0, h) + f(0, 0).$$

We will show that $\Delta(h)/h^2$ tends to both $D_{2,1} f(0, 0)$ and $D_{1,2} f(0, 0)$ as $h \to 0$.

Let $G(x) = f(x, h) - f(x, 0)$ and note that

$$\Delta(h) = G(h) - G(0). \tag{28}$$

By the one-dimensional Mean-Value Theorem we have

$$G(h) - G(0) = hG'(x_1) = h\{D_1 f(x_1, h) - D_1 f(x_1, 0)\}, \tag{29}$$

where x_1 lies between 0 and h. Since $D_1 f$ is differentiable at $(0, 0)$, we have the first-order Taylor formulas

$$D_1 f(x_1, h) = D_1 f(0, 0) + D_{1,1} f(0, 0)x_1 + D_{2,1} f(0, 0)h + (x_1^2 + h^2)^{1/2} E_1(h),$$

and

$$D_1 f(x_1, 0) = D_1 f(0, 0) + D_{1,1} f(0, 0)x_1 + |x_1| \, E_2(h),$$

where $E_1(h)$ and $E_2(h) \to 0$ as $h \to 0$. Using these in (29) and (28) we find

$$\Delta(h) = D_{2,1} f(0, 0)h^2 + E(h),$$

where $E(h) = h(x_1^2 + h^2)^{1/2} E_1(h) + h|x_1| E_2(h)$. Since $|x_1| \le |h|$, we have

$$0 \le |E(h)| \le \sqrt{2}\, h^2\, |E_1(h)| + h^2\, |E_2(h)|,$$

so

$$\lim_{h \to 0} \frac{\Delta(h)}{h^2} = D_{2,1} f(0, 0).$$

Applying the same procedure to the function $H(y) = f(h, y) - f(0, y)$ in place of $G(x)$, we find that

$$\lim_{h \to 0} \frac{\Delta(h)}{h^2} = D_{1,2} f(0, 0),$$

which completes the proof.

As a consequence of Theorems 12.11 and 12.12 we have:

Theorem 12.13. *If both partial derivatives $D_r\mathbf{f}$ and $D_k\mathbf{f}$ exist in an n-ball $B(\mathbf{c})$ and if both $D_{r,k}\mathbf{f}$ and $D_{k,r}\mathbf{f}$ are continuous at $\mathbf{c}$, then*

$$D_{r,k}\mathbf{f}(\mathbf{c}) = D_{k,r}\mathbf{f}(\mathbf{c}).$$

NOTE. We mention (without proof) another result which states that if $D_r\mathbf{f}$, $D_k\mathbf{f}$ and $D_{k,r}\mathbf{f}$ are continuous in an n-ball $B(\mathbf{c})$, then $D_{r,k}\mathbf{f}(\mathbf{c})$ exists and equals $D_{k,r}\mathbf{f}(\mathbf{c})$.

If f is a real-valued function of two variables, there are four second-order partial derivatives to consider; namely, $D_{1,1}f$, $D_{1,2}f$, $D_{2,1}f$, and $D_{2,2}f$. We have just shown that only three of these are distinct if f is suitably restricted.

The number of partial derivatives of order k which can be formed is 2^k. If all these derivatives are continuous in a neighborhood of the point (x, y), then certain of the mixed partials will be equal. Each mixed partial is of the form $D_{r_1, \ldots, r_k}f$, where each r_j is either 1 or 2. If we have two such mixed partials, $D_{r_1, \ldots, r_k}f$ and $D_{p_1, \ldots, p_k}f$, where the k-tuple $(r_1, \ldots, r_k)$ is a permutation of the k-tuple $(p_1, \ldots, p_k)$, then the two partials will be equal at (x, y) if all 2^k partials are continuous in a neighborhood of (x, y). This statement can be easily proved by mathematical induction, using Theorem 12.13 (which is the case $k = 2$). We omit the proof for general k. From this it follows that among the 2^k partial derivatives of order k, there are only $k + 1$ *distinct* partials in general, namely, those of the form $D_{r_1, \ldots, r_k}f$, where the k-tuple $(r_1, \ldots, r_k)$ assumes the following $k + 1$ forms:

$$(2, 2, \ldots, 2), \quad (1, 2, 2, \ldots, 2), \quad (1, 1, 2, \ldots, 2), \ldots,$$

$$(1, 1, \ldots, 1, 2), \quad (1, \ldots, 1).$$

Similar statements hold, of course, for functions of n variables. In this case, there are n^k partial derivatives of order k that can be formed. Continuity of all these partials at a point $\mathbf{x}$ implies that $D_{r_1, \ldots, r_k}f(\mathbf{x})$ is unchanged when the indices $r_1, \ldots, r_k$ are permuted. Each r_i is now a positive integer $\le n$.

12.14 TAYLOR'S FORMULA FOR FUNCTIONS FROM $\mathbf{R}^n$ TO $\mathbf{R}^1$

Taylor's formula (Theorem 5.19) can be extended to real-valued functions f defined on subsets of $\mathbf{R}^n$. In order to state the general theorem in a form which resembles the one-dimensional case, we introduce special symbols

$$f''(\mathbf{x}; \mathbf{t}), \quad f'''(\mathbf{x}; \mathbf{t}), \ldots, f^{(m)}(\mathbf{x}; \mathbf{t}),$$

for certain sums that arise in Taylor's formula. These play the role of higher-order directional derivatives, and they are defined as follows:

If $\mathbf{x}$ is a point in $\mathbf{R}^n$ where all second-order partial derivatives of f exist, and if $\mathbf{t} = (t_1, \ldots, t_n)$ is an arbitrary point in $\mathbf{R}^n$, we write

$$f''(\mathbf{x}; \mathbf{t}) = \sum_{i=1}^{n} \sum_{j=1}^{n} D_{i,j} f(\mathbf{x}) t_j t_i.$$

We also define

$$f'''(\mathbf{x}; \mathbf{t}) = \sum_{i=1}^{n} \sum_{j=1}^{n} \sum_{k=1}^{n} D_{i,j,k} f(\mathbf{x}) t_k t_j t_i$$

if all third-order partial derivatives exist at $\mathbf{x}$. The symbol $f^{(m)}(\mathbf{x}; \mathbf{t})$ is similarly defined if all mth-order partials exist.

These sums are analogous to the formula

$$f'(\mathbf{x}; \mathbf{t}) = \sum_{i=1}^{n} D_i f(\mathbf{x}) t_i$$

for the directional derivative of a function which is differentiable at $\mathbf{x}$.

Theorem 12.14 (Taylor's formula). *Assume that f and all its partial derivatives of order $<m$ are differentiable at each point of an open set S in $\mathbf{R}^n$. If $\mathbf{a}$ and $\mathbf{b}$ are two points of S such that $L(\mathbf{a}, \mathbf{b}) \subseteq S$, then there is a point $\mathbf{z}$ on the line segment $L(\mathbf{a}, \mathbf{b})$ such that*

$$f(\mathbf{b}) - f(\mathbf{a}) = \sum_{k=1}^{m-1} \frac{1}{k!} f^{(k)}(\mathbf{a}; \mathbf{b} - \mathbf{a}) + \frac{1}{m!} f^{(m)}(\mathbf{z}; \mathbf{b} - \mathbf{a}).$$

Proof. Since S is open, there is a $\delta > 0$ such that $\mathbf{a} + t(\mathbf{b} - \mathbf{a}) \in S$ for all real t in the interval $-\delta < t < 1 + \delta$. Define g on $(-\delta, 1 + \delta)$ by the equation

$$g(t) = f[\mathbf{a} + t(\mathbf{b} - \mathbf{a})].$$

Then $f(\mathbf{b}) - f(\mathbf{a}) = g(1) - g(0)$. We will prove the theorem by applying the one-dimensional Taylor formula to g, writing

$$g(1) - g(0) = \sum_{k=1}^{m-1} \frac{1}{k!} g^{(k)}(0) + \frac{1}{m!} g^{(m)}(\theta), \qquad \text{where } 0 < \theta < 1. \qquad (30)$$

Now g is a composite function given by $g(t) = f[\mathbf{p}(t)]$, where $\mathbf{p}(t) = \mathbf{a} + t(\mathbf{b} - \mathbf{a})$. The kth component of $\mathbf{p}$ has derivative $p'_k(t) = b_k - a_k$. Applying the chain rule, we see that $g'(t)$ exists in the interval $(-\delta, 1 + \delta)$ and is given by the formula

$$g'(t) = \sum_{j=1}^{n} D_j f[\mathbf{p}(t)](b_j - a_j) = f'(\mathbf{p}(t); \mathbf{b} - \mathbf{a}).$$

Again applying the chain rule, we obtain

$$g''(t) = \sum_{i=1}^{n} \sum_{j=1}^{n} D_{i,j} f[\mathbf{p}(t)](b_j - a_j)(b_i - a_i) = f''(\mathbf{p}(t); \mathbf{b} - \mathbf{a}).$$

Similarly, we find that $g^{(m)}(t) = f^{(m)}(\mathbf{p}(t); \mathbf{b} - \mathbf{a})$. When these are used in (30) we obtain the theorem, since the point $\mathbf{z} = \mathbf{a} + \theta(\mathbf{b} - \mathbf{a}) \in L(\mathbf{a}, \mathbf{b})$.

EXERCISES

Differentiable functions

12.1 Let S be an open subset of $\mathbf{R}^n$, and let $f: S \to \mathbf{R}$ be a real-valued function with finite partial derivatives $D_1 f, \ldots, D_n f$ on S. If f has a local maximum or a local minimum at a point $\mathbf{c}$ in S, prove that $D_k f(\mathbf{c}) = 0$ for each k.

12.2 Calculate all first-order partial derivatives and the directional derivative $f'(\mathbf{x}; \mathbf{u})$ for each of the real-valued functions defined on $\mathbf{R}^n$ as follows:

a) $f(\mathbf{x}) = \mathbf{a} \cdot \mathbf{x}$,   where $\mathbf{a}$ is a fixed vector in $\mathbf{R}^n$.

b) $f(\mathbf{x}) = \|\mathbf{x}\|^4$.

c) $f(\mathbf{x}) = \mathbf{x} \cdot L(\mathbf{x})$,   where $L: \mathbf{R}^n \to \mathbf{R}^n$ is a linear function.

d) $f(\mathbf{x}) = \sum_{i=1}^{n} \sum_{j=1}^{n} a_{ij} x_i x_j$,   where $a_{ij} = a_{ji}$.

12.3 Let $\mathbf{f}$ and $\mathbf{g}$ be functions with values in $\mathbf{R}^m$ such that the directional derivatives $\mathbf{f}'(\mathbf{c}; \mathbf{u})$ and $\mathbf{g}'(\mathbf{c}; \mathbf{u})$ exist. Prove that the sum $\mathbf{f} + \mathbf{g}$ and dot product $\mathbf{f} \cdot \mathbf{g}$ have directional derivatives given by

$$(\mathbf{f} + \mathbf{g})'(\mathbf{c}; \mathbf{u}) = \mathbf{f}'(\mathbf{c}; \mathbf{u}) + \mathbf{g}'(\mathbf{c}; \mathbf{u})$$

and

$$(\mathbf{f} \cdot \mathbf{g})'(\mathbf{c}; \mathbf{u}) = \mathbf{f}(\mathbf{c}) \cdot \mathbf{g}'(\mathbf{c}; \mathbf{u}) + \mathbf{g}(\mathbf{c}) \cdot \mathbf{f}'(\mathbf{c}; \mathbf{u}).$$

12.4 If $S \subseteq \mathbf{R}^n$, let $\mathbf{f}: S \to \mathbf{R}^m$ be a function with values in $\mathbf{R}^m$, and write $\mathbf{f} = (f_1, \ldots, f_m)$. Prove that $\mathbf{f}$ is differentiable at an interior point $\mathbf{c}$ of S if, and only if, each f_i is differentiable at $\mathbf{c}$.

12.5 Given n real-valued functions $f_1, \ldots, f_n$, each differentiable on an open interval (a, b) in $\mathbf{R}$. For each $\mathbf{x} = (x_1, \ldots, x_n)$ in the n-dimensional open interval

$$S = \{(x_1, \ldots, x_n) : a < x_k < b, \qquad k = 1, 2, \ldots, n\},$$

define $f(\mathbf{x}) = f_1(x_1) + \cdots + f_n(x_n)$. Prove that f is differentiable at each point of S and that

$$f'(\mathbf{x})(\mathbf{u}) = \sum_{i=1}^{n} f_i'(x_i) u_i, \qquad \text{where } \mathbf{u} = (u_1, \ldots, u_n).$$

12.6 Given n real-valued functions $f_1, \ldots, f_n$ defined on an open set S in $\mathbf{R}^n$. For each $\mathbf{x}$ in S, define $f(\mathbf{x}) = f_1(\mathbf{x}) + \cdots + f_n(\mathbf{x})$. Assume that for each $k = 1, 2, \ldots, n$, the following limit exists:

$$\lim_{\substack{\mathbf{y} \to \mathbf{x} \\ y_k \neq x_k}} \frac{f_k(\mathbf{y}) - f_k(\mathbf{x})}{y_k - x_k}.$$

Call this limit $a_k(\mathbf{x})$. Prove that f is differentiable at $\mathbf{x}$ and that

$$f'(\mathbf{x})(\mathbf{u}) = \sum_{k=1}^{n} a_k(\mathbf{x}) u_k \qquad \text{if } \mathbf{u} = (u_1, \ldots, u_n).$$

12.7 Let $\mathbf{f}$ and $\mathbf{g}$ be functions from $\mathbf{R}^n$ to $\mathbf{R}^m$. Assume that $\mathbf{f}$ is differentiable at $\mathbf{c}$, that $\mathbf{f}(\mathbf{c}) = \mathbf{0}$, and that $\mathbf{g}$ is continuous at $\mathbf{c}$. Let $h(\mathbf{x}) = \mathbf{g}(\mathbf{x}) \cdot \mathbf{f}(\mathbf{x})$. Prove that h is differentiable at $\mathbf{c}$ and that

$$h'(\mathbf{c})(\mathbf{u}) = \mathbf{g}(\mathbf{c}) \cdot \{\mathbf{f}'(\mathbf{c})(\mathbf{u})\} \qquad \text{if } \mathbf{u} \in \mathbf{R}^n.$$

12.8 Let $\mathbf{f} : \mathbf{R}^2 \to \mathbf{R}^3$ be defined by the equation

$$\mathbf{f}(x, y) = (\sin x \cos y, \quad \sin x \sin y, \quad \cos x \cos y).$$

Determine the Jacobian matrix $\mathbf{Df}(x, y)$.

12.9 Prove that there is no real-valued function f such that $f'(\mathbf{c}; \mathbf{u}) > 0$ for a fixed point $\mathbf{c}$ in $\mathbf{R}^n$ and every nonzero vector $\mathbf{u}$ in $\mathbf{R}^n$. Give an example such that $f'(\mathbf{c}; \mathbf{u}) > 0$ for a fixed direction $\mathbf{u}$ and every $\mathbf{c}$ in $\mathbf{R}^n$.

12.10 Let $f = u + iv$ be a complex-valued function such that the derivative $f'(c)$ exists for some complex c. Write $z = c + re^{i\alpha}$ (where α is real and fixed) and let $r \to 0$ in the difference quotient $[f(z) - f(c)]/(z - c)$ to obtain

$$f'(c) = e^{-i\alpha}[u'(c; \mathbf{a}) + iv'(c; \mathbf{a})],$$

where $\mathbf{a} = (\cos \alpha, \sin \alpha)$, and $u'(c; \mathbf{a})$ and $v'(c; \mathbf{a})$ are directional derivatives. Let $\mathbf{b} = (\cos \beta, \sin \beta)$, where $\beta = \alpha + \frac{1}{2}\pi$, and show by a similar argument that

$$f'(c) = e^{-i\alpha}[v'(c; \mathbf{b}) - iu'(c; \mathbf{b})].$$

Deduce that $u'(c; \mathbf{a}) = v'(c; \mathbf{b})$ and $v'(c; \mathbf{a}) = -u'(c; \mathbf{b})$. The Cauchy–Riemann equations (Theorem 5.22) are a special case.

Gradients and the chain rule

12.11 Let f be real-valued and differentiable at a point $\mathbf{c}$ in $\mathbf{R}^n$, and assume that $\|\nabla f(\mathbf{c})\| \neq 0$. Prove that there is one and only one unit vector $\mathbf{u}$ in $\mathbf{R}^n$ such that $|f'(\mathbf{c}; \mathbf{u})| = \|\nabla f(\mathbf{c})\|$, and that this is the unit vector for which $|f'(\mathbf{c}; \mathbf{u})|$ has its maximum value.

12.12 Compute the gradient vector $\nabla f(x, y)$ at those points (x, y) in $\mathbf{R}^2$ where it exists:

a) $f(x, y) = x^2 y^2 \log (x^2 + y^2)$ if $(x, y) \neq (0, 0)$, $f(0, 0) = 0$.

b) $f(x, y) = xy \sin \dfrac{1}{x^2 + y^2}$ if $(x, y) \neq (0, 0)$, $f(0, 0) = 0$.

12.13 Let f and g be real-valued functions defined on $\mathbf{R}^1$ with continuous second derivatives f'' and g''. Define

$$F(x, y) = f[x + g(y)] \text{ for each } (x, y) \text{ in } \mathbf{R}^2.$$

Find formulas for all partials of F of first and second order in terms of the derivatives of f and g. Verify the relation

$$(D_1F)(D_{1,2}F) = (D_2F)(D_{1,1}F).$$

12.14 Given a function f defined in $\mathbf{R}^2$. Let

$$F(r, \theta) = f(r \cos \theta, r \sin \theta).$$

a) Assume appropriate differentiability properties of f and show that

$$D_1F(r, \theta) = \cos \theta \, D_1f(x, y) + \sin \theta \, D_2f(x, y),$$

$$D_{1,1}F(r, \theta) = \cos^2 \theta D_{1,1}f(x, y) + 2 \sin \theta \cos \theta \, D_{1,2}f(x, y) + \sin^2 \theta D_{2,2}f(x, y),$$

where $x = r \cos \theta, y = r \sin \theta$.
b) Find similar formulas for D_2F, $D_{1,2}F$, and $D_{2,2}F$.
c) Verify the formula

$$\|\nabla f(r \cos \theta, r \sin \theta)\|^2 = [D_1F(r, \theta)]^2 + \frac{1}{r^2} [D_2F(r, \theta)]^2.$$

12.15 If f and g have gradient vectors $\nabla f(\mathbf{x})$ and $\nabla g(\mathbf{x})$ at a point $\mathbf{x}$ in $\mathbf{R}^n$ show that the product function h defined by $h(\mathbf{x}) = f(\mathbf{x})g(\mathbf{x})$ also has a gradient vector at $\mathbf{x}$ and that

$$\nabla h(\mathbf{x}) = f(\mathbf{x})\nabla g(\mathbf{x}) + g(\mathbf{x})\nabla f(\mathbf{x}).$$

State and prove a similar result for the quotient f/g.

12.16 Let f be a function having a derivative f' at each point in $\mathbf{R}^1$ and let g be defined on $\mathbf{R}^3$ by the equation

$$g(x, y, z) = x^2 + y^2 + z^2.$$

If h denotes the composite function $h = f \circ g$, show that

$$\|\nabla h(x, y, z)\|^2 = 4g(x, y, z)\{f'[g(x, y, z)]\}^2.$$

12.17 Assume f is differentiable at each point (x, y) in $\mathbf{R}^2$. Let g_1 and g_2 be defined on $\mathbf{R}^3$ by the equations

$$g_1(x, y, z) = x^2 + y^2 + z^2, \qquad g_2(x, y, z) = x + y + z,$$

and let $\mathbf{g}$ be the vector-valued function whose values (in $\mathbf{R}^2$) are given by

$$\mathbf{g}(x, y, z) = (g_1(x, y, z), g_2(x, y, z)).$$

Let h be the composite function $h = f \circ \mathbf{g}$ and show that

$$\|\nabla h\|^2 = 4(D_1f)^2g_1 + 4(D_1f)(D_2f)g_2 + 3(D_2f)^2.$$

12.18 Let f be defined on an open set S in $\mathbf{R}^n$. We say that f is homogeneous of degree p over S if $f(\lambda\mathbf{x}) = \lambda^p f(\mathbf{x})$ for every real λ and for every $\mathbf{x}$ in S for which $\lambda\mathbf{x} \in S$. If such a function is differentiable at $\mathbf{x}$, show that

$$\mathbf{x} \cdot \nabla f(\mathbf{x}) = pf(\mathbf{x}).$$

NOTE. This is known as *Euler's theorem* for homogeneous functions. *Hint.* For fixed $\mathbf{x}$, define $g(\lambda) = f(\lambda \mathbf{x})$ and compute $g'(1)$.

Also prove the converse. That is, show that if $\mathbf{x} \cdot \nabla f(\mathbf{x}) = pf(\mathbf{x})$ for all $\mathbf{x}$ in an open set S, then f must be homogeneous of degree p over S.

Mean-Value theorems

12.19 Let $\mathbf{f} : \mathbf{R} \to \mathbf{R}^2$ be defined by the equation $\mathbf{f}(t) = (\cos t, \sin t)$. Then $\mathbf{f}'(t)(u) = u(-\sin t, \cos t)$ for every real u. The Mean-Value formula

$$\mathbf{f}(y) - \mathbf{f}(x) = \mathbf{f}'(z)(y - x)$$

cannot hold when $x = 0$, $y = 2\pi$, since the left member is zero and the right member is a vector of length 2π. Nevertheless, Theorem 12.9 states that for every vector $\mathbf{a}$ in $\mathbf{R}^2$ there is a z in the interval $(0, 2\pi)$ such that

$$\mathbf{a} \cdot \{\mathbf{f}(y) - \mathbf{f}(x)\} = \mathbf{a} \cdot \{\mathbf{f}'(z)(y - x)\}.$$

Determine z in terms of $\mathbf{a}$ when $x = 0$ and $y = 2\pi$.

12.20 Let f be a real-valued function differentiable on a 2-ball $B(\mathbf{x})$. By considering the function

$$g(t) = f[ty_1 + (1 - t)x_1, y_2] + f[x_1, ty_2 + (1 - t)x_2]$$

prove that

$$f(\mathbf{y}) - f(\mathbf{x}) = (y_1 - x_1)D_1 f(z_1, y_2) + (y_2 - x_2)D_2 f(x_1, z_2),$$

where $z_1 \in L(x_1, y_1)$ and $z_2 \in L(x_2, y_2)$.

12.21 State and prove a generalization of the result in Exercise 12.20 for a real-valued function differentiable on an n-ball $B(\mathbf{x})$.

12.22 Let f be real-valued and assume that the directional derivative $f'(\mathbf{c} + t\mathbf{u}; \mathbf{u})$ exists for each t in the interval $0 \le t \le 1$. Prove that for some θ in the open interval $(0, 1)$ we have

$$f(\mathbf{c} + \mathbf{u}) - f(\mathbf{c}) = f'(\mathbf{c} + \theta\mathbf{u}; \mathbf{u}).$$

12.23 a) If f is real-valued and if the directional derivative $f'(\mathbf{x}; \mathbf{u}) = 0$ for every $\mathbf{x}$ in an n-ball $B(\mathbf{c})$ and every direction $\mathbf{u}$, prove that f is constant on $B(\mathbf{c})$.

b) What can you conclude about f if $f'(\mathbf{x}; \mathbf{u}) = 0$ for a fixed direction $\mathbf{u}$ and every $\mathbf{x}$ in $B(\mathbf{c})$?

Derivatives of higher order and Taylor's formula

12.24 For each of the following functions, verify that the mixed partial derivatives $D_{1,2}f$ and $D_{2,1}f$ are equal.

a) $f(x, y) = x^4 + y^4 - 4x^2y^2$.

b) $f(x, y) = \log (x^2 + y^2)$, $\quad (x, y) \ne (0, 0)$.

c) $f(x, y) = \tan (x^2/y)$, $\quad y \ne 0$.

12.25 Let f be a function of two variables. Use induction and Theorem 12.13 to prove that if the 2^k partial derivatives of f of order k are continuous in a neighborhood of a point (x, y), then all mixed partials of the form $D_{r_1,\ldots,r_k}f$ and $D_{p_1,\ldots,p_k}f$ will be equal at (x, y) if the k-tuple $(r_1, \ldots, r_k)$ contains the same number of ones as the k-tuple $(p_1, \ldots, p_k)$.

12.26 If f is a function of two variables having continuous partials of order k on some open set S in $\mathbf{R}^2$, show that

$$f^{(k)}(\mathbf{x}; \mathbf{t}) = \sum_{r=0}^{k} \binom{k}{r} t_1^r t_2^{k-r} D_{p_1, \ldots, \, p_k} f(\mathbf{x}), \qquad \text{if } \mathbf{x} \in S, \qquad \mathbf{t} = (t_1, t_2),$$

where in the rth term we have $p_1 = \cdots = p_r = 1$ and $p_{r+1} = \cdots = p_k = 2$. Use this result to give an alternative expression for Taylor's formula (Theorem 12.14) in the case when $n = 2$. The symbol $\binom{k}{r}$ is the binomial coefficient $k!/[r!\,(k-r)!]$.

12.27 Use Taylor's formula to express the following in powers of $(x - 1)$ and $(y - 2)$:

    a) $f(x, y) = x^3 + y^3 + xy^2,$ b) $f(x, y) = x^2 + xy + y^2.$

SUGGESTED REFERENCES FOR FURTHER STUDY

12.1 Apostol, T. M., *Calculus*, Vol. 2, 2nd ed. Xerox, Waltham, 1969.

12.2 Chaundy, T. W., *The Differential Calculus*. Clarendon Press, Oxford, 1935.

12.3 Woll, J. W., *Functions of Several Variables*. Harcourt Brace and World, New York, 1966.

CHAPTER 13

IMPLICIT FUNCTIONS
AND EXTREMUM PROBLEMS

13.1 INTRODUCTION

This chapter consists of two principal parts. The first part discusses an important theorem of analysis called the *implicit function theorem;* the second part treats extremum problems. Both parts use the theorems developed in Chapter 12.

The implicit function theorem in its simplest form deals with an equation of the form

$$f(x, t) = 0. \tag{1}$$

The problem is to decide whether this equation determines x as a function of t. If so, we have

$$x = g(t),$$

for some function g. We say that g is defined "implicitly" by (1).

The problem assumes a more general form when we have a system of several equations involving several variables and we ask whether we can solve these equations for some of the variables in terms of the remaining variables. This is the same type of problem as above, except that x and t are replaced by vectors, and f and g are replaced by vector-valued functions. Under rather general conditions, a solution always exists. The implicit function theorem gives a description of these conditions and some conclusions about the solution.

An important special case is the familiar problem in algebra of solving n linear equations of the form

$$\sum_{j=1}^{n} a_{ij}x_j = t_i \qquad (i = 1, 2, \ldots, n), \tag{2}$$

where the a_{ij} and t_i are considered as given numbers and $x_1, \ldots, x_n$ represent unknowns. In linear algebra it is shown that such a system has a unique solution if, and only if, the determinant of the coefficient matrix $A = [a_{ij}]$ is nonzero.

NOTE. The determinant of a square matrix $A = [a_{ij}]$ is denoted by det A or det $[a_{ij}]$. If det $[a_{ij}] \neq 0$, the solution of (2) can be obtained by Cramer's rule which expresses each x_k as a quotient of two determinants, say $x_k = A_k/D$, where $D = \det [a_{ij}]$ and A_k is the determinant of the matrix obtained by replacing the kth column of $[a_{ij}]$ by $t_1, \ldots, t_n$. (For a proof of Cramer's rule, see Reference 13.1, Theorem 3.14.) In particular, if each $t_i = 0$, then each $x_k = 0$.

Next we show that the system (2) can be written in the form (1). Each equation in (2) has the form

$$f_i(\mathbf{x}, \mathbf{t}) = 0 \qquad \text{where } \mathbf{x} = (x_1, \ldots, x_n), \quad \mathbf{t} = (t_1, \ldots, t_n),$$

and

$$f_i(\mathbf{x}, \mathbf{t}) = \sum_{j=1}^{n} a_{ij} x_j - t_i.$$

Therefore the system in (2) can be expressed as one vector equation $\mathbf{f}(\mathbf{x}, \mathbf{t}) = \mathbf{0}$, where $\mathbf{f} = (f_1, \ldots, f_n)$. If $D_j f_i$ denotes the partial derivative of f_i with respect to the jth coordinate x_j, then $D_j f_i(\mathbf{x}, \mathbf{t}) = a_{ij}$. Thus the coefficient matrix $A = [a_{ij}]$ in (2) is a Jacobian matrix. Linear algebra tells us that (2) has a unique solution if the determinant of this Jacobian matrix is nonzero.

In the general implicit function theorem, the nonvanishing of the determinant of a Jacobian matrix also plays a role. This comes about by approximating $\mathbf{f}$ by a linear function. The equation $\mathbf{f}(\mathbf{x}, \mathbf{t}) = \mathbf{0}$ gets replaced by a system of linear equations whose coefficient matrix is the Jacobian matrix of $\mathbf{f}$.

NOTATION. If $\mathbf{f} = (f_1, \ldots, f_n)$ and $\mathbf{x} = (x_1, \ldots, x_n)$, the Jacobian matrix $\mathbf{Df}(\mathbf{x}) = [D_j f_i(\mathbf{x})]$ is an $n \times n$ matrix. Its determinant is called a *Jacobian determinant* and is denoted by $J_{\mathbf{f}}(\mathbf{x})$. Thus,

$$J_{\mathbf{f}}(\mathbf{x}) = \det \mathbf{Df}(\mathbf{x}) = \det [D_j f_i(\mathbf{x})].$$

The notation

$$\frac{\partial(f_1, \ldots, f_n)}{\partial(x_1, \ldots, x_n)},$$

is also used to denote the Jacobian determinant $J_{\mathbf{f}}(\mathbf{x})$.

The next theorem relates the Jacobian determinant of a complex-valued function with its derivative.

Theorem 13.1. *If $f = u + iv$ is a complex-valued function with a derivative at a point z in $\mathbf{C}$, then $J_f(z) = |f'(z)|^2$.*

Proof. We have $f'(z) = D_1 u + iD_1 v$, so $|f'(z)|^2 = (D_1 u)^2 + (D_1 v)^2$. Also,

$$J_f(z) = \det \begin{bmatrix} D_1 u & D_2 u \\ D_1 v & D_2 v \end{bmatrix} = D_1 u \, D_2 v - D_1 v \, D_2 u = (D_1 u)^2 + (D_1 v)^2,$$

by the Cauchy–Riemann equations.

13.2 FUNCTIONS WITH NONZERO JACOBIAN DETERMINANT

This section gives some properties of functions with nonzero Jacobian determinant at certain points. These results will be used later in the proof of the implicit function theorem.

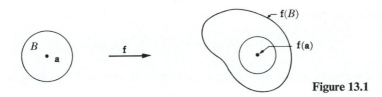

Figure 13.1

Theorem 13.2. *Let $B = B(\mathbf{a}; r)$ be an n-ball in $\mathbf{R}^n$, let ∂B denote its boundary,*

$$\partial B = \{\mathbf{x} : \|\mathbf{x} - \mathbf{a}\| = r\},$$

and let $\bar{B} = B \cup \partial B$ denote its closure. Let $\mathbf{f} = (f_1, \ldots, f_n)$ be continuous on $\bar{B}$, and assume that all the partial derivatives $D_j f_i(\mathbf{x})$ exist if $\mathbf{x} \in B$. Assume further that $\mathbf{f}(\mathbf{x}) \neq \mathbf{f}(\mathbf{a})$ if $\mathbf{x} \in \partial B$ and that the Jacobian determinant $J_{\mathbf{f}}(\mathbf{x}) \neq 0$ for each $\mathbf{x}$ in B. Then $\mathbf{f}(B)$, the image of B under $\mathbf{f}$, contains an n-ball with center at $\mathbf{f}(\mathbf{a})$.

Proof. Define a real-valued function g on ∂B as follows:

$$g(\mathbf{x}) = \|\mathbf{f}(\mathbf{x}) - \mathbf{f}(\mathbf{a})\| \qquad \text{if } \mathbf{x} \in \partial B.$$

Then $g(\mathbf{x}) > 0$ for each $\mathbf{x}$ in ∂B because $\mathbf{f}(\mathbf{x}) \neq \mathbf{f}(\mathbf{a})$ if $\mathbf{x} \in \partial B$. Also, g is continuous on ∂B since $\mathbf{f}$ is continuous on $\bar{B}$. Since ∂B is compact, g takes on its absolute minimum (call it m) somewhere on ∂B. Note that $m > 0$ since g is positive on ∂B. Let T denote the n-ball

$$T = B\left(\mathbf{f}(\mathbf{a}); \frac{m}{2}\right).$$

We will prove that $T \subseteq \mathbf{f}(B)$ and this will prove the theorem. (See Fig. 13.1.)

To do this we show that $\mathbf{y} \in T$ implies $\mathbf{y} \in \mathbf{f}(B)$. Choose a point $\mathbf{y}$ in T, keep $\mathbf{y}$ fixed, and define a new real-valued function h on $\bar{B}$ as follows:

$$h(\mathbf{x}) = \|\mathbf{f}(\mathbf{x}) - \mathbf{y}\| \qquad \text{if } \mathbf{x} \in \bar{B}.$$

Then h is continuous on the compact set $\bar{B}$ and hence attains its absolute minimum on $\bar{B}$. We will show that h attains its minimum somewhere in the open n-ball B. At the center we have $h(\mathbf{a}) = \|\mathbf{f}(\mathbf{a}) - \mathbf{y}\| < m/2$ since $\mathbf{y} \in T$. Hence the minimum value of h in $\bar{B}$ must also be $< m/2$. But at each point $\mathbf{x}$ on the boundary ∂B we have

$$h(\mathbf{x}) = \|\mathbf{f}(\mathbf{x}) - \mathbf{y}\| = \|\mathbf{f}(\mathbf{x}) - \mathbf{f}(\mathbf{a}) - (\mathbf{y} - \mathbf{f}(\mathbf{a}))\|$$

$$\geq \|\mathbf{f}(\mathbf{x}) - \mathbf{f}(\mathbf{a})\| - \|\mathbf{f}(\mathbf{a}) - \mathbf{y}\| > g(\mathbf{x}) - \frac{m}{2} \geq \frac{m}{2},$$

so the minimum of h cannot occur on the boundary ∂B. Hence there is an interior point $\mathbf{c}$ in B at which h attains its minimum. At this point the square of h also has a minimum. Since

$$h^2(\mathbf{x}) = \|\mathbf{f}(\mathbf{x}) - \mathbf{y}\|^2 = \sum_{r=1}^{n} [f_r(\mathbf{x}) - y_r]^2,$$

and since each partial derivative $D_k(h^2)$ must be zero at $\mathbf{c}$, we must have

$$\sum_{r=1}^{n} [f_r(\mathbf{c}) - y_r]D_k f_r(\mathbf{c}) = 0 \qquad \text{for } k = 1, 2, \ldots, n.$$

But this is a system of linear equations whose determinant $J_{\mathbf{f}}(\mathbf{c})$ is not zero, since $\mathbf{c} \in B$. Therefore $f_r(\mathbf{c}) = y_r$ for each r, or $\mathbf{f}(\mathbf{c}) = \mathbf{y}$. That is, $\mathbf{y} \in f(B)$. Hence $T \subseteq \mathbf{f}(B)$ and the proof is complete.

A function $f : S \to T$ from one metric space (S, d_S) to another (T, d_T) is called an *open mapping* if, for every open set A in S, the image $f(A)$ is open in T.

The next theorem gives a sufficient condition for a mapping to carry open sets onto open sets. (See also Theorem 13.5.)

Theorem 13.3. *Let A be an open subset of $\mathbf{R}^n$ and assume that $\mathbf{f} : A \to \mathbf{R}^n$ is continuous and has finite partial derivatives $D_j f_i$ on A. If $\mathbf{f}$ is one-to-one on A and if $J_{\mathbf{f}}(\mathbf{x}) \neq 0$ for each $\mathbf{x}$ in A, then $\mathbf{f}(A)$ is open.*

Proof. If $\mathbf{b} \in \mathbf{f}(A)$, then $\mathbf{b} = \mathbf{f}(\mathbf{a})$ for some $\mathbf{a}$ in A. There is an n-ball $B(\mathbf{a}; r) \subseteq A$ on which $\mathbf{f}$ satisfies the hypotheses of Theorem 13.2, so $\mathbf{f}(B)$ contains an n-ball with center at $\mathbf{b}$. Therefore, $\mathbf{b}$ is an interior point of $\mathbf{f}(A)$, so $\mathbf{f}(A)$ is open.

The next theorem shows that a function with continuous partial derivatives is locally one-to-one near a point where the Jacobian determinant does not vanish.

Theorem 13.4. *Assume that $\mathbf{f} = (f_1, \ldots, f_n)$ has continuous partial derivatives $D_j f_i$ on an open set S in $\mathbf{R}^n$, and that the Jacobian determinant $J_{\mathbf{f}}(\mathbf{a}) \neq 0$ for some point $\mathbf{a}$ in S. Then there is an n-ball $B(\mathbf{a})$ on which $\mathbf{f}$ is one-to-one.*

Proof. Let $\mathbf{Z}_1, \ldots, \mathbf{Z}_n$ be n points in S and let $\mathbf{Z} = (\mathbf{Z}_1; \ldots; \mathbf{Z}_n)$ denote that point in $\mathbf{R}^{n^2}$ whose first n components are the components of $\mathbf{Z}_1$, whose next n components are the components of $\mathbf{Z}_2$, and so on. Define a real-valued function h as follows:

$$h(\mathbf{Z}) = \det [D_j f_i(\mathbf{Z}_i)].$$

This function is continuous at those points $\mathbf{Z}$ in $\mathbf{R}^{n^2}$ where $h(\mathbf{Z})$ is defined because each $D_j f_i$ is continuous on S and a determinant is a polynomial in its n^2 entries. Let $\mathbf{Z}$ be the special point in $\mathbf{R}^{n^2}$ obtained by putting

$$\mathbf{Z}_1 = \mathbf{Z}_2 = \cdots = \mathbf{Z}_n = \mathbf{a}.$$

Then $h(\mathbf{Z}) = J_{\mathbf{f}}(\mathbf{a}) \neq 0$ and hence, by continuity, there is some n-ball $B(\mathbf{a})$ such that $\det [D_j f_i(\mathbf{Z}_i)] \neq 0$ if each $\mathbf{Z}_i \in B(\mathbf{a})$. We will prove that $\mathbf{f}$ is one-to-one on $B(\mathbf{a})$.

Assume the contrary. That is, assume that $\mathbf{f}(\mathbf{x}) = \mathbf{f}(\mathbf{y})$ for some pair of points $\mathbf{x} \neq \mathbf{y}$ in $B(\mathbf{a})$. Since $B(\mathbf{a})$ is convex, the line segment $L(\mathbf{x}, \mathbf{y}) \subseteq B(\mathbf{a})$ and we can apply the Mean-Value Theorem to each component of $\mathbf{f}$ to write

$$0 = f_i(\mathbf{y}) - f_i(\mathbf{x}) = \nabla f_i(\mathbf{Z}_i) \cdot (\mathbf{y} - \mathbf{x}) \qquad \text{for } i = 1, 2, \ldots, n,$$

where each $\mathbf{Z}_i \in L(\mathbf{x}, \mathbf{y})$ and hence $\mathbf{Z}_i \in B(\mathbf{a})$. (The Mean-Value Theorem is applicable because $\mathbf{f}$ is differentiable on S.) But this is a system of linear equations of the form

$$\sum_{k=1}^{n} (y_k - x_k)a_{ik} = 0 \qquad \text{with } a_{ik} = D_k f_i(\mathbf{Z}_i).$$

The determinant of this system is not zero, since $\mathbf{Z}_i \in B(\mathbf{a})$. Hence $y_k - x_k = 0$ for each k, and this contradicts the assumption that $\mathbf{x} \neq \mathbf{y}$. We have shown, therefore, that $\mathbf{x} \neq \mathbf{y}$ implies $\mathbf{f}(\mathbf{x}) \neq \mathbf{f}(\mathbf{y})$ and hence that $\mathbf{f}$ is one-to-one on $B(\mathbf{a})$.

NOTE. The reader should be cautioned that Theorem 13.4 is a *local* theorem and not a global theorem. The nonvanishing of $J_\mathbf{f}(\mathbf{a})$ guarantees that $\mathbf{f}$ is one-to-one on a neighborhood of $\mathbf{a}$. It does not follow that $\mathbf{f}$ is one-to-one on S, even when $J_\mathbf{f}(\mathbf{x}) \neq 0$ for every $\mathbf{x}$ in S. The following example illustrates this point. Let f be the complex-valued function defined by $f(z) = e^z$ if $z \in \mathbf{C}$. If $z = x + iy$ we have

$$J_f(z) = |f'(z)|^2 = |e^z|^2 = e^{2x}.$$

Thus $J_f(z) \neq 0$ for every z in $\mathbf{C}$. However, f is not one-to-one on $\mathbf{C}$ because $f(z_1) = f(z_2)$ for every pair of points z_1 and z_2 which differ by $2\pi i$.

The next theorem gives a global property of functions with nonzero Jacobian determinant.

Theorem 13.5. *Let A be an open subset of $\mathbf{R}^n$ and assume that $\mathbf{f} : A \to \mathbf{R}^n$ has continuous partial derivatives $D_j f_i$ on A. If $J_\mathbf{f}(\mathbf{x}) \neq 0$ for all $\mathbf{x}$ in A, then $\mathbf{f}$ is an open mapping.*

Proof. Let S be any open subset of A. If $\mathbf{x} \in S$ there is an n-ball $B(\mathbf{x})$ in which $\mathbf{f}$ is one-to-one (by Theorem 13.4). Therefore, by Theorem 13.3, the image $\mathbf{f}(B(\mathbf{x}))$ is open in $\mathbf{R}^n$. But we can write $S = \bigcup_{\mathbf{x} \in S} B(\mathbf{x})$. Applying $\mathbf{f}$ we find $\mathbf{f}(S) = \bigcup_{\mathbf{x} \in S} \mathbf{f}(B(\mathbf{x}))$, so $\mathbf{f}(S)$ is open.

NOTE. If a function $\mathbf{f} = (f_1, \ldots, f_n)$ has continuous partial derivatives on a set S, we say that $\mathbf{f}$ is *continuously differentiable* on S, and we write $\mathbf{f} \in C'$ on S. In view of Theorem 12.11, continuous differentiability at a point implies differentiability at that point.

Theorem 13.4 shows that a continuously differentiable function with a non-vanishing Jacobian at a point $\mathbf{a}$ has a local inverse in a neighborhood of $\mathbf{a}$. The next theorem gives some local differentiability properties of this local inverse function.

13.3 THE INVERSE FUNCTION THEOREM

Theorem 13.6. *Assume* $\mathbf{f} = (f_1, \ldots, f_n) \in C'$ *on an open set* S *in* $\mathbf{R}^n$, *and let* $T = \mathbf{f}(S)$. *If the Jacobian determinant* $J_{\mathbf{f}}(\mathbf{a}) \neq 0$ *for some point* $\mathbf{a}$ *in* S, *then there are two open sets* $X \subseteq S$ *and* $Y \subseteq T$ *and a uniquely determined function* $\mathbf{g}$ *such that*

a) $\mathbf{a} \in X$ *and* $\mathbf{f}(\mathbf{a}) \in Y$,

b) $Y = \mathbf{f}(X)$,

c) $\mathbf{f}$ *is one-to-one on* X,

d) $\mathbf{g}$ *is defined on* Y, $\mathbf{g}(Y) = X$, *and* $\mathbf{g}[\mathbf{f}(\mathbf{x})] = \mathbf{x}$ *for every* $\mathbf{x}$ *in* X,

e) $\mathbf{g} \in C'$ *on* Y.

Proof. The function $J_{\mathbf{f}}$ is continuous on S and, since $J_{\mathbf{f}}(\mathbf{a}) \neq 0$, there is an n-ball $B_1(\mathbf{a})$ such that $J_{\mathbf{f}}(\mathbf{x}) \neq 0$ for all $\mathbf{x}$ in $B_1(\mathbf{a})$. By Theorem 13.4, there is an n-ball $B(\mathbf{a}) \subseteq B_1(\mathbf{a})$ on which $\mathbf{f}$ is one-to-one. Let B be an n-ball with center at $\mathbf{a}$ and radius smaller than that of $B(\mathbf{a})$. Then, by Theorem 13.2, $\mathbf{f}(B)$ contains an n-ball with center at $\mathbf{f}(\mathbf{a})$. Denote this by Y and let $X = \mathbf{f}^{-1}(Y) \cap B$. Then X is open since both $\mathbf{f}^{-1}(Y)$ and B are open. (See Fig. 13.2.)

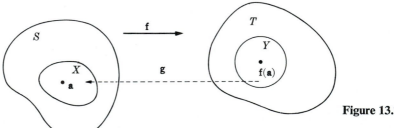

Figure 13.2

    The set $\bar{B}$ (the closure of B) is compact and $\mathbf{f}$ is one-to-one and continuous on $\bar{B}$. Hence, by Theorem 4.29, there exists a function $\mathbf{g}$ (the inverse function $\mathbf{f}^{-1}$ of Theorem 4.29) defined on $\mathbf{f}(\bar{B})$ such that $\mathbf{g}[\mathbf{f}(\mathbf{x})] = \mathbf{x}$ for all $\mathbf{x}$ in $\bar{B}$. Moreover, $\mathbf{g}$ is continuous on $\mathbf{f}(\bar{B})$. Since $X \subseteq \bar{B}$ and $Y \subseteq \mathbf{f}(\bar{B})$, this proves parts (a), (b), (c) and (d). The uniqueness of $\mathbf{g}$ follows from (d).

    Next we prove (e). For this purpose, define a real-valued function h by the equation $h(\mathbf{Z}) = \det [D_j f_i(\mathbf{Z}_i)]$, where $\mathbf{Z}_1, \ldots, \mathbf{Z}_n$ are n points in S, and $\mathbf{Z} = (\mathbf{Z}_1; \ldots; \mathbf{Z}_n)$ is the corresponding point in $\mathbf{R}^{n^2}$. Then, arguing as in the proof of Theorem 13.4, there is an n-ball $B_2(\mathbf{a})$ such that $h(\mathbf{Z}) \neq 0$ if each $\mathbf{Z}_i \in B_2(\mathbf{a})$. We can now assume that, in the earlier part of the proof, the n-ball $B(\mathbf{a})$ was chosen so that $B(\mathbf{a}) \subseteq B_2(\mathbf{a})$. Then $\bar{B} \subseteq B_2(\mathbf{a})$ and $h(\mathbf{Z}) \neq 0$ if each $\mathbf{Z}_i \in \bar{B}$.

    To prove (e), write $\mathbf{g} = (g_1, \ldots, g_n)$. We will show that each $g_k \in C'$ on Y. To prove that $D_r g_k$ exists on Y, assume $\mathbf{y} \in Y$ and consider the difference quotient $[g_k(\mathbf{y} + t\mathbf{u}_r) - g_k(\mathbf{y})]/t$, where $\mathbf{u}_r$ is the rth unit coordinate vector. (Since Y is open, $\mathbf{y} + t\mathbf{u}_r \in Y$ if t is sufficiently small.) Let $\mathbf{x} = \mathbf{g}(\mathbf{y})$ and let $\mathbf{x}' = \mathbf{g}(\mathbf{y} + t\mathbf{u}_r)$. Then both $\mathbf{x}$ and $\mathbf{x}'$ are in X and $\mathbf{f}(\mathbf{x}') - \mathbf{f}(\mathbf{x}) = t\mathbf{u}_r$. Hence $f_i(\mathbf{x}') - f_i(\mathbf{x})$ is 0 if $i \neq r$, and is t if $i = r$. By the Mean-Value Theorem we have

$$\frac{f_i(\mathbf{x}') - f_i(\mathbf{x})}{t} = \nabla f_i(\mathbf{Z}_i) \cdot \frac{\mathbf{x}' - \mathbf{x}}{t} \qquad \text{for } i = 1, 2, \ldots, n,$$

where each $\mathbf{Z}_i$ is on the line segment joining $\mathbf{x}$ and $\mathbf{x}'$; hence $\mathbf{Z}_i \in B$. The expression on the left is 1 or 0, according to whether $i = r$ or $i \neq r$. This is a system of n linear equations in n unknowns $(x'_j - x_j)/t$ and has a unique solution, since

$$\det [D_j f_i(\mathbf{Z}_i)] = h(\mathbf{Z}) \neq 0.$$

Solving for the kth unknown by Cramer's rule, we obtain an expression for $[g_k(\mathbf{y} + t\mathbf{u}_r) - g_k(\mathbf{y})]/t$ as a quotient of determinants. As $t \to 0$, the point $\mathbf{x} \to \mathbf{x}$, since $\mathbf{g}$ is continuous, and hence each $\mathbf{Z}_i \to \mathbf{x}$, since $\mathbf{Z}_i$ is on the segment joining $\mathbf{x}$ to $\mathbf{x}'$. The determinant which appears in the denominator has for its limit the number $\det [D_j f_i(\mathbf{x})] = J_{\mathbf{f}}(\mathbf{x})$, and this is nonzero, since $\mathbf{x} \in X$. Therefore, the following limit exists:

$$\lim_{t \to 0} \frac{g_k(\mathbf{y} + t\mathbf{u}_r) - g_k(\mathbf{y})}{t} = D_r g_k(\mathbf{y}).$$

This establishes the existence of $D_r g_k(\mathbf{y})$ for each $\mathbf{y}$ in Y and each $r = 1, 2, \ldots, n$. Moreover, this limit is a quotient of two determinants involving the derivatives $D_j f_i(\mathbf{x})$. Continuity of the $D_j f_i$ implies continuity of each partial $D_r g_k$. This completes the proof of (e).

NOTE. The foregoing proof also provides a method for computing $D_r g_k(\mathbf{y})$. In practice, the derivatives $D_r g_k$ can be obtained more easily (without recourse to a limiting process) by using the fact that, if $\mathbf{y} = \mathbf{f}(\mathbf{x})$, the product of the two Jacobian matrices $\mathbf{Df}(\mathbf{x})$ and $\mathbf{Dg}(\mathbf{y})$ is the identity matrix. When this is written out in detail it gives the following system of n^2 equations:

$$\sum_{k=1}^{n} D_k g_i(\mathbf{y}) D_j f_k(\mathbf{x}) = \begin{cases} 1 & \text{if } i = j, \\ 0 & \text{if } i \neq j. \end{cases}$$

For each fixed i, we obtain n linear equations as j runs through the values $1, 2, \ldots, n$. These can then be solved for the n unknowns, $D_1 g_i(\mathbf{y}), \ldots, D_n g_i(\mathbf{y})$, by Cramer's rule, or by some other method.

13.4 THE IMPLICIT FUNCTION THEOREM

The reader knows that the equation of a curve in the xy-plane can be expressed either in an "explicit" form, such as $y = f(x)$, or in an "implicit" form, such as $F(x, y) = 0$. However, if we are given an equation of the form $F(x, y) = 0$, this does not necessarily represent a function. (Take, for example, $x^2 + y^2 - 5 = 0$.) The equation $F(x, y) = 0$ *does* always represent a *relation*, namely, that set of all pairs (x, y) which satisfy the equation. The following question therefore presents itself quite naturally: When is the relation defined by $F(x, y) = 0$ also a function? In other words, when can the equation $F(x, y) = 0$ be solved explicitly for y in terms of x, yielding a unique solution? The implicit function theorem deals with this question *locally*. It tells us that, give a point (x_0, y_0) such that $F(x_0, y_0) = 0$, under certain conditions there will be a neighborhood of (x_0, y_0) such that *in this neighborhood* the relation defined by $F(x, y) = 0$ is also a function. The conditions are that F and D_2F be continuous in some neighborhood of (x_0, y_0) and that $D_2F(x_0, y_0) \neq 0$. In its more general form, the theorem treats, instead of one equation in two variables, a system of n equations in $n + k$ variables:

$$f_r(x_1, \ldots, x_n; t_1, \ldots, t_k) = 0 \qquad (r = 1, 2, \ldots, n).$$

This system can be solved for $x_1, \ldots, x_n$ in terms of $t_1, \ldots, t_k$, provided that certain partial derivatives are continuous and provided that the $n \times n$ Jacobian determinant $\partial(f_1, \ldots, f_n)/\partial(x_1, \ldots, x_n)$ is not zero.

For brevity, we shall adopt the following notation in this theorem: Points in $(n + k)$-dimensional space $\mathbf{R}^{n+k}$ will be written in the form $(\mathbf{x}; \mathbf{t})$, where

$$\mathbf{x} = (x_1, \ldots, x_n) \in \mathbf{R}^n \qquad \text{and} \qquad \mathbf{t} = (t_1, \ldots, t_k) \in \mathbf{R}^k.$$

Theorem 13.7 (Implicit function theorem). *Let $\mathbf{f} = (f_1, \ldots, f_n)$ be a vector-valued function defined on an open set S in $\mathbf{R}^{n+k}$ with values in $\mathbf{R}^n$. Suppose $\mathbf{f} \in C'$ on S. Let $(\mathbf{x}_0; \mathbf{t}_0)$ be a point in S for which $\mathbf{f}(\mathbf{x}_0; \mathbf{t}_0) = \mathbf{0}$ and for which the $n \times n$ determinant $\det [D_j f_i(\mathbf{x}_0; \mathbf{t}_0)] \neq 0$. Then there exists a k-dimensional open set T_0 containing $\mathbf{t}_0$ and one, and only one, vector-valued function $\mathbf{g}$, defined on T_0 and having values in $\mathbf{R}^n$, such that*

a) $\mathbf{g} \in C'$ *on* T_0,

b) $\mathbf{g}(\mathbf{t}_0) = \mathbf{x}_0$,

c) $\mathbf{f}(\mathbf{g}(\mathbf{t}); \mathbf{t}) = \mathbf{0}$ *for every* $\mathbf{t}$ *in* T_0.

Proof. We shall apply the inverse function theorem to a certain vector-valued function $\mathbf{F} = (F_1, \ldots, F_n; F_{n+1}, \ldots, F_{n+k})$ defined on S and having values in $\mathbf{R}^{n+k}$. The function $\mathbf{F}$ is defined as follows: For $1 \leq m \leq n$, let $F_m(\mathbf{x}; \mathbf{t}) = f_m(\mathbf{x}; \mathbf{t})$, and for $1 \leq m \leq k$, let $F_{n+m}(\mathbf{x}; \mathbf{t}) = t_m$. We can then write $\mathbf{F} = (\mathbf{f}; \mathbf{I})$, where $\mathbf{f} = (f_1, \ldots, f_n)$ and where $\mathbf{I}$ is the identity function defined by $\mathbf{I}(\mathbf{t}) = \mathbf{t}$ for each $\mathbf{t}$ in $\mathbf{R}^k$. The Jacobian $J_{\mathbf{F}}(\mathbf{x}; \mathbf{t})$ then has the same value as the $n \times n$ determinant $\det [D_j f_i(\mathbf{x}; \mathbf{t})]$ because the terms which appear in the last k rows and also in the last k columns of $J_{\mathbf{F}}(\mathbf{x}; \mathbf{t})$ form a $k \times k$ determinant with ones along the main diagonal and zeros elsewhere; the intersection of the first n rows and n columns consists of the determinant $\det [D_j f_i(\mathbf{x}; \mathbf{t})]$, and

$$D_i F_{n+j}(\mathbf{x}; \mathbf{t}) = 0 \qquad \text{for } 1 \leq i \leq n, \quad 1 \leq j \leq k.$$

Hence the Jacobian $J_{\mathbf{F}}(\mathbf{x}_0; \mathbf{t}_0) \neq 0$. Also, $\mathbf{F}(\mathbf{x}_0; \mathbf{t}_0) = (\mathbf{0}; \mathbf{t}_0)$. Therefore, by Theorem 13.6, there exist open sets X and Y containing $(\mathbf{x}_0; \mathbf{t}_0)$ and $(\mathbf{0}; \mathbf{t}_0)$, respectively, such that $\mathbf{F}$ is one-to-one on X, and $X = \mathbf{F}^{-1}(Y)$. Also, there exists a local inverse function $\mathbf{G}$, defined on Y and having values in X, such that

$$\mathbf{G}[\mathbf{F}(\mathbf{x}; t)] = (\mathbf{x}; t),$$

and such that $\mathbf{G} \in C'$ on Y.

Now $\mathbf{G}$ can be reduced to components as follows: $\mathbf{G} = (\mathbf{v}; \mathbf{w})$ where $\mathbf{v} = (v_1, \ldots, v_n)$ is a vector-valued function defined on Y with values in $\mathbf{R}^n$ and $\mathbf{w} = (w_1, \ldots, w_k)$ is also defined on Y but has values in $\mathbf{R}^k$. We can now determine $\mathbf{v}$ and $\mathbf{w}$ explicitly. The equation $\mathbf{G}[\mathbf{F}(\mathbf{x}; t)] = (\mathbf{x}; t)$, when written in terms of the components $\mathbf{v}$ and $\mathbf{w}$, gives us the two equations

$$\mathbf{v}[\mathbf{F}(\mathbf{x}; t)] = \mathbf{x} \qquad \text{and} \qquad \mathbf{w}[\mathbf{F}(\mathbf{x}; t)] = t.$$

But now, *every point* $(\mathbf{x}; t)$ in Y can be written uniquely in the form $(\mathbf{x}; t) = \mathbf{F}(\mathbf{x}'; t')$ for some $(\mathbf{x}'; t')$ in X, because $\mathbf{F}$ is one-to-one on X and the inverse image $\mathbf{F}^{-1}(Y)$ contains X. Furthermore, by the manner in which $\mathbf{F}$ was defined, when we write $(\mathbf{x}; t) = \mathbf{F}(\mathbf{x}'; t')$, we must have $t' = t$. Therefore,

$$\mathbf{v}(\mathbf{x}; t) = \mathbf{v}[\mathbf{F}(\mathbf{x}'; t)] = \mathbf{x}' \qquad \text{and} \qquad \mathbf{w}(\mathbf{x}; t) = \mathbf{w}[\mathbf{F}(\mathbf{x}'; t)] = t.$$

Hence the function $\mathbf{G}$ can be described as follows: Given a point $(\mathbf{x}; t)$ in Y, we have $\mathbf{G}(\mathbf{x}; t) = (\mathbf{x}'; t)$, where $\mathbf{x}'$ is that point in $\mathbf{R}^n$ such that $(\mathbf{x}; t) = \mathbf{F}(\mathbf{x}'; t)$. This statement implies that

$$\mathbf{F}[\mathbf{v}(\mathbf{x}; t); t] = (\mathbf{x}; t) \qquad \text{for every } (\mathbf{x}; t) \text{ in } Y.$$

Now we are ready to define the set T_0 and the function $\mathbf{g}$ in the theorem. Let

$$T_0 = \{t : t \in \mathbf{R}^k, \quad (\mathbf{0}; t) \in Y\},$$

and for each t in T_0 define $\mathbf{g}(t) = \mathbf{v}(\mathbf{0}; t)$. The set T_0 is open in $\mathbf{R}^k$. Moreover, $\mathbf{g} \in C'$ on T_0 because $\mathbf{G} \in C'$ on Y and the components of $\mathbf{g}$ are taken from the components of $\mathbf{G}$. Also,

$$\mathbf{g}(t_0) = \mathbf{v}(\mathbf{0}; t_0) = \mathbf{x}_0$$

because $(\mathbf{0}; t_0) = \mathbf{F}(\mathbf{x}_0; t_0)$. Finally, the equation $\mathbf{F}[\mathbf{v}(\mathbf{x}; t); t] = (\mathbf{x}; t)$, which holds for every $(\mathbf{x}; t)$ in Y, yields (by considering the components in $\mathbf{R}^n$) the equation $\mathbf{f}[\mathbf{v}(\mathbf{x}; t); t] = \mathbf{x}$. Taking $\mathbf{x} = \mathbf{0}$, we see that for every t in T_0, we have $\mathbf{f}[\mathbf{g}(t); t] = \mathbf{0}$, and this completes the proof of statements (a), (b), and (c). It remains to prove that there is only one such function $\mathbf{g}$. But this follows at once from the one-to-one character of $\mathbf{f}$. If there were another function, say $\mathbf{h}$, which satisfied (c), then we would have $\mathbf{f}[\mathbf{g}(t); t] = \mathbf{f}[\mathbf{h}(t); t]$, and this would imply $(\mathbf{g}(t); t) = (\mathbf{h}(t); t)$, or $\mathbf{g}(t) = \mathbf{h}(t)$ for every t in T_0.

13.5 EXTREMA OF REAL-VALUED FUNCTIONS OF ONE VARIABLE

In the remainder of this chapter we shall consider real-valued functions f with a view toward determining those points (if any) at which f has a local extremum, that is, either a local maximum or a local minimum.

We have already obtained one result in this connection for functions of one variable (Theorem 5.9). In that theorem we found that a necessary condition for a function f to have a local extremum at an interior point c of an interval is that $f'(c) = 0$, provided that $f'(c)$ exists. This condition, however, is not sufficient, as we can see by taking $f(x) = x^3$, $c = 0$. We now derive a sufficient condition.

Theorem 13.8. *For some integer $n \geq 1$, let f have a continuous nth derivative in the open interval (a, b). Suppose also that for some interior point c in (a, b) we have*

$$f'(c) = f''(c) = \cdots = f^{(n-1)}(c) = 0, \qquad but \qquad f^{(n)}(c) \neq 0.$$

Then for n even, f has a local minimum at c if $f^{(n)}(c) > 0$, and a local maximum at c if $f^{(n)}(c) < 0$. If n is odd, there is neither a local maximum nor a local minimum at c.

Proof. Since $f^{(n)}(c) \neq 0$, there exists an interval $B(c)$ such that for every x in $B(c)$, the derivative $f^{(n)}(x)$ will have the same sign as $f^{(n)}(c)$. Now by Taylor's formula (Theorem 5.19), for every x in $B(c)$ we have

$$f(x) - f(c) = \frac{f^{(n)}(x_1)}{n!} (x - c)^n, \qquad \text{where } x_1 \in B(c).$$

If n is even, this equation implies $f(x) \geq f(c)$ when $f^{(n)}(c) > 0$, and $f(x) \leq f(c)$ when $f^{(n)}(c) \leq 0$. If n is odd and $f^{(n)}(c) > 0$, then $f(x) > f(c)$ when $x > c$, but $f(x) < f(c)$ when $x < c$, and there can be no extremum at c. A similar statement holds if n is odd and $f^{(n)}(c) < 0$. This proves the theorem.

13.6 EXTREMA OF REAL-VALUED FUNCTIONS OF SEVERAL VARIABLES

We turn now to functions of several variables. Exercise 12.1 gives a necessary condition for a function to have a local maximum or a local minimum at an interior point $\mathbf{a}$ of an open set. The condition is that each partial derivative $D_k f(\mathbf{a})$ must be zero at that point. We can also state this in terms of directional derivatives by saying that $f'(\mathbf{a}; \mathbf{u})$ must be zero for every direction $\mathbf{u}$.

The converse of this statement is not true, however. Consider the following example of a function of two real variables:

$$f(x, y) = (y - x^2)(y - 2x^2).$$

Here we have $D_1 f(0, 0) = D_2 f(0, 0) = 0$. Now $f(0, 0) = 0$, but the function assumes both positive and negative values in every neighborhood of $(0, 0)$, so there is neither a local maximum nor a local minimum at $(0, 0)$. (See Fig. 13.3.)

This example illustrates another interesting phenomenon. If we take a fixed straight line through the origin and restrict the point (x, y) to move along this line toward $(0, 0)$, then the point will finally enter the region above the parabola $y = 2x^2$ (or below the parabola $y = x^2$) in which $f(x, y)$ becomes and stays positive for every $(x, y) \neq (0, 0)$. Therefore, along every such line, f has a minimum at $(0, 0)$, but the origin is not a local minimum in any two-dimensional neighborhood of $(0, 0)$.

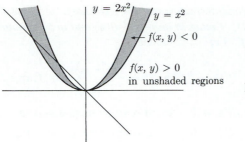

$y = 2x^2$

$y = x^2$

$f(x, y) < 0$

$f(x, y) > 0$
in unshaded regions

Figure 13.3

Definition 13.9. *If f is differentiable at* **a** *and if* $\nabla f(\mathbf{a}) = \mathbf{0}$, *the point* **a** *is called a stationary point of* **f**. *A stationary point is called a saddle point if every n-ball* $B(\mathbf{a})$ *contains points* **x** *such that* $f(\mathbf{x}) > f(\mathbf{a})$ *and other points such that* $f(\mathbf{x}) < f(\mathbf{a})$.

In the foregoing example, the origin is a saddle point of the function.

To determine whether a function of n variables has a local maximum, a local minimum, or a saddle point at a stationary point **a**, we must determine the algebraic sign of $f(\mathbf{x}) - f(\mathbf{a})$ for all **x** in a neighborhood of **a**. As in the one-dimensional case, this is done with the help of Taylor's formula (Theorem 12.14). Take $m = 2$ and $\mathbf{y} = \mathbf{a} + \mathbf{t}$ in Theorem 12.14. If the partial derivatives of f are differentiable on an n-ball $B(\mathbf{a})$ then

$$f(\mathbf{a} + \mathbf{t}) - f(\mathbf{a}) = \nabla f(\mathbf{a}) \cdot \mathbf{t} + \tfrac{1}{2} f''(\mathbf{z}; \mathbf{t}), \qquad (3)$$

where **z** lies on the line segment joining **a** and $\mathbf{a} + \mathbf{t}$, and

$$f''(\mathbf{z}; \mathbf{t}) = \sum_{i=1}^{n} \sum_{j=1}^{n} D_{i,j} f(\mathbf{z}) t_i t_j.$$

At a stationary point we have $\nabla f(\mathbf{a}) = \mathbf{0}$ so (3) becomes

$$f(\mathbf{a} + \mathbf{t}) - f(\mathbf{a}) = \tfrac{1}{2} f''(\mathbf{z}; \mathbf{t}).$$

Therefore, as $\mathbf{a} + \mathbf{t}$ ranges over $B(\mathbf{a})$, the algebraic sign of $f(\mathbf{a} + \mathbf{t}) - f(\mathbf{a})$ is determined by that of $f''(\mathbf{z}; \mathbf{t})$. We can write (3) in the form

$$f(\mathbf{a} + \mathbf{t}) - f(\mathbf{a}) = \tfrac{1}{2} f''(\mathbf{a}; \mathbf{t}) + \|\mathbf{t}\|^2 E(\mathbf{t}), \qquad (4)$$

where

$$\|\mathbf{t}\|^2 E(\mathbf{t}) = \tfrac{1}{2} f''(\mathbf{z}; \mathbf{t}) - \tfrac{1}{2} f''(\mathbf{a}; \mathbf{t}).$$

The inequality

$$\|\mathbf{t}\|^2 \, |E(\mathbf{t})| \le \frac{1}{2} \sum_{i=1}^{n} \sum_{j=1}^{n} |D_{i,j} f(\mathbf{z}) - D_{i,j} f(\mathbf{a})| \, \|\mathbf{t}\|^2,$$

shows that $E(\mathbf{t}) \to 0$ as $\mathbf{t} \to \mathbf{0}$ if the second-order partial derivatives of f are continuous at **a**. Since $\|\mathbf{t}\|^2 E(\mathbf{t})$ tends to zero faster than $\|\mathbf{t}\|^2$, it seems reasonable to expect that the algebraic sign of $f(\mathbf{a} + \mathbf{t}) - f(\mathbf{a})$ should be determined by that of $f''(\mathbf{a}; \mathbf{t})$. This is what is proved in the next theorem.

Theorem 13.10 (Second-derivative test for extrema). *Assume that the second-order partial derivatives $D_{i,j}f$ exist in an n-ball $B(\mathbf{a})$ and are continuous at $\mathbf{a}$, where $\mathbf{a}$ is a stationary point of f. Let*

$$Q(\mathbf{t}) = \tfrac{1}{2}f''(\mathbf{a}; \mathbf{t}) = \frac{1}{2}\sum_{i=1}^{n}\sum_{j=1}^{n} D_{i,j}f(\mathbf{a})t_i t_j. \qquad (5)$$

a) *If $Q(\mathbf{t}) > 0$ for all $\mathbf{t} \neq \mathbf{0}$, f has a relative minimum at $\mathbf{a}$.*

b) *If $Q(\mathbf{t}) < 0$ for all $\mathbf{t} \neq \mathbf{0}$, f has a relative maximum at $\mathbf{a}$.*

c) *If $Q(\mathbf{t})$ takes both positive and negative values, then f has a saddle point at $\mathbf{a}$.*

Proof. The function Q is continuous at each point $\mathbf{t}$ in $\mathbf{R}^n$. Let $S = \{\mathbf{t} : \|\mathbf{t}\| = 1\}$ denote the boundary of the n-ball $B(\mathbf{0}; 1)$. If $Q(\mathbf{t}) > 0$ for all $\mathbf{t} \neq \mathbf{0}$, then $Q(\mathbf{t})$ is positive on S. Since S is compact, Q has a minimum on S (call it m), and $m > 0$. Now $Q(c\mathbf{t}) = c^2 Q(\mathbf{t})$ for every real c. Taking $c = 1/\|\mathbf{t}\|$ where $\mathbf{t} \neq \mathbf{0}$ we see that $c\mathbf{t} \in S$ and hence $c^2 Q(\mathbf{t}) \geq m$, so $Q(\mathbf{t}) \geq m\|\mathbf{t}\|^2$. Using this in (4) we find

$$f(\mathbf{a} + \mathbf{t}) - f(\mathbf{a}) = Q(\mathbf{t}) + \|\mathbf{t}\|^2 E(\mathbf{t}) \geq m\,\|\mathbf{t}\|^2 + \|\mathbf{t}\|^2 E(\mathbf{t}).$$

Since $E(\mathbf{t}) \to 0$ as $\mathbf{t} \to \mathbf{0}$, there is a positive number r such that $|E(\mathbf{t})| < \tfrac{1}{2}m$ whenever $0 < \|\mathbf{t}\| < r$. For such $\mathbf{t}$ we have $0 \leq \|\mathbf{t}\|^2 |E(\mathbf{t})| < \tfrac{1}{2}m\|\mathbf{t}\|^2$, so

$$f(\mathbf{a} + \mathbf{t}) - f(\mathbf{a}) > m\|\mathbf{t}\|^2 - \tfrac{1}{2}m\|\mathbf{t}\|^2 = \tfrac{1}{2}m\|\mathbf{t}\|^2 > 0.$$

Therefore f has a relative minimum at $\mathbf{a}$, which proves (a). To prove (b) we use a similar argument, or simply apply part (a) to $-f$.

Finally, we prove (c). For each $\lambda > 0$ we have, from (4),

$$f(\mathbf{a} + \lambda\mathbf{t}) - f(\mathbf{a}) = Q(\lambda\mathbf{t}) + \lambda^2\|\mathbf{t}\|^2 E(\lambda\mathbf{t}) = \lambda^2\{Q(\mathbf{t}) + \|\mathbf{t}\|^2 E(\lambda\mathbf{t})\}.$$

Suppose $Q(\mathbf{t}) \neq 0$ for some $\mathbf{t}$. Since $E(\mathbf{y}) \to 0$ as $\mathbf{y} \to \mathbf{0}$, there is a positive r such that

$$\|\mathbf{t}\|^2 E(\lambda\mathbf{t}) < \tfrac{1}{2}|Q(\mathbf{t})| \qquad \text{if } 0 < \lambda < r.$$

Therefore, for each such λ the quantity $\lambda^2\{Q(\mathbf{t}) + \|\mathbf{t}\|^2 E(\lambda\mathbf{t})\}$ has the same sign as $Q(\mathbf{t})$. Therefore, if $0 < \lambda < r$, the difference $f(\mathbf{a} + \lambda\mathbf{t}) - f(\mathbf{a})$ has the same sign as $Q(\mathbf{t})$. Hence, if $Q(\mathbf{t})$ takes both positive and negative values, it follows that f has a saddle point at $\mathbf{a}$.

NOTE. A real-valued function Q defined on $\mathbf{R}^n$ by an equation of the type

$$Q(\mathbf{x}) = \sum_{i=1}^{n}\sum_{j=1}^{n} a_{ij}x_i x_j,$$

where $\mathbf{x} = (x_1, \ldots, x_n)$ and the a_{ij} are real is called a *quadratic form*. The form is called *symmetric* if $a_{ij} = a_{ji}$ for all i and j, *positive definite* if $\mathbf{x} \neq \mathbf{0}$ implies $Q(\mathbf{x}) > 0$, and *negative definite* if $\mathbf{x} \neq \mathbf{0}$ implies $Q(\mathbf{x}) < 0$.

In general, it is not easy to determine whether a quadratic form is positive or negative definite. One criterion, involving eigenvalues, is described in Reference

13.1, Theorem 9.5. Another, involving determinants, can be described as follows. Let $\Delta = \det [a_{ij}]$ and let Δ_k denote the determinant of the $k \times k$ matrix obtained by deleting the last $(n - k)$ rows and columns of $[a_{ij}]$. Also, put $\Delta_0 = 1$. From the theory of quadratic forms it is known that a necessary and sufficient condition for a symmetric form to be positive definite is that the $n + 1$ numbers $\Delta_0, \Delta_1, \ldots, \Delta_n$ be positive. The form is negative definite if, and only if, the same $n + 1$ numbers are alternately positive and negative. (See Reference 13.2, pp. 304–308.) The quadratic form which appears in (5) is symmetric because the mixed partials $D_{i,j}f(\mathbf{a})$ and $D_{j,i}f(\mathbf{a})$ are equal. Therefore, under the conditions of Theorem 13.10, we see that f has a local minimum at $\mathbf{a}$ if the $(n + 1)$ numbers $\Delta_0, \Delta_1, \ldots, \Delta_n$ are all positive, and a local maximum if these numbers are alternately positive and negative. The case $n = 2$ can be handled directly and gives the following criterion.

Theorem 13.11. *Let f be a real-valued function with continuous second-order partial derivatives at a stationary point $\mathbf{a}$ in $\mathbf{R}^2$. Let*

$$A = D_{1,1}f(\mathbf{a}), \qquad B = D_{1,2}f(\mathbf{a}), \qquad C = D_{2,2}f(\mathbf{a}),$$

and let

$$\Delta = \det \begin{bmatrix} A & B \\ B & C \end{bmatrix} = AC - B^2.$$

Then we have:

a) *If $\Delta > 0$ and $A > 0$, f has a relative minimum at $\mathbf{a}$.*

b) *If $\Delta > 0$ and $A < 0$, f has a relative maximum at $\mathbf{a}$.*

c) *If $\Delta < 0$, f has a saddle point at $\mathbf{a}$.*

Proof. In the two-dimensional case we can write the quadratic form in (5) as follows:

$$Q(x, y) = \tfrac{1}{2}\{Ax^2 + 2Bxy + Cy^2\}.$$

If $A \neq 0$, this can also be written as

$$Q(x, y) = \frac{1}{2A} \{(Ax + By)^2 + \Delta y^2\}.$$

If $\Delta > 0$, the expression in brackets is the sum of two squares, so $Q(x, y)$ has the same sign as A. Therefore, statements (a) and (b) follow at once from parts (a) and (b) of Theorem 13.10.

If $\Delta < 0$, the quadratic form is the product of two linear factors. Therefore, the set of points (x, y) such that $Q(x, y) = 0$ consists of two lines in the xy-plane intersecting at $(0, 0)$. These lines divide the plane into four regions; $Q(x, y)$ is positive in two of these regions and negative in the other two. Therefore f has a saddle point at $\mathbf{a}$.

NOTE. If $\Delta = 0$, there may be a local maximum, a local minimum, or a saddle point at $\mathbf{a}$.

13.7 EXTREMUM PROBLEMS WITH SIDE CONDITIONS

Consider the following type of extremum problem. Suppose that $f(x, y, z)$ represents the temperature at the point (x, y, z) in space and we ask for the maximum or minimum value of the temperature on a certain surface. If the equation of the surface is given explicitly in the form $z = h(x, y)$, then in the expression $f(x, y, z)$ we can replace z by $h(x, y)$ to obtain the temperature on the surface as a function of x and y alone, say $F(x, y) = f[x, y, h(x, y)]$. The problem is then reduced to finding the extreme values of F. However, in practice, certain difficulties arise. The equation of the surface might be given in an implicit form, say $g(x, y, z) = 0$, and it may be impossible, in practice, to solve this equation explicitly for z in terms of x and y, or even for x or y in terms of the remaining variables. The problem might be further complicated by asking for the extreme values of the temperature at those points which lie on a given *curve* in space. Such a curve is the intersection of two surfaces, say $g_1(x, y, z) = 0$ and $g_2(x, y, z) = 0$. If we could solve these two equations simultaneously, say for x and y in terms of z, then we could introduce these expressions into f and obtain a new function of z alone, whose extrema we would then seek. In general, however, this procedure cannot be carried out and a more practicable method must be sought. A very elegant and useful method for attacking such problems was developed by Lagrange.

Lagrange's method provides a *necessary* condition for an extremum and can be described as follows. Let $f(x_1, \ldots, x_n)$ be an expression whose extreme values are sought when the variables are restricted by a certain number of side conditions, say $g_1(x_1, \ldots, x_n) = 0, \ldots, g_m(x_1, \ldots, x_n) = 0$. We then form the linear combination

$$\phi(x_1, \ldots, x_n) = f(x_1, \ldots, x_n) + \lambda_1 g_1(x_1, \ldots, x_n) + \cdots + \lambda_m g_m(x_1, \ldots, x_n),$$

where $\lambda_1, \ldots, \lambda_m$ are m constants. We then differentiate ϕ with respect to each coordinate and consider the following system of $n + m$ equations:

$$D_r\phi(x_1, \ldots, x_n) = 0, \qquad r = 1, 2, \ldots, n,$$

$$g_k(x_1, \ldots, x_n) = 0, \qquad k = 1, 2, \ldots, m.$$

Lagrange discovered that if the point $(x_1, \ldots, x_n)$ is a solution of the extremum problem, then it will also satisfy this system of $n + m$ equations. In practice, one attempts to solve this system for the $n + m$ "unknowns," $\lambda_1, \ldots, \lambda_m$, and $x_1, \ldots, x_n$. The points $(x_1, \ldots, x_n)$ so obtained must then be tested to determine whether they yield a maximum, a minimum, or neither. The numbers $\lambda_1, \ldots, \lambda_m$, which are introduced only to help solve the system for $x_1, \ldots, x_n$, are known as *Lagrange's multipliers*. One multiplier is introduced for each side condition.

A complicated analytic criterion exists for distinguishing between maxima and minima in such problems. (See, for example, Reference 13.3.) However, this criterion is not very useful in practice and in any particular prolem it is usually easier to rely on some other means (for example, physical or geometrical considerations) to make this distinction.

The following theorem establishes the validity of Lagrange's method:

Theorem 13.12. *Let f be a real-valued function such that $f \in C'$ on an open set S in $\mathbf{R}^n$. Let $g_1, \ldots, g_m$ be m real-valued functions such that $\mathbf{g} = (g_1, \ldots, g_m) \in C'$ on S, and assume that $m < n$. Let X_0 be that subset of S on which $\mathbf{g}$ vanishes, that is,*

$$X_0 = \{\mathbf{x} : \mathbf{x} \in S, \mathbf{g}(\mathbf{x}) = \mathbf{0}\}.$$

Assume that $\mathbf{x}_0 \in X_0$ and assume that there exists an n-ball $B(\mathbf{x}_0)$ such that $f(\mathbf{x}) \leq f(\mathbf{x}_0)$ for all $\mathbf{x}$ in $X_0 \cap B(\mathbf{x}_0)$ or such that $f(\mathbf{x}) \geq f(\mathbf{x}_0)$ for all $\mathbf{x}$ in $X_0 \cap B(\mathbf{x}_0)$. Assume also that the m-rowed determinant $\det\left[D_j g_i(\mathbf{x}_0)\right] \neq 0$. Then there exist m real numbers $\lambda_1, \ldots, \lambda_m$ such that the following n equations are satisfied:

$$D_r f(\mathbf{x}_0) + \sum_{k=1}^{m} \lambda_k D_r g_k(\mathbf{x}_0) = 0 \qquad (r = 1, 2, \ldots, n). \tag{6}$$

NOTE. The n equations in (6) are equivalent to the following vector equation:

$$\nabla f(\mathbf{x}_0) + \lambda_1 \nabla g_1(\mathbf{x}_0) + \cdots + \lambda_m \nabla g_m(\mathbf{x}_0) = \mathbf{0}.$$

Proof. Consider the following system of m linear equations in the m unknowns $\lambda_1, \ldots, \lambda_m$:

$$\sum_{k=1}^{m} \lambda_k D_r g_k(\mathbf{x}_0) = -D_r f(\mathbf{x}_0) \qquad (r = 1, 2, \ldots, m).$$

This system has a unique solution since, by hypothesis, the determinant of the system is not zero. Therefore, the *first m equations* in (6) are satisfied. We must now verify that for this choice of $\lambda_1, \ldots, \lambda_m$, the remaining $n - m$ equations in (6) are also satisfied.

To do this, we apply the implicit function theorem. Since $m < n$, every point $\mathbf{x}$ in S can be written in the form $\mathbf{x} = (\mathbf{x}'; \mathbf{t})$, say, where $\mathbf{x}' \in \mathbf{R}^m$ and $\mathbf{t} \in \mathbf{R}^{n-m}$. In the remainder of this proof we will write $\mathbf{x}'$ for $(x_1, \ldots, x_m)$ and $\mathbf{t}$ for $(x_{m+1}, \ldots, x_n)$, so that $t_k = x_{m+k}$. In terms of the vector-valued function $\mathbf{g} = (g_1, \ldots, g_m)$, we can now write

$$\mathbf{g}(\mathbf{x}_0'; \mathbf{t}_0) = \mathbf{0} \qquad \text{if } \mathbf{x}_0 = (\mathbf{x}_0'; \mathbf{t}_0).$$

Since $\mathbf{g} \in C'$ on S, and since the determinant $\det\left[D_j g_i(\mathbf{x}_0'; \mathbf{t}_0)\right] \neq 0$, all the conditions of the implicit function theorem are satisfied. Therefore, there exists an $(n - m)$-dimensional neighborhood T_0 of $\mathbf{t}_0$ and a unique vector-valued function $\mathbf{h} = (h_1, \ldots, h_m)$, defined on T_0 and having values in $\mathbf{R}^m$ such that $\mathbf{h} \in C'$ on T_0, $\mathbf{h}(\mathbf{t}_0) = \mathbf{x}_0'$, and for every $\mathbf{t}$ in T_0, we have $\mathbf{g}[\mathbf{h}(\mathbf{t}); \mathbf{t}] = \mathbf{0}$. This amounts to saying that the system of m equations

$$g_1(x_1, \ldots, x_n) = 0, \ldots, g_m(x_1, \ldots, x_n) = 0,$$

can be solved for $x_1, \ldots, x_m$ in terms of $x_{m+1}, \ldots, x_n$, giving the solutions in the form $x_r = h_r(x_{m+1}, \ldots, x_n)$, $r = 1, 2, \ldots, m$. We shall now substitute these expressions for $x_1, \ldots, x_m$ into the expression $f(x_1, \ldots, x_n)$ and also into each expression $g_p(x_1, \ldots, x_n)$. That is to say, we define a new function F as follows:

$$F(x_{m+1}, \ldots, x_n) = f[h_1(x_{m+1}, \ldots, x_n), \ldots, h_m(x_{m+1}, \ldots, x_n); x_{m+1}, \ldots, x_n];$$

and we define m new functions $G_1, \ldots, G_m$ as follows:

$$G_p(x_{m+1}, \ldots, x_n) = g_p[h_1(x_{m+1}, \ldots, x_n), \ldots, h_m(x_{m+1}, \ldots, x_n); x_{m+1}, \ldots, x_n].$$

More briefly, we can write $F(\mathbf{t}) = f[\mathbf{H}(\mathbf{t})]$ and $G_p(\mathbf{t}) = g_p[\mathbf{H}(\mathbf{t})]$, where $\mathbf{H}(\mathbf{t}) = (\mathbf{h}(\mathbf{t}); \mathbf{t})$. Here $\mathbf{t}$ is restricted to lie in the set T_0.

Each function G_p so defined is identically zero on the set T_0 by the implicit function theorem. Therefore, each derivative $D_r G_p$ is also identically zero on T_0 and, in particular, $D_r G_p(\mathbf{t}_0) = 0$. But by the chain rule (Eq. 12.20), we can compute these derivatives as follows:

$$D_r G_p(\mathbf{t}_0) = \sum_{k=1}^{n} D_k g_p(\mathbf{x}_0) D_r H_k(\mathbf{t}_0) \qquad (r = 1, 2, \ldots, n - m).$$

But $H_k(\mathbf{t}) = h_k(\mathbf{t})$ if $1 \le k \le m$, and $H_k(\mathbf{t}) = x_k$ if $m + 1 \le k \le n$. Therefore, when $m + 1 \le k \le n$, we have $D_r H_k(\mathbf{t}) \equiv 0$ if $m + r \ne k$ and $D_r H_{m+r}(\mathbf{t}) = 1$ for every $\mathbf{t}$. Hence the above set of equations becomes

$$\sum_{k=1}^{m} D_k g_p(\mathbf{x}_0) D_r h_k(\mathbf{t}_0) + D_{m+r} g_p(\mathbf{x}_0) = 0 \qquad \begin{cases} p = 1, 2, \ldots, m, \\ r = 1, 2, \ldots, n - m. \end{cases} \qquad (7)$$

By continuity of $\mathbf{h}$, there is an $(n - m)$-ball $B(\mathbf{t}_0) \subseteq T_0$ such that $\mathbf{t} \in B(\mathbf{t}_0)$ implies $(\mathbf{h}(\mathbf{t}); \mathbf{t}) \in B(\mathbf{x}_0)$, where $B(\mathbf{x}_0)$ is the n-ball in the statement of the theorem. Hence, $\mathbf{t} \in B(\mathbf{t}_0)$ implies $(\mathbf{h}(\mathbf{t}); \mathbf{t}) \in X_0 \cap B(\mathbf{x}_0)$ and therefore, by hypothesis, we have either $F(\mathbf{t}) \le F(\mathbf{t}_0)$ for all $\mathbf{t}$ in $B(\mathbf{t}_0)$ or else we have $F(\mathbf{t}) \ge F(\mathbf{t}_0)$ for all $\mathbf{t}$ in $B(\mathbf{t}_0)$. That is, F has a local maximum or a local minimum at the interior point $\mathbf{t}_0$. Each partial derivative $D_r F(\mathbf{t}_0)$ must therefore be zero. If we use the chain rule to compute these derivatives, we find

$$D_r F(\mathbf{t}_0) = \sum_{k=1}^{n} D_k f(\mathbf{x}_0) D_r H_k(\mathbf{t}_0) \qquad (r = 1, \ldots, n - m),$$

and hence we can write

$$\sum_{k=1}^{m} D_k f(\mathbf{x}_0) D_r h_k(\mathbf{t}_0) + D_{m+r} f(\mathbf{x}_0) = 0 \qquad (r = 1, \ldots, n - m). \qquad (8)$$

If we now multiply (7) by λ_p, sum on p, and add the result to (8), we find

$$\sum_{k=1}^{m} \left[D_k f(\mathbf{x}_0) + \sum_{p=1}^{m} \lambda_p D_k g_p(\mathbf{x}_0) \right] D_r h_k(\mathbf{t}_0) + D_{m+r} f(\mathbf{x}_0) + \sum_{p=1}^{m} \lambda_p D_{m+r} g_p(\mathbf{x}_0) = 0,$$

for $r = 1, \ldots, n - m$. In the sum over k, the expression in square brackets vanishes because of the way $\lambda_1, \ldots, \lambda_m$ were defined. Thus we are left with

$$D_{m+r}f(\mathbf{x}_0) + \sum_{p=1}^{m} \lambda_p D_{m+r}g_p(\mathbf{x}_0) = 0 \qquad (r = 1, 2, \ldots, n - m),$$

and these are exactly the equations needed to complete the proof.

NOTE. In attempting the solution of a particular extremum problem by Lagrange's method, it is usually very easy to determine the system of equations (6) but, in general, it is not a simple matter to actually *solve* the system. Special devices can often be employed to obtain the extreme values of f directly from (6) without first finding the particular points where these extremes are taken on. The following example illustrates some of these devices:

Example. *A quadric surface with center at the origin has the equation*

$$Ax^2 + By^2 + Cz^2 + 2Dyz + 2Ezx + 2Fxy = 1.$$

Find the lengths of its semi-axes.

Solution. Let us write (x_1, x_2, x_3) instead of (x, y, z), and introduce the quadratic form

$$q(\mathbf{x}) = \sum_{j=1}^{3} \sum_{i=1}^{3} a_{ij}x_i x_j, \tag{9}$$

where $\mathbf{x} = (x_1, x_2, x_3)$ and the $a_{ij} = a_{ji}$ are chosen so that the equation of the surface becomes $q(\mathbf{x}) = 1$. (Hence the quadratic form is symmetric and positive definite.) The problem is equivalent to finding the extreme values of $f(\mathbf{x}) = \|\mathbf{x}\|^2 = x_1^2 + x_2^2 + x_3^2$ subject to the side condition $g(\mathbf{x}) = 0$, where $g(\mathbf{x}) = q(\mathbf{x}) - 1$. Using Lagrange's method, we introduce one multiplier and consider the vector equation

$$\nabla f(\mathbf{x}) + \lambda \nabla q(\mathbf{x}) = \mathbf{0} \tag{10}$$

(since $\nabla g = \nabla q$). In this particular case, both f and q are homogeneous functions of degree 2 and we can apply Euler's theorem (see Exercise 12.18) in (10) to obtain

$$\mathbf{x} \cdot \nabla f(\mathbf{x}) + \lambda \mathbf{x} \cdot \nabla q(\mathbf{x}) = 2f(\mathbf{x}) + 2\lambda q(\mathbf{x}) = 0.$$

Since $q(\mathbf{x}) = 1$ on the surface we find $\lambda = -f(\mathbf{x})$, and (10) becomes

$$t \, \nabla f(\mathbf{x}) - \nabla q(\mathbf{x}) = \mathbf{0}, \tag{11}$$

where $t = 1/f(\mathbf{x})$. (We cannot have $f(\mathbf{x}) = 0$ in this problem.) The vector equation (11) then leads to the following three equations for x_1, x_2, x_3:

$$(a_{11} - t)x_1 + \qquad a_{12}x_2 + \qquad a_{13}x_3 = 0,$$
$$a_{21}x_1 + (a_{22} - t)x_2 + \qquad a_{23}x_3 = 0,$$
$$a_{31}x_1 + \qquad a_{32}x_2 + (a_{33} - t)x_3 = 0.$$

Since $\mathbf{x} = \mathbf{0}$ cannot yield a solution to our problem, the determinant of this system must vanish. That is, we must have

$$\begin{vmatrix} a_{11} - t & a_{12} & a_{13} \\ a_{21} & a_{22} - t & a_{23} \\ a_{31} & a_{32} & a_{33} - t \end{vmatrix} = 0. \qquad (12)$$

Equation (12) is called the *characteristic equation* of the quadratic form in (9). In this case, the geometrical nature of the problem assures us that the three roots t_1, t_2, t_3 of this cubic must be real and positive. [Since $q(\mathbf{x})$ is symmetric and positive definite, the general theory of quadratic forms also guarantees that the roots of (12) are all real and positive. (See Reference 13.1, Theorem 9.5.)] The semi-axes of the quadric surface are $t_1^{-1/2}$, $t_2^{-1/2}$, $t_3^{-1/2}$.

EXERCISES

Jacobians

13.1 Let f be the complex-valued function defined for each complex $z \neq 0$ by the equation $f(z) = 1/\bar{z}$. Show that $J_f(z) = -|z|^{-4}$. Show that f is one-to-one and compute f^{-1} explicitly.

13.2 Let $\mathbf{f} = (f_1, f_2, f_3)$ be the vector-valued function defined (for every point (x_1, x_2, x_3) in $\mathbf{R}^3$ for which $x_1 + x_2 + x_3 \neq -1$) as follows:

$$f_k(x_1, x_2, x_3) = \frac{x_k}{1 + x_1 + x_2 + x_3} \qquad (k = 1, 2, 3).$$

Show that $J_{\mathbf{f}}(x_1, x_2, x_3) = (1 + x_1 + x_2 + x_3)^{-4}$. Show that $\mathbf{f}$ is one-to-one and compute $\mathbf{f}^{-1}$ explicitly.

13.3 Let $\mathbf{f} = (f_1, \ldots, f_n)$ be a vector-valued function defined in $\mathbf{R}^n$, suppose $\mathbf{f} \in C'$ on $\mathbf{R}^n$, and let $J_{\mathbf{f}}(\mathbf{x})$ denote the Jacobian determinant. Let $g_1, \ldots, g_n$ be n real-valued functions defined on $\mathbf{R}^1$ and having continuous derivatives $g_1', \ldots, g_n'$. Let $h_k(\mathbf{x}) = f_k[g_1(x_1), \ldots, g_n(x_n)]$, $k = 1, 2, \ldots, n$, and put $\mathbf{h} = (h_1, \ldots, h_n)$. Show that

$$J_{\mathbf{h}}(\mathbf{x}) = J_{\mathbf{f}}[g_1(x_1), \ldots, g_n(x_n)] g_1'(x_1) \cdots g_n'(x_n).$$

13.4 a) If $x(r, \theta) = r \cos \theta$, $y(r, \theta) = r \sin \theta$, show that

$$\frac{\partial(x, y)}{\partial(r, \theta)} = r.$$

b) If $x(r, \theta, \phi) = r \cos \theta \sin \phi$, $y(r, \theta, \phi) = r \sin \theta \sin \phi$, $z = r \cos \phi$, show that

$$\frac{\partial(x, y, z)}{\partial(r, \theta, \phi)} = -r^2 \sin \phi.$$

13.5 a) State conditions on f and g which will ensure that the equations $x = f(u, v)$, $y = g(u, v)$ can be solved for u and v in a neighborhood of (x_0, y_0). If the solutions are $u = F(x, y)$, $v = G(x, y)$, and if $J = \partial(f, g)/\partial(u, v)$, show that

$$\frac{\partial F}{\partial x} = \frac{1}{J} \frac{\partial g}{\partial v}, \qquad \frac{\partial F}{\partial y} = -\frac{1}{J} \frac{\partial f}{\partial v}, \qquad \frac{\partial G}{\partial x} = -\frac{1}{J} \frac{\partial g}{\partial u}, \qquad \frac{\partial G}{\partial y} = \frac{1}{J} \frac{\partial f}{\partial u}.$$

b) Compute J and the partial derivatives of F and G at $(x_0, y_0) = (1, 1)$ when $f(u, v) = u^2 - v^2$, $g(u, v) = 2uv$.

13.6 Let **f** and **g** be related as in Theorem 13.6. Consider the case $n = 3$ and show that we have

$$J_{\mathbf{f}}(\mathbf{x})D_1 g_i(\mathbf{y}) = \begin{vmatrix} \delta_{i,1} & D_1 f_2(\mathbf{x}) & D_1 f_3(\mathbf{x}) \\ \delta_{i,2} & D_2 f_2(\mathbf{x}) & D_2 f_3(\mathbf{x}) \\ \delta_{i,3} & D_3 f_2(\mathbf{x}) & D_3 f_3(\mathbf{x}) \end{vmatrix} \qquad (i = 1, 2, 3),$$

where $\mathbf{y} = \mathbf{f}(\mathbf{x})$ and $\delta_{i,j} = 0$ or 1 according as $i \neq j$ or $i = j$. Use this to deduce the formula

$$D_1 g_1 = \frac{\partial(f_2, f_3)}{\partial(x_2, x_3)} \bigg/ \frac{\partial(f_1, f_2, f_3)}{\partial(x_1, x_2, x_3)}.$$

There are similar expressions for the other eight derivatives $D_k g_i$.

13.7 Let $f = u + iv$ be a complex-valued function satisfying the following conditions: $u \in C'$ and $v \in C'$ on the open disk $A = \{z : |z| < 1\}$; f is continuous on the closed disk $\bar{A} = \{z : |z| \leq 1\}$; $u(x, y) = x$ and $v(x, y) = y$ whenever $x^2 + y^2 = 1$; the Jacobian $J_f(z) > 0$ if $z \in A$. Let $B = f(A)$ denote the image of A under f and prove that:

a) If X is an open subset of A, then $f(X)$ is an open subset of B.

b) B is an open disk of radius 1.

c) For each point $u_0 + iv_0$ in B, there is only a finite number of points z in A such that $f(z) = u_0 + iv_0$.

Extremum problems

13.8 Find and classify the extreme values (if any) of the functions defined by the following equations:

a) $f(x, y) = y^2 + x^2 y + x^4$,

b) $f(x, y) = x^2 + y^2 + x + y + xy$,

c) $f(x, y) = (x - 1)^4 + (x - y)^4$,

d) $f(x, y) = y^2 - x^3$.

13.9 Find the shortest distance from the point $(0, b)$ on the y-axis to the parabola $x^2 - 4y = 0$. Solve this problem using Lagrange's method and also without using Lagrange's method.

13.10 Solve the following geometric problems by Lagrange's method:

a) Find the shortest distance from the point (a_1, a_2, a_3) in $\mathbf{R}^3$ to the plane whose equation is $b_1 x_1 + b_2 x_2 + b_3 x_3 + b_0 = 0$.

b) Find the point on the line of intersection of the two planes

$$a_1 x_1 + a_2 x_2 + a_3 x_3 + a_0 = 0$$

and

$$b_1 x_1 + b_2 x_2 + b_3 x_3 + b_0 = 0$$

which is nearest the origin.

13.11 Find the maximum value of $|\sum_{k=1}^n a_k x_k|$, if $\sum_{k=1}^n x_k^2 = 1$, by using

    a) the Cauchy–Schwarz inequality.

    b) Lagrange's method.

13.12 Find the maximum of $(x_1 x_2 \cdots x_n)^2$ under the restriction

$$x_1^2 + \cdots + x_n^2 = 1.$$

Use the result to derive the following inequality, valid for positive real numbers $a_1, \ldots, a_n$:

$$(a_1 \cdots a_n)^{1/n} \le \frac{a_1 + \cdots + a_n}{n}.$$

13.13 If $f(\mathbf{x}) = x_1^k + \cdots + x_n^k$, $\mathbf{x} = (x_1, \ldots, x_n)$, show that a local extreme of f, subject to the condition $x_1 + \cdots + x_n = a$, is $a^k n^{1-k}$.

13.14 Show that all points (x_1, x_2, x_3, x_4) where $x_1^2 + x_2^2$ has a local extremum subject to the two side conditions $x_1^2 + x_3^2 + x_4^2 = 4$, $x_2^2 + 2x_3^2 + 3x_4^2 = 9$, are found among

$$(0, 0, \pm\sqrt{3}, \pm 1), \quad (0, \pm 1, +2, 0), \quad (\pm 1, 0, 0, \pm\sqrt{3}), \quad (\pm 2, \pm 3, 0, 0).$$

Which of these yield a local maximum and which yield a local minimum? Give reasons for your conclusions.

13.15 Show that the extreme values of $f(x_1, x_2, x_3) = x_1^2 + x_2^2 + x_3^2$, subject to the two side conditions

$$\sum_{j=1}^3 \sum_{i=1}^3 a_{ij} x_i x_j = 1 \qquad (a_{ij} = a_{ji})$$

and

$$b_1 x_1 + b_2 x_2 + b_3 x_3 = 0, \qquad (b_1, b_2, b_3) \ne (0, 0, 0),$$

are t_1^{-1}, t_2^{-1}, where t_1 and t_2 are the roots of the equation

$$\begin{vmatrix} b_1 & b_2 & b_3 & 0 \\ a_{11} - t & a_{12} & a_{13} & b_1 \\ a_{21} & a_{22} - t & a_{23} & b_2 \\ a_{31} & a_{32} & a_{33} - t & b_3 \end{vmatrix} = 0.$$

Show that this is a quadratic equation in t and give a geometric argument to explain why the roots t_1, t_2 are real and positive.

13.16 Let $\Delta = \det [x_{ij}]$ and let $\mathbf{X}_i = (x_{i1}, \ldots, x_{in})$. A famous theorem of Hadamard states that $|\Delta| \le d_1 \cdots d_n$, if $d_1, \ldots, d_n$ are n positive constants such that $\|\mathbf{X}_i\|^2 = d_i^2$ $(i = 1, 2, \ldots, n)$. Prove this by treating Δ as a function of n^2 variables subject to n constraints, using Lagrange's method to show that, when Δ has an extreme under these conditions, we must have

$$\Delta^2 = \begin{vmatrix} d_1^2 & 0 & 0 & \cdots & 0 \\ 0 & d_2^2 & 0 & \cdots & 0 \\ \vdots & \vdots & \vdots & & \vdots \\ 0 & 0 & 0 & \cdots & d_n^2 \end{vmatrix}.$$

SUGGESTED REFERENCES FOR FURTHER STUDY

13.1 Apostol, T. M., *Calculus*, Vol. 2, 2nd ed. Xerox, Waltham, 1969.

13.2 Gantmacher, F. R., *The Theory of Matrices*, Vol. 1. K. A. Hirsch, translator. Chelsea, New York, 1959.

13.3 Hancock, H., *Theory of Maxima and Minima*. Ginn, Boston, 1917.

CHAPTER 14

MULTIPLE RIEMANN INTEGRALS

14.1 INTRODUCTION

The Riemann integral $\int_a^b f(x)\,dx$ can be generalized by replacing the interval $[a, b]$ by an n-dimensional region in which f is defined and bounded. The simplest regions in $\mathbf{R}^n$ suitable for this purpose are n-dimensional intervals. For example, in $\mathbf{R}^2$ we take a rectangle I partitioned into subrectangles I_k and consider Riemann sums of the form $\sum f(x_k, y_k)A(I_k)$, where $(x_k, y_k) \in I_k$ and $A(I_k)$ denotes the area of I_k. This leads us to the concept of a double integral. Similarly, in $\mathbf{R}^3$ we use rectangular parallelepipeds subdivided into smaller parallelepipeds I_k and, by considering sums of the form $\sum f(x_k, y_k, z_k)V(I_k)$, where $(x_k, y_k, z_k) \in I_k$ and $V(I_k)$ is the volume of I_k, we are led to the concept of a triple integral. It is just as easy to discuss multiple integrals in $\mathbf{R}^n$, provided that we have a suitable generalization of the notions of area and volume. This "generalized volume" is called *measure* or *content* and is defined in the next section.

14.2 THE MEASURE OF A BOUNDED INTERVAL IN $\mathbf{R}^n$

Let $A_1, \ldots, A_n$ denote n general intervals in $\mathbf{R}^1$; that is, each A_k may be bounded, unbounded, open, closed, or half-open in $\mathbf{R}^1$. A set A in $\mathbf{R}^n$ of the form

$$A = A_1 \times \cdots \times A_n = \{(x_1, \ldots, x_n) : x_k \in A_k \quad \text{for } k = 1, 2, \ldots, n\},$$

is called a general n-dimensional interval. We also allow the degenerate case in which one or more of the intervals A_k consists of a single point.

If each A_k is open, closed, or bounded in $\mathbf{R}^1$, then A has the corresponding property in $\mathbf{R}^n$.

If each A_k is bounded, the n-dimensional measure (or n-measure) of A, denoted by $\mu(A)$, is defined by the equation

$$\mu(A) = \mu(A_1) \cdots \mu(A_n),$$

where $\mu(A_k)$ is the one-dimensional measure (length) of A_k. When $n = 2$, this is called the *area* of A, and when $n = 3$, it is called the *volume* of A. Note that $\mu(A) = 0$ if $\mu(A_k) = 0$ for some k.

We turn next to a discussion of Riemann integration in $\mathbf{R}^n$. The only essential difference between the case $n = 1$ and the case $n > 1$ is that the quantity $\Delta x_k = x_k - x_{k-1}$ which was used to measure the length of the subinterval

$[x_{k-1}, x_k]$ is replaced by the measure $\mu(I_k)$ of an n-dimensional subinterval. Since the work proceeds on exactly the same lines as the one-dimensional case, we shall omit many of the details in the discussions that follow.

14.3 THE RIEMANN INTEGRAL OF A BOUNDED FUNCTION DEFINED ON A COMPACT INTERVAL IN $\mathbf{R}^n$

Definition 14.1. *Let* $A = A_1 \times \cdots \times A_n$ *be a compact interval in* $\mathbf{R}^n$. *If* P_k *is a partition of* A_k, *the cartesian product*

$$P = P_1 \times \cdots \times P_n,$$

is said to be a partition of A. *If* P_k *divides* A_k *into* m_k *one-dimensional subintervals, then* P *determines a decomposition of* A *as a union of* $m_1 \cdots m_n$ *n-dimensional intervals (called subintervals of* P). *A partition* P' *of* A *is said to be finer than* P *if* $P \subseteq P'$. *The set of all partitions of* A *will be denoted by* $\mathscr{P}(A)$.

Figure 14.1 illustrates partitions of intervals in $\mathbf{R}^2$ and in $\mathbf{R}^3$.

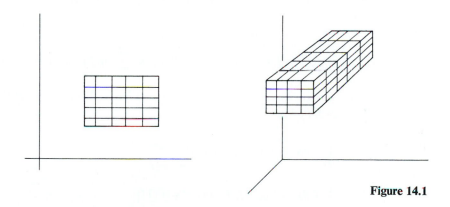

Figure 14.1

Definition 14.2. *Let* f *be defined and bounded on a compact interval* I *in* $\mathbf{R}^n$. *If* P *is a partition of* I *into* m *subintervals* $I_1, \ldots, I_m$ *and if* $\mathbf{t}_k \in I_k$, *a sum of the form*

$$S(P, f) = \sum_{k=1}^{m} f(\mathbf{t}_k)\mu(I_k),$$

is called a Riemann sum. We say f *is Riemann-integrable on* I *and we write* $f \in R$ *on* I, *whenever there exists a real number* A *having the following property: For every* $\varepsilon > 0$ *there exists a partition* P_ε *of* I *such that* P *finer than* P_ε *implies*

$$|S(P, f) - A| < \varepsilon,$$

for all Riemann sums $S(P, f)$. *When such a number* A *exists, it is uniquely determined and is denoted by*

$$\int_I f \, d\mathbf{x}, \qquad \int_I f(\mathbf{x}) \, d\mathbf{x}, \quad \text{or by} \quad \int_I f(x_1, \ldots, x_n) \, d(x_1, \ldots, x_n).$$

NOTE. For $n > 1$ the integral is called a *multiple* or *n-fold* integral. When $n = 2$ and 3, the terms *double* and *triple* integral are used. As in $\mathbf{R}^1$, the symbol $\mathbf{x}$ in $\int_I f(\mathbf{x}) \, d\mathbf{x}$ is a "dummy variable" and may be replaced by any other convenient symbol. The notation $\int_I f(x_1, \ldots, x_n) \, dx_1 \cdots dx_n$ is also used instead of $\int_I f(x_1, \ldots, x_n) \, d(x_1, \ldots, x_n)$. Double integrals are sometimes written with two integral signs and triple integrals with three such signs, thus:

$$\iint_I f(x, y) \, dx \, dy, \qquad \iiint_I f(x, y, z) \, dx \, dy \, dz.$$

Definition 14.3. *Let f be defined and bounded on a compact interval I in $\mathbf{R}^n$. If P is a partition of I into m subintervals $I_1, \ldots, I_m$, let*

$$m_k(f) = \inf \{f(\mathbf{x}) : \mathbf{x} \in I_k\}, \qquad M_k(f) = \sup \{f(\mathbf{x}) : \mathbf{x} \in I_k\}.$$

The numbers

$$U(P, f) = \sum_{k=1}^m M_k(f)\mu(I_k) \qquad and \qquad L(P, f) = \sum_{k=1}^m m_k(f)\mu(I_k),$$

are called upper and lower Riemann sums. The upper and lower Riemann integrals of f over I are defined as follows:

$$\overline{\int_I} f \, d\mathbf{x} = \inf \{U(P, f) : P \in \mathscr{P}(I)\},$$

$$\underline{\int_I} f \, d\mathbf{x} = \sup \{L(P, f) : P \in \mathscr{P}(I)\}.$$

The function f is said to satisfy Riemann's condition on I if, for every $\varepsilon > 0$, there exists a partition P_ε of I such that P finer than P_ε implies $U(P, f) - L(P, f) < \varepsilon$.

NOTE. As in the one-dimensional case, upper and lower integrals have the following properties:

a)
$$\overline{\int_I} (f + g) \, d\mathbf{x} \le \overline{\int_I} f \, d\mathbf{x} + \overline{\int_I} g \, d\mathbf{x},$$

$$\underline{\int_I} (f + g) \, d\mathbf{x} \ge \underline{\int_I} f \, d\mathbf{x} + \underline{\int_I} g \, d\mathbf{x}.$$

b) If an interval I is decomposed into a union of two nonoverlapping intervals I_1, I_2, then we have

$$\overline{\int_I} f\, d\mathbf{x} = \overline{\int_{I_1}} f\, d\mathbf{x} + \overline{\int_{I_2}} f\, d\mathbf{x} \quad \text{and} \quad \underline{\int_I} f\, d\mathbf{x} = \underline{\int_{I_1}} f\, d\mathbf{x} + \underline{\int_{I_2}} f\, d\mathbf{x}.$$

The proof of the following theorem is essentially the same as that of Theorem 7.19 and will be omitted.

Theorem 14.4. *Let f be defined and bounded on a compact interval I in $\mathbf{R}^n$. Then the following statements are equivalent:*

i) $f \in R$ on I.

ii) f satisfies Riemann's condition on I.

iii) $\underline{\int_I} f\, d\mathbf{x} = \overline{\int_I} f\, d\mathbf{x}$.

14.4 SETS OF MEASURE ZERO AND LEBESGUE'S CRITERION FOR EXISTENCE OF A MULTIPLE RIEMANN INTEGRAL

A subset T of $\mathbf{R}^n$ is said to be of n-measure zero if, for every $\varepsilon > 0$, T can be covered by a countable collection of n-dimensional intervals, the sum of whose n-measures is $< \varepsilon$.

As in the one-dimensional case, the union of a countable collection of sets of n-measure 0 is itself of n-measure 0. If $m < n$, every subset of $\mathbf{R}^m$, when considered as a subset of $\mathbf{R}^n$, has n-measure 0.

A property is said to hold almost everywhere on a set S in $\mathbf{R}^n$ if it holds everywhere on S except for a subset of n-measure 0.

Lebesgue's criterion for the existence of a Riemann integral in $\mathbf{R}^1$ has a straightforward extension to multiple integrals. The proof is analogous to that of Theorem 7.48.

Theorem 14.5. *Let f be defined and bounded on a compact interval I in $\mathbf{R}^n$. Then $f \in R$ on I if, and only if, the set of discontinuities of f in I has n-measure zero.*

14.5 EVALUATION OF A MULTIPLE INTEGRAL BY ITERATED INTEGRATION

From elementary calculus the reader has learned to evaluate certain double and triple integrals by successive integration with respect to each variable. For example, if f is a function of two variables continuous on a compact rectangle Q in the xy-plane, say $Q = \{(x, y) : a \le x \le b, c \le y \le d\}$, then for each fixed y in $[c, d]$ the function F defined by the equation $F(x) = f(x, y)$ is continuous (and hence integrable) on $[a, b]$. The value of the integral $\int_a^b F(x)\, dx$ depends on y and defines a new function G, where $G(y) = \int_a^b f(x, y)\,dx$. This function G is continuous (by Theorem 7.38), and hence integrable, on $[c, d]$. The integral $\int_c^d G(y)\,dy$ turns out to have the same value as the *double* integral $\int_Q f(x, y)\,d(x, y)$. That is, we have the equation

$$\int_Q f(x, y)\,d(x, y) = \int_c^d \left[\int_a^b f(x, y)\,dx \right] dy. \tag{1}$$

(This formula will be proved later.) The question now arises as to whether a similar result holds when f is merely integrable (and not necessarily continuous) on Q. We can see at once that certain difficulties are inevitable. For example, the inner integral $\int_a^b f(x, y)\,dx$ may not exist for certain values of y even though the double integral exists. In fact, if f is discontinuous at every point of the line segment $y = y_0$, $a \le x \le b$, then $\int_a^b f(x, y_0)\,dx$ will fail to exist. However, this line segment is a set whose 2-measure is zero and therefore does not affect the integrability of f on the whole rectangle Q. In a case of this kind we must use upper and lower integrals to obtain a suitable generalization of (1).

Theorem 14.6. *Let f be defined and bounded on a compact rectangle*

$$Q = [a, b] \times [c, d] \qquad \text{in } \mathbf{R}^2.$$

Then we have:

i) $\int_Q f\,d(x, y) \le \int_a^b \left[\underline{\int_c^d} f(x, y)\,dy \right] dx \le \overline{\int_a^b} \left[\underline{\int_c^d} f(x, y)\,dy \right] dx \le \overline{\int_Q} f\,d(x, y).$

ii) *Statement* (i) *holds with $\underline{\int_c^d}$ replaced by $\overline{\int_c^d}$ throughout.*

iii) $\int_Q f\,d(x, y) \le \int_c^d \left[\underline{\int_a^b} f(x, y)\,dx \right] dy \le \overline{\int_c^d} \left[\underline{\int_a^b} f(x, y)\,dx \right] dy \le \overline{\int_Q} f\,d(x, y).$

iv) *Statement* (iii) *holds with $\underline{\int_a^b}$ replaced by $\overline{\int_a^b}$ throughout.*

v) *When $\int_Q f(x, y)\,d(x, y)$ exists, we have*

$$\int_Q f(x, y)\,d(x, y) = \int_a^b \left[\underline{\int_c^d} f(x, y)\,dy \right] dx = \int_a^b \left[\overline{\int_c^d} f(x, y)\,dy \right] dx$$

$$= \int_c^d \left[\underline{\int_a^b} f(x, y)\,dx \right] dy = \int_c^d \left[\overline{\int_a^b} f(x, y)\,dx \right] dy.$$

Proof. To prove (i), define F by the equation

$$F(x) = \overline{\int_c^d} f(x, y)\,dy, \qquad \text{if } x \in [a, b].$$

Then $|F(x)| \le M(d - c)$, where $M = \sup \{|f(x, y)| : (x, y) \in Q\}$, and we can consider

$$\bar{I} = \overline{\int_a^b} F(x)\,dx = \overline{\int_a^b} \left[\overline{\int_c^d} f(x, y)\,dy \right] dx.$$

Similarly, we define

$$\underline{I} = \int_a^b F(x)\, dx = \int_a^b \left[\underline{\int_c^d} f(x, y)\, dy \right] dx.$$

Let $P_1 = \{x_0, x_1, \ldots, x_n\}$ be a partition of $[a, b]$ and let

$$P_2 = \{y_0, y_1, \ldots, y_m\},$$

be a partition of $[c, d]$. Then $P = P_1 \times P_2$ is a partition of Q into mn sub-rectangles Q_{ij} and we define

$$\bar{I}_{ij} = \int_{x_{i-1}}^{x_i} \left[\overline{\int_{y_{j-1}}^{y_j}} f(x, y)\, dy \right] dx, \qquad \underline{I}_{ij} = \int_{x_{i-1}}^{x_i} \left[\underline{\int_{y_{j-1}}^{y_j}} f(x, y)\, dy \right] dx.$$

Since we have

$$\overline{\int_c^d} f(x, y)\, dy = \sum_{j=1}^m \overline{\int_{y_{j-1}}^{y_j}} f(x, y)\, dy,$$

we can write

$$\int_a^b \left[\overline{\int_c^d} f(x, y)\, dy \right] dx \leq \sum_{j=1}^m \int_a^b \left[\overline{\int_{y_{j-1}}^{y_j}} f(x, y)\, dy \right] dx$$

$$= \sum_{j=1}^m \sum_{i=1}^n \int_{x_{i-1}}^{x_i} \left[\overline{\int_{y_{j-1}}^{y_j}} f(x, y)\, dy \right] dx.$$

That is, we have the inequality

$$\bar{I} \leq \sum_{j=1}^m \sum_{i=1}^n \bar{I}_{ij}.$$

Similarly, we find

$$\underline{I} \geq \sum_{j=1}^m \sum_{i=1}^n \underline{I}_{ij}.$$

If we write

$$m_{ij} = \inf \{f(x, y) : (x, y) \in Q_{ij}\},$$

and

$$M_{ij} = \sup \{f(x, y) : (x, y) \in Q_{ij}\},$$

then from the inequality $m_{ij} \leq f(x, y) \leq M_{ij}$, $(x, y) \in Q_{ij}$, we obtain

$$m_{ij}(y_j - y_{j-1}) \leq \int_{y_{j-1}}^{y_j} f(x, y)\, dy \leq M_{ij}(y_j - y_{j-1}).$$

This, in turn, implies

$$m_{ij}\mu(Q_{ij}) \leq \int_{x_{i-1}}^{x_i} \left[\int_{y_{j-1}}^{\overline{y_j}} f(x, y) \, dy \right] dx$$

$$\leq \int_{x_{i-1}}^{x_i} \left[\int_{\underline{y_{j-1}}}^{y_j} f(x, y) \, dy \right] dx \leq M_{ij}\mu(Q_{ij}).$$

Summing on i and j and using the above inequalities, we get

$$L(P, f) \leq \underline{I} \leq \overline{I} \leq U(P, f).$$

Since this holds for all partitions P of Q, we must have

$$\underline{\int_Q} f \, d(x, y) \leq \underline{I} \leq \overline{I} \leq \overline{\int_Q} f \, d(x, y).$$

This proves statement (i).

It is clear that the preceding proof could also be carried out if the function F were originally defined by the formula

$$F(x) = \int_c^d f(x, y) \, dy,$$

and hence (ii) follows by the same argument.

Statements (iii) and (iv) can be similarly proved by interchanging the roles of x and y. Finally, statement (v) is an immediate consequence of statements (i) through (iv).

As a corollary, we have the formula mentioned earlier:

$$\int_Q f(x, y) \, d(x, y) = \int_a^b \left[\int_c^d f(x, y) \, dy \right] dx = \int_c^d \left[\int_a^b f(x, y) \, dx \right] dy,$$

which is valid when f is continuous on Q. This is often called *Fubini's theorem*.

NOTE. The existence of the iterated integrals

$$\int_a^b \left[\int_c^d f(x, y) \, dy \right] dx \qquad \text{and} \qquad \int_c^d \left[\int_a^b f(x, y) \, dx \right] dy,$$

does not imply the existence of $\int_Q f(x, y) \, d(x, y)$. A counter example is given in Exercise 14.7.

Before commenting on the analog of Theorem 14.6 in $\mathbf{R}^n$, we first introduce some further notation and terminology. If $k \leq n$, the set of $\mathbf{x}$ in $\mathbf{R}^n$ for which $x_k = 0$ is called the *coordinate hyperplane* $\prod_k$. Given a set S in $\mathbf{R}^n$, the *projection* S_k of S on $\prod_k$ is defined to be the image of S under that mapping whose value at each point $(x_1, x_2, \ldots, x_n)$ in S is $(x_1, \ldots, x_{k-1}, 0, x_{k+1}, \ldots, x_n)$. It is easy to

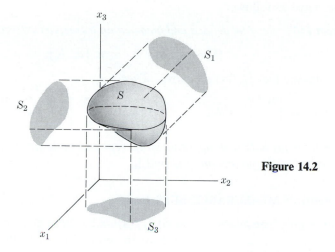

Figure 14.2

show that such a mapping is continuous on S. It follows that if S is compact, each projection S_k is compact. Also, if S is connected, each S_k is connected. Projections in $\mathbf{R}^3$ are illustrated in Fig. 14.2.

A theorem entirely analogous to Theorem 14.6 holds for n-fold integrals. It will suffice to indicate how the extension goes when $n = 3$. In this case, f is defined and bounded on a compact interval $Q = [a_1, b_1] \times [a_2, b_2] \times [a_3, b_3]$ in $\mathbf{R}^3$ and statement (i) of Theorem 14.6 is replaced by

$$\underline{\int_Q} f \, d\mathbf{x} \le \int_{a_1}^{b_1} \left[\underline{\int_{Q_1}} f \, d(x_2, x_3) \right] dx_1$$

$$\le \int_{a_1}^{\overline{b_1}} \left[\overline{\int_{Q_1}} f \, d(x_2, x_3) \right] dx_1 \le \overline{\int_Q} f \, d\mathbf{x}, \tag{2}$$

where Q_1 is the projection of Q on the coordinate plane $\prod_1$. When $\int_Q f(\mathbf{x}) \, d\mathbf{x}$ exists, the analog of part (v) of Theorem 14.6 is the formula

$$\int_Q f(\mathbf{x}) \, d\mathbf{x} = \int_{a_1}^{b_1} \left[\overline{\int_{Q_1}} f \, d(x_2, x_3) \right] dx_1 = \int_{Q_1} \left[\overline{\int_{a_1}^{b_1}} f \, dx_1 \right] d(x_2, x_3). \tag{3}$$

As in Theorem 14.6, similar statements hold with appropriate replacements of upper integrals by lower integrals, and there are also analogous formulas for the projections Q_2 and Q_3.

The reader should have no difficulty in stating analogous results for n-fold integrals (they can be proved by the method used in Theorem 14.6). The special case in which the n-fold integral $\int_Q f(\mathbf{x}) \, d\mathbf{x}$ exists is of particular importance and can be stated as follows:

Theorem 14.7. *Let f be defined and bounded on a compact interval*

$$Q = [a_1, b_1] \times \cdots \times [a_n, b_n],$$

in $\mathbf{R}^n$. *Assume that* $\int_Q f(\mathbf{x}) \, d\mathbf{x}$ *exists. Then*

$$\int_Q f \, d\mathbf{x} = \int_{a_1}^{b_1} \left[\overline{\int_{Q_1}} f \, d(x_2, \ldots, x_n) \right] dx_1 = \int_{Q_1} \left[\overline{\int_{a_1}^{b_1}} f \, dx_1 \right] d(x_2, \ldots, x_n).$$

Similar formulas hold with upper integrals replaced by lower integrals and with Q_1 *replaced by* Q_k, *the projection of* Q *on* $\prod_k$.

14.6 JORDAN-MEASURABLE SETS IN $\mathbf{R}^n$

Up to this point the multiple integral $\int_I f(\mathbf{x}) \, d\mathbf{x}$ has been defined only for intervals I. This, of course, is too restrictive for the applications of integration. It is not difficult to extend the definition to encompass more general sets called *Jordan-measurable* sets. These are discussed in this section. The definition makes use of the boundary of a set S in $\mathbf{R}^n$. We recall that a point $\mathbf{x}$ in $\mathbf{R}^n$ is called a *boundary point* of S if every n-ball $B(\mathbf{x})$ contains a point in S and also a point not in S. The set of all boundary points of S is called the *boundary* of S and is denoted by ∂S. (See Section 3.16.)

Definition 14.8. *Let S be a subset of a compact interval I in* $\mathbf{R}^n$. *For every partition P of I define* $\underline{J}(P, S)$ *to be the sum of the measures of those subintervals of P which contain only interior points of S and let* $\bar{J}(P, S)$ *be the sum of the measures of those subintervals of P which contain points of* $S \cup \partial S$. *The numbers*

$$\underline{c}(S) = \sup \{\underline{J}(P, S) : P \in \mathscr{P}(I)\},$$

$$\bar{c}(S) = \inf \{\bar{J}(P, S) : P \in \mathscr{P}(I)\},$$

are called, respectively, the (n-dimensional) inner and outer Jordan content of S. The set S is said to be Jordan-measurable if $\underline{c}(S) = \bar{c}(S)$, *in which case this common value is called the Jordan content of S, denoted by* $c(S)$.

It is easy to verify that $\underline{c}(S)$ and $\bar{c}(S)$ depend only on S and not on the interval I which contains S. Also, $0 \le \underline{c}(S) \le \bar{c}(S)$.

If S has content zero, then $\underline{c}(S) = \bar{c}(S) = 0$. Hence, for every $\varepsilon > 0$, S can be covered by a finite collection of intervals, the sum of whose measures is $<\varepsilon$. Note that content zero is described in terms of *finite* coverings, whereas measure zero is described in terms of *countable* coverings. Any set with content zero also has measure zero, but the converse is not necessarily true.

Every compact interval Q is Jordan-measurable and its content, $c(Q)$, is equal to its measure, $\mu(Q)$. If $k < n$, the n-dimensional content of every bounded set in $\mathbf{R}^k$ is zero.

Jordan-measurable sets S in $\mathbf{R}^2$ are also said to have *area* $c(S)$. In this case, the sums $\underline{J}(P, S)$ and $\bar{J}(P, S)$ represent approximations to the area from the "inside"

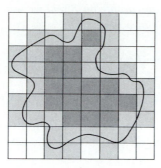

Figure 14.3

and the "outside" of S, respectively. This is illustrated in Fig. 14.3, where the lightly shaded rectangles are counted in $\bar{J}(P, S)$, the heavily shaded rectangles in $\underline{J}(P, S)$. For sets in $\mathbf{R}^3$, $c(S)$ is also called the *volume* of S.

The next theorem shows that a bounded set has Jordan content if, and only if, its boundary isn't too "thick."

Theorem 14.9. *Let S be a bounded set in $\mathbf{R}^n$ and let ∂S denote its boundary. Then we have*

$$\bar{c}(\partial S) = \bar{c}(S) - \underline{c}(S).$$

Hence, S is Jordan-measurable if, and only if, ∂S has content zero.

Proof. Let I be a compact interval containing S and ∂S. Then for every partition P of I we have

$$\bar{J}(P, \partial S) = \bar{J}(P, S) - \underline{J}(P, S).$$

Therefore, $\bar{J}(P, \partial S) \geq \bar{c}(S) - \underline{c}(S)$ and hence $\bar{c}(\partial S) \geq \bar{c}(S) - \underline{c}(S)$. To obtain the reverse inequality, let $\varepsilon > 0$ be given, choose P_1 so that $\bar{J}(P_1, S) < \bar{c}(S) + \varepsilon/2$ and choose P_2 so that $\underline{J}(P_2, S) > \underline{c}(S) - \varepsilon/2$. Let $P = P_1 \cup P_2$. Since refinement increases the inner sums $\underline{J}$ and decreases the outer sums $\bar{J}$, we find

$$\bar{c}(\partial S) \leq \bar{J}(P, \partial S) = \bar{J}(P, S) - \underline{J}(P, S) \leq \bar{J}(P_1, S) - \underline{J}(P_2, S)$$

$$< \bar{c}(S) - \underline{c}(S) + \varepsilon.$$

Since ε is arbitrary, this means that $\bar{c}(\partial S) \leq \bar{c}(S) - \underline{c}(S)$. Therefore, $\bar{c}(\partial S) = \bar{c}(S) - \underline{c}(S)$ and the proof is complete.

14.7 MULTIPLE INTEGRATION OVER JORDAN-MEASURABLE SETS

Definition 14.10. *Let f be defined and bounded on a bounded Jordan-measurable set S in $\mathbf{R}^n$. Let I be a compact interval containing S and define g on I as follows:*

$$g(\mathbf{x}) = \begin{cases} f(\mathbf{x}) & \text{if } \mathbf{x} \in S, \\ 0 & \text{if } \mathbf{x} \in I - S. \end{cases}$$

Then f is said to be Riemann-integrable on S and we write f ∈ R on S, whenever the integral $\int_I g(\mathbf{x})\, d\mathbf{x}$ exists. We also write

$$\int_S f(\mathbf{x})\, d\mathbf{x} = \int_I g(\mathbf{x})\, d\mathbf{x}.$$

The upper and lower integrals $\overline{\int}_S f(\mathbf{x})\, d\mathbf{x}$ and $\underline{\int}_S f(\mathbf{x})\, d\mathbf{x}$ are similarly defined.

NOTE. By considering the Riemann sums which approximate $\int_I g(\mathbf{x})\, d\mathbf{x}$, it is easy to see that the integral $\int_S f(\mathbf{x})\, d\mathbf{x}$ does not depend on the choice of the interval I used to enclose S.

A necessary and sufficient condition for the existence of $\int_S f(\mathbf{x})\, d\mathbf{x}$ can now be given.

Theorem 14.11. *Let S be a Jordan-measurable set in $\mathbf{R}^n$, and let f be defined and bounded on S. Then f ∈ R on S if, and only if, the discontinuities of f in S form a set of measure zero.*

Proof. Let I be a compact interval containing S and let $g(\mathbf{x}) = f(\mathbf{x})$ when $\mathbf{x} \in S$, $g(\mathbf{x}) = 0$ when $\mathbf{x} \in I - S$. The discontinuities of f will be discontinuities of g. However, g may also have discontinuities at some or all of the boundary points of S. Since S is Jordan measurable, Theorem 14.9 tells us that $c(\partial S) = 0$. Therefore, $g \in R$ on I if, and only if, the discontinuities of f form a set of measure zero.

14.8 JORDAN CONTENT EXPRESSED AS A RIEMANN INTEGRAL

Theorem 14.12. *Let S be a compact Jordan-measurable set in $\mathbf{R}^n$. Then the integral $\int_S 1$ exists and we have*

$$c(S) = \int_S 1.$$

Proof. Let I be a compact interval containing S and let χ_S denote the characteristic function of S. That is,

$$\chi_S(\mathbf{x}) = \begin{cases} 1 & \text{if } \mathbf{x} \in S, \\ 0 & \text{if } \mathbf{x} \in I - S. \end{cases}$$

The discontinuities of χ_S in I are the boundary points of S and these form a set of content zero, so the integral $\int_I \chi_S$ exists, and hence $\int_S 1$ exists.

Let P be a partition of I into subintervals $I_1, \ldots, I_m$, and let

$$A = \{k : I_k \cap S \text{ is nonempty}\}.$$

If $k \in A$, we have

$$M_k(\chi_S) = \sup \{\chi_S(\mathbf{x}) : \mathbf{x} \in I_k\} = 1,$$

and $M_k(\chi_S) = 0$ if $k \notin A$, so

$$U(P, \chi_S) = \sum_{k=1}^{m} M_k(\chi_S)\mu(I_k) = \sum_{k \in A} \mu(I_k) = \bar{J}(P, \chi_S).$$

Since this holds for all partitions, we have $\bar{\int_I} \chi_S = \bar{c}(S) = c(S)$. But

$$\bar{\int_I} \chi_S = \int_I \chi_S \quad \text{so} \quad c(S) = \int_I \chi_S = \int_S 1.$$

14.9 ADDITIVE PROPERTY OF THE RIEMANN INTEGRAL

The next theorem shows that the integral is additive with respect to sets having Jordan content.

Theorem 14.13. *Assume $f \in R$ on a Jordan-measurable set S in $\mathbf{R}^n$. Suppose $S = A \cup B$, where A and B are Jordan-measurable but have no interior points in common. Then $f \in R$ on A, $f \in R$ on B, and we have*

$$\int_S f(\mathbf{x}) \, d\mathbf{x} = \int_A f(\mathbf{x}) \, d\mathbf{x} + \int_B f(\mathbf{x}) \, d\mathbf{x}. \qquad (4)$$

Proof. Let I be a compact interval containing S and define g as follows:

$$g(\mathbf{x}) = \begin{cases} f(\mathbf{x}) & \text{if } \mathbf{x} \in S, \\ 0 & \text{if } \mathbf{x} \in I - S. \end{cases}$$

The existence of $\int_A f(\mathbf{x}) \, d\mathbf{x}$ and $\int_B f(\mathbf{x}) \, d\mathbf{x}$ is an easy consequence of Theorem 14.11. To prove (4), let P be a partition of I into m subintervals $I_1, \ldots, I_m$ and form a Riemann sum

$$S(P, g) = \sum_{k=1}^{m} g(\mathbf{t}_k)\mu(I_k).$$

If S_A denotes that part of the sum arising from those subintervals containing points of A, and if S_B is similarly defined, we can write

$$S(P, g) = S_A + S_B - S_C,$$

where S_C contains those terms coming from subintervals which contain both points of A and points of B. In particular, all points common to the two boundaries ∂A and ∂B will fall in this third class. But now S_A is a Riemann sum approximating the integral $\int_A f(\mathbf{x}) \, d\mathbf{x}$, and S_B is a Riemann sum approximating $\int_B f(\mathbf{x}) \, d\mathbf{x}$. Since $c(\partial A \cap \partial B) = 0$, it follows that $|S_C|$ can be made arbitrarily small when P is sufficiently fine. The equation in the theorem is an easy consequence of these remarks.

NOTE. Formula (4) also holds for upper and lower integrals.

For sets S whose structure is relatively simple, Theorem 14.6 can be used to obtain formulas for evaluating double integrals by iterated integration. These formulas are given in the next theorem.

Theorem 14.14. *Let ϕ_1 and ϕ_2 be two continuous functions defined on $[a, b]$ such that $\phi_1(x) \leq \phi_2(x)$ for each x in $[a, b]$. Let S be the compact set in $\mathbf{R}^2$ given by*

$$S = \{(x, y) : a \leq x \leq b, \phi_1(x) \leq y \leq \phi_2(x)\}.$$

If $f \in R$ on S, we have

$$\int_S f(x, y)\, d(x, y) = \int_a^b \left[\int_{\phi_1(x)}^{\phi_2(x)} f(x, y)\, dy \right] dx.$$

NOTE. The set S is Jordan-measurable because its boundary has content zero. (See Exercise 14.9.)

Analogous statements hold for n-fold integrals. The extensions are too obvious to require further comment.

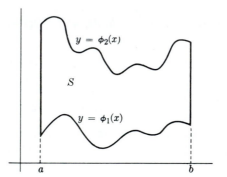

Figure 14.4

Figure 14.4 illustrates the type of region described in the theorem. For sets which can be decomposed into a finite number of Jordan-measurable regions of this type, we can apply iterated integration to each separate part and add the results in accordance with Theorem 14.13.

14.10 MEAN-VALUE THEOREM FOR MULTIPLE INTEGRALS

As in the one-dimensional case, multiple integrals satisfy a mean value property. This can be obtained as an easy consequence of the following theorem, the proof of which is left as an exercise.

Theorem 14.15. *Assume $f \in R$ and $g \in R$ on a Jordan-measurable set S in $\mathbf{R}^n$. If $f(\mathbf{x}) \leq g(\mathbf{x})$ for each $\mathbf{x}$ in S, then we have*

$$\int_S f(\mathbf{x})\, d\mathbf{x} \leq \int_S g(\mathbf{x})\, d\mathbf{x}.$$

Theorem 14.16 (Mean-Value Theorem for multiple integrals). *Assume that $g \in R$ and $f \in R$ on a Jordan-measurable set S in $\mathbf{R}^n$ and suppose that $g(\mathbf{x}) \geq 0$ for each $\mathbf{x}$ in S. Let $m = \inf f(S)$, $M = \sup f(S)$. Then there exists a real number λ in the interval $m \leq \lambda \leq M$ such that*

$$\int_S f(\mathbf{x})g(\mathbf{x}) \, d\mathbf{x} = \lambda \int_S g(\mathbf{x}) \, d\mathbf{x}. \tag{5}$$

In particular, we have

$$mc(S) \leq \int_S f(\mathbf{x}) \, d\mathbf{x} \leq Mc(S). \tag{6}$$

NOTE. If, in addition, S is connected and f is continuous on S, then $\lambda = f(\mathbf{x}_0)$ for some $\mathbf{x}_0$ in S (by Theorem 4.38.) and (5) becomes

$$\int_S f(\mathbf{x})g(\mathbf{x}) \, d\mathbf{x} = f(\mathbf{x}_0) \int_S g(\mathbf{x}) \, d\mathbf{x}. \tag{7}$$

In particular, (7) implies $\int_S f(\mathbf{x}) \, d\mathbf{x} = f(\mathbf{x}_0)c(S)$, where $\mathbf{x}_0 \in S$.

Proof. Since $g(\mathbf{x}) \geq 0$, we have $mg(\mathbf{x}) \leq f(\mathbf{x}) \, g(\mathbf{x}) \leq Mg(\mathbf{x})$ for each $\mathbf{x}$ in S. By Theorem 14.15, we can write

$$m \int_S g(\mathbf{x}) \, d\mathbf{x} \leq \int_S f(\mathbf{x})g(\mathbf{x}) \, d\mathbf{x} \leq M \int_S g(\mathbf{x}) \, d\mathbf{x}.$$

If $\int_S g(\mathbf{x}) \, d\mathbf{x} = 0$, (5) holds for every λ. If $\int_S g(\mathbf{x}) \, d\mathbf{x} > 0$, (5) holds with $\lambda = \int_S f(\mathbf{x})g(\mathbf{x}) \, d\mathbf{x} / \int_S g(\mathbf{x}) \, d\mathbf{x}$. Taking $g(\mathbf{x}) \equiv 1$, we obtain (6).

We can use (6) to prove that the integrand f can be disturbed on a set of content zero without affecting the value of the integral. In fact, we have the following theorem:

Theorem 14.17. *Assume that $f \in R$ on a Jordan-measurable set S in $\mathbf{R}^n$. Let T be a subset of S having n-dimensional Jordan content zero. Let g be a function, defined and bounded on S, such that $g(\mathbf{x}) = f(\mathbf{x})$ when $\mathbf{x} \in S - T$. Then $g \in R$ on S and*

$$\int_S f(\mathbf{x}) \, d\mathbf{x} = \int_S g(\mathbf{x}) \, d\mathbf{x}.$$

Proof. Let $h = f - g$. Then $\int_S h(\mathbf{x}) \, d\mathbf{x} = \int_T h(\mathbf{x}) \, d\mathbf{x} + \int_{S-T} h(\mathbf{x}) \, d\mathbf{x}$. However, $\int_T h(\mathbf{x}) \, d\mathbf{x} = 0$ because of (6), and $\int_{S-T} h(\mathbf{x}) \, d\mathbf{x} = 0$ since $h(\mathbf{x}) = 0$ for each $\mathbf{x}$ in $S - T$.

NOTE. This theorem suggests a way of extending the definition of the Riemann integral $\int_S f(\mathbf{x}) \, d\mathbf{x}$ for functions which may not be defined and bounded on the *whole* of S. In fact, let S be a bounded set in $\mathbf{R}^n$ having Jordan content and let T be a subset of S having content zero. If f is defined and bounded on $S - T$ and if $\int_{S-T} f(\mathbf{x})\, d\mathbf{x}$ exists, we agree to write

$$\int_S f(\mathbf{x})\, d\mathbf{x} = \int_{S-T} f(\mathbf{x})\, d\mathbf{x},$$

and to say that f is Riemann-integrable on S. In view of the theorem just proved, this is essentially the same as extending the domain of definition of f to the whole of S by defining f on T in such a way that it remains bounded.

EXERCISES

Multiple integrals

14.1 If $f_1 \in R$ on $[a_1, b_1], \ldots, f_n \in R$ on $[a_n, b_n]$, prove that

$$\int_S f_1(x_1) \cdots f_n(x_n)\, d(x_1, \ldots, x_n) = \left(\int_{a_1}^{b_1} f_1(x_1)\, dx_1 \right) \cdots \left(\int_{a_n}^{b_n} f_n(x_n)\, dx_n \right),$$

where $S = [a_1, b_1] \times \cdots \times [a_n, b_n]$.

14.2 Let f be defined and bounded on a compact rectangle $Q = [a, b] \times [c, d]$ in $\mathbf{R}^2$. Assume that for each fixed y in $[c, d]$, $f(x, y)$ is an increasing function of x, and that for each fixed x in $[a, b]$, $f(x, y)$ is an increasing function of y. Prove that $f \in R$ on Q.

14.3 Evaluate each of the following double integrals.

a) $\iint\limits_Q \sin^2 x \sin^2 y\, dx\, dy$,   where $Q = [0, \pi] \times [0, \pi]$.

b) $\iint\limits_Q |\cos (x + y)|\, dx\, dy$,   where $Q = [0, \pi] \times [0, \pi]$.

c) $\iint\limits_Q [x + y]\, dx\, dy$,   where $Q = [0, 2] \times [0, 2]$, and $[t]$ is the greatest

integer $\leq t$.

14.4 Let $Q = [0, 1] \times [0, 1]$ and calculate $\iint_Q f(x, y)\, dx\, dy$ in each case.
a) $f(x, y) = 1 - x - y$ if $x + y \leq 1$, $f(x, y) = 0$ otherwise.
b) $f(x, y) = x^2 + y^2$ if $x^2 + y^2 \leq 1$, $f(x, y) = 0$ otherwise.
c) $f(x, y) = x + y$ if $x^2 \leq y \leq 2x^2$, $f(x, y) = 0$ otherwise.

14.5 Define f on the square $Q = [0, 1] \times [0, 1]$ as follows:

$$f(x, y) =\cdot \begin{cases} 1 & \text{if } x \text{ is rational,} \\ 2y & \text{if } x \text{ is irrational.} \end{cases}$$

a) Prove that $\int_0^t f(x, y)\, dy$ exists for $0 \leq t \leq 1$ and that

$$\int_0^1 \left[\int_0^t f(x, y)\, dy \right] dx = t^2,$$

and

$$\overline{\int_0^1} \left[\int_0^t f(x, y)\, dy \right] dx = t.$$

This shows that $\int_0^1 [\int_0^1 f(x, y)\, dy]\, dx$ exists and equals 1.

b) Prove that $\int_0^1 [\int_0^1 f(x, y)\, dx]\, dy$ exists and find its value.

c) Prove that the double integral $\int_Q f(x, y)\, d(x, y)$ does not exist.

14.6 Define f on the square $Q = [0, 1] \times [0, 1]$ as follows:

$$f(x, y) = \begin{cases} 0 & \text{if at least one of } x, y \text{ is irrational,} \\ 1/n & \text{if } y \text{ is rational and } x = m/n, \end{cases}$$

where m and n are relatively prime integers, $n > 0$. Prove that

$$\int_0^1 f(x, y)\, dx = \int_0^1 \left[\int_0^1 f(x, y)\, dx \right] dy = \int_Q f(x, y)\, d(x, y) = 0$$

but that $\int_0^1 f(x, y)\, dy$ does not exist for rational x.

14.7 If p_k denotes the kth prime number, let

$$S(p_k) = \left\{ \left(\frac{n}{p_k}, \frac{m}{p_k} \right) : n = 1, 2, \ldots, p_k - 1, \quad m = 1, 2, \ldots, p_k - 1 \right\},$$

let $S = \bigcup_{k=1}^{\infty} S(p_k)$, and let $Q = [0, 1] \times [0, 1]$.

a) Prove that S is dense in Q (that is, the closure of S contains Q) but that any line parallel to the coordinate axes contains at most a finite subset of S.

b) Define f on Q as follows:

$$f(x, y) = 0 \quad \text{if } (x, y) \in S, \qquad f(x, y) = 1 \quad \text{if } (x, y) \in Q - S.$$

Prove that $\int_0^1 [\int_0^1 f(x, y)\, dy]\, dx = \int_0^1 [\int_0^1 f(x, y)\, dx]\, dy = 1$, but that the double integral $\int_Q f(x, y)\, d(x, y)$ does not exist.

Jordan content

14.8 Let S be a bounded set in $\mathbf{R}^n$ having at most a finite number of accumulation points. Prove that $c(S) = 0$.

14.9 Let f be a continuous real-valued function defined on $[a, b]$. Let S denote the graph of f, that is, $S = \{(x, y) : y = f(x), a \leq x \leq b\}$. Prove that S has two-dimensional Jordan content zero.

14.10 Let Γ be a rectifiable curve in $\mathbf{R}^n$. Prove that Γ has n-dimensional Jordan content zero.

14.11 Let f be a nonnegative function defined on a set S in $\mathbf{R}^n$. The *ordinate set* of f over S is defined to be the following subset of $\mathbf{R}^{n+1}$:

$$\{(x_1, \ldots, x_n, x_{n+1}) : (x_1, \ldots, x_n) \in S, \quad 0 \leq x_{n+1} \leq f(x_1, \ldots, x_n)\}.$$

If S is a Jordan-measurable region in $\mathbf{R}^n$ and if f is continuous on S, prove that the ordinate set of f over S has $(n + 1)$-dimensional Jordan content whose value is

$$\int_S f(x_1, \ldots, x_n)\, d(x_1, \ldots, x_n).$$

Interpret this problem geometrically when $n = 1$ and $n = 2$.

14.12 Assume that $f \in R$ on S and suppose $\int_S f(\mathbf{x})\, d\mathbf{x} = 0$. ($S$ is a subset of $\mathbf{R}^n$). Let $A = \{\mathbf{x} : \mathbf{x} \in S, f(\mathbf{x}) < 0\}$ and assume that $c(A) = 0$. Prove that there exists a set B of measure zero such that $f(\mathbf{x}) = 0$ for each $\mathbf{x}$ in $S - B$.

14.13 Assume that $f \in R$ on S, where S is a region in $\mathbf{R}^n$ and f is continuous on S. Prove that there exists an *interior* point $\mathbf{x}_0$ of S such that

$$\int_S f(\mathbf{x})\, d\mathbf{x} = f(\mathbf{x}_0)c(S).$$

14.14 Let f be continuous on a rectangle $Q = [a, b] \times [c, d]$. For each interior point (x_1, x_2) in Q, define

$$F(x_1, x_2) = \int_a^{x_1} \left(\int_c^{x_2} f(x, y)\, dy \right) dx.$$

Prove that $D_{1,2}F(x_1, x_2) = D_{2,1}F(x_1, x_2) = f(x_1, x_2)$.

14.15 Let T denote the following triangular region in the plane:

$$T = \left\{ (x, y): 0 \le \frac{x}{a} + \frac{y}{b} \le 1 \right\}, \qquad \text{where } a > 0, b > 0.$$

Assume that f has a continuous second-order partial derivative $D_{1,2}f$ on T. Prove that there is a point (x_0, y_0) on the segment joining $(a, 0)$ and $(0, b)$ such that

$$\int_T D_{1,2}f(x, y)\, d(x, y) = f(0, 0) - f(a, 0) + aD_1 f(x_0, y_0).$$

SUGGESTED REFERENCES FOR FURTHER STUDY

14.1 Apostol, T. M., *Calculus*, Vol. 2, 2nd ed. Xerox, Waltham, 1969.

14.2 Kestelman, H., *Modern Theories of Integration*. Oxford University Press, 1937.

14.3 Rogosinski, W. W., *Volume and Integral*. Wiley, New York, 1952.

CHAPTER 15

MULTIPLE LEBESGUE INTEGRALS

15.1 INTRODUCTION

The Lebesgue integral was described in Chapter 10 for functions defined on subsets of $\mathbf{R}^1$. The method used there can be generalized to provide a theory of Lebesgue integration for functions defined on subsets of n-dimensional space $\mathbf{R}^n$. The resulting integrals are called *multiple integrals*. When $n = 2$ they are called *double integrals*, and when $n = 3$ they are called *triple integrals*.

As in the one-dimensional case, multiple Lebesgue integration is an extension of multiple Riemann integration. It permits more general functions as integrands, it treats unbounded as well as bounded functions, and it encompasses more general sets as regions of integration.

The basic definitions and the principal convergence theorems are completely analogous to the one-dimensional case. However, there is one new feature that does not appear in $\mathbf{R}^1$. A multiple integral in $\mathbf{R}^n$ can be evaluated by calculating a succession of n one-dimensional integrals. This result, called *Fubini's Theorem*, is one of the principal concerns of this chapter.

As in the one-dimensional case we define the integral first for step functions, then for a larger class (called upper functions) which contains limits of certain increasing sequences of step functions, and finally for an even larger class, the Lebesgue-integrable functions. Since the development proceeds on exactly the same lines as in the one-dimensional case, we shall omit most of the details of the proofs.

We recall some of the concepts introduced in Chapter 14. If $I = I_1 \times \cdots \times I_n$ is a bounded interval in $\mathbf{R}^n$, the n-measure of I is defined by the equation

$$\mu(I) = \mu(I_1) \cdots \mu(I_n),$$

where $\mu(I_k)$ is the one-dimensional measure, or length, of I_k.

A subset T of $\mathbf{R}^n$ is said to be of n-measure 0 if, for every $\varepsilon > 0$, T can be covered by a countable collection of n-dimensional intervals, the sum of whose n-measures is $< \varepsilon$.

A property is said to hold almost everywhere on a set S in $\mathbf{R}^n$ if it holds everywhere on S except for a subset of n-measure 0. For example, if $\{f_n\}$ is a sequence of functions, we say $f_n \to f$ almost everywhere on S if $\lim_{n \to \infty} f_n(\mathbf{x}) = f(\mathbf{x})$ for all $\mathbf{x}$ in S except for those $\mathbf{x}$ in a subset of n-measure 0.

15.2 STEP FUNCTIONS AND THEIR INTEGRALS

Let I be a compact interval in $\mathbf{R}^n$, say

$$I = I_1 \times \cdots \times I_n,$$

where each I_k is a compact subinterval of $\mathbf{R}^1$. If P_k is a partition of I_k, the cartesian product $P = P_1 \times \cdots \times P_n$ is called a partition of I. If P_k decomposes I_k into m_k one-dimensional subintervals, then P decomposes I into $m = m_1 \cdots m_k$ n-dimensional subintervals, say $J_1, \ldots, J_m$.

A function s defined on I is called a *step function* if a partition P of I exists such that s is constant on the interior of each subinterval J_k, say

$$s(\mathbf{x}) = c_k \qquad \text{if } \mathbf{x} \in \text{int } J_k.$$

The integral of s over I is defined by the equation

$$\int_I s = \sum_{k=1}^m c_k \mu(J_k). \qquad (1)$$

Now let G be a general n-dimensional interval, that is, an interval in $\mathbf{R}^n$ which need not be compact. A function s is called a step function on G if there is a compact n-dimensional subinterval I of G such that s is a step function on I and $s(\mathbf{x}) = 0$ if $\mathbf{x} \in G - I$. The integral of s over G is defined by the formula

$$\int_G s = \int_I s,$$

where the integral over I is given by (1). As in the one-dimensional case the integral is independent of the choice of I.

15.3 UPPER FUNCTIONS AND LEBESGUE-INTEGRABLE FUNCTIONS

Upper functions and Lebesgue-integrable functions are defined exactly as in the one-dimensional case.

A real-valued function f defined on an interval I in $\mathbf{R}^n$ is called an upper function on I, and we write $f \in U(I)$, if there exists an increasing sequence of step functions $\{s_n\}$ such that

a) $s_n \to f$ almost everywhere on I,
and

b) $\lim_{n \to \infty} \int_I s_n$ exists.

The sequence $\{s_n\}$ is said to generate f. The integral of f over I is defined by the equation

$$\int_I f = \lim_{n \to \infty} \int_I s_n. \qquad (2)$$

We denote by $L(I)$ the set of all functions f of the form $f = u - v$, where $u \in U(I)$ and $v \in U(I)$. Each function f in $L(I)$ is said to be Lebesgue-integrable on I, and its integral is defined by the equation

$$\int_I f = \int_I u - \int_I v.$$

Since these definitions are completely analogous to the one-dimensional case, it is not surprising to learn that many of the theorems derived from these definitions are also valid. In particular, Theorems 10.5, 10.6, 10.7, 10.9, 10.10, 10.11, 10.13, 10.14, 10.16, 10.17(a) and (c), 10.18, and 10.19 are all valid for multiple integrals. Theorem 10.17(b), which describes the behavior of an integral under expansion or contraction of the interval of integration, needs to be modified as follows:

If $f \in L(I)$ and if $g(\mathbf{x}) = f(\mathbf{x}/c)$, where $c > 0$, then $g \in L(cI)$ and

$$\int_{cI} g = c^n \int_I f.$$

In other words, expansion of the interval by a positive factor c has the effect of multiplying the integral by c^n, where n is the dimension of the space.

The Levi convergence theorems (Theorems 10.22 through 10.26), and the Lebesgue dominated convergence theorem (Theorem 10.27) and its consequences (Theorems 10.28, 10.29, and 10.30) are also valid for multiple integrals.

NOTATION. The integral $\int_I f$ is also denoted by

$$\int_I f(\mathbf{x})\, dx \qquad \text{or} \qquad \int_I f(x_1, \ldots, x_n) d(x_1, \ldots, x_n).$$

The notation $\int_I f(x_1, \ldots, x_n)\, dx_1 \cdots dx_n$ is also used. Double integrals are sometimes written with two integral signs, and triple integrals with three such signs, thus:

$$\iint_I f(x, y)\, dx\, dy, \qquad \iiint_I f(x, y, z)\, dx\, dy\, dz.$$

15.4 MEASURABLE FUNCTIONS AND MEASURABLE SETS IN R^n

A real-valued function f defined on an interval I in $\mathbf{R}^n$ is called *measurable* on I, and we write $f \in M(I)$, if there exists a sequence of step functions $\{s_n\}$ on I such that

$$\lim_{n \to \infty} s_n(\mathbf{x}) = f(\mathbf{x}) \qquad a.e. \text{ on } I.$$

The properties of measurable functions described in Theorems 10.35, 10.36, and 10.37 are also valid in this more general setting.

A subset S of $\mathbf{R}^n$ is called measurable if its characteristic function χ_S is measurable. If, in addition, χ_S is Lebesgue-integrable on $\mathbf{R}^n$, then the n-measure $\mu(S)$ of the set S is defined by the equation

$$\mu(S) = \int_{\mathbf{R}^n} \chi_S.$$

If χ_S is measurable but not in $L(\mathbf{R}^n)$, we define $\mu(S) = +\infty$. The function μ so defined is called n-dimensional Lebesgue measure.

The properties of measure described in Theorems 10.44 through 10.47 are also valid for n-dimensional Lebesgue measure. Also, the Lebesgue integral can be defined for arbitrary subsets of $\mathbf{R}^n$ by the method used in Section 10.19.

We emphasize in particular the countably additive property of Lebesgue measure described in Theorem 10.47:

If $\{A_1, A_2, \ldots\}$ is a countable disjoint collection of measurable sets in $\mathbf{R}^n$, then the union $\bigcup_{i=1}^{\infty} A_i$ is measurable and

$$\mu\left(\bigcup_{i=1}^{\infty} A_i\right) = \sum_{i=1}^{\infty} \mu(A_i).$$

The next theorem shows that every open subset of $\mathbf{R}^n$ is measurable.

Theorem 15.1. *Every open set S in $\mathbf{R}^n$ can be expressed as the union of a countable disjoint collection of bounded cubes whose closure is contained in S. Therefore S is measurable. Moreover, if S is bounded, then $\mu(S)$ is finite.*

Proof. Fix an integer $m \geq 1$ and consider all half-open intervals in $\mathbf{R}^1$ of the form

$$\left(\frac{k}{2^m}, \frac{k+1}{2^m}\right] \qquad \text{for } k = 0, \pm 1, \pm 2, \ldots$$

All the intervals are of length 2^{-m}, and they form a countable disjoint collection whose union is $\mathbf{R}^1$. The cartesian product of n such intervals is an n-dimensional cube of edge-length 2^{-m}. Let F_m denote the collection of all these cubes. Then F_m is a countable disjoint collection whose union is $\mathbf{R}^n$. Note that the cubes in F_{m+1} are obtained by bisecting the edges of those in F_m. Therefore, if Q_m is a cube in F_m and if Q_{m+1} is a cube in F_{m+1}, then either $Q_{m+1} \subseteq Q_m$, or Q_{m+1} and Q_m are disjoint.

Now we extract a subcollection G_m from F_m as follows. If $m = 1$, G_1 consists of all cubes in F_1 whose closure lies in S. If $m = 2$, G_2 consists of all cubes in F_2 whose closure lies in S but not in any of the cubes in G_1. If $m = 3$, G_3 consists of all cubes in F_3 whose closure lies in S but not in any of the cubes in G_1 or G_2, and so on. The construction is illustrated in Fig. 15.1 where S is a quarter of an open disk in $\mathbf{R}^2$. The blank square is in G_1, the lightly shaded ones are in G_2, and the darker ones are in G_3.

Now let

$$T = \bigcup_{m=1}^{\infty} \bigcup_{Q \in G_m} Q.$$

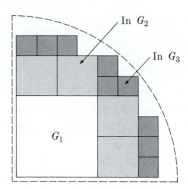

Figure 15.1

That is, T is the union of all the cubes in $G_1, G_2, \ldots$ We will prove that $S = T$ and this will prove the theorem because T is a countable disjoint collection of cubes whose closure lies in S. Now $T \subseteq S$ because each Q in G_m is a subset of S. Hence we need only show that $S \subseteq T$.

Let $\mathbf{p} = (p_1, \ldots, p_n)$ be a point in S. Since S is open, there is a cube with center $\mathbf{p}$ and edge-length $\delta > 0$, which lies in S. Choose m so that $2^{-m} < \delta/2$. Then for each i we have

$$p_i - \frac{\delta}{2} < p_i - \frac{1}{2^m} < p_i < p_i + \frac{1}{2^m} < p_i + \frac{\delta}{2}.$$

Now choose k_i, so that

$$\frac{k_i}{2^m} < p_i \le \frac{k_i + 1}{2^m},$$

and let Q be the Cartesian product of the intervals $(k_i 2^{-m}, (k_i + 1)2^{-m}]$ for $i = 1, 2, \ldots, n$. Then $\mathbf{p} \in Q$ for some cube Q in F_m. If m is the smallest integer with this property, then $Q \in G_m$, so $\mathbf{p} \in T$. Hence $S \subseteq T$. The statements about the measurability of S follow at once from the countably additive property of Lebesgue measure.

NOTE. If S is measurable, so is $\mathbf{R}^n - S$ because $\chi_{\mathbf{R}^n - S} = 1 - \chi_S$. Therefore, every closed subset of $\mathbf{R}^n$ is measurable.

15.5 FUBINI'S REDUCTION THEOREM FOR THE DOUBLE INTEGRAL OF A STEP FUNCTION

Up to this point, Lebesgue theory in $\mathbf{R}^n$ is completely analogous to the one-dimensional case. New ideas are required when we come to Fubini's theorem for calculating a multiple integral in $\mathbf{R}^n$ by iterated lower-dimensional integrals. To better understand what is needed, we consider first the two-dimensional case.

Let us recall the corresponding result for multiple Riemann integrals. If $I = [a, b] \times [c, d]$ is a compact interval in $\mathbf{R}^2$ and if f is Riemann-integrable on I, then we have the following reduction formula (from part (v) of Theorem 14.6):

$$\int_I f(x, y) \, d(x, y) = \int_c^d \left[\underline{\int_a^b} f(x, y) \, dx \right] dy. \tag{3}$$

There is a companion formula with the lower integral $\underline{\int_a^b}$ replaced by the upper integral $\overline{\int_a^b}$, and there are two similar formulas with the order of integration reversed. The upper and lower integrals are needed here because the hypothesis of Riemann-integrability on I is not strong enough to ensure the existence of the one-dimensional Riemann integral $\int_a^b f(x, y) \, dx$. This difficulty does not arise in the Lebesgue theory. Fubini's theorem for double Lebesgue integrals gives us the reduction formulas

$$\int_I f(x, y) \, d(x, y) = \int_c^d \left[\int_a^b f(x, y) \, dx \right] dy = \int_a^b \left[\int_c^d f(x, y) \, dy \right] dx,$$

under the sole hypothesis that f is Lebesgue-integrable on I. We will show that the inner integrals always exist as Lebesgue integrals. This is another example illustrating how Lebesgue theory overcomes difficulties inherent in the Riemann theory.

In this section we prove Fubini's theorem for step functions, and in a later section we extend it to arbitrary Lebesgue integrable functions.

Theorem 15.2. (Fubini's theorem for step functions). *Let s be a step function on $\mathbf{R}^2$. Then for each fixed y in $\mathbf{R}^1$ the integral $\int_{\mathbf{R}^1} s(x, y) \, dx$ exists and, as a function of y, is Lebesgue-integrable on $\mathbf{R}^1$. Moreover, we have*

$$\iint_{\mathbf{R}^2} s(x, y) \, d(x, y) = \int_{\mathbf{R}^1} \left[\int_{\mathbf{R}^1} s(x, y) \, dx \right] dy. \tag{4}$$

Similarly, for each fixed x in $\mathbf{R}^1$ the integral $\int_{\mathbf{R}^1} s(x, y) \, dy$ exists and, as a function of x, is Lebesgue-integrable on $\mathbf{R}^1$. Also, we have

$$\iint_{\mathbf{R}^2} s(x, y) \, d(x, y) = \int_{\mathbf{R}^1} \left[\int_{\mathbf{R}^1} s(x, y) \, dy \right] dx. \tag{5}$$

Proof. This theorem can be derived from the reduction formula (3) for Riemann integrals, but we prefer to give a direct proof independent of the Riemann theory.

There is a compact interval $I = [a, b] \times [c, d]$ such that s is a step function on I and $s(x, y) = 0$ if $(x, y) \in \mathbf{R}^2 - I$. There is a partition of I into mn sub-rectangles $I_{ij} = [x_{i-1}, x_i] \times [y_{j-i}, y_j]$ such that s is constant on the interior of I_{ij}, say

$$s(x, y) = c_{ij} \qquad \text{if } (x, y) \in \text{int } I_{ij}.$$

Then

$$\iint\limits_{I_{ij}} s(x,\, y)\, d(x,\, y) = c_{ij}(x_i - x_{i-1})(y_j - y_{j-1}) = \int_{y_{j-1}}^{y_j} \left[\int_{x_{i-1}}^{x_i} s(x,\, y)\, dx \right] dy.$$

Summing on i and j we find

$$\iint\limits_{I} s(x,\, y)\, d(x,\, y) = \int_c^d \left[\int_a^b s(x,\, y)\, dx \right] dy.$$

Since s vanishes outside I, this proves (4), and a similar argument proves (5).

To extend Fubini's theorem to Lebesgue-integrable functions we need some further results concerning sets of measure zero. These are discussed in the next section.

15.6 SOME PROPERTIES OF SETS OF MEASURE ZERO

Theorem 15.3. *Let S be a subset of $\mathbf{R}^n$. Then S has n-measure 0 if, and only if, there exists a countable collection of n-dimensional intervals $\{J_1, J_2, \ldots\}$, the sum of whose n-measures is finite, such that each point in S belongs to J_k for infinitely many k.*

Proof. Assume first that S has n-measure 0. Then, for every $m \geq 1$, S can be covered by a countable collection of n-dimensional intervals $\{I_{m,1}, I_{m,2}, \ldots\}$, the sum of whose n-measures is $< 2^{-m}$. The set A consisting of all intervals $I_{m,k}$ for $m = 1, 2, \ldots$, and $k = 1, 2, \ldots$, is a countable collection which covers S, and the sum of the n-measures of all these intervals is $< \sum_{m=1}^{\infty} 2^{-m} = 1$. Moreover, if $\mathbf{a} \in S$ then, for each m, $\mathbf{a} \in I_{m,k}$ for some k. Therefore if we write $A = \{J_1, J_2, \ldots\}$, we see that $\mathbf{a}$ belongs to J_k for infinitely many k.

Conversely, assume that there is a countable collection of n-dimensional intervals $\{J_1, J_2, \ldots\}$ such that the series $\sum_{k=1}^{\infty} \mu(J_k)$ converges and such that each point in S belongs to J_k for infinitely many k. Given $\varepsilon > 0$, there is an integer N such that

$$\sum_{k=N}^{\infty} \mu(J_k) < \varepsilon.$$

Each point of S lies in the set $\bigcup_{k=N}^{\infty} J_k$, so $S \subseteq \bigcup_{k=N}^{\infty} J_k$. Thus, S has been covered by a countable collection of intervals, the sum of whose n-measures is $< \varepsilon$, so S has n-measure 0.

Definition 15.4. *If S is an arbitrary subset of $\mathbf{R}^2$, and if $(x, y) \in \mathbf{R}^2$, we denote by S_y and S^x the following subsets of $\mathbf{R}^1$:*

$$S_y = \{x : x \in \mathbf{R}^1 \quad and \quad (x, y) \in S\},$$
$$S^x = \{y : y \in \mathbf{R}^1 \quad and \quad (x, y) \in S\}.$$

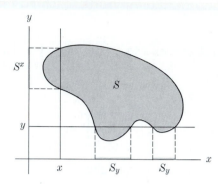

Figure 15.2

Examples are shown in Fig. 15.2. Geometrically, S_y is the projection on the x-axis of a horizontal cross section of S; and S^x is the projection on the y-axis of a vertical cross section of S.

Theorem 15.5. *If S is a subset of $\mathbf{R}^2$ with 2-measure 0, then S_y has 1-measure 0 for almost all y in $\mathbf{R}^1$, and S^x has 1-measure 0 for almost all x in $\mathbf{R}^1$.*

Proof. We will prove that S_y has 1-measure 0 for almost all y in $\mathbf{R}^1$. The proof makes use of Theorem 15.3.

Since S has 2-measure 0, by Theorem 15.3 there is a countable collection of rectangles $\{I_k\}$ such that the series

$$\sum_{k=1}^{\infty} \mu(I_k) \qquad \text{converges,} \qquad\qquad (6)$$

and such that every point (x, y) of S belongs to I_k for infinitely many k. Write $I_k = X_k \times Y_k$, where X_k and Y_k are subintervals of $\mathbf{R}^1$. Then

$$\mu(I_k) = \mu(X_k)\mu(Y_k) = \mu(X_k) \int_{\mathbf{R}^1} \chi_{Y_k} = \int_{\mathbf{R}^1} \mu(X_k)\,\chi_{Y_k},$$

where χ_{Y_k} is the characteristic function of the interval Y_k. Let $g_k = \mu(X_k)\chi_{Y_k}$. Then (6) implies that the series

$$\sum_{k=1}^{\infty} \int_{\mathbf{R}^1} g_k \qquad \text{converges.}$$

Now $\{g_k\}$ is a sequence of nonnegative functions in $L(\mathbf{R}^1)$ such that the series $\sum_{k=1}^{\infty} \int_{\mathbf{R}^1} g_k$ converges. Therefore, by the Levi theorem (Theorem 10.25), the series $\sum_{k=1}^{\infty} g_k$ converges almost everywhere on $\mathbf{R}^1$. In other words, there is a subset T of $\mathbf{R}^1$ of 1-measure 0 such that the series

$$\sum_{k=1}^{\infty} \mu(X_k)\chi_{Y_k}(y) \qquad \text{converges for all } y \text{ in } \mathbf{R}^1 - T. \qquad\qquad (7)$$

Take a point y in $\mathbf{R}^1 - T$, keep y fixed and consider the set S_y. We will prove that S_y has 1-measure zero.

We can assume that S_y is nonempty; otherwise the result is trivial. Let

$$A(y) = \{X_k : y \in Y_k, \quad k = 1, 2, \ldots\}.$$

Then $A(y)$ is a countable collection of one-dimensional intervals which we relabel as $\{J_1, J_2, \ldots\}$. The sum of the lengths of all the intervals J_k converges because of (7). If $x \in S_y$, then $(x, y) \in S$ so $(x, y) \in I_k = X_k \times Y_k$ for infinitely many k, and hence $x \in J_k$ for infinitely many k. By the one-dimensional version of Theorem 15.3 it follows that S_y has 1-measure zero. This shows that S_y has 1-measure zero for almost all y in $\mathbf{R}^1$, and a similar argument proves that S^x has 1-measure zero for almost all x in $\mathbf{R}^1$.

15.7 FUBINI'S REDUCTION THEOREM FOR DOUBLE INTEGRALS

Theorem 15.6. *Assume f is Lebesgue-integrable on $\mathbf{R}^2$. Then we have:*

a) *There is a set T of 1-measure 0 such that the Lebesgue integral $\int_{\mathbf{R}^1} f(x, y)\, dx$ exsits for all y in $\mathbf{R}^1 - T$.*

b) *The function G defined on $\mathbf{R}^1$ by the equation*

$$G(y) = \begin{cases} \int_{\mathbf{R}^1} f(x, y)\, dx & \text{if } y \in \mathbf{R}^1 - T, \\ 0 & \text{if } y \in T, \end{cases}$$

is Lebesgue-integrable on $\mathbf{R}^1$.

c) $\iint_{\mathbf{R}^2} f = \int_{\mathbf{R}^1} G(y)\, dy.$ *That is,*

$$\iint_{\mathbf{R}^2} f(x, y)\, d(x, y) = \int_{\mathbf{R}^1}\left[\int_{\mathbf{R}^1} f(x, y)\, dx\right] dy.$$

NOTE. There is a corresponding result which concludes that

$$\iint_{\mathbf{R}^2} f(x, y)\, d(x, y) = \int_{\mathbf{R}^1}\left[\int_{\mathbf{R}^1} f(x, y)\, dy\right] dx.$$

Proof. We have already proved the theorem for step functions. We prove it next for upper functions. If $f \in U(\mathbf{R}^2)$ there is an increasing sequence of step functions $\{s_n\}$ such that $s_n(x, y) \to f(x, y)$ for all (x, y) in $\mathbf{R}^2 - S$, where S is a set of 2-measure 0; also,

$$\lim_{n \to \infty} \iint_{\mathbf{R}^2} s_n(x, y)\, d(x, y) = \iint_{\mathbf{R}^2} f(x, y)\, d(x, y).$$

Now $(x, y) \in \mathbf{R}^2 - S$ if, and only if, $x \in \mathbf{R}^1 - S_y$. Hence

$$s_n(x, y) \to f(x, y) \qquad \text{if } x \in \mathbf{R}^1 - S_y. \tag{8}$$

Let $t_n(y) = \int_{\mathbf{R}^1} s_n(x, y) \, dx$. This integral exists for each real y and is an integrable function of y. Moreover, by Theorem 15.2 we have

$$\int_{\mathbf{R}^1} t_n(y) \, dy = \int_{\mathbf{R}^1} \left[\int_{\mathbf{R}^1} s_n(x, y) \, dx \right] dy = \iint_{\mathbf{R}^2} s_n(x, y) \, d(x, y) \le \iint_{\mathbf{R}^2} f.$$

Since the sequence $\{t_n\}$ is increasing, the last inequality shows that $\lim_{n \to \infty} \int_{\mathbf{R}^1} t_n(y) \, dy$ exists. Therefore, by the Levi theorem (Theorem 10.24) there is a function t in $L(\mathbf{R}^1)$ such that $t_n \to t$ almost everywhere on $\mathbf{R}^1$. In other words, there is a set T_1 of 1-measure 0 such that $t_n(y) \to t(y)$ if $y \in \mathbf{R}^1 - T_1$. Moreover,

$$\int_{\mathbf{R}^1} t(y) \, dy = \lim_{n \to \infty} \int_{\mathbf{R}^1} t_n(y) \, dy.$$

Again, since $\{t_n\}$ is increasing, we have

$$t_n(y) = \int_{\mathbf{R}^1} s_n(x, y) \, dx \le t(y) \qquad \text{if } y \in \mathbf{R}^1 - T_1.$$

Applying the Levi theorem to $\{s_n\}$ we find that if $y \in \mathbf{R}^1 - T_1$ there is a function g in $L(\mathbf{R}^1)$ such that $s_n(x, y) \to g(x, y)$ for x in $\mathbf{R}^1 - A$, where A is a set of 1-measure 0. (The set A depends on y.) Comparing this with (8) we see that if $y \in \mathbf{R}^1 - T_1$ then

$$g(x, y) = f(x, y) \qquad \text{if } x \in \mathbf{R}^1 - (A \cup S_y). \tag{9}$$

But A has 1-measure 0 and S_y has 1-measure 0 for almost all y, say for all y in $\mathbf{R}^1 - T_2$, where T_2 has 1-measure 0. Let $T = T_1 \cup T_2$. Then T has 1-measure 0. If $y \in \mathbf{R}^1 - T$, the set $A \cup S_y$ has 1-measure 0 and (9) holds. Since the integral $\int_{\mathbf{R}^1} g(x, y) \, dx$ exists if $y \in \mathbf{R}^1 - T$ it follows that the integral $\int_{\mathbf{R}^1} f(x, y) \, dx$ also exists if $y \in \mathbf{R}^1 - T$. This proves (a). Also, if $y \in \mathbf{R}^1 - T$ we have

$$\int_{\mathbf{R}^1} f(x, y) \, dx = \int_{\mathbf{R}^1} g(x, y) \, dx = \lim_{n \to \infty} \int_{\mathbf{R}^1} s_n(x, y) \, dx = t(y). \tag{10}$$

Since $t \in L(\mathbf{R}^1)$, this proves (b). Finally, we have

$$\int_{\mathbf{R}^1} t(y) \, dy = \int_{\mathbf{R}^1} \lim_{n \to \infty} t_n(y) \, dy = \lim_{n \to \infty} \int_{\mathbf{R}^1} t_n(y) \, dy$$

$$= \lim_{n \to \infty} \int_{\mathbf{R}^1} \left[\int_{\mathbf{R}^1} s_n(x, y) \, dx \right] dy = \lim_{n \to \infty} \iint_{\mathbf{R}^2} s_n(x, y) \, d(x, y)$$

$$= \iint_{\mathbf{R}^2} f(x, y) \, d(x, y).$$

Comparing this with (10) we obtain (c). This proves Fubini's theorem for upper functions.

To prove it for Lebesgue-integrable functions we write $f = u - v$, where $u \in L(\mathbf{R}^2)$ and $v \in L(\mathbf{R}^2)$ and we obtain

$$\iint_{\mathbf{R}^2} f = \iint_{\mathbf{R}^2} u - \iint_{\mathbf{R}^2} v = \int_{\mathbf{R}^1} \left[\int_{\mathbf{R}^1} u(x, y) \, dx \right] dy - \int_{\mathbf{R}^1} \left[\int_{\mathbf{R}^1} v(x, y) \, dx \right] dy$$

$$= \int_{\mathbf{R}^1} \left[\int_{\mathbf{R}^1} \{ u(x, y) - v(x, y) \} \, dx \right] dy = \int_{\mathbf{R}^1} \left[\int_{\mathbf{R}^1} f(x, y) \, dx \right] dy.$$

As an immediate corollary of Theorem 15.6 and the two-dimensional analog of Theorem 10.11 we obtain:

Theorem 15.7. *Assume that f is defined and bounded on a compact rectangle $I = [a, b] \times [c, d]$, and that f is continuous almost everywhere on I. Then $f \in L(I)$ and we have*

$$\iint_I f(x, y) \, d(x, y) = \int_c^d \left[\int_a^b f(x, y) \, dx \right] dy = \int_a^b \left[\int_c^d f(x, y) \, dy \right] dx.$$

NOTE. The one-dimensional integral $\int_a^b f(x, y) \, dx$ exists for almost all y in $[c, d]$ as a Lebesgue integral. It need not exist as a Riemann integral. A similar remark applies to the integral $\int_c^d f(x, y) \, dy$. In the Riemann theory, the inner integrals in the reduction formula must be replaced by upper or lower integrals. (See Theorem 14.6, part (v).)

There is, of course, an extension of Fubini's theorem to higher-dimensional integrals. If f is Lebesgue-integrable on $\mathbf{R}^{m+k}$ the analog of Theorem 15.6 concludes that

$$\int_{\mathbf{R}^{m+k}} f = \int_{\mathbf{R}^k} \left[\int_{\mathbf{R}^m} f(\mathbf{x}; \mathbf{y}) \, d\mathbf{x} \right] d\mathbf{y} = \int_{\mathbf{R}^m} \left[\int_{\mathbf{R}^k} f(\mathbf{x}; \mathbf{y}) \, d\mathbf{y} \right] d\mathbf{x}.$$

Here we have written a point in $\mathbf{R}^{m+k}$ as $(\mathbf{x}; \mathbf{y})$, where $\mathbf{x} \in \mathbf{R}^m$ and $\mathbf{y} \in \mathbf{R}^k$. This can be proved by an extension of the method used to prove the two-dimensional case, but we shall omit the details.

15.8 THE TONELLI–HOBSON TEST FOR INTEGRABILITY

Which functions are Lebesgue-integrable on $\mathbf{R}^2$? The next theorem gives a useful sufficient condition for integrability. Its proof makes use of Fubini's theorem.

Theorem 15.8. *Assume that f is measurable on $\mathbf{R}^2$ and assume that at least one of the two iterated integrals*

$$\int_{\mathbf{R}^1} \left[\int_{\mathbf{R}^1} |f(x, y)| \, dx \right] dy \qquad or \qquad \int_{\mathbf{R}^1} \left[\int_{\mathbf{R}^1} |f(x, y)| \, dy \right] dx,$$

exists. Then we have:

a) $f \in L(\mathbf{R}^2)$.

b) $$\iint_{\mathbf{R}^2} f = \int_{\mathbf{R}^1} \left[\int_{\mathbf{R}^1} f(x, y) \, dx \right] dy = \int_{\mathbf{R}^1} \left[\int_{\mathbf{R}^1} f(x, y) \, dy \right] dx.$$

Proof. Part (b) follows from part (a) because of Fubini's theorem. We will also use Fubini's theorem to prove part (a). Assume that the iterated integral $\int_{\mathbf{R}^1} \left[\int_{\mathbf{R}^1} |f(x, y)| \, dx \right] dy$ exists. Let $\{s_n\}$ denote the increasing sequence of nonnegative step functions defined as follows:

$$s_n(x, y) = \begin{cases} n & \text{if } |x| \le n \text{ and } |y| \le n, \\ 0 & \text{otherwise.} \end{cases}$$

Let $f_n(x, y) = \min \{s_n(x, y), |f(x, y)|\}$. Both s_n and $|f|$ are measurable so f_n is measurable. Also, we have $0 \le f_n(x, y) \le s_n(x, y)$, so f_n is dominated by a Lebesgue-integrable function. Therefore, $f_n \in L(\mathbf{R}^2)$. Hence we can apply Fubini's theorem to f_n along with the inequality $0 \le f_n(x, y) \le |f(x, y)|$ to obtain

$$\iint_{\mathbf{R}^2} f_n = \int_{\mathbf{R}^1} \left[\int_{\mathbf{R}^1} f_n(x, y) \, dx \right] dy \le \int_{\mathbf{R}^1} \left[\int_{\mathbf{R}^1} |f(x, y)| \, dx \right] dy.$$

Since $\{f_n\}$ is increasing, this shows that the limit $\lim_{n \to \infty} \iint_{\mathbf{R}^2} f_n$ exists. By the Levi theorem (the two-dimensional analog of Theorem 10.24), $\{f_n\}$ converges almost everywhere on $\mathbf{R}^2$ to a limit function in $L(\mathbf{R}^2)$. But $f_n(x, y) \to |f(x, y)|$ as $n \to \infty$, so $|f| \in L(\mathbf{R}^2)$. Since f is measurable, it follows that $f \in L(\mathbf{R}^2)$. This proves (a). The proof is similar if the other iterated integral exists.

15.9 COORDINATE TRANSFORMATIONS

One of the most important results in the theory of multiple integration is the formula for making a change of variables. This is an extension of the formula

$$\int_{g(c)}^{g(d)} f(x) \, dx = \int_c^d f[g(t)]g'(t) \, dt,$$

which was proved in Theorem 7.36 for Riemann integrals under the assumption that g has a continuous derivative g' on an interval $T = [c, d]$ and that f is continuous on the image $g(T)$.

Consider the special case in which g' is never zero (hence of constant sign) on T. If g' is positive on T, then g is increasing, so $g(c) < g(d)$, $g(T) = [g(c), g(d)]$, and the above formula can be written as follows:

$$\int_{g(T)} f(x) \, dx = \int_T f[g(t)]g'(t) \, dt.$$

On the other hand, if g' is negative on T, then $g(T) = [g(d), g(c)]$ and the above formula becomes

$$\int_{g(T)} f(x)\, dx = -\int_{T} f[g(t)]g'(t)\, dt.$$

Both cases are included, therefore, in the *single* formula

$$\int_{g(T)} f(x)\, dx = \int_{T} f[g(t)]\, |g'(t)|\, dt. \tag{11}$$

Equation (11) is also valid when $c > d$, and it is in this form that the result will be generalized to multiple integrals. The function g which transforms the variables must be replaced by a vector-valued function called a *coordinate transformation* which is defined as follows.

Definition 15.9. *Let T be an open subset of $\mathbf{R}^n$. A vector-valued function $\mathbf{g} : T \to \mathbf{R}^n$ is called a coordinate transformation on T if it has the following three properties:*

a) $\mathbf{g} \in C'$ *on T.*

b) $\mathbf{g}$ *is one-to-one on T.*

c) *The Jacobian determinant $J_{\mathbf{g}}(\mathbf{t}) = \det \mathbf{Dg}(\mathbf{t}) \neq 0$ for all $\mathbf{t}$ in T.*

NOTE. A coordinate transformation is sometimes called a *diffeomorphism*.

Property (a) states that $\mathbf{g}$ is continuously differentiable on T. From Theorem 13.4 we know that a continuously differentiable function is locally one-to-one near each point where its Jacobian determinant does not vanish. Property (b) assumes that $\mathbf{g}$ is *globally* one-to-one on T. This guarantees the existence of a global inverse $\mathbf{g}^{-1}$ which is defined and one-to-one on the image $\mathbf{g}(T)$. Properties (a) and (c) together imply that $\mathbf{g}$ is an open mapping (by Theorem 13.5). Also, $\mathbf{g}^{-1}$ is continuously differentiable on $\mathbf{g}(T)$ (by Theorem 13.6).

Further properties of coordinate transformations will be deduced from the following multiplicative property of Jacobian determinants.

Theorem 15.10 (Multiplication theorem for Jacobian determinants). *Assume that $\mathbf{g}$ is differentiable on an open set T in $\mathbf{R}^n$ and that $\mathbf{h}$ is differentiable on the image $\mathbf{g}(T)$. Then the composition $\mathbf{k} = \mathbf{h} \circ \mathbf{g}$ is differentiable on T, and for every $\mathbf{t}$ in T we have*

$$J_{\mathbf{k}}(\mathbf{t}) = J_{\mathbf{h}}[\mathbf{g}(\mathbf{t})]J_{\mathbf{g}}(\mathbf{t}). \tag{12}$$

Proof. The chain rule (Theorem 12.7) tells us that the composition $\mathbf{k}$ is differentiable on T, and the matrix form of the chain rule tells us that the corresponding Jacobian matrices are related as follows:

$$\mathbf{Dk}(\mathbf{t}) = \mathbf{Dh}[\mathbf{g}(\mathbf{t})]\mathbf{Dg}(\mathbf{t}). \tag{13}$$

From the theory of determinants we know that $\det (AB) = \det A \det B$, so (13) implies (12).

This theorem shows that if **g** is a coordinate transformation on T and if **h** is a coordinate transformation on $\mathbf{g}(T)$, then the composition **k** is a coordinate transformation on T. Also, if $\mathbf{h} = \mathbf{g}^{-1}$, then

$$\mathbf{k(t)} = \mathbf{t} \quad \text{for all } \mathbf{t} \text{ in } T, \qquad \text{and} \qquad J_{\mathbf{k}}(\mathbf{t}) = 1,$$

so $J_{\mathbf{h}}[\mathbf{g(t)}]J_{\mathbf{g}}(\mathbf{t}) = 1$ and $\mathbf{g}^{-1}$ is a coordinate transformation on $\mathbf{g}(T)$.

A coordinate transformation **g** and its inverse $\mathbf{g}^{-1}$ set up a one-to-one correspondence between the open subsets of T and the open subsets of $\mathbf{g}(T)$, and also between the compact subsets of T and the compact subsets of $\mathbf{g}(T)$. The following examples are commonly used coordinate transformations.

Example 1. *Polar coordinates in* $\mathbf{R}^2$. In this case we take

$$T = \{(t_1, t_2) : t_1 > 0, 0 < t_2 < 2\pi\},$$

and we let $\mathbf{g} = (g_1, g_2)$ be the function defined on T as follows:

$$g_1(\mathbf{t}) = t_1 \cos t_2, \qquad g_2(\mathbf{t}) = t_1 \sin t_2.$$

It is customary to denote the components of **t** by (r, θ) rather than (t_1, t_2). The coordinate transformation **g** maps each point (r, θ) in T onto the point (x, y) in $\mathbf{g}(T)$ given by the familiar formulas

$$x = r \cos \theta, \qquad y = r \sin \theta.$$

The image $\mathbf{g}(T)$ is the set $\mathbf{R}^2 - \{(x, 0) : x \geq 0\}$, and the Jacobian determinant is

$$J_{\mathbf{g}}(\mathbf{t}) = \begin{vmatrix} \cos \theta & \sin \theta \\ -r \sin \theta & r \cos \theta \end{vmatrix} = r.$$

Example 2. *Cylindrical coordinates in* $\mathbf{R}^3$. Here we write $\mathbf{t} = (r, \theta, z)$ and we take

$$T = \{(r, \theta, z) : r > 0, \ 0 < \theta < 2\pi, \ -\infty < z < +\infty\}.$$

The coordinate transformation **g** maps each point (r, θ, z) in T onto the point (x, y, z) in $\mathbf{g}(T)$ given by the equations

$$x = r \cos \theta, \qquad y = r \sin \theta, \qquad z = z.$$

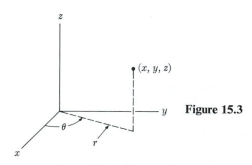

Figure 15.3

The image $g(T)$ is the set $\mathbf{R}^3 - \{(x, 0, 0) : x \geq 0\}$, and the Jacobian determinant is given by

$$J_g(\mathbf{t}) = \begin{vmatrix} \cos \theta & \sin \theta & 0 \\ -r \sin \theta & r \cos \theta & 0 \\ 0 & 0 & 1 \end{vmatrix} = r.$$

The geometric significance of r, θ, and z is shown in Fig. 15.3.

Example 3. *Spherical coordinates in* $\mathbf{R}^3$. In this case we write $\mathbf{t} = (\rho, \theta, \varphi)$ and we take

$$T = \{(\rho, \theta, \varphi) : \rho > 0, \quad 0 < \theta < 2\pi, \quad 0 < \varphi < \pi\}.$$

The coordinate transformation g maps each point (ρ, θ, φ) in T onto the point (x, y, z) in $g(T)$ given by the equations

$$x = \rho \cos \theta \sin \varphi, \qquad y = \rho \sin \theta \sin \varphi, \qquad z = \rho \cos \varphi.$$

The image $g(T)$ is the set $\mathbf{R}^3 - [\{(x, 0, 0) : x \geq 0\} \cup \{(0, 0, z) : z \in \mathbf{R}\}]$, and the Jacobian determinant is

$$J_g(\mathbf{t}) = \begin{vmatrix} \cos \theta \sin \varphi & \sin \theta \sin \varphi & \cos \varphi \\ -\rho \sin \theta \sin \varphi & \rho \cos \theta \sin \varphi & 0 \\ \rho \cos \theta \cos \varphi & \rho \sin \theta \cos \varphi & -\rho \sin \varphi \end{vmatrix} = -\rho^2 \sin \varphi.$$

The geometric significance of ρ, θ, and φ is shown in Fig. 15.4.

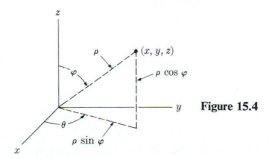

Figure 15.4

Example 4. *Linear transformations in* $\mathbf{R}^n$. Let $g : \mathbf{R}^n \to \mathbf{R}^n$ be a linear transformation represented by a matrix $(a_{ij}) = m(\mathbf{g})$, so that

$$\mathbf{g}(\mathbf{t}) = \left(\sum_{j=1}^n a_{1j} t_j, \ldots, \sum_{j=1}^n a_{nj} t_j \right).$$

Then $\mathbf{g} = (g_1, \ldots, g_n)$ where $g_i(\mathbf{t}) = \sum_{j=1}^n a_{ij} t_j$, and the Jacobian matrix is

$$\mathbf{Dg}(\mathbf{t}) = (D_j g_i(\mathbf{t})) = (a_{ij}).$$

Thus the Jacobian determinant $J_g(\mathbf{t})$ is constant, and equals $\det (a_{ij})$, the determinant of the matrix (a_{ij}). We also call this the determinant of $\mathbf{g}$ and we write

$$\det \mathbf{g} = \det (a_{ij}).$$

A linear transformation **g** which is one-to-one on $\mathbf{R}^n$ is called *nonsingular*. We shall use the following elementary facts concerning nonsingular transformations from $\mathbf{R}^n$ to $\mathbf{R}^n$. (Proofs can be found in any text on linear algebra; see also Reference 14.1.)

A linear transformation **g** is nonsingular if, and only if, its matrix $A = m(\mathbf{g})$ has an inverse A^{-1} such that $AA^{-1} = I$, where I is the identity matrix (the matrix of the identity transformation), in which case A is also called nonsingular. An $n \times n$ matrix A is nonsingular if, and only if, det $A \neq 0$. Thus, a linear function **g** is a coordinate transformation if, and only if, det $\mathbf{g} \neq 0$.

Every nonsingular **g** can be expressed as a composition of three special types of nonsingular transformations called *elementary transformations*, which we refer to as *types a, b,* and *c*. They are defined as follows:

Type a: $\mathbf{g}_a(t_1, \ldots, t_k, \ldots, t_n) = (t_1, \ldots, \lambda t_k, \ldots, t_n)$, where $\lambda \neq 0$. In other words, $\mathbf{g}_a$ multiplies one component of **t** by a nonzero scalar λ. In particular, $\mathbf{g}_a$ maps the unit coordinate vectors as follows:

$$\mathbf{g}_a(\mathbf{u}_k) = \lambda \mathbf{u}_k \quad \text{for some } k, \qquad \mathbf{g}_a(\mathbf{u}_i) = \mathbf{u}_i \quad \text{for all } i \neq k.$$

The matrix of $\mathbf{g}_a$ can be obtained by multiplying the entries in the kth row of the identity matrix by λ. Also, det $\mathbf{g}_a = \lambda$.

Type b: $\mathbf{g}_b(t_1, \ldots, t_k, \ldots, t_n) = (t_1, \ldots, t_k + t_j, \ldots, t_n)$, where $j \neq k$. Thus, $\mathbf{g}_b$ replaces one component of **t** by itself plus another. In particular, $\mathbf{g}_b$ maps the coordinate vectors as follows:

$$\mathbf{g}_b(\mathbf{u}_k) = \mathbf{u}_k + \mathbf{u}_j \quad \text{for some fixed } k \text{ and } j, \quad k \neq j,$$

$$\mathbf{g}_b(\mathbf{u}_i) = \mathbf{u}_i \quad \text{for all } i \neq k.$$

The matrix $\mathbf{g}_b$ can be obtained from the identity matrix by replacing the kth row of I by the kth row of I plus the jth row of I. Also, det $\mathbf{g}_b = 1$.

Type c: $\mathbf{g}_c(t_1, \ldots, t_i, \ldots, t_j, \ldots, t_n) = (t_1, \ldots, t_j, \ldots, t_i, \ldots, t_n)$, where $i \neq j$. That is, $\mathbf{g}_c$ interchanges the ith and jth components of **t** for some i and j with $i \neq j$. In particular, $\mathbf{g}(\mathbf{u}_i) = \mathbf{u}_j$, $\mathbf{g}(\mathbf{u}_j) = \mathbf{u}_i$, and $\mathbf{g}(\mathbf{u}_k) = \mathbf{u}_k$ for all $k \neq i$, $k \neq j$. The matrix of $\mathbf{g}_c$ is the identity matrix with the ith and jth rows interchanged. In this case det $\mathbf{g}_c = -1$.

The inverse of an elementary transformation is another of the same type. The matrix of an elementary transformation is called an *elementary matrix*. Every nonsingular matrix A can be transformed to the identity matrix I by multiplying A on the left by a succession of elementary matrices. (This is the familiar Gauss–Jordan process of linear algebra.) Thus,

$$I = T_1 T_2 \cdots T_r A,$$

where each T_k is an elementary matrix, Hence,

$$A = T_r^{-1} \cdots T_2^{-1} T_1^{-1}.$$

If $A = m(\mathbf{g})$, this gives a corresponding factorization of $\mathbf{g}$ as a composition of elementary transformations.

15.10 THE TRANSFORMATION FORMULA FOR MULTIPLE INTEGRALS

The rest of this chapter is devoted to a proof of the following transformation formula for multiple integrals.

Theorem 15.11. *Let T be an open subset of $\mathbf{R}^n$ and let $\mathbf{g}$ be a coordinate transformation on T. Let f be a real-valued function defined on the image $\mathbf{g}(T)$ and assume that the Lebesgue integral $\int_{\mathbf{g}(T)} f(\mathbf{x})\,d\mathbf{x}$ exists. Then the Lebesgue integral $\int_T f[\mathbf{g}(\mathbf{t})]\,|J_{\mathbf{g}}(\mathbf{t})|\,d\mathbf{t}$ also exists and we have*

$$\int_{\mathbf{g}(T)} f(\mathbf{x})\,d\mathbf{x} = \int_T f[\mathbf{g}(\mathbf{t})]\,|J_{\mathbf{g}}(\mathbf{t})|\,d\mathbf{t}. \tag{14}$$

The proof of Theorem 15.11 is divided into three parts. Part 1 shows that the formula holds for every linear coordinate transformation $\boldsymbol{\alpha}$. As a corollary we obtain the relation

$$\mu[\boldsymbol{\alpha}(A)] = |\det \boldsymbol{\alpha}|\,\mu(A),$$

for every subset A of $\mathbf{R}^n$ with finite Lebesgue measure. In part 2 we consider a general coordinate transformation $\mathbf{g}$ and show that (14) holds when f is the characteristic function of a compact cube. This gives us

$$\mu(K) = \int_{\mathbf{g}^{-1}(K)} |J_{\mathbf{g}}(\mathbf{t})|\,d\mathbf{t}, \tag{15}$$

for every compact cube K in $\mathbf{g}(T)$. This is the lengthiest part of the proof. In part 3 we use Equation (15) to deduce (14) in its general form.

15.11 PROOF OF THE TRANSFORMATION FORMULA FOR LINEAR COORDINATE TRANSFORMATIONS

Theorem 15.12. *Let $\boldsymbol{\alpha} : \mathbf{R}^n \to \mathbf{R}^n$ be a linear coordinate transformation. If the Lebesgue integral $\int_{\mathbf{R}^n} f(\mathbf{x})\,d\mathbf{x}$ exists, then the Lebesgue integral $\int_{\mathbf{R}^n} f[\boldsymbol{\alpha}(\mathbf{t})]\,|J_{\boldsymbol{\alpha}}(\mathbf{t})|\,d\mathbf{t}$ also exists, and the two integrals are equal.*

Proof. First we note that if the theorem is true for $\boldsymbol{\alpha}$ and $\boldsymbol{\beta}$, then it is also true for the composition $\boldsymbol{\gamma} = \boldsymbol{\alpha} \circ \boldsymbol{\beta}$ because

$$\int_{\mathbf{R}^n} f(\mathbf{x})\,d\mathbf{x} = \int_{\mathbf{R}^n} f[\boldsymbol{\alpha}(\mathbf{t})]\,|J_{\boldsymbol{\alpha}}(\mathbf{t})|\,d\mathbf{t} = \int_{\mathbf{R}^n} f(\boldsymbol{\alpha}[\boldsymbol{\beta}(\mathbf{t})])\,|J_{\boldsymbol{\alpha}}[\boldsymbol{\beta}(\mathbf{t})]|\,|J_{\boldsymbol{\beta}}(\mathbf{t})|\,d\mathbf{t}$$

$$= \int_{\mathbf{R}^n} f[\boldsymbol{\gamma}(\mathbf{t})]\,|J_{\boldsymbol{\gamma}}(\mathbf{t})|\,d\mathbf{t},$$

since $J_{\boldsymbol{\gamma}}(\mathbf{t}) = J_{\boldsymbol{\alpha}}[\boldsymbol{\beta}(\mathbf{t})]\,J_{\boldsymbol{\beta}}(\mathbf{t})$.

Therefore, since every nonsingular linear transformation α is a composition of elementary transformations, it suffices to prove the theorem for every elementary transformation. It also suffices to assume $f \geq 0$.

Suppose α is of *type a*. For simplicity, assume that α multiplies the last component of $\mathbf{t}$ by a nonzero scalar λ, say

$$\alpha(t_1, \ldots, t_n) = (t_1, \ldots, t_{n-1}, \lambda t_n).$$

Then $|J_\alpha(\mathbf{t})| = |\det \alpha| = |\lambda|$. We apply Fubini's theorem to write the integral of f over $\mathbf{R}^n$ as the iteration of an $(n-1)$-dimensional integral over $\mathbf{R}^{n-1}$ and a one-dimensional integral over $\mathbf{R}^1$. For the integral over $\mathbf{R}^1$ we use Theorem 10.17(b) and (c), and we obtain

$$\int_{\mathbf{R}^n} f(\mathbf{x})\, d\mathbf{x} = \int_{\mathbf{R}^{n-1}} \left[\int_{-\infty}^{\infty} f(x_1, \ldots, x_n)\, dx_n \right] dx_1 \cdots dx_{n-1}$$

$$= \int_{\mathbf{R}^{n-1}} \left[|\lambda| \int_{-\infty}^{\infty} f(x_1, \ldots, x_{n-1}, \lambda t_n)\, dt_n \right] dx_1 \cdots dx_{n-1}$$

$$= \int_{\mathbf{R}^{n-1}} \left[\int_{-\infty}^{\infty} f[\alpha(\mathbf{t})]\, |J_\alpha(\mathbf{t})|\, dt_n \right] dt_1 \cdots dt_{n-1}$$

$$= \int_{\mathbf{R}^n} f[\alpha(\mathbf{t})]\, |J_\alpha(\mathbf{t})|\, d\mathbf{t},$$

where in the last step we use the Tonelli-Hobson theorem. This proves the theorem if α is of *type a*. If α is of *type b*, the proof is similar except that we use Theorem 10.17(a) in the one-dimensional integral. In this case $|J_\alpha(\mathbf{t})| = 1$. Finally, if α is of *type c* we simply use Fubini's theorem to interchange the order of integration over the ith and jth coordinates. Again, $|J_\alpha(\mathbf{t})| = 1$ in this case.

As an immediate corollary we have:

Theorem 15.13. *If $\alpha : \mathbf{R}^n \to \mathbf{R}^n$ is a linear coordinate transformation and if A is any subset of $\mathbf{R}^n$ such that the Lebesgue integral $\int_{\alpha(A)} f(\mathbf{x})\, d\mathbf{x}$ exists, then the Lebesgue integral $\int_A f[\alpha(\mathbf{t})]\, |J_\alpha(\mathbf{t})|\, d\mathbf{t}$ also exists, and the two are equal.*

Proof. Let $\tilde{f}(\mathbf{x}) = f(\mathbf{x})$ if $\mathbf{x} \in \alpha(A)$, and let $\tilde{f}(\mathbf{x}) = 0$ otherwise. Then

$$\int_{\alpha(A)} f(\mathbf{x})\, d\mathbf{x} = \int_{\mathbf{R}^n} \tilde{f}(\mathbf{x})\, d\mathbf{x} = \int_{\mathbf{R}^n} \tilde{f}[\alpha(\mathbf{t})]\, |J_\alpha(\mathbf{t})|\, d\mathbf{t} = \int_A f[\alpha(\mathbf{t})]\, |J_\alpha(\mathbf{t})|\, d\mathbf{t}.$$

As a corollary of Theorem 15.13 we have the following relation between the measure of A and the measure of $\alpha(A)$.

Theorem 15.14. *Let* $\alpha : \mathbf{R}^n \to \mathbf{R}^n$ *be a linear coordinate transformation. If A is a subset of $\mathbf{R}^n$ with finite Lebesgue measure $\mu(A)$, then $\alpha(A)$ also has finite Lebesgue measure and*

$$\mu[\alpha(A)] = |\det \alpha| \, \mu(A). \tag{16}$$

Proof. Write $A = \alpha^{-1}(B)$, where $B = \alpha(A)$. Since α^{-1} is also a coordinate transformation, we find

$$\mu(A) = \int_A d\mathbf{x} = \int_{\alpha^{-1}(B)} d\mathbf{x} = \int_B |\det \alpha^{-1}| \, d\mathbf{t} = |\det \alpha^{-1}| \, \mu(B).$$

This proves (16) since $B = \alpha(A)$ and $\det(\alpha^{-1}) = (\det \alpha)^{-1}$.

Theorem 15.15. *If A is a compact Jordan-measurable subset of $\mathbf{R}^n$, then for any linear coordinate transformation $\alpha : \mathbf{R}^n \to \mathbf{R}^n$ the image $\alpha(A)$ is a compact Jordan-measurable set and its content is given by*

$$c[\alpha(A)] = |\det \alpha| \, c(A).$$

Proof. The set $\alpha(A)$ is compact because α is continuous on A. To prove the theorem we argue as in the proof of Theorem 15.14. In this case, however, all the integrals exist both as Lebesgue integrals and as Riemann integrals.

15.12 PROOF OF THE TRANSFORMATION FORMULA FOR THE CHARACTERISTIC FUNCTION OF A COMPACT CUBE

This section contains part 2 of the proof of Theorem 15.11. Throughout the section we assume that **g** is a coordinate transformation on an open set T in $\mathbf{R}^n$. Our purpose is to prove that

$$\mu(K) = \int_{\mathbf{g}^{-1}(K)} |J_{\mathbf{g}}(T)| \, d\mathbf{t},$$

for every compact cube K in T. The auxiliary results needed to prove this formula are labelled as lemmas.

To help simplify the details, we introduce some convenient notation. Instead of the usual Euclidean metric for $\mathbf{R}^n$ we shall use the metric d given by

$$d(\mathbf{x}, \mathbf{y}) = \max_{1 \le i \le n} |x_i - y_i|.$$

This metric was introduced in Example 9, Section 3.13. In this section only we shall write $\|\mathbf{x} - \mathbf{y}\|$ for $d(\mathbf{x}, \mathbf{y})$.

With this metric, a ball $B(\mathbf{a}; r)$ with center $\mathbf{a}$ and radius r is an n-dimensional cube with center $\mathbf{a}$ and edge-length $2r$; that is, $B(\mathbf{a}; r)$ is the cartesian product of n one-dimensional intervals, each of length $2r$. The measure of such a cube is $(2r)^n$, the product of the edge-lengths.

If $\alpha : \mathbf{R}^n \to \mathbf{R}^n$ is a linear transformation represented by a matrix (a_{ij}), so that

$$\alpha(\mathbf{x}) = \left(\sum_{j=1}^{n} a_{1j}x_j, \ldots, \sum_{j=1}^{n} a_{nj}x_j \right),$$

then

$$\|\alpha(\mathbf{x})\| = \max_{1 \le i \le n} \left| \sum_{j=1}^{n} a_{ij}x_j \right| \le \|\mathbf{x}\| \max_{1 \le i \le n} \sum_{j=1}^{n} |a_{ij}|. \tag{17}$$

We also define

$$\|\alpha\| = \max_{1 \le i \le n} \sum_{j=1}^{n} |a_{ij}|. \tag{18}$$

This defines a metric $\|\alpha - \beta\|$ on the space of all linear transformations from $\mathbf{R}^n$ to $\mathbf{R}^n$. The first lemma gives some properties of this metric.

Lemma 1. *Let α and β denote linear transformations from $\mathbf{R}^n$ to $\mathbf{R}^n$. Then we have:*

a) $\|\alpha\| = \|\alpha(\mathbf{x})\|$ *for some $\mathbf{x}$ with $\|\mathbf{x}\| = 1$.*

b) $\|\alpha(\mathbf{x})\| \le \|\alpha\| \, \|\mathbf{x}\|$ *for all $\mathbf{x}$ in $\mathbf{R}^n$.*

c) $\|\alpha \circ \beta\| \le \|\alpha\| \, \|\beta\|$.

d) $\|\mathbf{I}\| = 1$, *where $\mathbf{I}$ is the identity transformation.*

Proof. Suppose that $\max_{1 \le i \le n} \sum_{j=1}^{n} |a_{ij}|$ is attained for $i = p$. Take $x_p = 1$ if $a_{pj} \ge 0$, $x_p = -1$ if $a_{pj} < 0$, and $x_j = 0$ if $j \ne p$. Then $\|\mathbf{x}\| = 1$ and $\|\alpha\| = \|\alpha(\mathbf{x})\|$, which proves (a).

Part (b) follows at once from (17) and (18). To prove (c) we use (b) to write

$$\|(\alpha \circ \beta)(\mathbf{x})\| = \|\alpha(\beta(\mathbf{x}))\| \le \|\alpha\| \, \|\beta(\mathbf{x})\| \le \|\alpha\| \, \|\beta\| \, \|\mathbf{x}\|.$$

Taking $\mathbf{x}$ with $\|\mathbf{x}\| = 1$ so that $\|(\alpha \circ \beta)(\mathbf{x})\| = \|\alpha \circ \beta\|$, we obtain (c).

Finally, if $\mathbf{I}$ is the identity transformation, then each sum $\sum_{j=1}^{n} |a_{ij}| = 1$ in (18) so $\|\mathbf{I}\| = 1$.

The coordinate transformation $\mathbf{g}$ is differentiable on T, so for each $\mathbf{t}$ in T the total derivative $\mathbf{g}'(\mathbf{t})$ is a linear transformation from $\mathbf{R}^n$ to $\mathbf{R}^n$ represented by the Jacobian matrix $\mathbf{Dg}(\mathbf{t}) = (D_j g_i(\mathbf{t}))$. Therefore, taking $\alpha = \mathbf{g}'(\mathbf{t})$ in (18), we find

$$\|\mathbf{g}'(\mathbf{t})\| = \max_{1 \le i \le n} \sum_{j=1}^{n} |D_j g_i(\mathbf{t})|.$$

We note that $\|\mathbf{g}'(\mathbf{t})\|$ is a continuous function of $\mathbf{t}$ since all the partial derivatives $D_j g_i$ are continuous on T.

If Q is a compact subset of T, each function $D_j g_i$ is bounded on Q; hence $\|\mathbf{g}'(\mathbf{t})\|$ is also bounded on Q, and we define

$$\lambda_{\mathbf{g}}(Q) = \sup_{\mathbf{t} \in Q} \|\mathbf{g}'(\mathbf{t})\| = \sup_{\mathbf{t} \in Q} \left\{ \max_{1 \le i \le n} \sum_{j=1}^{n} |D_j g_i(\mathbf{t})| \right\}. \tag{19}$$

The next lemma states that the image $\mathbf{g}(Q)$ of a cube Q of edge-length $2r$ lies in another cube of edge-length $2r\lambda_{\mathbf{g}}(Q)$.

Lemma 2. *Let* $Q = \{\mathbf{x} : \|\mathbf{x} - \mathbf{a}\| \leq r\}$ *be a compact cube of edge-length* $2r$ *lying in* T. *Then for each* $\mathbf{x}$ *in* Q *we have*

$$\|\mathbf{g}(\mathbf{x}) - \mathbf{g}(\mathbf{a})\| \leq r\lambda_{\mathbf{g}}(Q). \tag{20}$$

Therefore $\mathbf{g}(Q)$ *lies in a cube of edge-length* $2r\lambda_{\mathbf{g}}(Q)$.

Proof. By the Mean-Value theorem for real-valued functions we have

$$g_i(\mathbf{x}) - g_i(\mathbf{a}) = \nabla g_i(\mathbf{z}_i) \cdot (\mathbf{x} - \mathbf{a}) = \sum_{j=1}^{n} D_j g_i(\mathbf{z}_i)(x_j - a_j),$$

where $\mathbf{z}_i$ lies on the line segment joining $\mathbf{x}$ and $\mathbf{a}$. Therefore

$$|g_i(\mathbf{x}) - g_i(\mathbf{a})| \leq \sum_{j=1}^{n} |D_j g_i(\mathbf{z}_i)| \, |x_j - a_j| \leq \|\mathbf{x} - \mathbf{a}\| \sum_{j=1}^{n} |D_j g_i(\mathbf{z}_i)| \leq r\lambda_{\mathbf{g}}(Q),$$

and this implies (20).

NOTE. Inequality (20) shows that $\mathbf{g}(Q)$ lies inside a cube of content

$$(2r\lambda_{\mathbf{g}}(Q))^n = \{\lambda_{\mathbf{g}}(Q)\}^n c(Q).$$

Lemma 3. *If* A *is any compact Jordan-measurable subset of* T, *then* $\mathbf{g}(A)$ *is a compact Jordan-measurable subset of* $\mathbf{g}(T)$.

Proof. The compactness of $\mathbf{g}(A)$ follows from the continuity of $\mathbf{g}$. Since A is Jordan-measurable, its boundary ∂A has content zero. Also, $\partial(\mathbf{g}(A)) = \mathbf{g}(\partial A)$, since $\mathbf{g}$ is one-to-one and continuous. Therefore, to complete the proof, it suffices to show that $\mathbf{g}(\partial A)$ has content zero.

Given $\varepsilon > 0$, there is a finite number of open intervals $A_1, \ldots, A_m$ lying in T, the sum of whose measures is $< \varepsilon$, such that $\partial A \subseteq \bigcup_{i=1}^{m} A_i$. By Theorem 15.1, this union can also be expressed as a union $U(\varepsilon)$ of a countable disjoint collection of cubes, the sum of whose measures is $< \varepsilon$. If $\varepsilon < 1$ we can assume that each cube in $U(\varepsilon)$ is contained in $U(1)$. (If not, intersect the cubes in $U(\varepsilon)$ with $U(1)$ and apply Theorem 15.1 again.) Since ∂A is compact, a finite subcollection of the cubes in $U(\varepsilon)$ covers ∂A, say $Q_1, \ldots, Q_k$. By Lemma 2, the image $\mathbf{g}(\bar{Q}_i)$ lies in a cube of measure $\{\lambda_{\mathbf{g}}(\bar{Q}_i)\}^n c(Q_i)$. Let $\lambda = \lambda_{\mathbf{g}}(\overline{U(1)})$. Then $\lambda_{\mathbf{g}}(\bar{Q}_i) \leq \lambda$ since $\bar{Q}_i \subseteq \overline{U(1)}$. Thus $\mathbf{g}(\partial A)$ is covered by a finite number of cubes, the sum of whose measures does not exceed $\lambda^n \sum_{i=1}^{k} c(Q_i) < \varepsilon\lambda^n$. Since this holds for every $\varepsilon < 1$, it follows that $\mathbf{g}(\partial A)$ has Jordan content 0, so $\mathbf{g}(A)$ is Jordan-measurable.

The next lemma relates the content of a cube Q with that of its image $\mathbf{g}(Q)$.

Lemma 4. *Let Q be a compact cube in T and let $\mathbf{h} = \alpha \circ \mathbf{g}$, where $\alpha : \mathbf{R}^n \to \mathbf{R}^n$ is any nonsingular linear transformation. Then*

$$c[\mathbf{g}(Q)] \leq |\det \alpha|^{-1} \{\lambda_\mathbf{h}(Q)\}^n c(Q). \tag{21}$$

Proof. From Lemma 2 we have $c[\mathbf{g}(Q)] \leq \{\lambda_\mathbf{g}(Q)\}^n c(Q)$. Applying this inequality to the coordinate transformation $\mathbf{h}$, we find

$$c[\mathbf{h}(Q)] \leq \{\lambda_\mathbf{h}(Q)\}^n c(Q).$$

But by Theorem 15.15 we have $c[\mathbf{h}(Q)] = c[\alpha(\mathbf{g}(Q))] = |\det \alpha| \, c[\mathbf{g}(Q)]$, so

$$c[\mathbf{g}(Q)] = |\det \alpha|^{-1} c[\mathbf{h}(Q)] \leq |\det \alpha|^{-1} \{\lambda_\mathbf{h}(Q)\}^n c(Q).$$

Lemma 5. *Let Q be a compact cube in T. Then for every $\varepsilon > 0$, there is a $\delta > 0$ such that if $\mathbf{t} \in Q$ and $\mathbf{a} \in Q$ we have*

$$\|\mathbf{g}'(\mathbf{a})^{-1} \circ \mathbf{g}'(\mathbf{t})\| < 1 + \varepsilon \qquad \text{whenever } \|\mathbf{t} - \mathbf{a}\| < \delta. \tag{22}$$

Proof. The function $\|\mathbf{g}'(\mathbf{t})^{-1}\|$ is continuous and hence bounded on Q, say $\|\mathbf{g}'(\mathbf{t})^{-1}\| < M$ for all $\mathbf{t}$ in Q where $M > 0$. By the continuity of $\|\mathbf{g}'(\mathbf{t})\|$, there is a $\delta > 0$ such that

$$\|\mathbf{g}'(\mathbf{t}) - \mathbf{g}'(\mathbf{a})\| < \frac{\varepsilon}{M} \qquad \text{whenever } \|\mathbf{t} - \mathbf{a}\| < \delta.$$

If $\mathbf{I}$ denotes the identity transformation, then

$$\mathbf{g}'(\mathbf{a})^{-1} \circ \mathbf{g}'(\mathbf{t}) - \mathbf{I}(\mathbf{t}) = \mathbf{g}'(\mathbf{a})^{-1} \circ \{\mathbf{g}'(\mathbf{t}) - \mathbf{g}'(\mathbf{a})\},$$

so if $\|\mathbf{t} - \mathbf{a}\| < \delta$ we have

$$\|\mathbf{g}'(\mathbf{a})^{-1} \circ \mathbf{g}'(\mathbf{t}) - \mathbf{I}(\mathbf{t})\| \leq \|\mathbf{g}'(\mathbf{a})^{-1}\| \, \|\mathbf{g}'(\mathbf{t}) - \mathbf{g}'(\mathbf{a})\| < M \frac{\varepsilon}{M} = \varepsilon.$$

The triangle inequality gives us $\|\alpha\| \leq \|\beta\| + \|\alpha - \beta\|$. Taking

$$\alpha = \mathbf{g}'(\mathbf{a})^{-1} \circ \mathbf{g}'(\mathbf{t}) \qquad \text{and} \qquad \beta = \mathbf{I}(\mathbf{t}),$$

we obtain (22).

Lemma 6. *Let Q be a compact cube in T. Then we have*

$$c[\mathbf{g}(Q)] \leq \int_Q |J_\mathbf{g}(\mathbf{t})| \, d\mathbf{t}.$$

Proof. The integral on the right exists as a Riemann integral because the integrand is continuous and bounded on Q. Therefore, given $\varepsilon > 0$, there is a partition P_ε of Q such that for every Riemann sum $S(P, |J_\mathbf{g}|)$ with P finer than P_ε we have

$$\left| S(P, |J_\mathbf{g}|) - \int_Q |J_\mathbf{g}(\mathbf{t})| \, d\mathbf{t} \right| < \varepsilon.$$

Take such a partition P into a finite number of cubes $Q_1, \ldots, Q_m$, each of which has edge-length $< \delta$, where δ is the number (depending on ε) given by Lemma 5. Let $\mathbf{a}_i$ denote the center of Q_i and apply Lemma 4 to Q_i with $\alpha = \mathbf{g}'(\mathbf{a}_i)^{-1}$ to obtain the inequality

$$c[\mathbf{g}(Q_i)] \leq |\det \mathbf{g}'(\mathbf{a}_i)| \{\lambda_\mathbf{h}(Q_i)\}^n c(Q_i), \qquad (23)$$

where $\mathbf{h} = \alpha \circ \mathbf{g}$. By the chain rule we have $\mathbf{h}'(\mathbf{t}) = \alpha'(\mathbf{x}) \circ \mathbf{g}'(\mathbf{t})$, where $\mathbf{x} = \mathbf{g}(\mathbf{t})$. But $\alpha'(\mathbf{x}) = \alpha$ since α is a linear function, so

$$\mathbf{h}'(\mathbf{t}) = \alpha \circ \mathbf{g}'(\mathbf{t}) = \mathbf{g}'(\mathbf{a}_i)^{-1} \circ \mathbf{g}'(\mathbf{t}).$$

But by Lemma 5 we have $\|\mathbf{h}'(\mathbf{t})\| < 1 + \varepsilon$ if $\mathbf{t} \in Q_i$, so

$$\lambda_\mathbf{h}(Q_i) = \sup_{\mathbf{t} \in Q_i} \|\mathbf{h}'(\mathbf{t})\| \leq 1 + \varepsilon.$$

Thus (23) gives us

$$c[\mathbf{g}(Q_i)] \leq |\det \mathbf{g}'(\mathbf{a}_i)| (1 + \varepsilon)^n c(Q_i).$$

Summing over all i, we find

$$c[\mathbf{g}(Q)] \leq (1 + \varepsilon)^n \sum_{i=1}^{m} |\det \mathbf{g}'(\mathbf{a}_i)| c(Q_i).$$

Since $\det \mathbf{g}'(\mathbf{a}_i) = J_\mathbf{g}(\mathbf{a}_i)$, the sum on the right is a Riemann sum $S(P, |J_\mathbf{g}|)$, and since $S(P, |J_\mathbf{g}|) < \int_Q |J_\mathbf{g}(\mathbf{t})| \, d\mathbf{t} + \varepsilon$, we find

$$c[\mathbf{g}(Q)] \leq (1 + \varepsilon)^n \left\{ \int_Q |J_\mathbf{g}(\mathbf{t})| \, d\mathbf{t} + \varepsilon \right\}.$$

But ε is arbitrary, so this implies $c[\mathbf{g}(Q)] \leq \int_Q |J_\mathbf{g}(\mathbf{t})| \, d\mathbf{t}$.

Lemma 7. *Let K be a compact cube in $\mathbf{g}(T)$. Then*

$$\mu(K) \leq \int_{\mathbf{g}^{-1}(K)} |J_\mathbf{g}(\mathbf{t})| \, d\mathbf{t}. \qquad (24)$$

Proof. The integral exists as a Riemann integral because the integrand is continuous on the compact set $\mathbf{g}^{-1}(K)$. Also, by Lemma 3, the integral over $\mathbf{g}^{-1}(K)$ is equal to that over the interior of $\mathbf{g}^{-1}(K)$. By Theorem 15.1 we can write

$$\text{int } \mathbf{g}^{-1}(K) = \bigcup_{i=1}^{\infty} A_i,$$

where $\{A_1, A_2, \ldots\}$ is a countable disjoint collection of cubes whose closure lies in the interior of $\mathbf{g}^{-1}(K)$. Thus, int $\mathbf{g}^{-1}(K) = \bigcup_{i=1}^{\infty} Q_i$ where each Q_i is the closure of A_i. Since the integral in (24) is also a Lebesgue integral, we can use countable additivity along with Lemma 6 to write

$$\int_{\mathbf{g}^{-1}(K)} |J_\mathbf{g}(\mathbf{t})| \, d\mathbf{t} = \sum_{i=1}^{\infty} \int_{Q_i} |J_\mathbf{g}(\mathbf{t})| \, d\mathbf{t} \geq \sum_{i=1}^{\infty} \mu[\mathbf{g}(Q_i)] = \mu\left(\bigcup_{i=1}^{\infty} \mathbf{g}(Q_i) \right) = \mu(K).$$

Lemma 8. *Let K be a compact cube in $\mathbf{g}(T)$. Then for any nonnegative upper function f which is bounded on K, the integral $\int_{\mathbf{g}^{-1}(K)} f[\mathbf{g}(\mathbf{t})] |J_{\mathbf{g}}(\mathbf{t})| \, d\mathbf{t}$ exists, and we have the inequality*

$$\int_K f(\mathbf{x}) \, d\mathbf{x} \leq \int_{\mathbf{g}^{-1}(K)} f[\mathbf{g}(\mathbf{t})] |J_{\mathbf{g}}(\mathbf{t})| \, d\mathbf{t}. \tag{25}$$

Proof. Let s be any nonnegative step function on K. Then there is a partition of K into a finite number of cubes $K_1, \ldots, K_r$ such that s is constant on the interior of each K_i, say $s(\mathbf{x}) = a_i \geq 0$ if $\mathbf{x} \in \text{int } K_i$. Apply (24) to each cube K_i, multiply by a_i and add, to obtain

$$\int_K s(\mathbf{x}) \, d\mathbf{x} \leq \int_{\mathbf{g}^{-1}(K)} s[\mathbf{g}(\mathbf{t})] |J_{\mathbf{g}}(\mathbf{t})| \, d\mathbf{t}. \tag{26}$$

Now let $\{s_k\}$ be an increasing sequence of nonnegative step functions which converges almost everywhere on K to the upper function f. Then (26) holds for each s_k, and we let $k \to \infty$ to obtain (25). The existence of the integral on the right follows from the Lebesgue bounded convergence theorem since both $f[\mathbf{g}(\mathbf{t})]$ and $|J_{\mathbf{g}}(\mathbf{t})|$ are bounded on the compact set $\mathbf{g}^{-1}(K)$.

Theorem 15.16. *Let K be a compact cube in $\mathbf{g}(T)$. Then we have*

$$\mu(K) = \int_{\mathbf{g}^{-1}(K)} |J_{\mathbf{g}}(\mathbf{t})| \, d\mathbf{t}. \tag{27}$$

Proof. In view of Lemma 7, it suffices to prove the inequality

$$\int_{\mathbf{g}^{-1}(K)} |J_{\mathbf{g}}(\mathbf{t})| \, d\mathbf{t} \leq \mu(K). \tag{28}$$

As in the proof of Lemma 7, we write

$$\text{int } \mathbf{g}^{-1}(K) = \bigcup_{i=1}^{\infty} A_i = \bigcup_{i=1}^{\infty} Q_i,$$

where $\{A_1, A_2, \ldots\}$ is a countable disjoint collection of cubes and Q_i is the closure of A_i. Then

$$\int_{\mathbf{g}^{-1}(K)} |J_{\mathbf{g}}(\mathbf{t})| \, d\mathbf{t} = \sum_{i=1}^{\infty} \int_{Q_i} |J_{\mathbf{g}}(\mathbf{t})| \, d\mathbf{t}. \tag{29}$$

Now we apply Lemma 8 to each integral $\int_{Q_i} |J_{\mathbf{g}}(\mathbf{t})| \, d\mathbf{t}$, taking $f = |J_{\mathbf{g}}|$ and using the coordinate transformation $\mathbf{h} = \mathbf{g}^{-1}$. This gives us the inequality

$$\int_{Q_i} |J_{\mathbf{g}}(\mathbf{t})| \, d\mathbf{t} \leq \int_{\mathbf{g}(Q_i)} |J_{\mathbf{g}}[\mathbf{h}(\mathbf{u})]| \, |J_{\mathbf{h}}(\mathbf{u})| \, d\mathbf{u} = \int_{\mathbf{g}(Q_i)} d\mathbf{u} = \mu[\mathbf{g}(Q_i)],$$

which, when used in (29) gives (28).

15.13 COMPLETION OF THE PROOF OF THE TRANSFORMATION FORMULA

Now it is relatively easy to complete the proof of the formula

$$\int_{g(T)} f(\mathbf{x})\, d\mathbf{x} = \int_{T} f[\mathbf{g}(\mathbf{t})]\, |J_{\mathbf{g}}(\mathbf{t})|\, d\mathbf{t}, \tag{30}$$

under the conditions stated in Theorem 15.11. That is, we assume that T is an open subset of $\mathbf{R}^n$, that $\mathbf{g}$ is a coordinate transformation on T, and that the integral on the left of (30) exists. We are to prove that the integral on the right also exists and that the two are equal. This will be deduced from the special case in which the integral on the left is extended over a cube K.

Theorem 15.17. *Let K be a compact cube in $\mathbf{g}(T)$ and assume the Lebesgue integral $\int_K f(\mathbf{x})\, d\mathbf{x}$ exists. Then the Lebesgue integral $\int_{\mathbf{g}^{-1}(K)} f[\mathbf{g}(\mathbf{t})]\, |J_{\mathbf{g}}(\mathbf{t})|\, d\mathbf{t}$ also exists, and the two are equal.*

Proof. It suffices to prove the theorem when f is an upper function on K. Then there is an increasing sequence of step functions $\{s_k\}$ such that $s_k \to f$ almost everywhere on K. By Theorem 15.16 we have

$$\int_K s_k(\mathbf{x})\, d\mathbf{x} = \int_{\mathbf{g}^{-1}(K)} s_k[\mathbf{g}(\mathbf{t})]\, |J_{\mathbf{g}}(\mathbf{t})|\, d\mathbf{t},$$

for each step function s_k. When $k \to \infty$, we have $\int_K s_k(\mathbf{x})\, d\mathbf{x} \to \int_K f(\mathbf{x})\, d\mathbf{x}$. Now let

$$f_k(\mathbf{t}) = \begin{cases} s_k[\mathbf{g}(\mathbf{t})]\, |J_{\mathbf{g}}(\mathbf{t})| & \text{if } \mathbf{t} \in \mathbf{g}^{-1}(K), \\ 0 & \text{if } \mathbf{t} \in \mathbf{R}^n - \mathbf{g}^{-1}(K). \end{cases}$$

Then

$$\int_{\mathbf{R}^n} f_k(\mathbf{t})\, d\mathbf{t} = \int_{\mathbf{g}^{-1}(K)} s_k[\mathbf{g}(\mathbf{t})]\, |J_{\mathbf{g}}(\mathbf{t})|\, d\mathbf{t} = \int_K s_k(\mathbf{x})\, d\mathbf{x},$$

so

$$\lim_{k \to \infty} \int_{\mathbf{R}^n} f_k(\mathbf{t})\, d\mathbf{t} = \lim_{k \to \infty} \int_K s_k(\mathbf{x})\, d\mathbf{x} = \int_K f(\mathbf{x})\, d\mathbf{x}.$$

By the Levi theorem (the analog of Theorem 10.24), the sequence $\{f_k\}$ converges almost everywhere on $\mathbf{R}^n$ to a function in $L(\mathbf{R}^n)$. Since we have

$$\lim_{k \to \infty} f_k(\mathbf{t}) = \begin{cases} f[\mathbf{g}(\mathbf{t})]\, |J_{\mathbf{g}}(\mathbf{t})| & \text{if } \mathbf{t} \in \mathbf{g}^{-1}(K), \\ 0 & \text{if } \mathbf{t} \in \mathbf{R}^n - \mathbf{g}^{-1}(K), \end{cases}$$

almost everywhere on $\mathbf{R}^n$, it follows that the integral $\int_{\mathbf{g}^{-1}(K)} f[\mathbf{g}(\mathbf{t})]\, |J_{\mathbf{g}}(\mathbf{t})|\, d\mathbf{t}$ exists and equals $\int_K f(\mathbf{x})\, d\mathbf{x}$. This completes the proof of Theorem 15.17.

Proof of Theorem 15.11. Now assume that the integral $\int_{\mathbf{g}(T)} f(\mathbf{x})\, d\mathbf{x}$ exists. Since $\mathbf{g}(T)$ is open, we can write

$$\mathbf{g}(T) = \bigcup_{i=1}^{\infty} A_i,$$

where $\{A_1, A_2, \ldots\}$ is a countable disjoint collection of cubes whose closure lies in $\mathbf{g}(T)$. Let K_i denote the closure of A_i. Using countable additivity and Theorem 15.17 we have

$$\int_{\mathbf{g}(T)} f(\mathbf{x})\, d\mathbf{x} = \sum_{i=1}^{\infty} \int_{K_i} f(\mathbf{x})\, d\mathbf{x}$$

$$= \sum_{i=1}^{\infty} \int_{\mathbf{g}^{-1}(K_i)} f[\mathbf{g}(\mathbf{t})]\, |J_{\mathbf{g}}(\mathbf{t})|\, d\mathbf{t}$$

$$= \int_{T} f[\mathbf{g}(\mathbf{t})]\, |J_{\mathbf{g}}(\mathbf{t})|\, d\mathbf{t}.$$

EXERCISES

15.1 If $f \in L(T)$, where T is the triangular region in $\mathbf{R}^2$ with vertices at $(0, 0)$, $(1, 0)$, and $(0, 1)$, prove that

$$\int_{T} f(x, y)\, d(x, y) = \int_{0}^{1} \left[\int_{0}^{x} f(x, y)\, dy \right] dx = \int_{0}^{1} \left[\int_{y}^{1} f(x, y)\, dx \right] dy.$$

15.2 For fixed c, $0 < c < 1$, define f on $\mathbf{R}^2$ as follows:

$$f(x, y) = \begin{cases} (1 - y)^c/(x - y)^c & \text{if } 0 \le y < x, 0 < x < 1, \\ 0 & \text{otherwise.} \end{cases}$$

Prove that $f \in L(\mathbf{R}^2)$ and calculate the double integral $\int_{\mathbf{R}^2} f(x, y)\, d(x, y)$.

15.3 Let S be a measurable subset of $\mathbf{R}^2$ with finite measure $\mu(S)$. Using the notation of Definition 15.4, prove that

$$\mu(S) = \int_{-\infty}^{\infty} \mu(S^x)\, dx = \int_{-\infty}^{\infty} \mu(S_y)\, dy.$$

15.4 Let $f(x, y) = e^{-xy} \sin x \sin y$ if $x \ge 0$, $y \ge 0$, and let $f(x, y) = 0$ otherwise. Prove that both iterated integrals

$$\int_{\mathbf{R}^1} \left[\int_{\mathbf{R}^1} f(x, y)\, dx \right] dy \qquad \text{and} \qquad \int_{\mathbf{R}^1} \left[\int_{\mathbf{R}^1} f(x, y)\, dy \right] dx$$

exist and are equal, but that the double integral of f over $\mathbf{R}^2$ does not exist. Also, explain why this does not contradict the Tonelli-Hobson test (Theorem 15.8).

15.5 Let $f(x, y) = (x^2 - y^2)/(x^2 + y^2)^2$ for $0 \leq x \leq 1, 0 < y \leq 1$, and let $f(0, 0) = 0$. Prove that both iterated integrals

$$\int_0^1 \left[\int_0^1 f(x, y) \, dy \right] dx \quad \text{and} \quad \int_0^1 \left[\int_0^1 f(x, y) \, dx \right] dy$$

exist but are not equal. This shows that f is not Lebesgue-integrable on $[0, 1] \times [0, 1]$.

15.6 Let $I = [0, 1] \times [0, 1]$, let $f(x, y) = (x - y)/(x + y)^3$ if $(x, y) \in I$, $(x, y) \neq (0, 0)$, and let $f(0, 0) = 0$. Prove that $f \notin L(I)$ by considering the iterated integrals

$$\int_0^1 \left[\int_0^1 f(x, y) \, dy \right] dx \quad \text{and} \quad \int_0^1 \left[\int_0^1 f(x, y) \, dx \right] dy.$$

15.7 Let $I = [0, 1] \times [1, +\infty)$ and let $f(x, y) = e^{-xy} - 2e^{-2xy}$ if $(x, y) \in I$. Prove that $f \notin L(I)$ by considering the iterated integrals

$$\int_0^1 \left[\int_1^\infty f(x, y) \, dy \right] dx \quad \text{and} \quad \int_1^\infty \left[\int_0^1 f(x, y) \, dx \right] dy.$$

15.8 The following formulas for transforming double and triple integrals occur in elementary calculus. Obtain them as consequences of Theorem 15.11 and give restrictions on T and T' for validity of these formulas.

a) $\displaystyle\iint_T f(x, y) \, dx \, dy = \iint_{T'} f(r \cos \theta, r \sin \theta) r \, dr \, d\theta.$

b) $\displaystyle\iiint_T f(x, y, z) \, dx \, dy \, dz = \iiint_{T'} f(r \cos \theta, r \sin \theta, z) r \, dr \, d\theta \, dz.$

c) $\displaystyle\iiint_T f(x, y, z) \, dx \, dy \, dz$

$$= \iiint_{T'} f(\rho \cos \theta \sin \varphi, \rho \sin \theta \sin \varphi, \rho \cos \varphi) \rho^2 \sin \varphi \, d\rho \, d\theta \, d\varphi.$$

15.9 a) Prove that $\int_{\mathbf{R}^2} e^{-(x^2 + y^2)} \, d(x, y) = \pi$ by transforming the integral to polar coordinates.

b) Use part (a) to prove that $\int_{-\infty}^\infty e^{-x^2} \, dx = \sqrt{\pi}$.

c) Use part (b) to prove that $\int_{\mathbf{R}^n} e^{-\|x\|^2} \, d(x_1, \dots, x_n) = \pi^{n/2}$.

d) Use part (b) to calculate $\int_{-\infty}^\infty e^{-tx^2} \, dx$ and $\int_{-\infty}^\infty x^2 e^{-tx^2} \, dx$, $t > 0$.

15.10 Let $V_n(a)$ denote the n-measure of the n-ball $B(0; a)$ of radius a. This exercise outlines a proof of the formula

$$V_n(a) = \frac{\pi^{n/2} a^n}{\Gamma(\frac{1}{2}n + 1)}.$$

a) Use a linear change of variable to prove that $V_n(a) = a^n V_n(1)$.

b) Assume $n \geq 3$, express the integral for $V_n(1)$ as the iteration of an $(n - 2)$-fold integral and a double integral, and use part (a) for an $(n - 2)$-ball to obtain the formula

$$V_n(1) = V_{n-2}(1) \int_0^{2\pi} \left[\int_0^1 (1 - r^2)^{n/2-1} r \, dr \right] d\theta = V_{n-2}(1) \frac{2\pi}{n}.$$

c) From the recursion formula in (b) deduce that

$$V_n(1) = \frac{\pi^{n/2}}{\Gamma(\frac{1}{2}n + 1)}.$$

15.11 Refer to Exercise 15.10 and prove that

$$\int_{B(0;1)} x_k^2 \, d(x_1, \ldots, x_n) = \frac{V_n(1)}{n + 2}$$

for each $k = 1, 2, \ldots, n$.

15.12 Refer to Exercise 15.10 and express the integral for $V_n(1)$ as the iteration of an $(n - 1)$-fold integral and a one-dimensional integral, to obtain the recursion formula

$$V_n(1) = 2V_{n-1}(1) \int_0^1 (1 - x^2)^{(n-1)/2} \, dx.$$

Put $x = \cos t$ in the integral, and use the formula of Exercise 15.10 to deduce that

$$\int_0^{\pi/2} \cos^n t \, dt = \frac{\sqrt{\pi} \, \Gamma(\frac{1}{2}n + \frac{1}{2})}{2 \, \Gamma(\frac{1}{2}n + 1)}$$

15.13 If $a > 0$, let $S_n(a) = \{(x_1, \ldots, x_n) : |x_1| + \cdots + |x_n| \leq a\}$, and let $V_n(a)$ denote the n-measure of $S_n(a)$. This exercise outlines a proof of the formula $V_n(a) = 2^n a^n/n!$.

a) Use a linear change of variable to prove that $V_n(a) = a^n V_n(1)$.

b) Assume $n \geq 2$, express the integral for $V_n(1)$ as an iteration of a one-dimensional integral and an $(n - 1)$-fold integral, use (a) to show that

$$V_n(1) = V_{n-1}(1) \int_{-1}^1 (1 - |x|)^{n-1} \, dx = 2V_{n-1}(1)/n,$$

and deduce that $V_n(1) = 2^n/n!$.

15.14 If $a > 0$ and $n \geq 2$, let $S_n(a)$ denote the following set in $\mathbf{R}^n$:

$$S_n(a) = \{(x_1, \ldots, x_n) : |x_i| + |x_n| \leq a \quad \text{for each } i = 1, \ldots, n - 1\}.$$

Let $V_n(a)$ denote the n-measure of $S_n(a)$. Use a method suggested by Exercise 15.13 to prove that $V_n(a) = 2^n a^n/n$.

15.15 Let $Q_n(a)$ denote the "first quadrant" of the n-ball $B(0:a)$ given by

$$Q_n(a) = \{(x_1, \ldots, x_n) : \|x\| \leq a \quad \text{and} \quad 0 \leq x_i \leq a \quad \text{for each } i = 1, 2, \ldots, n\}.$$

Let $f(\mathbf{x}) = x_1 \cdots x_n$ and prove that

$$\int_{Q_n(a)} f(\mathbf{x}) \, d\mathbf{x} = \frac{a^{2n}}{2^n n!}.$$

SUGGESTED REFERENCES FOR FURTHER STUDY

15.1 Asplund, E., and Bungart, L., *A First Course in Integration*. Holt, Rinehart, and Winston, New York, 1966.

15.2 Bartle, R., *The Elements of Integration*. Wiley, New York, 1966.

15.3 Kestelman, H., *Modern Theories of Integration*. Oxford University Press, 1937.

15.4 Korevaar, J., *Mathematical Methods*, Vol. 1. Academic Press, New York, 1968.

15.5 Riesz, F., and Sz.-Nagy, B., *Functional Analysis*. L. Boron, translator. Ungar, New York, 1955.

CHAPTER 16

CAUCHY'S THEOREM
AND THE
RESIDUE CALCULUS

16.1 ANALYTIC FUNCTIONS

The concept of derivative for functions of a complex variable was introduced in Chapter 5 (Section 5.15). The most important functions in complex variable theory are those which possess a continuous derivative at each point of an open set. These are called *analytic functions*.

Definition 16.1. *Let $f = u + iv$ be a complex-valued function defined on an open set S in the complex plane $\mathbf{C}$. Then f is said to be analytic on S if the derivative f' exists and is continuous* at every point of S.*

NOTE. If T is an arbitrary subset of $\mathbf{C}$ (not necessarily open), the terminology "f is analytic on T" is used to mean that f is analytic on some open set containing T. In particular, f is analytic at a point z if there is an open disk about z on which f is analytic.

It is possible for a function to have a derivative at a point without being analytic at the point. For example, if $f(z) = |z|^2$, then f has a derivative at 0 but at no other point of $\mathbf{C}$.

Examples of analytic functions were encountered in Chapter 5. If $f(z) = z^n$ (where n is a positive integer), then f is analytic everywhere in $\mathbf{C}$ and its derivative is $f'(z) = nz^{n-1}$. When n is a negative integer, the equation $f(z) = z^n$ if $z \neq 0$ defines a function analytic everywhere except at 0. Polynomials are analytic everywhere in $\mathbf{C}$, and rational functions are analytic everywhere except at points where the denominator vanishes. The exponential function, defined by the formula $e^z = e^x(\cos y + i \sin y)$, where $z = x + iy$, is analytic everywhere in $\mathbf{C}$ and is equal to its derivative. The complex sine and cosine functions (being linear combinations of exponentials) are also analytic everywhere in $\mathbf{C}$.

Let $f(z) = \operatorname{Log} z$ if $z \neq 0$, where $\operatorname{Log} z$ denotes the principal logarithm of z (see Definition 1.53). Then f is analytic everywhere in $\mathbf{C}$ except at those points $z = x + iy$ for which $x \leq 0$ and $y = 0$. At these points, the principal logarithm fails to be continuous. Analyticity at the other points is easily shown by verifying

---

* It can be shown that the existence of f' on S automatically implies continuity of f' on S (a fact discovered by Goursat in 1900). Hence an analytic function can be defined as one which merely possesses a derivative everywhere on S. However, we shall include continuity of f' as part of the definition of analyticity, since this allows some of the proofs to run more smoothly.

that the real and imaginary parts of f satisfy the Cauchy–Riemann equations (Theorem 12.6).

We shall see later that analyticity at a point z puts severe restrictions on a function. It implies the existence of all higher derivatives in a neighborhood of z and also guarantees the existence of a convergent power series which represents the function in a neighborhood of z. This is in marked contrast to the behavior of real-valued functions, where it is possible to have existence and continuity of the first derivative without existence of the second derivative.

16.2 PATHS AND CURVES IN THE COMPLEX PLANE

Many fundamental properties of analytic functions are most easily deduced with the help of integrals taken along curves in the complex plane. These are called *contour integrals* (or *complex line integrals*) and they are discussed in the next section. This section lists some terminology used for different types of curves, such as those in Fig. 16.1.

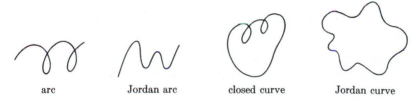

arc   Jordan arc   closed curve   Jordan curve

Figure 16.1

We recall that a *path* in the complex plane is a complex-valued function γ, continuous on a compact interval $[a, b]$. The image of $[a, b]$ under γ (the graph of γ) is said to be a *curve* described by γ and it is said to join the points $\gamma(a)$ and $\gamma(b)$.

If $\gamma(a) \neq \gamma(b)$, the curve is called an *arc* with endpoints $\gamma(a)$ and $\gamma(b)$.

If γ is one-to-one on $[a, b]$, the curve is called a *simple arc* or a *Jordan arc*.

If $\gamma(a) = \gamma(b)$, the curve is called a *closed curve*. If $\gamma(a) = \gamma(b)$ and if γ is one-to-one on the half-open interval $[a, b)$, the curve is called a *simple closed curve*, or a *Jordan curve*.

The path γ is called *rectifiable* if it has finite arc length, as defined in Section 6.10. We recall that γ is rectifiable if, and only if, γ is of bounded variation on $[a, b]$. (See Section 7.27 and Theorem 6.17.)

A path γ is called *piecewise smooth* if it has a bounded derivative γ' which is continuous everywhere on $[a, b]$ except (possibly) at a finite number of points. At these exceptional points it is required that both right- and left-hand derivatives exist. Every piecewise smooth path is rectifiable and its arc length is given by the integral $\int_a^b |\gamma'(t)| \, dt$.

A piecewise smooth closed path will be called a *circuit*.

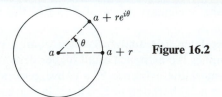

Figure 16.2

Definition 16.2. *If $a \in \mathbf{C}$ and $r > 0$, the path γ defined by the equation*

$$\gamma(\theta) = a + re^{i\theta}, \qquad 0 \le \theta \le 2\pi,$$

is called a positively oriented circle with center at a and radius r.

NOTE. The geometric meaning of $\gamma(\theta)$ is shown in Fig. 16.2. As θ varies from 0 to 2π, the point $\gamma(\theta)$ moves counterclockwise around the circle.

16.3 CONTOUR INTEGRALS

Contour integrals will be defined in terms of complex Riemann–Stieltjes integrals, discussed in Section 7.27.

Definition 16.3. *Let γ be a path in the complex plane with domain $[a, b]$, and let f be a complex-valued function defined on the graph of γ. The contour integral of f along γ, denoted by $\int_\gamma f$, is defined by the equation*

$$\int_\gamma f = \int_a^b f[\gamma(t)] \, d\gamma(t),$$

whenever the Riemann–Stieltjes integral on the right exists.

NOTATION. We also write

$$\int_\gamma f(z) \, dz \qquad \text{or} \qquad \int_{\gamma(a)}^{\gamma(b)} f(z) \, dz,$$

for the integral. The dummy symbol z can be replaced by any other convenient symbol. For example, $\int_\gamma f(z) \, dz = \int_\gamma f(w) \, dw$.

If γ is rectifiable, then a sufficient condition for the existence of $\int_\gamma f$ is that f be continuous on the graph of γ (Theorem 7.27).

The effect of replacing γ by an equivalent path (as defined in Section 6.12) is, at worst, a change in sign. In fact, we have:

Theorem 16.4. *Let γ and δ be equivalent paths describing the same curve Γ. If $\int_\gamma f$ exists, then $\int_\delta f$ also exists. Moreover, we have*

$$\int_\gamma f = \int_\delta f,$$

if γ and δ trace out Γ in the same direction, whereas

$$\int_\gamma f = - \int_\delta f,$$

if γ and δ trace out Γ in opposite directions.

Proof. Suppose $\delta(t) = \gamma[u(t)]$ where u is strictly monotonic on $[c, d]$. From the change-of-variable formula for Riemann–Stieltjes integrals (Theorem 7.7) we have

$$\int_{u(c)}^{u(d)} f[\gamma(t)] \, d\gamma(t) = \int_c^d f[\delta(t)] \, d\delta(t) = \int_\delta f. \qquad (1)$$

If u is increasing then $u(c) = a$, $u(d) = b$ and (1) becomes $\int_\gamma f = \int_\delta f$.
If u is decreasing then $u(c) = b$, $u(d) = a$ and (1) becomes $-\int_\gamma f = \int_\delta f$.

The reader can easily verify the following additive properties of contour integrals.

Theorem 16.5. *Let γ be a path with domain $[a, b]$.*

i) *If the integrals $\int_\gamma f$ and $\int_\gamma g$ exist, then the integral $\int_\gamma (\alpha f + \beta g)$ exists for every pair of complex numbers α, β, and we have*

$$\int_\gamma (\alpha f + \beta g) = \alpha \int_\gamma f + \beta \int_\gamma g.$$

ii) *Let γ_1 and γ_2 denote the restrictions of γ to $[a, c]$ and $[c, b]$, respectively, where $a < c < b$. If two of the three integrals in (2) exist, then the third also exists and we have*

$$\int_\gamma f = \int_{\gamma_1} f + \int_{\gamma_2} f. \qquad (2)$$

In practice, most paths of integration are rectifiable. For such paths the following theorem is often used to estimate the absolute value of a contour integral.

Theorem 16.6. *Let γ be a rectifiable path of length $\Lambda(\gamma)$. If the integral $\int_\gamma f$ exists, and if $|f(z)| \leq M$ for all z on the graph of γ, then we have the inequality*

$$\left| \int_\gamma f \right| \leq M \Lambda(\gamma).$$

Proof. We simply observe that all Riemann–Stieltjes sums which occur in the definition of $\int_a^b f[\gamma(t)] \, d\gamma(t)$ have absolute value not exceeding $M\Lambda(\gamma)$.

Contour integrals taken over piecewise smooth curves can be expressed as Riemann integrals. The following theorem is an easy consequence of Theorem 7.8.

Theorem 16.7. *Let γ be a piecewise smooth path with domain $[a, b]$. If the contour integral $\int_\gamma f$ exists, we have*

$$\int_\gamma f = \int_a^b f[\gamma(t)] \, \gamma'(t) \, dt.$$

16.4 THE INTEGRAL ALONG A CIRCULAR PATH AS A FUNCTION OF THE RADIUS

Consider a circular path γ of radius $r \geq 0$ and center a, given by

$$\gamma(\theta) = a + re^{i\theta}, \qquad 0 \leq \theta \leq 2\pi.$$

In this section we study the integral $\int_\gamma f$ as a function of the radius r.

Let $\varphi(r) = \int_\gamma f$. Since $\gamma'(\theta) = ire^{i\theta}$, Theorem 16.7 gives us

$$\varphi(r) = \int_0^{2\pi} f(a + re^{i\theta}) ire^{i\theta} \, d\theta. \tag{3}$$

As r varies over an interval $[r_1, r_2]$, where $0 \leq r_1 < r_2$, the points $\gamma(\theta)$ trace out an *annulus* which we denote by $A(a; r_1, r_2)$. (See Fig. 16.3.) Thus,

$$A(a; r_1, r_2) = \{z : r_1 \leq |z - a| \leq r_2\}.$$

If $r_1 = 0$ the annulus is a closed disk of radius r_2. If f is continuous on the annulus, then φ is continuous on the interval $[r_1, r_2]$. If f is analytic on the annulus, then φ is differentiable on $[r_1, r_2]$. The next theorem shows that φ is *constant* on $[r_1, r_2]$ if f is analytic everywhere on the annulus except possibly on a finite subset, provided that f is continuous on this subset.

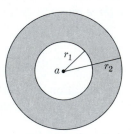

Figure 16.3

Theorem 16.8. *Assume f is analytic on the annulus $A(a; r_1, r_2)$, except possibly at a finite number of points. At these exceptional points assume that f is continuous. Then the function φ defined by (3) is constant on the interval $[r_1, r_2]$. Moreover, if $r_1 = 0$ the constant is 0.*

Proof. Let $z_1, \ldots, z_n$ denote the exceptional points where f fails to be analytic. Label these points according to increasing distances from the center, say

$$|z_1 - a| \leq |z_2 - a| \leq \cdots \leq |z_n - a|,$$

and let $R_k = |z_k - a|$. Also, let $R_0 = r_1$, $R_{n+1} = r_2$.

The union of the intervals $[R_k, R_{k+1}]$ for $k = 0, 1, 2, \ldots, n$ is the interval $[r_1, r_2]$. We will show that φ is constant on each interval $[R_k, R_{k+1}]$. We write (3) in the form

$$\varphi(r) = \int_0^{2\pi} g(r, \theta)\, d\theta, \qquad \text{where } g(r, \theta) = f(a + re^{i\theta})ire^{i\theta}.$$

An easy application of the chain rule shows that we have

$$\frac{\partial g}{\partial \theta} = ir \frac{\partial g}{\partial r}. \tag{4}$$

(The reader should verify this formula.) Continuity of f' implies continuity of the partial derivatives $\partial g / \partial r$ and $\partial g / \partial \theta$. Therefore, on each open interval (R_k, R_{k+1}), we can calculate $\varphi'(r)$ by differentiation under the integral sign (Theorem 7.40) and then use (4) and the second fundamental theorem of calculus (Theorem 7.34) to obtain

$$\varphi'(r) = \int_0^{2\pi} \frac{\partial g}{\partial r}\, d\theta = \frac{1}{ir} \int_0^{2\pi} \frac{\partial g}{\partial \theta}\, d\theta = \frac{1}{ir} \{g(r, 2\pi) - g(r, 0)\} = 0.$$

Applying Theorem 12.10, we see that φ is constant on each open subinterval (R_k, R_{k+1}). By continuity, φ is constant on each closed subinterval $[R_k, R_{k+1}]$ and hence on their union $[r_1, r_2]$. From (3) we see that $\varphi(r) \to 0$ as $r \to 0$ so the constant value of φ is 0 if $r_1 = 0$.

16.5 CAUCHY'S INTEGRAL THEOREM FOR A CIRCLE

The following special case of Theorem 16.8 is of particular importance.

Theorem 16.9 (Cauchy's integral theorem for a circle). *If f is analytic on a disk $B(a; R)$ except possibly for a finite number of points at which it is continuous, then*

$$\int_\gamma f = 0,$$

for every circular path γ with center at a and radius $r < R$.

Proof. Choose r_2 so that $r < r_2 < R$ and apply Theorem 16.8 with $r_1 = 0$.

NOTE. There is a more general form of Cauchy's integral theorem in which the circular path γ is replaced by a more general closed path. These more general paths will be introduced through the concept of *homotopy*.

16.6 HOMOTOPIC CURVES

Figure 16.4 shows three arcs having the same endpoints A and B and lying in an open region D. Arc 1 can be continuously deformed into arc 2 through a collection of intermediate arcs, each of which lies in D. Two arcs with this property are said

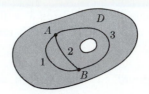

Figure 16.4

to be *homotopic* in *D*. Arc 1 cannot be so deformed into arc 3 (because of the hole separating them) so they are not homotopic in *D*.

   In this section we give a formal definition of homotopy. Then we show that, if *f* is analytic in *D*, the contour integral of *f* from *A* to *B* has the same value along any two homotopic paths in *D*. In other words, the value of a contour integral $\int_A^B f$ is unaltered under a continuous deformation of the path, provided the intermediate contours remain within the region of analyticity of *f*. This property of contour integrals is of utmost importance in the applications of complex integration.

Definition 16.10. *Let γ_0 and γ_1 be two paths with a common domain $[a, b]$. Assume that either*

a) *γ_0 and γ_1 have the same endpoints: $\gamma_0(a) = \gamma_1(a)$ and $\gamma_0(b) = \gamma_1(b)$, or*

b) *γ_0 and γ_1 are both closed paths: $\gamma_0(a) = \gamma_0(b)$ and $\gamma_1(a) = \gamma_1(b)$.*

*Let D be a subset of **C** containing the graphs of γ_0 and γ_1. Then γ_0 and γ_1 are said to be homotopic in D if there exists a function h, continuous on the rectangle $[0, 1] \times [a, b]$, and with values in D, such that*

1) $h(0, t) = \gamma_0(t)$ *if $t \in [a, b]$,*

2) $h(1, t) = \gamma_1(t)$ *if $t \in [a, b]$.*

In addition we require that for each s in $[0, 1]$ we have

3a) $h(s, a) = \gamma_0(a)$ and $h(s, b) = \gamma_0(b)$,   *in case (a);*

or

3b) $h(s, a) = h(s, b)$,   *in case (b).*

The function h is called a homotopy.

   The concept of homotopy has a simple geometric interpretation. For each fixed *s* in $[0, 1]$, let $\gamma_s(t) = h(s, t)$. Then γ_s can be regarded as an intermediate moving path which starts from γ_0 when $s = 0$ and ends at γ_1 when $s = 1$.

Example 1. *Homotopy to a point.* If γ_1 is a constant function, so that its graph is a single point, and if γ_0 is homotopic to γ_1 in *D*, we say that γ_0 is *homotopic to a point* in *D*.

Example 2. *Linear homotopy.* If, for each *t* in $[a, b]$, the line segment joining $\gamma_0(t)$ and $\gamma_1(t)$ lies in *D*, then γ_0 and γ_1 are homotopic in *D* because the function

$$h(s, t) = s\gamma_1(t) + (1 - s)\gamma_0(t)$$

serves as a homotopy. In this case we say that γ_0 and γ_1 are *linearly homotopic* in D. In particular, any two paths with domain $[a, b]$ are linearly homotopic in $\mathbf{C}$ (the complex plane) or, more generally, in any convex set containing their graphs.

NOTE. Homotopy is an equivalence relation.

The next theorem shows that between any two homotopic paths we can interpolate a finite number of intermediate polygonal paths, each of which is linearly homotopic to its neighbor.

Theorem 16.11 (Polygonal interpolation theorem). *Let γ_0 and γ_1 be homotopic paths in an open set D. Then there exist a finite number of paths $\alpha_0, \alpha_1, \ldots, \alpha_n$ such that:*

a) $\alpha_0 = \gamma_0$ *and* $\alpha_n = \gamma_1$,

b) α_j *is a polygonal path for* $1 \leq j \leq n - 1$,

c) α_j *is linearly homotopic in D to* α_{j+1} *for* $0 \leq j \leq n - 1$.

Proof. Since γ_0 and γ_1 are homotopic in D, there is a homotopy h satisfying the conditions in Definition 16.10. Consider partitions

$$\{s_0, s_1, \ldots, s_n\} \quad \text{of } [0, 1] \quad \text{and} \quad \{t_0, t_1, \ldots, t_n\} \quad \text{of } [a, b],$$

into n equal parts, choosing n so large that the image of each rectangle $[s_j, s_{j+1}] \times [t_k, t_{k+1}]$ under h is contained in an open disk D_{jk} contained in D. (The reader should verify that this is possible because of uniform continuity of h.)

On the intermediate path γ_{s_j} given by

$$\gamma_{s_j}(t) = h(s_j, t) \qquad \text{for } 0 < j < n,$$

we inscribe a polygonal path α_j with vertices at the points $h(s_j, t_k)$. That is,

$$\alpha_j(t_k) = h(s_j, t_k) \qquad \text{for } k = 0, 1, \ldots, n,$$

and α_j is linear on each subinterval $[t_k, t_{k+1}]$ for $0 \leq k \leq n - 1$. We also define $\alpha_0 = \gamma_0$ and $\alpha_n = \gamma_1$. (An example is shown in Fig. 16.5.)

The four vertices $\alpha_j(t_k)$, $\alpha_j(t_{k+1})$, $\alpha_{j+1}(t_k)$, and $\alpha_{j+1}(t_{k+1})$ all lie in the disk D_{jk}. Since D_{jk} is convex, the line segments joining them also lie in D_{jk} and hence the points

$$s\alpha_{j+1}(t) + (1 - s)\alpha_j(t), \tag{5}$$

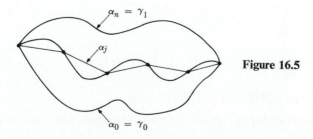

Figure 16.5

lie in D_{jk} for each (s, t) in $[0, 1] \times [t_k, t_{k+1}]$. Therefore the points (5) lie in D for all (s, t) in $[0, 1] \times [a, b]$, so α_{j+1} is linearly homotopic to α_j in D.

16.7 INVARIANCE OF CONTOUR INTEGRALS UNDER HOMOTOPY

Theorem 16.12. *Assume f is analytic on an open set D, except possibly for a finite number of points where it is continuous. If γ_0 and γ_1 are piecewise smooth paths which are homotopic in D we have*

$$\int_{\gamma_0} f = \int_{\gamma_1} f.$$

Proof. First we consider the case in which γ_0 and γ_1 are *linearly* homotopic. For each s in $[0, 1]$ let

$$\gamma_s(t) = s\gamma_1(t) + (1 - s)\gamma_0(t) \qquad \text{if } t \in [a, b].$$

Then γ_s is piecewise smooth and its graph lies in D. Write

$$\gamma_s(t) = \gamma_0(t) + s\alpha(t), \qquad \text{where } \alpha(t) = \gamma_1(t) - \gamma_0(t),$$

and define

$$\varphi(s) = \int_{\gamma_s} f = \int_a^b f[\gamma_s(t)] \, d\gamma_0(t) + s \int_a^b f[\gamma_s(t)] \, d\alpha(t),$$

for $0 \le s \le 1$. We wish to prove that $\varphi(0) = \varphi(1)$. We will in fact prove that φ is constant on $[0, 1]$.

We use Theorem 7.40 to calculate $\varphi'(s)$ by differentiation under the integral sign. Since

$$\frac{\partial}{\partial s} \gamma_s(t) = \alpha(t),$$

this gives us

$$\varphi'(s) = \int_a^b f'[\gamma_s(t)]\alpha(t) \, d\gamma_0(t) + s \int_a^b f'[\gamma_s(t)]\alpha(t) \, d\alpha(t) + \int_a^b f[\gamma_s(t)] \, d\alpha(t)$$

$$= \int_a^b \alpha(t)f'[\gamma_s(t)] \, d\gamma_s(t) + \int_a^b f[\gamma_s(t)] \, d\alpha(t)$$

$$= \int_a^b \alpha(t)f'[\gamma_s(t)]\gamma_s'(t) \, dt + \int_a^b f[\gamma_s(t)] \, d\alpha(t)$$

$$= \int_a^b \alpha(t) \, d\{f[\gamma_s(t)]\} + \int_a^b f[\gamma_s(t)] \, d\alpha(t)$$

$$= \alpha(b)f[\gamma_s(b)] - \alpha(a)f[\gamma_s(a)],$$

by the formula for integration by parts (Theorem 7.6). But, as the reader can easily verify, the last expression vanishes because γ_0 and γ_1 are homotopic, so $\varphi'(s) = 0$ for all s in $[0, 1]$. Therefore φ is constant on $[0, 1]$. This proves the theorem when γ_0 and γ_1 are linearly homotopic in D.

If they are homotopic in D under a general homotopy h, we interpolate polygonal paths α_j as described in Theorem 16.11. Since each polygonal path is piecewise smooth, we can repeatedly apply the result just proved to obtain

$$\int_{\gamma_0} f = \int_{\alpha_0} f = \int_{\alpha_1} f = \cdots = \int_{\alpha_n} f = \int_{\gamma_1} f.$$

16.8 GENERAL FORM OF CAUCHY'S INTEGRAL THEOREM

The general form of Cauchy's theorem referred to earlier can now be easily deduced from Theorems 16.9 and 16.12. We remind the reader that a *circuit* is a piecewise smooth closed path.

Theorem 16.13 (Cauchy's integral theorem for circuits homotopic to a point). Assume *f is analytic on an open set D, except possibly for a finite number of points at which we assume f is continuous. Then for every circuit γ which is homotopic to a point in D we have*

$$\int_{\gamma} f = 0.$$

Proof. Since γ is homotopic to a point in D, γ is also homotopic to a circular path δ in D with arbitrarily small radius. Therefore $\int_{\gamma} f = \int_{\delta} f$, and $\int_{\delta} f = 0$ by Theorem 16.9.

Definition 16.14. An open connected set D is called simply connected if every closed path in D is homotopic to a point in D.

Geometrically, a simply connected region is one without holes. Cauchy's theorem shows that, *in a simply connected region D the integral of an analytic function is zero around any circuit in D.*

16.9 CAUCHY'S INTEGRAL FORMULA

The next theorem reveals a remarkable property of analytic functions. It relates the value of an analytic function at a point with the values on a closed curve not containing the point.

Theorem 16.15 (Cauchy's integral formula). Assume f is analytic on an open set D, *and let γ be any circuit which is homotopic to a point in D. Then for any point z in D which is not on the graph of γ we have*

$$\int_{\gamma} \frac{f(w)}{w - z}\, dw = f(z) \int_{\gamma} \frac{1}{w - z}\, dw. \tag{6}$$

Proof. Define a new function g on D as follows:

$$g(w) = \begin{cases} \dfrac{f(w) - f(z)}{w - z} & \text{if } w \neq z \\ f'(z) & \text{if } w = z. \end{cases}$$

Then g is analytic at each point $w \neq z$ in D and, at the point z itself, g is continuous. Applying Cauchy's integral theorem to g we have $\int_\gamma g = 0$ for every circuit γ homotopic to a point in D. But if z is not on the graph of γ we can write

$$\int_\gamma g = \int_\gamma \frac{f(w) - f(z)}{w - z}\, dw = \int_\gamma \frac{f(w)}{w - z}\, dw - f(z) \int_\gamma \frac{1}{w - z}\, dw,$$

which proves (6).

NOTE. The same proof shows that (6) is also valid if there is a finite subset T of D on which f is not analytic, provided that f is continuous on T and z is not in T.

The integral $\int_\gamma (w - z)^{-1}\, dw$ which appears in (6) plays an important role in complex integration theory and is discussed further in the next section. We can easily calculate its value for a circular path.

Example. If γ is a positively oriented circular path with center at z and radius r, we can write $\gamma(\theta) = z + re^{i\theta}, 0 \leq \theta \leq 2\pi$. Then $\gamma'(\theta) = ire^{i\theta} = i\{\gamma(\theta) - z\}$, and we find

$$\int_\gamma \frac{dw}{w - z} = \int_0^{2\pi} \frac{\gamma'(\theta)}{\gamma(\theta) - z}\, d\theta = \int_0^{2\pi} i\, d\theta = 2\pi i.$$

NOTE. In this case Cauchy's integral formula (6) takes the form

$$2\pi i f(z) = \int_\gamma \frac{f(w)}{w - z}\, dw.$$

Again writing $\gamma(\theta) = z + re^{i\theta}$, we can put this in the form

$$f(z) = \frac{1}{2\pi} \int_0^{2\pi} f(z + re^{i\theta})\, d\theta. \tag{7}$$

This can be interpreted as a *Mean-Value Theorem* expressing the value of f at the center of a disk as an average of its values at the boundary of the disk. The function f is assumed to be analytic on the closure of the disk, except possibly for a finite subset on which it is continuous.

16.10 THE WINDING NUMBER OF A CIRCUIT WITH RESPECT TO A POINT

Theorem 16.16. *Let γ be a circuit and let z be a point not on the graph of γ. Then there is an integer n (depending on γ and on z) such that*

$$\int_\gamma \frac{dw}{w - z} = 2\pi i n. \tag{8}$$

Proof. Suppose γ has domain $[a, b]$. By Theorem 16.7 we can express the integral in (8) as a Riemann integral,

$$\int_\gamma \frac{dw}{w - z} = \int_a^b \frac{\gamma'(t)\, dt}{\gamma(t) - z} .$$

Define a complex-valued function on the interval $[a, b]$ by the equation

$$F(x) = \int_a^x \frac{\gamma'(t)\, dt}{\gamma(t) - z} \qquad \text{if } a \le x \le b.$$

To prove the theorem we must show that $F(b) = 2\pi i n$ for some integer n. Now F is continuous on $[a, b]$ and has a derivative

$$F'(x) = \frac{\gamma'(x)}{\gamma(x) - z} ,$$

at each point of continuity of γ'. Therefore the function G defined by

$$G(t) = e^{-F(t)}\{\gamma(t) - z\} \qquad \text{if } t \in [a, b],$$

is also continuous on $[a, b]$. Moreover, at each point of continuity of γ' we have

$$G'(t) = e^{-F(t)}\gamma'(t) - F'(t)e^{-F(t)}\{\gamma(t) - z\} = 0.$$

Therefore $G'(t) = 0$ for each t in $[a, b]$ except (possibly) for a finite number of points. By continuity, G is constant throughout $[a, b]$. Hence, $G(b) = G(a)$. In other words, we have

$$e^{-F(b)}\{\gamma(b) - z\} = \gamma(a) - z.$$

Since $\gamma(b) = \gamma(a) \ne z$ we find

$$e^{-F(b)} = 1,$$

which implies $F(b) = 2\pi i n$, where n is an integer. This completes the proof.

Definition 16.17. *If γ is a circuit whose graph does not contain z, then the integer n defined by (8) is called the* winding number *(or* index*) of γ with respect to z, and is denoted by* $n(\gamma, z)$. *Thus,*

$$n(\gamma, z) = \frac{1}{2\pi i} \int_\gamma \frac{dw}{w - z} .$$

NOTE. Cauchy's integral formula (6) can now be restated in the form

$$n(\gamma, z)f(z) = \frac{1}{2\pi i} \int_\gamma \frac{f(w)}{w - z}\, dw.$$

The term "winding number" is used because $n(\gamma, z)$ gives a mathematically precise way of counting the number of times the point $\gamma(t)$ "winds around" the point z as t varies over the interval $[a, b]$. For example, if γ is a positively oriented circle given by $\gamma(\theta) = z + re^{i\theta}$, where $0 \leq \theta \leq 2\pi$, we have already seen that the winding number is 1. This is in accord with the physical interpretation of the point $\gamma(\theta)$ moving once around a circle in the positive direction as θ varies from 0 to 2π. If θ varies over the interval $[0, 2\pi n]$, the point $\gamma(\theta)$ moves n times around the circle in the positive direction and an easy calculation shows that the winding number is n. On the other hand, if $\delta(\theta) = z + re^{-i\theta}$ for $0 \leq \theta \leq 2\pi n$, then $\delta(\theta)$ moves n times around the circle in the opposite direction and the winding number is $-n$. Such a path δ is said to be *negatively oriented*.

16.11 THE UNBOUNDEDNESS OF THE SET OF POINTS WITH WINDING NUMBER ZERO

Let Γ denote the graph of a circuit γ. Since Γ is a compact set, its complement $\mathbf{C} - \Gamma$ is an open set which, by Theorem 4.44, is a countable union of disjoint open regions (the components of $\mathbf{C} - \Gamma$). If we consider the components as subsets of the extended plane $\mathbf{C}^*$, exactly one of these contains the ideal point ∞. In other words, one and only one of the components of $\mathbf{C} - \Gamma$ is unbounded. The next theorem shows that the winding number $n(\gamma, z)$ is 0 for each z in the unbounded component.

Theorem 16.18. Let γ be a circuit with graph Γ. Divide the set $\mathbf{C} - \Gamma$ into two subsets:

$$E = \{z : n(\gamma, z) = 0\} \qquad and \qquad I = \{z : n(\gamma, z) \neq 0\}.$$

Then both E and I are open. Moreover, E is unbounded and I is bounded.

Proof. Define a function g on $\mathbf{C} - \Gamma$ by the formula

$$g(z) = n(\gamma, z) = \frac{1}{2\pi i} \int_\gamma \frac{dw}{w - z}.$$

By Theorem 7.38, g is continuous on $\mathbf{C} - \Gamma$ and, since $g(z)$ is always an integer, it follows that g is constant on each component of $\mathbf{C} - \Gamma$. Therefore both E and I are open since each is a union of components of $\mathbf{C} - \Gamma$.

Let U denote the unbounded component of $\mathbf{C} - \Gamma$. If we prove that E contains U this will show that E is unbounded and that I is bounded. Let K be a constant such that $|\gamma(t)| < K$ for all t in the domain of γ, and let c be a point in U such that $|c| > K + \Lambda(\gamma)$ where $\Lambda(\gamma)$ is the length of γ. Then we have

$$\left| \frac{1}{\gamma(t) - c} \right| \leq \frac{1}{|c| - |\gamma(t)|} < \frac{1}{|c| - K}.$$

Estimating the integral for $n(\gamma, c)$ by Theorem 16.6 we find

$$0 \leq |g(c)| \leq \frac{\Lambda(\gamma)}{|c| - K} < 1.$$

Since $g(c)$ is an integer we must have $g(c) = 0$, so g has the constant value 0 on U. Hence E contains the point c, so E contains all of U.

There is a general theorem, called the *Jordan curve theorem*, which states that if Γ is a Jordan curve (simple closed curve) described by γ, then each of the sets E and I in Theorem 16.18 is connected. In other words, a Jordan curve Γ divides $\mathbf{C} - \Gamma$ into exactly *two* components E and I having Γ as their common boundary. The set I is called the *inner* (or *interior*) *region* of Γ, and its points are said to be *inside* Γ. The set E is called the *outer* (or *exterior*) *region* of Γ, and its points are said to be *outside* Γ.

Although the Jordan curve theorem is intuitively evident and easy to prove for certain familiar Jordan curves such as circles, triangles, and rectangles, the proof for an arbitrary Jordan curve is by no means simple. (Proofs can be found in References 16.3 and 16.5.)

We shall not need the Jordan curve theorem to prove any of the theorems in this chapter. However, the reader should realize that the Jordan curves occurring in the ordinary applications of complex integration theory are usually made up of a finite number of line segments and circular arcs, and for such examples it is usually quite obvious that $\mathbf{C} - \Gamma$ consists of exactly two components. For points z inside such curves the winding number $n(\gamma, z)$ is $+1$ or -1 because γ is homotopic in I to some circular path δ with center z, so $n(\gamma, z) = n(\delta, z)$, and $n(\delta, z)$ is $+1$ or -1 depending on whether the circular path δ is positively or negatively oriented. For this reason we say that a Jordan circuit γ is *positively oriented* if, for some z inside Γ we have $n(\gamma, z) = +1$, and *negatively oriented* if $n(\gamma, z) = -1$.

16.12 ANALYTIC FUNCTIONS DEFINED BY CONTOUR INTEGRALS

Cauchy's integral formula, which states that

$$n(\gamma, z)f(z) = \frac{1}{2\pi i} \int_\gamma \frac{f(w)}{w - z}\, dw,$$

has many important consequences. Some of these follow from the next theorem which treats integrals of a slightly more general type in which the integrand $f(w)/(w - z)$ is replaced by $\varphi(w)/(w - z)$, where φ is merely continuous and not necessarily analytic, and γ is any rectifiable path, not necessarily a circuit.

Theorem 16.19. *Let γ be a rectifiable path with graph Γ. Let φ be a complex-valued function which is continuous on Γ, and let f be defined on $\mathbf{C} - \Gamma$ by the equation*

$$f(z) = \int_\gamma \frac{\varphi(w)}{w - z}\, dw \qquad \text{if } z \notin \Gamma.$$

Then f has the following properties:

a) *For each point a in $\mathbf{C} - \Gamma$, f has a power-series representation*

$$f(z) = \sum_{n=0}^\infty c_n(z - a)^n, \tag{9}$$

where

$$c_n = \int_\gamma \frac{\varphi(w)}{(w - a)^{n+1}} \, dw \qquad for \ n = 0, 1, 2, \ldots \tag{10}$$

b) *The series in* (a) *has a positive radius of convergence* $\geq R$, *where*

$$R = \inf \{|w - a| : w \in \Gamma\}. \tag{11}$$

c) *The function f has a derivative of every order n on* $\mathbf{C} - \Gamma$ *given by*

$$f^{(n)}(z) = n! \int_\gamma \frac{\varphi(w)}{(w - z)^{n+1}} \, dw \qquad if \ z \notin \Gamma. \tag{12}$$

Proof. First we note that the number R defined by (11) is positive because the function $g(w) = |w - a|$ has a minimum on the compact set Γ, and this minimum is not zero since $a \notin \Gamma$. Thus, R is the distance from a to the nearest point of Γ. (See Fig. 16.6.)

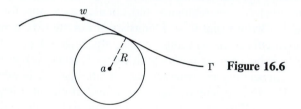

Figure 16.6

To prove (a) we begin with the identity

$$\frac{1}{1 - t} = \sum_{n=0}^{k} t^n + \frac{t^{k+1}}{1 - t}, \tag{13}$$

valid for all $t \neq 1$. We take $t = (z - a)/(w - a)$ where $|z - a| < R$ and $w \in \Gamma$. Then $1/(1 - t) = (w - a)/(w - z)$. Multiplying (13) by $\varphi(w)/(w - a)$ and integrating along γ, we find

$$f(z) = \int_\gamma \frac{\varphi(w)}{w - z} \, dw$$

$$= \sum_{n=0}^{k} (z - a)^n \int_\gamma \frac{\varphi(w)}{(w - a)^{n+1}} \, dw + \int_\gamma \frac{\varphi(w)}{w - z} \left(\frac{z - a}{w - a}\right)^{k+1} dw$$

$$= \sum_{n=0}^{k} c_n (z - a)^n + E_k,$$

where c_n is given by (10) and E_k is given by

$$E_k = \int_\gamma \frac{\varphi(w)}{w - z} \left(\frac{z - a}{w - a}\right)^{k+1} dw. \tag{14}$$

Now we show that $E_k \to 0$ as $k \to \infty$ by estimating the integrand in (14). We have

$$\left|\frac{z-a}{w-a}\right| \le \frac{|z-a|}{R} \quad \text{and} \quad \frac{1}{|w-z|} = \frac{1}{|w-a+a-z|} \le \frac{1}{R-|a-z|}.$$

Let $M = \max\{|\varphi(w)| : w \in \Gamma\}$, and let $\Lambda(\gamma)$ denote the length of γ. Then (14) gives us

$$|E_k| \le \frac{M\,\Lambda(\gamma)}{R-|a-z|}\left(\frac{|z-a|}{R}\right)^{k+1}.$$

Since $|z-a| < R$ we find that $E_k \to 0$ as $k \to \infty$. This proves (a) and (b).

Applying Theorem 9.23 to (9) we find that f has derivatives of every order on the disk $B(a; R)$ and that $f^{(n)}(a) = n!c_n$. Since a is an arbitrary point of $\mathbf{C} - \Gamma$ this proves (c).

NOTE. The series in (9) may have a radius of convergence greater than R, in which case it may or may not represent f at more distant points.

16.13 POWER-SERIES EXPANSIONS FOR ANALYTIC FUNCTIONS

A combination of Cauchy's integral formula with Theorem 16.19 gives us:

Theorem 16.20. *Assume f is analytic on an open set S in $\mathbf{C}$, and let a be any point of S. Then all derivatives $f^{(n)}(a)$ exist, and f can be represented by the convergent power series*

$$f(z) = \sum_{n=0}^{\infty} \frac{f^{(n)}(a)}{n!}(z-a)^n, \tag{15}$$

in every disk $B(a; R)$ whose closure lies in S. Moreover, for every $n \ge 0$ we have

$$f^{(n)}(a) = \frac{n!}{2\pi i}\int_\gamma \frac{f(w)}{(w-a)^{n+1}}\,dw, \tag{16}$$

where γ is any positively oriented circular path with center at a and radius $r < R$.

NOTE. The series in (15) is known as the *Taylor expansion* of f about a. Equation (16) is called *Cauchy's integral formula* for $f^{(n)}(a)$.

Proof. Let γ be a circuit homotopic to a point in S, and let Γ be the graph of γ. Define g on $\mathbf{C} - \Gamma$ by the equation

$$g(z) = \int_\gamma \frac{f(w)}{w-z}\,dw \quad \text{if } z \notin \Gamma.$$

If $z \in B(a; R)$, Cauchy's integral formula tells us that $g(z) = 2\pi i n(\gamma, z)f(z)$. Hence,

$$n(\gamma, z)f(z) = \frac{1}{2\pi i}\int_\gamma \frac{f(w)}{w-z}\,dw \quad \text{if } |z-a| < R.$$

Now let $\gamma(\theta) = a + re^{i\theta}$, where $|z - a| < r < R$ and $0 \le \theta \le 2\pi$. Then $n(\gamma, z) = 1$, so by applying Theorem 16.19 to $\varphi(w) = f(w)/(2\pi i)$ we find a series representation

$$f(z) = \sum_{n=0}^{\infty} c_n(z - a)^n,$$

convergent for $|z - a| < R$, where $c_n = f^{(n)}(a)/n!$. Also, part (c) of Theorem 16.19 gives (16).

Theorems 16.20 and 9.23 together tell us that a necessary and sufficient condition for a complex-valued function f to be analytic at a point a is that f be representable by a power series in some neighborhood of a. When such a power series exists, its radius of convergence is at least as large as the radius of any disk $B(a)$ which lies in the region of analyticity of f. Since the circle of convergence cannot contain any points in its interior where f fails to be analytic, it follows that the radius of convergence is exactly equal to the distance from a to the nearest point at which f fails to be analytic.

This observation gives us a deeper insight concerning power-series expansions for real-valued functions of a real variable. For example, let $f(x) = 1/(1 + x^2)$ if x is real. This function is defined everywhere in $\mathbf{R}^1$ and has derivatives of every order at each point in $\mathbf{R}^1$. Also, it has a power-series expansion about the origin, namely,

$$\frac{1}{1 + x^2} = 1 - x^2 + x^4 - x^6 + \cdots$$

However, this representation is valid only in the open interval $(-1, 1)$. From the standpoint of real-variable theory, there is nothing in the behavior of f which explains this. But when we examine the situation in the complex plane, we see at once that the function $f(z) = 1/(1 + z^2)$ is analytic everywhere in $\mathbf{C}$ except at the points $z = \pm i$. Therefore the radius of convergence of the power-series expansion about 0 must equal 1, the distance from 0 to i and to $-i$.

Examples. The following power series expansions are valid for all z in $\mathbf{C}$:

a) $e^z = \displaystyle\sum_{n=0}^{\infty} \frac{z^n}{n!}$,

b) $\sin z = \displaystyle\sum_{n=0}^{\infty} \frac{(-1)^n z^{2n+1}}{(2n + 1)!}$,

c) $\cos z = \displaystyle\sum_{n=0}^{\infty} \frac{(-1)^n z^{2n}}{(2n)!}$.

16.14 CAUCHY'S INEQUALITIES. LIOUVILLE'S THEOREM

If f is analytic on a closed disk $B(a; R)$, Cauchy's integral formula (16) shows that

$$f^{(n)}(a) = \frac{n!}{2\pi i} \int_{\gamma} \frac{f(w)}{(w - a)^{n+1}} \, dw,$$

where γ is any positively oriented circular path with center a and radius $r < R$.

We can write $\gamma(\theta) = a + re^{i\theta}, 0 \le \theta \le 2\pi$, and put this in the form

$$f^{(n)}(a) = \frac{n!}{2\pi r^n} \int_0^{2\pi} f(a + re^{i\theta}) e^{-in\theta} \, d\theta. \tag{17}$$

This formula expresses the nth derivative at a as a weighted average of the values of f on a circle with center at a. The special case $n = 0$ was obtained earlier in Section 16.9.

Now, let $M(r)$ denote the maximum value of $|f|$ on the graph of γ. Estimating the integral in (17), we immediately obtain *Cauchy's inequalities*:

$$|f^{(n)}(a)| \le \frac{M(r)n!}{r^n} \qquad (n = 0, 1, 2, \ldots). \tag{18}$$

The next theorem is an easy consequence of the case $n = 1$.

Theorem 16.21 (Liouville's theorem). *If f is analytic everywhere on* **C** *and bounded on* **C**, *then f is constant.*

Proof. Suppose $|f(z)| \le M$ for all z in **C**. Then Cauchy's inequality with $n = 1$ gives us $|f'(a)| \le M/r$ for every $r > 0$. Letting $r \to +\infty$, we find $f'(a) = 0$ for every a in **C** and hence, by Theorem 5.23, f is constant.

NOTE. A function analytic everywhere on **C** is called an *entire* function. Examples are polynomials, the sine and cosine, and the exponential. Liouville's theorem states that *every bounded entire function is constant.*

Liouville's theorem leads to a simple proof of the Fundamental Theorem of Algebra.

Theorem 16.22 (Fundamental Theorem of Algebra). *Every polynomial of degree* $n \ge 1$ *has a zero.*

Proof. Let $P(z) = a_0 + a_1 z + \cdots + a_n z^n$, where $n \ge 1$ and $a_n \ne 0$. We assume that P has no zero and prove that P is constant. Let $f(z) = 1/P(z)$. Then f is analytic everywhere on **C** since P is never zero. Also, since

$$P(z) = z^n \left(\frac{a_0}{z^n} + \frac{a_1}{z^{n-1}} + \cdots + \frac{a_{n-1}}{z} + a_n \right),$$

we see that $|P(z)| \to +\infty$ as $|z| \to +\infty$, so $f(z) \to 0$ as $|z| \to +\infty$. Therefore f is bounded on **C** so, by Liouville's theorem, f and hence P is constant.

16.15 ISOLATION OF THE ZEROS OF AN ANALYTIC FUNCTION

If f is analytic at a and if $f(a) = 0$, the Taylor expansion of f about a has constant term zero and hence assumes the following form:

$$f(z) = \sum_{n=1}^{\infty} c_n(z - a)^n.$$

This is valid for each z in some disk $B(a)$. If f is identically zero on this disk [that is, if $f(z) = 0$ for every z in $B(a)$], then each $c_n = 0$, since $c_n = f^{(n)}(a)/n!$. If f is not identically zero on this neighborhood, there will be a first nonzero coefficient c_k in the expansion, in which case the point a is said to be a *zero of order* k. We will prove next that there is a neighborhood of a which contains no further zeros of f. This property is described by saying that the zeros of an analytic function are *isolated*.

Theorem 16.23. *Assume that f is analytic on an open set S in $\mathbf{C}$. Suppose $f(a) = 0$ for some point a in S and assume that f is not identically zero on any neighborhood of a. Then there exists a disk $B(a)$ in which f has no further zeros.*

Proof. The Taylor expansion about a becomes $f(z) = (z - a)^k g(z)$, where $k \geq 1$,

$$g(z) = c_k + c_{k+1}(z - a) + \cdots, \qquad \text{and} \qquad g(a) = c_k \neq 0.$$

Since g is continuous at a, there is a disk $B(a) \subseteq S$ on which g does not vanish. Therefore, $f(z) \neq 0$ for all $z \neq a$ in $B(a)$.

This theorem has several important consequences. For example, we can use it to show that a function which is analytic on an open region S cannot be zero on any nonempty open subset of S without being identically zero throughout S. We recall that an open region is an open *connected* set. (See Definitions 4.34 and 4.45.)

Theorem 16.24. *Assume that f is analytic on an open region S in $\mathbf{C}$. Let A denote the set of those points z in S for which there exists a disk $B(z)$ on which f is identically zero, and let $B = S - A$. Then one of the two sets A or B is empty and the other one is S itself.*

Proof. We have $S = A \cup B$, where A and B are disjoint sets. The set A is open by its very definition. If we prove that B is also open, it will follow from the connectedness of S that at least one of the two sets A or B is empty.

To prove B is open, let a be a point of B and consider the two possibilities: $f(a) \neq 0$, $f(a) = 0$. If $f(a) \neq 0$, there is a disk $B(a) \subseteq S$ on which f does not vanish. Each point of this disk must therefore belong to B. Hence, a is an interior point of B if $f(a) \neq 0$. But, if $f(a) = 0$, Theorem 16.23 provides us with a disk $B(a)$ containing no further zeros of f. This means that $B(a) \subseteq B$. Hence, in either case, a is an interior point of B. Therefore, B is open and one of the two sets A or B must be empty.

16.16 THE IDENTITY THEOREM FOR ANALYTIC FUNCTIONS

Theorem 16.25. *Assume that f is analytic on an open region S in $\mathbf{C}$. Let T be a subset of S having an accumulation point a in S. If $f(z) = 0$ for every z in T, then $f(z) = 0$ for every z in S.*

Proof. There exists an infinite sequence $\{z_n\}$, whose terms are points of T, such that $\lim_{n \to \infty} z_n = a$. By continuity, $f(a) = \lim_{n \to \infty} f(z_n) = 0$. We will prove next that there is a neighborhood of a on which f is identically zero. Suppose there is no such neighborhood. Then Theorem 16.23 tells us that there must be a disk $B(a)$ on which $f(z) \neq 0$ if $z \neq a$. But this is impossible, since every disk $B(a)$ contains points of T other than a. Therefore there must be a neighborhood of a on which f vanishes identically. Hence the set A of Theorem 16.24 cannot be empty. Therefore, $A = S$, and this means $f(z) = 0$ for every z in S.

As a corollary we have the following important result, sometimes referred to as the *identity theorem for analytic functions*:

Theorem 16.26. *Let f and g be analytic on an open region S in **C**. If T is a subset of S having an accumulation point a in S, and if $f(z) = g(z)$ for every z in T, then $f(z) = g(z)$ for every z in S.*

Proof. Apply Theorem 16.25 to $f - g$.

16.17 THE MAXIMUM AND MINIMUM MODULUS OF AN ANALYTIC FUNCTION

The absolute value or modulus $|f|$ of an analytic function f is a real-valued non-negative function. The theorems of this section refer to maxima and minima of $|f|$.

Theorem 16.27 (Local maximum modulus principle). *Assume f is analytic and not constant on an open region S. Then $|f|$ has no local maxima in S. That is, every disk $B(a; R)$ in S contains points z such that $|f(z)| > |f(a)|$.*

Proof. We assume there is a disk $B(a; R)$ in S in which $|f(z)| \leq |f(a)|$ and prove that f is constant on S. Consider the concentric disk $B(a; r)$ with $0 < r \leq R$. From Cauchy's integral formula, as expressed in (7), we have

$$|f(a)| \leq \frac{1}{2\pi} \int_0^{2\pi} |f(a + re^{i\theta})| \, d\theta. \tag{19}$$

Now $|f(a + re^{i\theta})| \leq |f(a)|$ for all θ. We show next that we cannot have *strict* inequality $|f(a + re^{i\theta})| < |f(a)|$ for any θ. Otherwise, by continuity we would have $|f(a + re^{i\theta})| \leq |f(a)| - \varepsilon$ for some $\varepsilon > 0$ and all θ in some subinterval I of $[0, 2\pi]$ of positive length h, say. Let $J = [0, 2\pi] - I$. Then J has measure $2\pi - h$, and (19) gives us

$$2\pi|f(a)| \leq \int_I |f(a + re^{i\theta})| \, d\theta + \int_J |f(a + re^{i\theta})| \, d\theta$$

$$\leq h\{|f(a)| - \varepsilon\} + (2\pi - h)\,|f(a)| = 2\pi\,|f(a)| - h\varepsilon < 2\pi\,|f(a)|.$$

Thus we get the contradiction $|f(a)| < |f(a)|$. This shows that if $r \leq R$, we cannot have strict inequality $|f(a + re^{i\theta})| < |f(a)|$ for any θ. Hence $|f(z)| = |f(a)|$ for every z in $B(a; R)$. Therefore $|f|$ is constant on this disk so, by Theorem 5.23, f itself is constant on this disk. By the identity theorem, f is constant on S.

Theorem 16.28 (Absolute maximum modulus principle). *Let T be a compact subset of the complex plane* **C**. *Assume f is continuous on T and analytic on the interior of T. Then the absolute maximum of $|f|$ on T is attained on ∂T, the boundary of T.*

Proof. Since T is compact, $|f|$ attains its absolute maximum somewhere on T, say at a. If $a \in \partial T$ there is nothing to prove. If $a \in \text{int } T$, let S be the component of int T containing a. Since $|f|$ has a local maximum at a, Theorem 16.27 implies that f is constant on S. By continuity, f is constant on $\partial S \subseteq T$, so the maximum value, $|f(a)|$, is attained on ∂S. But $\partial S \subseteq \partial T$ (Why?) so the maximum is attained on ∂T.

Theorem 16.29 (Minimum modulus principle). *Assume f is analytic and not constant on an open region S. If $|f|$ has a local minimum in S at a, then $f(a) = 0$.*

Proof. If $f(a) \neq 0$ we apply Theorem 16.27 to $g = 1/f$. Then g is analytic in some open disk $B(a; R)$ and $|g|$ has a local maximum at a. Therefore g and hence f is constant on this disk and therefore on S, contradicting the hypothesis.

16.18 THE OPEN MAPPING THEOREM

Nonconstant analytic functions are open mappings; that is, they map open sets onto open sets. We prove this as an application of the minimum modulus principle.

Theorem 16.30 (Open mapping theorem). *If f is analytic and not constant on an open region S, then f is open.*

Proof. Let A be any open subset of S. We are to prove that $f(A)$ is open. Take any b in $f(A)$ and write $b = f(a)$, where $a \in A$. First we note that a is an isolated point of the inverse-image $f^{-1}(\{b\})$. (If not, by the identity theorem f would be constant on S.) Hence there is some disk $B = B(a; r)$ whose closure $\bar{B}$ lies in A and contains no point of $f^{-1}(\{b\})$ except a. Since $f(\bar{B}) \subseteq f(A)$ the proof will be complete if we show that $f(\bar{B})$ contains a disk with center at b.

Let ∂B denote the boundary of B, $\partial B = \{z : |z - a| = r\}$. Then $f(\partial B)$ is a compact set which does not contain b. Hence the number m defined by

$$m = \inf \{|f(z) - b| : z \in \partial B\},$$

is positive. We will show that $f(\bar{B})$ contains the disk $B(b; m/2)$. To do this, we take any w in $B(b; m/2)$ and show that $w = f(z_0)$ for some z_0 in $\bar{B}$.

Let $g(z) = f(z) - w$ if $z \in \bar{B}$. We will prove that $g(z_0) = 0$ for some z_0 in $\bar{B}$. Now $|g|$ is continuous on $\bar{B}$ and, since $\bar{B}$ is compact, there is a point z_0 in $\bar{B}$ at which $|g|$ attains its minimum. Since $a \in \bar{B}$, this implies

$$|g(z_0)| \leq |g(a)| = |f(a) - w| = |b - w| < \frac{m}{2}.$$

But if $z \in \partial B$, we have

$$|g(z)| = |f(z) - b + b - w| \geq |f(z) - b| - |w - b| > m - \frac{m}{2} = \frac{m}{2}.$$

Hence, $z_0 \notin \partial B$ so z_0 is an interior point of $\bar{B}$. In other words, $|g|$ has a local minimum at z_0. Since g is analytic and not constant on B, the minimum modulus principle shows that $g(z_0) = 0$ and the proof is complete.

16.19 LAURENT EXPANSIONS FOR FUNCTIONS ANALYTIC IN AN ANNULUS

Consider two functions f_1 and g_1, both analytic at a point a, with $g_1(a) = 0$. Then we have power-series expansions

$$g_1(z) = \sum_{n=1}^{\infty} b_n(z - a)^n, \qquad \text{for } |z - a| < r_1,$$

and

$$f_1(z) = \sum_{n=0}^{\infty} c_n(z - a)^n, \qquad \text{for } |z - a| < r_2. \tag{20}$$

Let f_2 denote the composite function given by

$$f_2(z) = g_1 \left(\frac{1}{z - a} + a \right).$$

Then f_2 is defined and analytic in the region $|z - a| > r_1$ and is represented there by the convergent series

$$f_2(z) = \sum_{n=1}^{\infty} b_n(z - a)^{-n}, \qquad \text{for } |z - a| > r_1. \tag{21}$$

Now if $r_1 < r_2$, the series in (20) and (21) will have a region of convergence in common, namely the set of z for which

$$r_1 < |z - a| < r_2.$$

In this region, the interior of the annulus $A(a; r_1, r_2)$, both f_1 and f_2 are analytic and their sum $f_1 + f_2$ is given by

$$f_1(z) + f_2(z) = \sum_{n=0}^{\infty} c_n(z - a)^n + \sum_{n=1}^{\infty} b_n(z - a)^{-n}.$$

The sum on the right is written more briefly as

$$\sum_{n=-\infty}^{\infty} c_n(z - a)^n,$$

where $c_{-n} = b_n$ for $n = 1, 2, \ldots$ A series of this type, consisting of both positive and negative powers of $z - a$, is called a *Laurent series*. We say it converges if both parts converge separately.

Every convergent Laurent series represents an analytic function in the interior of the annulus $A(a; r_1, r_2)$. Now we will prove that, conversely, every function f which is analytic on an annulus can be represented in the interior of the annulus by a convergent Laurent series.

Theorem 16.31. *Assume that f is analytic on an annulus $A(a; r_1, r_2)$. Then for every interior point z of this annulus we have*

$$f(z) = f_1(z) + f_2(z),$$ (22)

where

$$f_1(z) = \sum_{n=0}^{\infty} c_n(z-a)^n \quad and \quad f_2(z) = \sum_{n=1}^{\infty} c_{-n}(z-a)^{-n}.$$

The coefficients are given by the formulas

$$c_n = \frac{1}{2\pi i} \int_\gamma \frac{f(w)}{(w-a)^{n+1}} \, dw \quad (n = 0, \pm 1, \pm 2, \ldots),$$ (23)

where γ is any positively oriented circular path with center at a and radius r, with $r_1 < r < r_2$. The function f_1 (called the regular part of f at a) is analytic on the disk $B(a; r_2)$. The function f_2 (called the principal part of f at a) is analytic outside the closure of the disk $B(a; r_1)$.

Proof. Choose an interior point z of the annulus, keep z fixed, and define a function g on $A(a; r_1, r_2)$ as follows:

$$g(w) = \begin{cases} \dfrac{f(w) - f(z)}{w - z} & \text{if } w \neq z \\ f'(z) & \text{if } w = z. \end{cases}$$

Then g is analytic at w if $w \neq z$ and g is continuous at z. Let

$$\varphi(r) = \int_{\gamma_r} g(w) \, dw,$$

where γ_r is a positively oriented circular path with center a and radius r, with $r_1 \leq r \leq r_2$. By Theorem 16.8, $\varphi(r_1) = \varphi(r_2)$ so

$$\int_{\gamma_1} g(w) \, dw = \int_{\gamma_2} g(w) \, dw,$$ (24)

where $\gamma_1 = \gamma_{r_1}$ and $\gamma_2 = \gamma_{r_2}$. Since z is not on the graph of γ_1 or of γ_2, in each of these integrals we can write

$$g(w) = \frac{f(w)}{w-z} - \frac{f(z)}{w-z}.$$

Substituting this in (24) and transposing terms, we find

$$f(z)\left\{\int_{\gamma_2} \frac{1}{w-z} \, dw - \int_{\gamma_1} \frac{1}{w-z} \, dw\right\} = \int_{\gamma_2} \frac{f(w)}{w-z} \, dw - \int_{\gamma_1} \frac{f(w)}{w-z} \, dw.$$ (25)

But $\int_{\gamma_1} (w-z)^{-1} \, dw = 0$ since the integrand is analytic on the disk $B(a; r_1)$, and $\int_{\gamma_2} (w - z)^{-1}\, dw = 2\pi i$ since $n(\gamma_2, z) = 1$. Therefore, (25) gives us the equation

$$f(z) = f_1(z) + f_2(z),$$

where

$$f_1(z) = \frac{1}{2\pi i} \int_{\gamma_2} \frac{f(w)}{w - z}\, dw \qquad \text{and} \qquad f_2(z) = -\frac{1}{2\pi i} \int_{\gamma_1} \frac{f(w)}{w - z}\, dw.$$

By Theorem 16.19, f_1 is analytic on the disk $B(a; r_2)$ and hence we have a Taylor expansion

$$f_1(z) = \sum_{n=0}^{\infty} c_n (z - a)^n \qquad \text{for } |z - a| < r_2,$$

where

$$c_n = \frac{1}{2\pi i} \int_{\gamma_2} \frac{f(w)}{(w - a)^{n+1}}\, dw. \tag{26}$$

Moreover, by Theorem 16.8, the path γ_2 can be replaced by γ_r for any r in the interval $r_1 \le r \le r_2$.

To find a series expansion for $f_2(z)$, we argue as in the proof of Theorem 16.19, using the identity (13) with $t = (w - a)/(z - a)$. This gives us

$$\frac{1}{1 - (w - a)/(z - a)} = \sum_{n=0}^{k} \left(\frac{w - a}{z - a}\right)^n + \left(\frac{w - a}{z - a}\right)^{k+1} \left(\frac{z - a}{z - w}\right). \tag{27}$$

If w is on the graph of γ_1, we have $|w - a| = r_1 < |z - a|$, so $|t| < 1$. Now we multiply (27) by $-f(w)/(z - a)$, integrate along γ_1, and let $k \to \infty$ to obtain

$$f_2(z) = \sum_{n=1}^{\infty} b_n (z - a)^{-n} \qquad \text{for } |z - a| > r_1$$

where

$$b_n = \frac{1}{2\pi i} \int_{\gamma_1} \frac{f(w)}{(w - a)^{1-n}}\, dw. \tag{28}$$

By Theorem 16.8, the path γ_1 can be replaced by γ_r for any r in $[r_1, r_2]$. If we take the same path γ_r in both (28) and (26) and if we write c_{-n} for b_n, both formulas can be combined into one as indicated in (23). Since z was an arbitrary interior point of the annulus, this completes the proof.

NOTE. Formula (23) shows that a function can have at most one Laurent expansion in a given annulus.

16.20 ISOLATED SINGULARITIES

A disk $B(a; r)$ minus its center, that is, the set $B(a; r) - \{a\}$, is called a *deleted neighborhood* of a and is denoted by $B'(a; r)$ or $B'(a)$.

Definition 16.32. *A point a is called an isolated singularity of f if*

a) *f is analytic on a deleted neighborhood of a,*

and

b) *f is not analytic at a.*

NOTE. *f* need not be defined at *a*.

If *a* is an isolated singularity of *f*, there is an annulus $A(a; r_1, r_2)$ on which *f* is analytic. Hence *f* has a uniquely determined Laurent expansion, say

$$f(z) = \sum_{n=0}^{\infty} c_n(z - a)^n + \sum_{n=1}^{\infty} c_{-n}(z - a)^{-n}. \tag{29}$$

Since the inner radius r_1 can be arbitrarily small, (29) is valid in the deleted neighborhood $B'(a; r_2)$. The singularity *a* is classified into one of three types (depending on the form of the principal part) as follows:

If no negative powers appear in (29), that is, if $c_{-n} = 0$ for every $n = 1, 2, \ldots$, the point *a* is called a *removable singularity*. In this case, $f(z) \to c_0$ as $z \to a$ and the singularity can be removed by defining *f* at *a* to have the value $f(a) = c_0$. (See Example 1 below.)

If only a finite number of negative powers appear, that is, if $c_{-n} \neq 0$ for some *n* but $c_{-m} = 0$ for every $m > n$, the point *a* is said to be a *pole* of order *n*. In this case, the principal part is simply a finite sum, namely,

$$\frac{c_{-1}}{z - a} + \frac{c_{-2}}{(z - a)^2} + \cdots + \frac{c_{-n}}{(z - a)^n}.$$

A pole of order 1 is usually called a *simple* pole. If there is a pole at *a*, then $|f(z)| \to \infty$ as $z \to a$.

Finally, if $c_{-n} \neq 0$ for infinitely many values of *n*, the point *a* is called an *essential singularity*. In this case, $f(z)$ does not tend to a limit as $z \to a$.

Example 1. *Removable singularity.* Let $f(z) = (\sin z)/z$ if $z \neq 0$, $f(0) = 0$. This function is analytic everywhere except at 0. (It is discontinuous at 0, since $(\sin z)/z \to 1$ as $z \to 0$.) The Laurent expansion about 0 has the form

$$\frac{\sin z}{z} = 1 - \frac{z^2}{3!} + \frac{z^4}{5!} - + \cdots.$$

Since no negative powers of *z* appear, the point 0 is a removable singularity. If we redefine *f* to have the value 1 at 0, the modified function becomes analytic at 0.

Example 2. *Pole.* Let $f(z) = (\sin z)/z^5$ if $z \neq 0$. The Laurent expansion about 0 is

$$\frac{\sin z}{z^5} = z^{-4} - \frac{1}{3!} z^{-2} + \frac{1}{5!} - \frac{1}{7!} z^2 + \cdots.$$

In this case, the point 0 is a pole of order 4. Note that nothing has been said about the value of *f* at 0.

Example 3. *Essential singularity.* Let $f(z) = e^{1/z}$ if $z \neq 0$. The point 0 is an essential singularity, since

$$e^{1/z} = 1 + z^{-1} + \frac{1}{2!} z^{-2} + \cdots + \frac{1}{n!} z^{-n} + \cdots$$

Theorem 16.33. *Assume that f is analytic on an open region S in $\mathbf{C}$ and define g by the equation $g(z) = 1/f(z)$ if $f(z) \neq 0$. Then f has a zero of order k at a point a in S if, and only if, g has a pole of order k at a.*

Proof. If f has a zero of order k at a, there is a deleted neighborhood $B'(a)$ in which f does not vanish. In the neighborhood $B(a)$ we have $f(z) = (z - a)^k h(z)$, where $h(z) \neq 0$ if $z \in B(a)$. Hence, $1/h$ is analytic in $B(a)$ and has an expansion

$$\frac{1}{h(z)} = b_0 + b_1(z - a) + \cdots, \qquad \text{where } b_0 = \frac{1}{h(a)} \neq 0.$$

Therefore, if $z \in B'(a)$, we have

$$g(z) = \frac{1}{(z - a)^k h(z)} = \frac{b_0}{(z - a)^k} + \frac{b_1}{(z - a)^{k-1}} + \cdots,$$

and hence a is a pole of order k for g. The converse is similarly proved.

16.21 THE RESIDUE OF A FUNCTION AT AN ISOLATED SINGULAR POINT

If a is an isolated singular point of f, there is a deleted neighborhood $B'(a)$ on which f has a Laurent expansion, say

$$f(z) = \sum_{n=0}^{\infty} c_n(z - a)^n + \sum_{n=1}^{\infty} c_{-n}(z - a)^{-n}. \tag{30}$$

The coefficient c_{-1} which multiplies $(z - a)^{-1}$ is called the *residue* of f at a and is denoted by the symbol

$$c_{-1} = \operatorname*{Res}_{z=a} f(z).$$

Formula (23) tells us that

$$\int_\gamma f(z)\, dz = 2\pi i \operatorname*{Res}_{z=a} f(z), \tag{31}$$

if γ is any positively oriented circular path with center at a whose graph lies in the disk $B(a)$.

In many cases it is relatively easy to evaluate the residue at a point without the use of integration. For example, if a is a simple pole, we can use formula (30) to obtain

$$\operatorname*{Res}_{z=a} f(z) = \lim_{z \to a} (z - a)f(z). \tag{32}$$

Similarly, if a is a pole of order 2, it is easy to show that

$$\operatorname*{Res}_{z=a} f(z) = g'(a), \qquad \text{where } g(z) = (z - a)^2 f(z).$$

In cases like this, where the residue can be computed very easily, (31) gives us a simple method for evaluating contour integrals around circuits.

Cauchy was the first to exploit this idea and he developed it into a powerful method known as the *residue calculus*. It is based on the *Cauchy residue theorem* which is a generalization of (31).

16.22 THE CAUCHY RESIDUE THEOREM

Theorem 16.34. *Let f be analytic on an open region S except for a finite number of isolated singularities $z_1, \ldots, z_n$ in S. Let γ be a circuit which is homotopic to a point in S, and assume that none of the singularities lies on the graph of γ. Then we have*

$$\int_{\gamma} f(z) \, dz = 2\pi i \sum_{k=1}^{n} n(\gamma, z_k) \operatorname*{Res}_{z=z_k} f(z), \qquad (33)$$

where $n(\gamma, z_k)$ is the winding number of γ with respect to z_k.

Proof. The proof is based on the following formula, where m denotes an integer (positive, negative, or zero):

$$\int_{\gamma} (z - z_k)^m \, dz = \begin{cases} 2\pi i n(\gamma, z_k) & \text{if } m = -1, \\ 0 & \text{if } m \neq -1. \end{cases} \qquad (34)$$

The formula for $m = -1$ is just the definition of the winding number $n(\gamma, z_k)$. Let $[a, b]$ denote the domain of γ. If $m \neq -1$, let $g(t) = \{\gamma(t) - z_k\}^{m+1}$ for t in $[a, b]$. Then we have

$$\int_{\gamma} (z - z_k)^m \, dz = \int_a^b \{\gamma(t) - z_k\}^m \gamma'(t) \, dt = \frac{1}{m + 1} \int_a^b g'(t) \, dt$$

$$= \frac{1}{m + 1} \{g(b) - g(a)\} = 0,$$

since $g(b) = g(a)$. This proves (34).

To prove the residue theorem, let f_k denote the principal part of f at the point z_k. By Theorem 16.31, f_k is analytic everywhere in $\mathbf{C}$ except at z_k. Therefore $f - f_1$ is analytic in S except at $z_2, \ldots, z_n$. Similarly, $f - f_1 - f_2$ is analytic in S except at $z_3, \ldots, z_n$ and, by induction, we find that $f - \sum_{k=1}^n f_k$ is analytic everywhere in S. Therefore, by Cauchy's integral theorem, $\int_{\gamma} (f - \sum_{k=1}^n f_k) = 0$, or

$$\int_{\gamma} f = \sum_{k=1}^{n} \int_{\gamma} f_k.$$

Now we express f_k as a Laurent series about z_k and integrate this series term by term, using (34) and the definition of residue to obtain (33).

NOTE. If γ is a positively oriented Jordan curve with graph Γ, then $n(\gamma, z_k) = 1$ for each z_k inside Γ, and $n(\gamma, z_k) = 0$ for each z_k outside Γ. In this case, the integral of f along γ is $2\pi i$ times the sum of the residues at those singularities lying inside Γ.

Some of the applications of the Cauchy residue theorem are given in the next few sections.

16.23 COUNTING ZEROS AND POLES IN A REGION

If f is analytic or has a pole at a, and if f is not identically 0, the Laurent expansion about a has the form

$$f(z) = \sum_{n=m}^{\infty} c_n(z - a)^n,$$

where $c_m \neq 0$. If $m > 0$ there is a zero at a of order m; if $m < 0$ there is a pole at a of order $-m$, and if $m = 0$ there is neither a zero nor a pole at a.

NOTE. We also write $m(f; a)$ for m to emphasize that m depends on both f and a.

Theorem 16.35. *Let f be a function, not identically zero, which is analytic on an open region S, except possibly for a finite number of poles. Let γ be a circuit which is homotopic to a point in S and whose graph contains no zero or pole of f. Then we have*

$$\frac{1}{2\pi i} \int_\gamma \frac{f'(z)}{f(z)} \, dz = \sum_{a \in S} n(\gamma, a) m(f; a), \tag{35}$$

where the sum on the right contains only a finite number of nonzero terms.

NOTE. If γ is a positively oriented Jordan curve with graph Γ, then $n(\gamma, a) = 1$ for each a inside Γ and (35) is usually written in the form

$$\frac{1}{2\pi i} \int_\gamma \frac{f'(z)}{f(z)} \, dz = N - P, \tag{36}$$

where N denotes the number of zeros and P the number of poles of f inside Γ, each counted as often as its order indicates.

Proof. Suppose that in a deleted neighborhood of a point a we have $f(z) = (z - a)^m g(z)$, where g is analytic at a and $g(a) \neq 0$, m being an integer (positive or negative). Then there is a deleted neighborhood of a on which we can write

$$\frac{f'(z)}{f(z)} = \frac{m}{z - a} + \frac{g'(z)}{g(z)},$$

the quotient g'/g being analytic at a. This equation tells us that a zero of f of order m is a simple pole of f'/f with residue m. Similarly, a pole of f of order m is a simple pole of f'/f with residue $-m$. This fact, used in conjunction with Cauchy's residue theorem, yields (35).

16.24 EVALUATION OF REAL-VALUED INTEGRALS BY MEANS OF RESIDUES

Cauchy's residue theorem can sometimes be used to evaluate real-valued Riemann integrals. There are several techniques available, depending on the form of the integral. We shall describe briefly two of these methods.

The first method deals with integrals of the form $\int_0^{2\pi} R(\sin\theta, \cos\theta)\, d\theta$, where R is a rational function* of two variables.

Theorem 16.36. *Let R be a rational function of two variables and let*

$$f(z) = R\left(\frac{z^2 - 1}{2iz}, \frac{z^2 + 1}{2z}\right),$$

whenever the expression on the right is finite. Let γ denote the positively oriented unit circle with center at 0. Then

$$\int_0^{2\pi} R(\sin\theta, \cos\theta)\, d\theta = \int_\gamma \frac{f(z)}{iz}\, dz, \tag{37}$$

provided that f has no poles on the graph of γ.

Proof. Since $\gamma(\theta) = e^{i\theta}$ with $0 \le \theta \le 2\pi$, we have

$$\gamma'(\theta) = i\gamma(\theta), \quad \frac{\gamma(\theta)^2 - 1}{2i\gamma(\theta)} = \sin\theta, \quad \frac{\gamma(\theta)^2 + 1}{2\gamma(\theta)} = \cos\theta,$$

and (37) follows at once from Theorem 16.7.

NOTE. To evaluate the integral on the right of (37), we need only compute the residues of the integrand at those poles which lie inside the unit circle.

Example. Evaluate $I = \int_0^{2\pi} d\theta/(a + \cos\theta)$, where a is real, $|a| > 1$. Applying (37), we find

$$I = -2i \int_\gamma \frac{dz}{z^2 + 2az + 1}.$$

The integrand has simple poles at the roots of the equation $z^2 + 2az + 1 = 0$. These are the points

$$z_1 = -a + \sqrt{a^2 - 1},$$

$$z_2 = -a - \sqrt{a^2 - 1}.$$

---

* A function P defined on $\mathbf{C} \times \mathbf{C}$ by an equation of the form

$$P(z_1, z_2) = \sum_{m=0}^{p} \sum_{n=0}^{q} a_{m,n} z_1^m z_2^n$$

is called a *polynomial in two variables*. The coefficients $a_{m,n}$ may be real or complex. The quotient of two such polynomials is called a *rational function of two variables*.

The corresponding residues R_1 and R_2 are given by

$$R_1 = \lim_{z \to z_1} \frac{z - z_1}{z^2 + 2az + 1} = \frac{1}{z_1 - z_2},$$

$$R_2 = \lim_{z \to z_2} \frac{z - z_2}{z^2 + 2az + 1} = \frac{1}{z_2 - z_1}.$$

If $a > 1$, z_1 is inside the unit circle, z_2 is outside, and $I = 4\pi/(z_1 - z_2) = 2\pi/\sqrt{a^2 - 1}$. If $a < -1$, z_2 is inside, z_1 is outside, and we get $I = -2\pi/\sqrt{a^2 - 1}$.

Many improper integrals can be dealt with by means of the following theorem:

Theorem 16.37. *Let $T = \{x + iy : y \geq 0\}$ denote the upper half-plane. Let S be an open region in $\mathbf{C}$ which contains T and suppose f is analytic on S, except, possibly, for a finite number of poles. Suppose further that none of these poles is on the real axis. If*

$$\lim_{R \to +\infty} \int_0^\pi f(Re^{i\theta}) \, Re^{i\theta} \, d\theta = 0, \tag{38}$$

then

$$\lim_{R \to +\infty} \int_{-R}^R f(x) \, dx = 2\pi i \sum_{k=1}^n \operatorname*{Res}_{z=z_k} f(z). \tag{39}$$

where $z_1, \ldots, z_n$ are the poles of f which lie in T.

Proof. Let γ be the positively oriented path formed by taking a portion of the real axis from $-R$ to R and a semicircle in T having $[-R, R]$ as its diameter, where R is taken large enough to enclose all the poles $z_1, \ldots, z_n$. Then

$$2\pi i \sum_{k=1}^n \operatorname*{Res}_{z=z_k} f(z) = \int_\gamma f(z) \, dz = \int_{-R}^R f(x) \, dx + i \int_0^\pi f(Re^{i\theta}) \, Re^{i\theta} \, d\theta.$$

When $R \to +\infty$, the last integral tends to zero by (38) and we obtain (39).

NOTE. Equation (38) is automatically satisfied if f is the quotient of two poly-nomials, say $f = P/Q$, provided that the degree of Q exceeds the degree of P by at least 2. (See Exercise 16.36.)

Example. To evaluate $\int_{-\infty}^\infty dx/(1 + x^4)$, let $f(z) = 1/(z^4 + 1)$. Then $P(z) = 1$, $Q(z) = 1 + z^4$, and hence (38) holds. The poles of f are the roots of the equation $1 + z^4 = 0$. These are z_1, z_2, z_3, z_4, where

$$z_k = e^{(2k-1)\pi i/4} \qquad (k = 1, 2, 3, 4).$$

Of these, only z_1 and z_2 lie in the upper half-plane. The residue at z_1 is

$$\operatorname*{Res}_{z=z_1} f(z) = \lim_{z \to z_1} (z - z_1) f(z) = \frac{1}{(z_1 - z_2)(z_1 - z_3)(z_1 - z_4)} = \frac{e^{-\pi i/4}}{4i}.$$

Similarly, we find $\text{Res}_{z=z_2} f(z) = (1/4i)e^{\pi i/4}$. Therefore,

$$\int_{-\infty}^{\infty} \frac{dx}{1 + x^4} = \frac{2\pi i}{4i}(e^{-\pi i/4} + e^{\pi i/4}) = \pi \cos\frac{\pi}{4} = \frac{\pi}{2}\sqrt{2}.$$

16.25 EVALUATION OF GAUSS'S SUM BY RESIDUE CALCULUS

The residue theorem is often used to evaluate sums by integration. We illustrate with a famous example called *Gauss's sum* $G(n)$, defined by the formula

$$G(n) = \sum_{r=0}^{n-1} e^{2\pi i r^2/n}, \tag{40}$$

where $n \geq 1$. This sum occurs in various parts of the Theory of Numbers. For small values of n it can easily be computed from its definition. For example, we have

$$G(1) = 1, \quad G(2) = 0, \quad G(3) = i\sqrt{3}, \quad G(4) = 2(1 + i).$$

Although each term of the sum has absolute value 1, the sum itself has absolute value 0, $\sqrt{n}$, or $\sqrt{2n}$. In fact, Gauss proved the remarkable formula

$$G(n) = \tfrac{1}{2}\sqrt{n}(1 + i)(1 + e^{-\pi i n/2}), \tag{41}$$

for every $n \geq 1$. A number of different proofs of (41) are known. We will deduce (41) by considering a more general sum $S(a, n)$ introduced by Dirichlet,

$$S(a, n) = \sum_{r=0}^{n-1} e^{\pi i a r^2/n},$$

where n and a are positive integers. If $a = 2$, then $S(2, n) = G(n)$. Dirichlet proved (41) as a corollary of a reciprocity law for $S(a, n)$ which can be stated as follows:

Theorem 16.38. *If the product na is even, we have*

$$S(a, n) = \sqrt{\frac{n}{a}}\left(\frac{1 + i}{\sqrt{2}}\right)\overline{S(n, a)}, \tag{42}$$

where the bar denotes the complex conjugate.

NOTE. To deduce Gauss's formula (41), we take $a = 2$ in (42), and observe that $\overline{S(n, 2)} = 1 + e^{-\pi i n/2}$.

Proof. The proof given here is particularly instructive because it illustrates several techniques used in complex analysis. Some minor computational details are left as exercises for the reader.

Let g be the function defined by the equation

$$g(z) = \sum_{r=0}^{n-1} e^{\pi i a(z+r)^2/n}. \tag{43}$$

Then g is analytic everywhere, and $g(0) = S(a, n)$. Since na is even we find

$$g(z + 1) - g(z) = e^{\pi i a z^2/n}(e^{2\pi i a z} - 1) = e^{\pi i a z^2/n}(e^{2\pi i z} - 1) \sum_{m=0}^{a-1} e^{2\pi i m z},$$

(Exercise 16.41). Now define f by the equation

$$f(z) = g(z)/(e^{2\pi i z} - 1).$$

Then f is analytic everywhere except for a first-order pole at each integer, and f satisfies the equation

$$f(z + 1) = f(z) + \varphi(z), \tag{44}$$

where

$$\varphi(z) = e^{\pi i a z^2/n} \sum_{m=0}^{a-1} e^{2\pi i m z}. \tag{45}$$

The function φ is analytic everywhere.

At $z = 0$ the residue of f is $g(0)/(2\pi i)$ (Exercise 16.41), and hence

$$S(a, n) = g(0) = 2\pi i \operatorname*{Res}_{z=0} f(z) = \int_\gamma f(z)\, dz, \tag{46}$$

where γ is any positively oriented simple closed path whose graph contains only the pole $z = 0$ in its interior region. We will choose γ so that it describes a parallelogram with vertices A, $A + 1$, $B + 1$, B, where

$$A = -\tfrac{1}{2} - Re^{\pi i/4} \qquad \text{and} \qquad B = -\tfrac{1}{2} + Re^{\pi i/4},$$

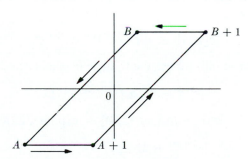

Figure 16.7

as shown in Fig. 16.7. Integrating f along γ we have

$$\int_\gamma f = \int_A^{A+1} f + \int_{A+1}^{B+1} f + \int_{B+1}^{B} f + \int_B^A f.$$

In the integral $\int_{A+1}^{B+1} f$ we make the change of variable $w = z + 1$ and then use (44) to get

$$\int_{A+1}^{B+1} f(w)\, dw = \int_A^B f(z + 1)\, dz = \int_A^B f(z)\, dz + \int_A^B \varphi(z)\, dz.$$

Therefore (46) becomes

$$S(a, n) = \int_A^B \varphi(z) \, dz + \int_A^{A+1} f(z) \, dz - \int_B^{B+1} f(z) \, dz. \qquad (47)$$

Now we show that the integrals along the horizontal segments from A to $A + 1$ and from B to $B + 1$ tend to 0 as $R \to +\infty$. To do this we estimate the integrand on these segments. We write

$$|f(z)| = \frac{|g(z)|}{|e^{2\pi i z} - 1|}, \qquad (48)$$

and estimate the numerator and denominator separately.

On the segment joining B to $B + 1$ we let

$$\gamma(t) = t + Re^{\pi i/4}, \qquad \text{where } -\tfrac{1}{2} \le t \le \tfrac{1}{2}.$$

From (43) we find

$$|g[\gamma(t)]| \le \sum_{r=0}^{n-1} \left| \exp \left\{ \frac{\pi i a(t + Re^{\pi i/4} + r)^2}{n} \right\} \right|, \qquad (49)$$

where $\exp z = e^z$. The expression in braces has real part (Exercise 16.41)

$$-\pi a(\sqrt{2}tR + R^2 + \sqrt{2}rR)/n.$$

Since $|e^{x+iy}| = e^x$ and $\exp \{-\pi a\sqrt{2}rR/n\} \le 1$, each term in (49) has absolute value not exceeding $\exp \{-\pi aR^2/n\} \exp \{-\sqrt{2}\pi atR/n\}$. But $-\tfrac{1}{2} \le t \le \tfrac{1}{2}$, so we obtain the estimate

$$|g[\gamma(t)]| \le n \, e^{\pi\sqrt{2}aR/(2n)} \, e^{-\pi aR^2/n}.$$

For the denominator in (48) we use the triangle inequality in the form

$$|e^{2\pi i z} - 1| \ge \big| |e^{2\pi i z}| - 1 \big|.$$

Since $|\exp \{2\pi i\gamma(t)\}| = \exp \{-2\pi R \sin (\pi/4)\} = \exp \{-\sqrt{2}\pi R\}$, we find

$$|e^{2\pi i \gamma(t)} - 1| \ge 1 - e^{-\sqrt{2}\pi R}.$$

Therefore on the line segment joining B to $B + 1$ we have the estimate

$$|f(z)| \le \frac{n e^{\pi\sqrt{2}aR/(2n)} \, e^{-\pi aR^2/n}}{1 - e^{-\sqrt{2}\pi R}} = o(1) \qquad \text{as } R \to +\infty.$$

Here $o(1)$ denotes a function of R which tends to 0 as $R \to +\infty$.

A similar argument shows that the integrand tends to 0 on the segment joining A to $A + 1$ as $R \to +\infty$. Since the length of the path of integration is 1 in each case, this shows that the second and third integrals on the right of (47) tend to 0

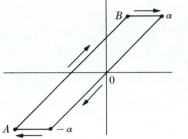

Figure 16.8

as $R \to +\infty$. Therefore we can write (47) in the form

$$S(a, n) = \int_A^B \varphi(z) \, dz + o(1) \qquad \text{as } R \to +\infty. \tag{50}$$

To deal with the integral $\int_A^B \varphi$ we apply Cauchy's theorem, integrating φ around the parallelogram with vertices A, B, α, $-\alpha$, where $\alpha = B + \frac{1}{2} = Re^{\pi i/4}$. (See Fig. 16.8.) Since φ is analytic everywhere, its integral around this parallelogram is 0, so

$$\int_A^B \varphi + \int_B^\alpha \varphi + \int_\alpha^{-\alpha} \varphi + \int_{-\alpha}^A \varphi = 0. \tag{51}$$

Because of the exponential factor $e^{\pi i a z^2/n}$ in (45), an argument similar to that given above shows that the integral of φ along each horizontal segment $\to 0$ as $R \to +\infty$. Therefore (51) gives us

$$\int_A^B \varphi = \int_{-\alpha}^\alpha \varphi + o(1) \qquad \text{as } R \to +\infty,$$

and (50) becomes

$$S(a, n) = \int_{-\alpha}^\alpha \varphi(z) \, dz + o(1) \qquad \text{as } R \to +\infty, \tag{52}$$

where $\alpha = Re^{\pi i/4}$. Using (45) we find

$$\int_{-\alpha}^\alpha \varphi(z) \, dz = \sum_{m=0}^{a-1} \int_{-\alpha}^\alpha e^{\pi i a z^2/n} e^{2\pi i m z} \, dz = \sum_{m=0}^{a-1} e^{-\pi i n m^2/a} I(a, m, n, R),$$

where

$$I(a, m, n, R) = \int_{-\alpha}^\alpha \exp\left\{ \frac{\pi i a}{n} \left(z + \frac{nm}{a} \right)^2 \right\} dz.$$

Applying Cauchy's theorem again to the parallelogram with vertices $-\alpha$, α, $\alpha - nm/a$, $-\alpha - nm/a$, we find as before that the integrals along the horizontal segments $\to 0$ as $R \to +\infty$, so

$$I(a, m, n, R) = \int_{-\alpha - nm/a}^{\alpha - mn/a} \exp\left\{ \frac{\pi i a}{n} \left(z + \frac{nm}{a} \right)^2 \right\} dz + o(1) \qquad \text{as } R \to +\infty.$$

The change of variable $w = \sqrt{a/n}(z + nm/a)$ puts this into the form

$$I(a, m, n, R) = \sqrt{\frac{n}{a}} \int_{-\alpha\sqrt{a/n}}^{\alpha\sqrt{a/n}} e^{\pi i w^2} \, dw + o(1) \qquad \text{as } R \to +\infty.$$

Letting $R \to +\infty$ in (52), we find

$$S(a, n) = \sum_{m=0}^{a-1} e^{-\pi i n m^2/a} \sqrt{\frac{n}{a}} \lim_{R \to +\infty} \int_{-R\sqrt{a/n}e^{\pi i/4}}^{R\sqrt{a/n}e^{\pi i/4}} e^{\pi i w^2} \, dw. \qquad (53)$$

By writing $T = \sqrt{a/n}R$, we see that the last limit is equal to

$$\lim_{T \to +\infty} \int_{-Te^{\pi i/4}}^{Te^{\pi i/4}} e^{\pi i w^2} \, dw = I.$$

say, where I is a number independent of a and n. Therefore (53) gives us

$$S(a, n) = \sqrt{\frac{n}{a}} I S(n, a). \qquad (54)$$

To evaluate I we take $a = 1$ and $n = 2$ in (54). Then $S(1, 2) = 1 + i$ and $S(2, 1) = 1$, so (54) implies $I = (1 + i)/\sqrt{2}$, and (54) reduces to (42).

16.26 APPLICATION OF THE RESIDUE THEOREM TO THE INVERSION FORMULA FOR LAPLACE TRANSFORMS

The following theorem is, in many cases, the easiest method for evaluating the limit which appears in the inversion formula for Laplace transforms. (See Exercise 11.38.)

Theorem 16.39. *Let F be a function analytic everywhere in* $\mathbf{C}$ *except, possibly, for a finite number of poles. Suppose there exist three positive constants M, b, c such that*

$$|F(z)| < \frac{M}{|z|^c} \qquad \text{whenever } |z| \geq b.$$

Let a be a positive number such that the vertical line $x = a$ contains no poles of F and let $z_1, \ldots, z_n$ denote the poles of F which lie to the left of this line. Then, for each real $t > 0$, we have

$$\lim_{T \to +\infty} \int_{-T}^{T} e^{(a+iv)t} F(a + iv) \, dv = 2\pi \sum_{k=1}^{n} \operatorname*{Res}_{z=z_k} \{e^{zt} F(z)\}. \qquad (55)$$

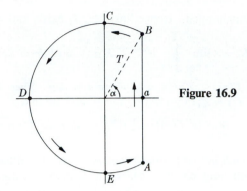

Figure 16.9

Proof. We apply Cauchy's residue theorem to the positively oriented path Γ shown in Fig. 16.9, where the radius T of the circular part is taken large enough to enclose all the poles of F which lie to the left of the line $x = a$, and also $T > b$. The residue theorem gives us

$$\int_{\Gamma} e^{zt} F(z) \, dz = 2\pi i \sum_{k=1}^{n} \operatorname*{Res}_{z=z_k} \{e^{zt} F(z)\}. \tag{56}$$

Now write

$$\int_{\Gamma} = \int_{A}^{B} + \int_{B}^{C} + \int_{C}^{D} + \int_{D}^{E} + \int_{E}^{A},$$

where A, B, C, D, E are the points indicated in Fig. 16.9, and denote these integrals by I_1, I_2, I_3, I_4, I_5. We will prove that $I_k \to 0$ as $T \to +\infty$ when $k > 1$.

First, we have

$$|I_2| < \frac{M}{T^c} \int_{\alpha}^{\pi/2} e^{tT \cos \theta} \, T \, d\theta \le \frac{Me^{at}}{T^{c-1}} \left(\frac{\pi}{2} - \alpha \right) = \frac{Me^{at}}{T^c} \, T \arcsin \left(\frac{a}{T} \right).$$

Since $T \arcsin (a/T) \to a$ as $T \to +\infty$, it follows that $I_2 \to 0$ as $T \to +\infty$. In the same way we prove $I_5 \to 0$ as $T \to +\infty$.

Next, consider I_3. We have

$$|I_3| < \frac{M}{T^{c-1}} \int_{\pi/2}^{\pi} e^{tT \cos \theta} \, d\theta = \frac{M}{T^{c-1}} \int_{0}^{\pi/2} e^{-tT \sin \varphi} \, d\varphi.$$

But $\sin \varphi \ge 2\varphi/\pi$ if $0 \le \varphi \le \pi/2$, and hence

$$|I_3| < \frac{M}{T^{c-1}} \int_{0}^{\pi/2} e^{-2tT\varphi/\pi} \, d\varphi = \frac{\pi M}{2tT^c} (1 - e^{-tT}) \to 0 \qquad \text{as } T \to +\infty.$$

Similarly, we find $I_4 \to 0$ as $T \to +\infty$. But as $T \to +\infty$ the righthand side of

(56) remains unchanged. Hence $\lim_{T \to +\infty} I_1$ exists and we have

$$\lim_{T \to +\infty} I_1 = \lim_{T \to +\infty} \int_{-T}^{T} e^{(a+iv)t} F(a + iv) \, i \, dv = 2\pi i \sum_{k=1}^{n} \operatorname*{Res}_{z=z_k} \{ e^{zt} F(z) \}.$$

Example. Let $F(z) = z/(z^2 + \alpha^2)$, where α is real. Then F has simple poles at $\pm i\alpha$. Since $z/(z^2 + \alpha^2) = \frac{1}{2}[1/(z + i\alpha) + 1/(z - i\alpha)]$, we find

$$\operatorname*{Res}_{z=i\alpha} \{ e^{zt} F(z) \} = \tfrac{1}{2} e^{i\alpha t}, \qquad \operatorname*{Res}_{z=-i\alpha} \{ e^{zt} F(t) \} = \tfrac{1}{2} e^{-i\alpha t}.$$

Therefore the limit in (55) has the value $2\pi i \cos \alpha t$. From Exercise 11.38 we see that the function f, continuous on $(0, +\infty)$, whose Laplace transform is F, is given by $f(t) = \cos \alpha t$.

16.27 CONFORMAL MAPPINGS

An analytic function f will map two line segments, intersecting at a point c, into two curves intersecting at $f(c)$. In this section we show that the tangent lines to these curves intersect at the same angle as the given line segments if $f'(c) \neq 0$.

This property is geometrically obvious for linear functions. For example, suppose $f(z) = z + b$. This represents a translation which moves every line parallel to itself, and it is clear that angles are preserved. Another example is $f(z) = az$, where $a \neq 0$. If $|a| = 1$, then $a = e^{i\alpha}$ and this represents a rotation about the origin through an angle α. If $|a| \neq 1$, then $a = Re^{i\alpha}$ and f represents a rotation composed with a stretching (if $R > 1$) or a contraction (if $R < 1$). Again, angles are preserved. A general linear function $f(z) = az + b$ with $a \neq 0$ is a composition of these types and hence also preserves angles.

In the general case, differentiability at c means that we have a linear approximation near c, say $f(z) = f(c) + f'(c)(z - c) + o(z - c)$, and if $f'(c) \neq 0$ we can expect angles to be preserved near c.

To formalize these ideas, let γ_1 and γ_2 be two piecewise smooth paths with respective graphs Γ_1 and Γ_2, intersecting at c. Suppose that γ_1 is one-to-one on an interval containing t_1, and that γ_2 is one-to-one on an interval containing t_2, where $\gamma_1(t_1) = \gamma_2(t_2) = c$. Assume also that $\gamma_1'(t_1) \neq 0$ and $\gamma_2'(t_2) \neq 0$. The difference

$$\arg [\gamma_2'(t_2)] - \arg [\gamma_1'(t_1)],$$

is called the *angle* from Γ_1 to Γ_2 at c.

Now assume that $f'(c) \neq 0$. Then (by Theorem 13.4) there is a disk $B(c)$ on which f is one-to-one. Hence the composite functions

$$w_1(t) = f[\gamma_1(t)] \qquad \text{and} \qquad w_2(t) = f[\gamma_2(t)],$$

will be locally one-to-one near t_1 and t_2, respectively, and will describe arcs C_1 and C_2 intersecting at $f(c)$. (See Fig. 16.10.) By the chain rule we have

$$w_1'(t_1) = f'(c)\gamma_1'(t_1) \neq 0 \qquad \text{and} \qquad w_2'(t_2) = f'(c)\gamma_2'(t_2) \neq 0.$$

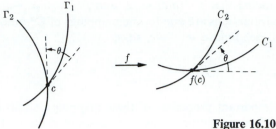

Figure 16.10

Therefore, by Theorem 1.48 there exist integers n_1 and n_2 such that

$$\arg\,[w_1'(t_1)] = \arg\,[f'(c)] + \arg\,[\gamma_1'(t_1)] + 2\pi n_1,$$
$$\arg\,[w_2'(t_2)] = \arg\,[f'(c)] + \arg\,[\gamma_2'(t_2)] + 2\pi n_2,$$

so the angle from C_1 to C_2 at $f(c)$ is equal to the angle from Γ_1 to Γ_2 at c plus an integer multiple of 2π. For this reason we say that f *preserves angles* at c. Such a function is also said to be *conformal* at c.

Angles are not preserved at points where the derivative is zero. For example, if $f(z) = z^2$, a straight line through the origin making an angle α with the real axis is mapped by f onto a straight line making an angle 2α with the real axis. In general, when $f'(c) = 0$, the Taylor expansion of f assumes the form

$$f(z) - f(c) = (z - c)^k[a_k + a_{k+1}(z - c) + \cdots],$$

where $k \geq 2$. Using this equation, it is easy to see that angles between curves intersecting at c are multiplied by a factor k under the mapping f.

Among the important examples of conformal mappings are the *Möbius transformations*. These are functions f defined as follows: If a, b, c, d are four complex numbers such that $ad - bc \neq 0$, we define

$$f(z) = \frac{az + b}{cz + d}, \tag{57}$$

whenever $cz + d \neq 0$. It is convenient to define f everywhere on the extended plane $\mathbf{C}^*$ by setting $f(-d/c) = \infty$ and $f(\infty) = a/c$. (If $c = 0$, these last two equations are to be replaced by the single equation $f(\infty) = \infty$.) Now (57) can be solved for z in terms of $f(z)$ to get

$$z = \frac{-df(z) + b}{cf(z) - a}.$$

This means that the inverse function f^{-1} exists and is given by

$$f^{-1}(z) = \frac{-dz + b}{cz - a},$$

with the understanding that $f^{-1}(a/c) = \infty$ and $f^{-1}(\infty) = -d/c$. Thus we see that Möbius transformations are one-to-one mappings of C^* onto itself. They are also conformal at each finite $z \neq -d/c$, since

$$f'(z) = \frac{bc - ad}{(cz + d)^2} \neq 0.$$

One of the most important properties of these mappings is that they map circles onto circles (including straight lines as special cases of circles). The proof of this is sketched in Exercise 16.46. Further properties of Möbius transformations are also described in the exercises near the end of the chapter.

EXERCISES

Complex integration; Cauchy's integral formulas

16.1 Let γ be a piecewise smooth path with domain $[a, b]$ and graph Γ. Assume that the integral $\int_\gamma f$ exists. Let S be an open region containing Γ and let g be a function such that $g'(z)$ exists and equals $f(z)$ for each z on Γ. Prove that

$$\int_\gamma f = \int_\gamma g' = g(B) - g(A), \qquad \text{where } A = \gamma(a) \text{ and } B = \gamma(b).$$

In particular, if γ is a circuit, then $A = B$ and the integral is 0. *Hint.* Apply Theorem 7.34 to each interval of continuity of γ'.

16.2 Let γ be a positively oriented circular path with center 0 and radius 2. Verify each of the following by using one of Cauchy's integral formulas.

a) $\displaystyle\int_\gamma \frac{e^z}{z} \, dz = 2\pi i.$

b) $\displaystyle\int_\gamma \frac{e^z}{z^3} \, dz = \pi i.$

c) $\displaystyle\int_\gamma \frac{e^z}{z^4} \, dz = \frac{\pi i}{3}.$

d) $\displaystyle\int_\gamma \frac{e^z}{z - 1} \, dz = 2\pi i e.$

e) $\displaystyle\int_\gamma \frac{e^z}{z(z - 1)} \, dz = 2\pi i(e - 1).$

f) $\displaystyle\int_\gamma \frac{e^z}{z^2(z - 1)} \, dz = 2\pi i(e - 2).$

16.3 Let $f = u + iv$ be analytic on a disk $B(a; R)$. If $0 < r < R$, prove that

$$f'(a) = \frac{1}{\pi r} \int_0^{2\pi} u(a + re^{i\theta})e^{-i\theta} \, d\theta.$$

16.4 a) Prove the following stronger version of Liouville's theorem: *If f is an entire function such that $\lim_{z \to \infty} |f(z)/z| = 0$, then f is a constant.*

b) What can you conclude about an entire function which satisfies an inequality of the form $|f(z)| \leq M|z|^c$ for every complex z, where $c > 0$?

16.5 Assume that f is analytic on $B(0; R)$. Let γ denote the positively oriented circle with center at 0 and radius r, where $0 < r < R$. If a is inside γ, show that

$$f(a) = \frac{1}{2\pi i} \int_\gamma f(z) \left\{ \frac{1}{z - a} - \frac{1}{z - r^2/\bar{a}} \right\} dz.$$

If $a = Ae^{i\alpha}$, show that this reduces to the formula

$$f(a) = \frac{1}{2\pi} \int_0^{2\pi} \frac{(r^2 - A^2)f(re^{i\theta})}{r^2 - 2rA\cos(\alpha - \theta) + A^2} \, d\theta.$$

By equating the real parts of this equation we obtain an expression known as *Poisson's integral formula*.

16.6 Assume that f is analytic on the closure of the disk $B(0; 1)$. If $|a| < 1$, show that

$$(1 - |a|^2)f(a) = \frac{1}{2\pi i} \int_\gamma f(z) \frac{1 - z\bar{a}}{z - a} \, dz,$$

where γ is the positively oriented unit circle with center at 0. Deduce the inequality

$$(1 - |a|^2)|f(a)| \leq \frac{1}{2\pi} \int_0^{2\pi} |f(e^{i\theta})| \, d\theta.$$

16.7 Let $f(z) = \sum_{n=0}^{\infty} 2^n z^n / 3^n$ if $|z| < 3/2$, and let $g(z) = \sum_{n=0}^{\infty} (2z)^{-n}$ if $|z| > \frac{1}{2}$. Let γ be the positively oriented circular path of radius 1 and center 0, and define $h(a)$ for $|a| \neq 1$ as follows:

$$h(a) = \frac{1}{2\pi i} \int_\gamma \left(\frac{f(z)}{z - a} + \frac{a^2 g(z)}{z^2 - az} \right) dz.$$

Prove that

$$h(a) = \begin{cases} \dfrac{3}{3 - 2a} & \text{if } |a| < 1, \\[2ex] \dfrac{2a^2}{1 - 2a} & \text{if } |a| > 1. \end{cases}$$

Taylor expansions

16.8 Define f on the disk $B(0; 1)$ by the equation $f(z) = \sum_{n=0}^{\infty} z^n$. Find the Taylor expansion of f about the point $a = \frac{1}{2}$ and also about the point $a = -\frac{1}{2}$. Determine the radius of convergence in each case.

16.9 Assume that f has the Taylor expansion $f(z) = \sum_{n=0}^{\infty} a(n)z^n$, valid in $B(0; R)$. Let

$$g(z) = \frac{1}{p} \sum_{k=0}^{p-1} f(ze^{2\pi i k/p}).$$

Prove that the Taylor expansion of g consists of every pth term in that of f. That is, if $z \in B(0; R)$ we have

$$g(z) = \sum_{n=0}^{\infty} a(pn) z^{pn}.$$

16.10 Assume that f has the Taylor expansion $f(z) = \sum_{n=0}^{\infty} a_n z^n$, valid in $B(0; R)$. Let $s_n(z) = \sum_{k=0}^{n} a_k z^k$. If $0 < r < R$ and if $|z| < r$, show that

$$s_n(z) = \frac{1}{2\pi i} \int_\gamma \frac{f(w)}{w^{n+1}} \frac{w^{n+1} - z^{n+1}}{w - z} \, dw,$$

where γ is the positively oriented circle with center at 0 and radius r.

16.11 Given the Taylor expansions $f(z) = \sum_{n=0}^{\infty} a_n z^n$ and $g(z) = \sum_{n=0}^{\infty} b_n z^n$, valid for $|z| \le R_1$ and $|z| < R_2$, respectively. Prove that if $|z| < R_1 R_2$ we have

$$\frac{1}{2\pi i} \int_\gamma \frac{f(w)}{w} g\left(\frac{z}{w}\right) dw = \sum_{n=0}^{\infty} a_n b_n z^n,$$

where γ is the positively oriented circle of radius R_1 with center at 0.

16.12 Assume that f has the Taylor expansion $f(z) = \sum_{n=0}^{\infty} a_n (z - a)^n$, valid in $B(a; R)$.

    a) If $0 \le r < R$, deduce *Parseval's identity*:

$$\frac{1}{2\pi} \int_0^{2\pi} |f(a + re^{i\theta})|^2 \, d\theta = \sum_{n=0}^{\infty} |a_n|^2 r^{2n}.$$

    b) Use (a) to deduce the inequality $\sum_{n=0}^{\infty} |a_n|^2 r^{2n} \le M(r)^2$, where $M(r)$ is the maximum of $|f|$ on the circle $|z - a| = r$.

    c) Use (b) to give another proof of the local maximum modulus principle (Theorem 16.27).

16.13 Prove *Schwarz's lemma: Let f be analytic on the disk $B(0; 1)$. Suppose that $f(0) = 0$ and $|f(z)| \le 1$ if $|z| < 1$. Then*

$$|f'(0)| \le 1 \qquad and \qquad |f(z)| \le |z|, \qquad if \; |z| < 1.$$

If $|f'(0)| = 1$ or if $|f(z_0)| = |z_0|$ for at least one z_0 in $B'(0; 1)$, then

$$f(z) = e^{i\alpha} z, \qquad where \; \alpha \; is \; real.$$

Hint. Apply the maximum-modulus theorem to g, where $g(0) = f'(0)$ and $g(z) = f(z)/z$ if $z \ne 0$.

Laurent expansions, singularities, residues

16.14 Let f and g be analytic on an open region S. Let γ be a Jordan circuit with graph Γ such that both Γ and its inner region lie within S. Suppose that $|g(z)| < |f(z)|$ for every z on Γ.

    a) Show that

$$\frac{1}{2\pi i} \int_\gamma \frac{f'(z) + g'(z)}{f(z) + g(z)} \, dz = \frac{1}{2\pi i} \int_\gamma \frac{f'(z)}{f(z)} \, dz.$$

    Hint.   Let $m = \inf \{|f(z)| - |g(z)| : z \in \Gamma\}$. Then $m > 0$ and hence

$$|f(z) + tg(z)| \ge m > 0$$

for each t in $[0, 1]$ and each z on Γ. Now let

$$\phi(t) = \frac{1}{2\pi i} \int_\gamma \frac{f'(z) + tg'(z)}{f(z) + tg(z)} \, dz, \qquad if \; 0 \le t \le 1.$$

Then ϕ is continuous, and hence constant, on $[0, 1]$. Thus, $\phi(0) = \phi(1)$.

b) Use (a) to prove that f and $f + g$ have the same number of zeros inside Γ (*Rouché's theorem*).

16.15 Let p be a polynomial of degree n, say $p(z) = a_0 + a_1 z + \cdots + a_n z^n$, where $a_n \neq 0$. Take $f(z) = a_n z^n$, $g(z) = p(z) - f(z)$ in Rouché's theorem, and prove that p has exactly n zeros in $\mathbf{C}$.

16.16 Let f be analytic on the closure of the disk $B(0; 1)$ and suppose $|f(z)| < 1$ if $|z| = 1$. Show that there is one, and only one, point z_0 in $B(0; 1)$ such that $f(z_0) = z_0$. *Hint*. Use Rouché's theorem.

16.17 Let $p_n(z)$ denote the nth partial sum of the Taylor expansion $e^z = \sum_{n=0}^{\infty} z^n/n!$. Using Rouché's theorem (or otherwise), prove that for every $r > 0$ there exists an N (depending on r) such that $n \geq N$ implies $p_n(z) \neq 0$ for every z in $B(0; r)$.

16.18 If $a > e$, find the number of zeros of the function $f(z) = e^z - az^n$ which lie inside the circle $|z| = 1$.

16.19 Give an example of a function which has all the following properties, or else explain why there is no such function: f is analytic everywhere in $\mathbf{C}$ except for a pole of order 2 at 0 and simple poles at i and $-i$; $f(z) = f(-z)$ for all z; $f(1) = 1$; the function $g(z) = f(1/z)$ has a zero of order 2 at $z = 0$; and $\text{Res}_{z=i} f(z) = 2i$.

16.20 Show that each of the following Laurent expansions is valid in the region indicated:

a) $\dfrac{1}{(z-1)(2-z)} = \displaystyle\sum_{n=0}^{\infty} \dfrac{z^n}{2^{n+1}} + \sum_{n=1}^{\infty} \dfrac{1}{z^n}$ if $1 < |z| < 2$.

b) $\dfrac{1}{(z-1)(2-z)} = \displaystyle\sum_{n=2}^{\infty} \dfrac{1 - 2^{n-1}}{z^n}$ if $|z| > 2$.

16.21 For each fixed t in $\mathbf{C}$, define $J_n(t)$ to be the coefficient of z^n in the Laurent expansion

$$e^{(z - 1/z)t/2} = \sum_{n=-\infty}^{\infty} J_n(t) z^n.$$

Show that for $n \geq 0$ we have

$$J_n(t) = \frac{1}{\pi} \int_0^{\pi} \cos(t \sin \theta - n\theta) \, d\theta$$

and that $J_{-n}(t) = (-1)^n J_n(t)$. Deduce the power series expansion

$$J_n(t) = \sum_{k=0}^{\infty} \frac{(-1)^k (\tfrac{1}{2}t)^{n+2k}}{k! \, (n+k)!} \quad (n \geq 0).$$

The function J_n is called the *Bessel function* of order n.

16.22 Prove *Riemann's theorem: If z_0 is an isolated singularity of f and if $|f|$ is bounded on some deleted neighborhood $B'(z_0)$, then z_0 is a removable singularity.* *Hint*. Estimate the integrals for the coefficients a_n in the Laurent expansion of f and show that $a_n = 0$ for each $n < 0$.

16.23 Prove the *Casorati–Weierstrass theorem: Assume that z_0 is an essential singularity of f and let c be an arbitrary complex number. Then, for every $\varepsilon > 0$ and every disk $B(z_0)$, there exists a point z in $B(z_0)$ such that $|f(z) - c| < \varepsilon$.* *Hint*. Assume that the theorem is false and arrive at a contradiction by applying Exercise 16.22 to g, where $g(z) = 1/[f(z) - c]$.

16.24 *The point at infinity.* A function f is said to be analytic at ∞ if the function g defined by the equation $g(z) = f(1/z)$ is analytic at the origin. Similarly, we say that f has a zero, a pole, a removable singularity, or an essential singularity at ∞ if g has a zero, a pole, etc., at 0. Liouville's theorem states that a function which is analytic everywhere in $\mathbf{C}^*$ must be a constant. Prove that

a) f is a polynomial if, and only if, the only singularity of f in $\mathbf{C}^*$ is a pole at ∞, in which case the order of the pole is equal to the degree of the polynomial.

b) f is a rational function if, and only if, f has no singularities in $\mathbf{C}^*$ other than poles.

16.25 Derive the following "short cuts" for computing residues:

a) If a is a first order pole for f, then

$$\operatorname*{Res}_{z=a} f(z) = \lim_{z \to a} (z - a)f(z).$$

b) If a is a pole of order 2 for f, then

$$\operatorname*{Res}_{z=a} f(z) = g'(a), \qquad \text{where } g(z) = (z - a)^2 f(z).$$

c) Suppose f and g are both analytic at a, with $f(a) \neq 0$ and a a first-order zero for g. Show that

$$\operatorname*{Res}_{z=a} \frac{f(z)}{g(z)} = \frac{f(a)}{g'(a)}, \qquad \operatorname*{Res}_{z=a} \frac{f(z)}{[g(z)]^2} = \frac{f'(a)g'(a) - f(a)g''(a)}{[g'(a)]^3}.$$

d) If f and g are as in (c), except that a is a second-order zero for g, then

$$\operatorname*{Res}_{z=a} \frac{f(z)}{g(z)} = \frac{6f'(a)g''(a) - 2f(a)g'''(a)}{3[g''(a)]^2}.$$

16.26 Compute the residues at the poles of f if

a) $f(z) = \dfrac{ze^z}{z^2 - 1}$,

b) $f(z) = \dfrac{e^z}{z(z - 1)^2}$,

c) $f(z) = \dfrac{\sin z}{z \cos z}$,

d) $f(z) = \dfrac{1}{1 - e^z}$,

e) $f(z) = \dfrac{1}{1 - z^n}$ (where n is a positive integer).

16.27 If $\gamma(a; r)$ denotes the positively oriented circle with center at a and radius r, show that

a) $\displaystyle\int_{\gamma(0;4)} \frac{3z - 1}{(z + 1)(z - 3)} \, dz = 6\pi i$,

b) $\displaystyle\int_{\gamma(0;2)} \frac{2z}{z^2 + 1} \, dz = 4\pi i$,

c) $\displaystyle\int_{\gamma(0;2)} \frac{z^3}{z^4 - 1} \, dz = 2\pi i$,

d) $\displaystyle\int_{\gamma(2;1)} \frac{e^z}{(z - 2)^2} \, dz = 2\pi i e^2$.

Evaluate the integrals in Exercises 16.28 through 16.35 by means of residues.

16.28 $\displaystyle\int_0^{2\pi} \frac{dt}{(a + b\cos t)^2} = \frac{2\pi a}{(a^2 - b^2)^{3/2}}$ if $0 < b < a$.

16.29 $\displaystyle\int_0^{2\pi} \frac{\cos 2t\, dt}{1 - 2a\cos t + a^2} = \frac{2\pi a^2}{1 - a^2}$ if $a^2 < 1$.

16.30 $\displaystyle\int_0^{2\pi} \frac{(1 + \cos 3t)\, dt}{1 - 2a\cos t + a^2} = \frac{\pi(a^2 - a + 1)}{1 - a}$ if $0 < a < 1$.

16.31 $\displaystyle\int_0^{2\pi} \frac{\sin^2 t\, dt}{a + b\cos t} = \frac{2\pi(a - \sqrt{a^2 - b^2})}{b^2}$ if $0 < b < a$.

16.32 $\displaystyle\int_{-\infty}^{\infty} \frac{1}{x^2 + x + 1}\, dx = \frac{2\pi\sqrt{3}}{3}$.

16.33 $\displaystyle\int_{-\infty}^{\infty} \frac{x^6}{(1 + x^4)^2}\, dx = \frac{3\pi\sqrt{2}}{16}$.

16.34 $\displaystyle\int_0^{\infty} \frac{x^2}{(x^2 + 4)^2(x^2 + 9)}\, dx = \frac{\pi}{200}$.

16.35 a) $\displaystyle\int_0^{\infty} \frac{x}{1 + x^5}\, dx = \frac{\pi}{5}\bigg/ \sin\frac{2\pi}{5}$.

Hint. Integrate $z/(1 + z^5)$ around the boundary of the circular sector $S = \{re^{i\theta} : 0 \le r \le R,\ 0 \le \theta \le 2\pi/5\}$, and let $R \to \infty$.

b) $\displaystyle\int_0^{\infty} \frac{x^{2m}}{1 + x^{2n}}\, dx = \frac{\pi}{2n}\bigg/ \sin\left(\frac{2m + 1}{2n}\pi\right),$ m, n integers, $0 < m < n$.

16.36 Prove that formula (38) holds if f is the quotient of two polynomials, say $f = P/Q$, where the degree of Q exceeds that of P by 2 or more.

16.37 Prove that formula (38) holds if $f(z) = e^{imz}P(z)/Q(z)$, where $m > 0$ and P and Q are polynomials such that the degree of Q exceeds that of P by 1 or more. This makes it possible to evaluate integrals of the form

$$\int_{-\infty}^{\infty} e^{imx}\frac{P(x)}{Q(x)}\, dx$$

by the method described in Theorem 16.37.

16.38 Use the method suggested in Exercise 16.37 to evaluate the following integrals:

a) $\displaystyle\int_0^{\infty} \frac{\sin mx}{x(a^2 + x^2)}\, dx = \frac{\pi}{2a^2}(1 - e^{-am})$ if $m \ge 0,\ a > 0$.

b) $\displaystyle\int_0^{\infty} \frac{\cos mx}{x^4 + a^4}\, dx = \frac{\pi}{2a^3}e^{-ma/\sqrt{2}}\sin\left(\frac{ma}{\sqrt{2}} + \frac{\pi}{4}\right)$ if $m > 0,\ a > 0$.

16.39 Let $w = e^{2\pi i/3}$ and let γ be a positively oriented circle whose graph does not pass through 1, w, or w^2. (The numbers 1, w, w^2 are the cube roots of 1.) Prove that the integral

$$\int_\gamma \frac{(z + 1)}{z^3 - 1} \, dz$$

is equal to $2\pi i(m + nw)/3$, where m and n are integers. Determine the possible values of m and n and describe how they depend on γ.

16.40 Let γ be a positively oriented circle with center 0 and radius $< 2\pi$. If a is complex and n is an integer, let

$$I(n, a) = \frac{1}{2\pi i} \int_\gamma \frac{z^{n-1}e^{az}}{1 - e^z} \, dz.$$

Prove that

$$I(0, a) = \tfrac{1}{2} - a, \qquad I(1, a) = -1, \qquad \text{and} \qquad I(n, a) = 0 \quad \text{if } n > 1.$$

Calculate $I(-n, a)$ in terms of Bernoulli polynomials when $n \geq 1$ (see Exercise 9.38).

16.41 This exercise requests some of the details of the proof of Theorem 16.38. Let

$$g(z) = \sum_{r=0}^{n-1} e^{\pi i a(z+r)^2/n}, \qquad f(z) = g(z)/(e^{2\pi i z} - 1),$$

where a and n are positive integers with na even. Prove that:

a) $g(z + 1) - g(z) = e^{\pi i a z^2/n}(e^{2\pi i z} - 1)\sum_{m=0}^{a-1} e^{2\pi i m z}$.

b) $\text{Res}_{z=0} f(z) = g(0)/(2\pi i)$.

c) The real part of $i(t + Re^{\pi i/4} + r)^2$ is $-(\sqrt{2}tR + R^2 + \sqrt{2}rR)$.

One-to-one analytic functions

16.42 Let S be an open subset of $\mathbf{C}$ and assume that f is analytic and one-to-one on S. Prove that:

a) $f'(z) \neq 0$ for each z in S. (Hence f is conformal at each point of S.)

b) If g is the inverse of f, then g is analytic on $f(S)$ and $g'(w) = 1/f'(g(w))$ if $w \in f(S)$.

16.43 Let $f: \mathbf{C} \to \mathbf{C}$ be analytic and one-to-one on $\mathbf{C}$. Prove that $f(z) = az + b$, where $a \neq 0$. What can you conclude if f is one-to-one on $\mathbf{C}^*$ and analytic on $\mathbf{C}^*$ except possibly for a finite number of poles?

16.44 If f and g are Möbius transformations, show that the composition $f \circ g$ is also a Möbius transformation.

16.45 Describe geometrically what happens to a point z when it is carried into $f(z)$ by the following special Möbius transformations:

    a) $f(z) = z + b$ (Translation).

    b) $f(z) = az$, where $a > 0$ (Stretching or contraction).

    c) $f(z) = e^{i\alpha}z$, where α is real   (Rotation).

    d) $f(z) = 1/z$ (Inversion).

16.46 If $c \neq 0$, we have

$$\frac{az + b}{cz + d} = \frac{a}{c} + \frac{bc - ad}{c(cz + d)}.$$

Hence every Möbius transformation can be expressed as a composition of the special cases described in Exercise 16.45. Use this fact to show that Möbius transformations carry circles into circles (where straight lines are considered as special cases of circles).

16.47 a) Show that all Möbius transformations which map the upper half-plane $T = \{x + iy : y \geq 0\}$ onto the closure of the disk $B(0; 1)$ can be expressed in the form $f(z) = e^{i\alpha}(z - a)/(z - \bar{a})$, where α is real and $a \in T$.

   b) Show that a and α can always be chosen to map any three given points of the real axis onto any three given points on the unit circle.

16.48 Find all Möbius transformations which map the right half-plane

$$S = \{x + iy : x \geq 0\}$$

onto the closure of $B(0; 1)$.

16.49 Find all Möbius transformations which map the closure of $B(0; 1)$ onto itself.

16.50 The *fixed points* of a Möbius transformation

$$f(z) = \frac{az + b}{cz + d} \qquad (ad - bc \neq 0)$$

are those points z for which $f(z) = z$. Let $D = (d - a)^2 + 4bc$.

   a) Determine all fixed points when $c = 0$.

   b) If $c \neq 0$ and $D \neq 0$, prove that f has exactly 2 fixed points z_1 and z_2 (both finite) and that they satisfy the equation

$$\frac{f(z) - z_1}{f(z) - z_2} = Re^{i\theta}\frac{z - z_1}{z - z_2}, \qquad \text{where } R > 0 \text{ and } \theta \text{ is real.}$$

   c) If $c \neq 0$ and $D = 0$, prove that f has exactly one fixed point z_1 and that it satisfies the equation

$$\frac{1}{f(z) - z_1} = \frac{1}{z - z_1} + C \qquad \text{for some } C \neq 0.$$

   d) Given any Möbius transformation, investigate the successive images of a given point w. That is, let

$$w_1 = f(w), \qquad w_2 = f(w_1), \qquad \dots, \qquad w_n = f(w_{n-1}), \qquad \dots,$$

and study the behavior of the sequence $\{w_n\}$. Consider the special case a, b, c, d real, $ad - bc = 1$.

MISCELLANEOUS EXERCISES

16.51 Determine all complex z such that

$$z = \sum_{n=2}^{\infty} \sum_{k=1}^{n} e^{2\pi ikz/n}.$$

16.52 If $f(z) = \sum_{n=0}^{\infty} a_n z^n$ is an entire function such that $|f(re^{i\theta})| \leq Me^{r^k}$ for all $r > 0$, where $M > 0$ and $k > 0$, prove that

$$|a_n| \leq \frac{Me^{n/k}}{(n/k)^{n/k}} \qquad \text{for } n \geq 1.$$

16.53 Assume f is analytic on a deleted neighborhood $B'(0; a)$. Prove that $\lim_{z \to 0} f(z)$ exists (possibly infinite) if, and only if, there exists an integer n and a function g, analytic on $B(0; a)$, with $g(0) \neq 0$, such that $f(z) = z^n g(z)$ in $B'(0; a)$.

16.54 Let $p(z) = \sum_{k=0}^{n} a_k z^k$ be a polynomial of degree n with real coefficients satisfying

$$a_0 > a_1 > \cdots > a_{n-1} > a_n > 0.$$

Prove that $p(z) = 0$ implies $|z| > 1$. *Hint.* Consider $(1 - z)p(z)$.

16.55 A function f, defined on a disk $B(a; r)$, is said to have a zero of infinite order at a if, for every integer $k > 0$, there is a function g_k, analytic at a, such that $f(z) = (z - a)^k g_k(z)$ on $B(a; r)$. If f has a zero of infinite order at a, prove that $f = 0$ everywhere in $B(a; r)$.

16.56 Prove Morera's theorem: *If f is continuous on an open region S in $\mathbf{C}$ and if $\int_\gamma f = 0$ for every polygonal circuit γ in S, then f is analytic on S.*

SUGGESTED REFERENCES FOR FURTHER STUDY

16.1   Ahlfors, L. V., *Complex Analysis*, 2nd ed. McGraw-Hill, New York, 1966.

16.2   Carathéodory, C., *Theory of Functions of a Complex Variable*, 2 vols. F. Steinhardt, translator. Chelsea, New York, 1954.

16.3   Estermann, T., *Complex Numbers and Functions*. Athlone Press, London, 1962.

16.4   Heins, M., *Complex Function Theory*. Academic Press, New York, 1968.

16.5   Heins, M., *Selected Topics in the Classical Theory of Functions of a Complex Variable*. Holt, Rinehart, and Winston, New York, 1962.

16.6   Knopp, K., *Theory of Functions*, 2 vols. F. Bagemihl, translator. Dover, New York, 1945.

16.7   Saks, S., and Zygmund, A., *Analytic Functions*, 2nd ed. E. J. Scott, translator. *Monografie Matematyczne* **28**, Warsaw, 1965.

16.8   Sansone, G., and Gerretsen, J., *Lectures on the Theory of Functions of a Complex Variable*, 2 vols. P. Noordhoff, Gröningen, 1960.

16.9   Titchmarsh, E. C., *Theory of Functions*, 2nd ed. Oxford University Press, 1939.

INDEX OF SPECIAL SYMBOLS

det $[a_{ij}]$, determinant of matrix $[a_{ij}]$, 367

$J_{\mathbf{f}}$, Jacobian determinant of $\mathbf{f}$, 368

$\mathbf{f} \in C'$, the components of $\mathbf{f}$ have continuous first-order partials, 371

$\displaystyle\int_I f(\mathbf{x})\, d\mathbf{x}$, multiple integral, 389, 407

$\underline{c}(S)$, $\bar{c}(S)$, inner (outer) Jordan content of S, 396

$c(S)$, Jordan content of S, 396

$\displaystyle\int_\gamma f$, contour integral of f along γ, 436

$A(a; r_1, r_2)$, annulus with center a, 438

$n(\gamma, z)$, winding number of a circuit γ with respect to z, 445

$B'(a)$, $B'(a; r)$, deleted neighborhood of a, 457

$\displaystyle\operatorname*{Res}_{z=a} f(z)$, residue of f at a, 459

INDEX

Abel, Neils Henrik, (1802–1829), 194, 245, 248
Abel, limit theorem, 245
  partial summation formula, 194
  test for convergence of series, 194, 248 (Ex. 9.13)
Absolute convergence, of products, 208
  of series, 189
Absolute value, 13, 18
Absolutely continuous function, 139
Accumulation point, 52, 62
Additive function, 45 (Ex. 2.22)
Additivity of Lebesgue measure, 291
Adherent point, 52, 62
Algebraic number, 45 (Ex. 2.15)
Almost everywhere, 172, 391
Analytic function, 434
Annulus, 438
Approximation theorem of Weierstrass, 322
Arc, 88, 435
Archimedean property of real numbers, 10
Arc length, 134
Arcwise connected set, 88
Area (content) of a plane region, 396
Argand, Jean-Robert (1768–1822), 17
Argument of complex number, 21
Arithmetic mean, 205
Arzelà, Cesare (1847–1912), 228, 273
Arzelà's theorem, 228, 273
Associative law, 2, 16
Axioms for real numbers, 1, 2, 9

Ball, in a metric space, 61
  in $\mathbf{R}^n$, 49
Basis vectors, 49
Bernoulli, James (1654–1705), 251, 338, 478
Bernoulli, numbers, 251 (Ex. 9.38)
  periodic functions, 338 (Ex. 11.18)
  polynomials, 251 (Ex. 9.38), 478 (Ex. 16.40)

Bernstein, Sergei Natanovic (1880–   ), 242
Bernstein's theorem, 242
Bessel, Friedrich Wilhelm (1784–1846), 309, 475
Bessel function, 475 (Ex. 16.21)
Bessel inequality, 309
Beta function, 331
Binary system, 225
Binomial series, 244
Bolzano, Bernard (1781–1848), 54, 85
Bolzano's theorem, 85
Bolzano-Weierstrass theorem, 54
Bonnet, Ossian (1819–1892), 165
Bonnet's theorem, 165
Borel, Émile (1871–1938), 58
Bound, greatest lower, 9
  least upper, 9
  lower, 8
  uniform, 221
  upper, 8
Boundary, of a set, 64
  point, 64
Bounded, away from zero, 130
  convergence, 227, 273
  function, 83
  set, 54, 63
  variation, 128

Cantor, Georg (1845–1918), 8, 32, 56, 67, 180, 312
Cantor intersection theorem, 56
Cantor–Bendixon theorem, 67 (Ex. 3.25)
Cantor set, 180 (Ex. 7.32)
Cardinal number, 38
Carleson, Lennart, 312
Cartesian product, 33
Casorati–Weierstrass theorem, 475 (Ex. 16.23)
Cauchy, Augustin-Louis (1789–1857), 14, 73, 118, 177, 183, 207, 222